Makers of Modern Strategy

Makers of Modern Strategy

Military Thought from Machiavelli to Hitler

EDITED BY

EDWARD MEAD EARLE

WITH THE COLLABORATION OF
GORDON A. CRAIG AND
FELIX GILBERT

PRINCETON
PRINCETON UNIVERSITY PRESS

COPYRIGHT, 1941, BY PRINCETON UNIVERSITY PRESS
LONDON: OXFORD UNIVERSITY PRESS

Second Printing 1945
Third Printing 1948
Fourth Printing 1953
Fifth Printing 1957
Sixth Printing 1960

PRINTED IN THE UNITED STATES OF AMERICA

Contents

PAGE

INTRODUCTION
    EDWARD MEAD EARLE   vii

SECTION I. THE ORIGINS OF MODERN WAR: FROM THE SIXTEENTH
    TO THE EIGHTEENTH CENTURY

Chapter 1. Machiavelli: The Renaissance of the Art of War
    FELIX GILBERT   3

Chapter 2. Vauban: The Impact of Science on War
    HENRY GUERLAC   26

Chapter 3. Frederick the Great, Guibert, Bülow: From Dynastic
    to National War
    R. R. PALMER   49

SECTION II. THE CLASSICS OF THE NINETEENTH CENTURY: INTER-
    PRETERS OF NAPOLEON

Chapter 4. Jomini
    CRANE BRINTON, GORDON A. CRAIG, AND FELIX GILBERT   77

Chapter 5. Clausewitz
    H. ROTHFELS   93

SECTION III. FROM THE NINETEENTH CENTURY TO THE FIRST
    WORLD WAR

Chapter 6. Adam Smith, Alexander Hamilton, Friedrich List:
    The Economic Foundations of Military Power
    EDWARD MEAD EARLE   117

Chapter 7. Engels and Marx: Military Concepts of the Social Rev-
    olutionaries
    SIGMUND NEUMANN   155

Chapter 8. Moltke and Schlieffen: The Prussian-German School
    HAJO HOLBORN   172

Chapter 9. Du Picq and Foch: The French School
    STEFAN T. POSSONY AND ETIENNE MANTOUX   206

Chapter 10. Bugeaud, Galliéni, Lyautey: The Development of
    French Colonial Warfare
    JEAN GOTTMANN   234

Chapter 11. Delbrück: The Military Historian
    GORDON A. CRAIG   260

Section IV. From the First to the Second World War

Chapter 12. Churchill, Lloyd George, Clemenceau: The Emergence
    of the Civilian
        Harvey A. DeWeerd   287

Chapter 13. Ludendorff: The German Concept of Total War
        Hans Speier   306

Chapter 14. Lenin, Trotsky, Stalin: Soviet Concepts of War
        Edward Mead Earle   322

Chapter 15. Maginot and Liddell Hart: The Doctrine of Defense
        Irving M. Gibson   365

Chapter 16. Haushofer: The Geopoliticians
        Derwent Whittlesey   388

Section V. Sea and Air War

Chapter 17. Mahan: Evangelist of Sea Power
        Margaret Tuttle Sprout   415

Chapter 18. Continental Doctrines of Sea Power
        Theodore Ropp   446

Chapter 19. Japanese Naval Strategy
        Alexander Kiralfy   457

Chapter 20. Douhet, Mitchell, Seversky: Theories of Air War-
    fare
        Edward Warner   485

Epilogue

Hitler: The Nazi Concept of War
        Edward Mead Earle   504

Who's Who of Contributors   517
Editor's Note and Acknowledgments   519
Bibliographical Notes   521
Index   548

Introduction

BY EDWARD MEAD EARLE

WHEN war comes it dominates our lives. As an American novelist wrote in 1861, it is like a great tempest which blows upon us all, mingles with the church organ, whistles through the streets, steals into our firesides, clinks glasses in barrooms, lifts the gray hairs of statesmen, invades the classrooms of our colleges, rustles the thumbed pages of our scholars. It offers inescapable tests of our allegiances—of work and duty, private loves and public devotions, personal preferences and social ties. Moreover, Mr. Walter Millis points out, "It challenges virtually every other institution of society—the justice and equity of its economy, the adequacy of its political systems, the energy of its productive plant, the bases, wisdom and purposes of its foreign policy. There is no aspect of our existence . . . which is not touched, modified, perhaps completely altered by the imperatives of war."

But war is not an act of God. It grows directly out of things which individuals, statesmen, and nations do or fail to do. It is, in short, the consequence of national policies or lack of policies. And once the nation's destiny is submitted to the terrible arbitrament of war, victory or defeat likewise ensues from what we do or fail to do.

These truths being self-evident, it would be folly to leave the comprehension of war policies to soldiers alone or statesmen alone or to soldiers and statesmen together. A national strategy must be formulated by the President and the General Staff and implemented by acts of Congress, for in such matters the nation cannot be turned into a debating society. However, the strategy determined upon can succeed only if it has the support of enlightened and determined citizens; they must dedicate to the success of that strategy their lives, their fortunes, and their sacred honor. Democracies in time of war require great leadership and usually produce it in the heroic figures of men like Washington, Lincoln, Lloyd George, Wilson, Clemenceau, Churchill, and Franklin Roosevelt, but the wellsprings of such leadership come from deep in the heart, the will, and the conscience of the people. Even the private soldiers and the junior officers of an embattled democratic society must know the purposes for which they risk their lives. It was a Prussian officer, Steuben, who laid down the fundamental principle that an American army must understand the rational bases of discipline and action. It was a great democrat, Thomas Jefferson, who proposed that military affairs be made a fundamental part of American education.

Winston Churchill appreciates these basic truths. For throughout the present war his magnificent speeches have told the British people why their sons and fathers have been dying on far-flung battlefields. His address to the Congress of the United States, May 20, 1943, served a similar purpose; it explained to Americans for the first time the underlying reasons for defeating Germany before Japan, and it won their support for a strategy on which public opinion formerly was sharply divided.

It is the purpose of this book, on a broader canvas and over a longer period of time, to explain the manner in which the strategy of modern war has developed, in the conviction that a knowledge of the best military thought will enable Anglo-Saxon readers to comprehend the causes of war and the fundamental principles which govern the conduct of war. We believe that eternal vigilance in such matters is the price of liberty. We believe, too, that if we are to have a durable peace we must have a clear understanding of the role which armed force plays in international society. And we have not always had this understanding. As Mr. Gordon Craig points out in Chapter 11, the greatest military historians of our time felt under constant necessity to apologize for their interest in military problems because dislike for war and ignorance concerning its role in human affairs have led peaceful peoples everywhere to deprecate its significance in history and ignore its portentous meaning for our future. For it is not force, in itself, which is wrong but the purposes to which force is sometimes put. As Pascal said almost three centuries ago, we must realize that: "Justice without force is impotent. Force without justice tyrannical. We must combine justice with force." The American nation is in the process of becoming the foremost military power of modern times. The manner in which we use this great power is momentous for ourselves and for the world.

Strategy deals with war, preparation for war, and the waging of war. Narrowly defined, it is the art of military command, of projecting and directing a campaign. It is different from tactics—which is the art of handling forces in battle—in much the same way that an orchestra is different from its individual instruments. Until about the end of the eighteenth century strategy consisted of the body of stratagems and tricks of war—*ruses de guerre*—by which a general sought to deceive the enemy and win victory. But as war and society have become more complicated—and war, it must be remembered, is an inherent part of society—strategy has of necessity required increasing consideration of nonmilitary factors, economic, psychological, moral, political, and technological. Strategy, therefore, is not merely a concept of wartime, but is an inherent element of statecraft at all times. Only the most restricted terminology would now define strategy as the art of military command. In the present-day world, then, strategy is the art of controlling and utilizing the resources of a nation— or a coalition of nations—including its armed forces, to the end that its vital interests shall be effectively promoted and secured against enemies, actual, potential, or merely presumed. The highest type of strategy—sometimes called grand strategy—is that which so integrates the policies and armaments of the nation that the resort to war is either rendered unnecessary or is undertaken with the maximum chance of victory. It is in this broader sense that the word strategy is used in this volume.

Because modern strategy has so many nonmilitary phases, a great many civilians march across these pages: the professor Adam Smith, the economist

Friedrich List, the social revolutionaries Marx and Engels, the historians Delbrück and Oman, the industrialist Rathenau, the journalist Trotsky, the politicians Lloyd George and Clemenceau. On the other hand, many gallant soldiers are missing from the story. Marlborough, Lee, Jackson, Wellington, Blücher, Grant, Sherman are not here, either because they were more tacticians than strategists or because they bequeathed to posterity no coherent statement of strategical doctrine. Napoleon—undoubtedly the greatest general of modern times and one of the foremost strategists of all time—does not have a chapter in his own right. His campaigns are eloquent testimony to his genius. But Napoleon recorded his strategy on the battlefield, not on the printed page (if we except his trite maxims); hence he is represented here by his interpreters Clausewitz and Jomini.

Only two American professional officers are discussed in this book—Admiral Mahan and General Mitchell. This is small representation for a people which has been preoccupied with war, to a greater or lesser degree, since the first colonists landed on our shores. The reason is, of course, that our significant contributions to warfare have been in the fields of tactics and technology, rather than strategy. When Americans are compelled to wage war, they can successfully match wits, ingenuity, and determination with those to whom war is more congenial. Our forefathers of colonial and revolutionary days developed the fundamental tactics of camouflage and taking cover; demonstrated the value of accuracy, economy, and concentration of fire; raised to a fine art the practice of impeding the enemy's advance by felling trees, weakening bridges, and otherwise "scorching the earth." The tactics of our Civil War were for generations the object of careful study by European staffs: the movement of troops by rail, mining and sapping, trench warfare, and aerial observation, among other things.

In the field of military technology, we introduced to the world the rifle with interchangeable parts, the machine gun, the balloon, the tractor for tanks, the parachute, the dive bomber, the submarine, and the airplane. Being mechanically minded and possessed of almost religious faith in the machine, we were first to adapt mass production to war. And in a very special sense we are the fathers of military aviation. Not only did the Wright brothers invent the airplane, but another favorite American child, the gasoline engine, has made possible the development of the airplane from a primitive thing to the powerful four-motored bomber. But we have not produced a Clausewitz or a Vauban. Mahan is our only military theorist of comparable reputation.

The authors of the following chapters do not necessarily agree with the editors and do not always agree with one another. Nevertheless, certain well-defined themes run through the story from Machiavelli to Hitler. Among these are the concept of lightning war and the battle of annihilation; the war of maneuver *vs.* the war of position; the relationship between war and social institutions and between economic strength and military power; psychology and morale as weapons of war; the role of discipline in the army; the question of the professional army *vs.* the militia. Military techniques are responsible for a kind of international freemasonry of war, so that the development of strategy cuts across national lines. So do ideas and ideologies—mercantilism; free trade; liberty, equality, fraternity; totalitarianism; socialism; pacifism—which are related to the causes and conduct of war.

But the national factors in strategy frequently are the determining factors. In part they grow out of differences in the character and psychology of peoples, as well as their standards of value and their outlook on life—what the Germans call their *Weltanschauung*. In part they are the consequence of political, social, and economic institutions. Even more they are the political and military expression of geographical situation and national tradition. In any case, they are readily discernible. It seems clear, for example, that since 1870 the Germans have thought in terms of aggressive warfare and military annihilation of the enemy, the British (and more recently the French) in terms of defensive warfare and the long-run consequences of economic attrition.

Diplomacy and strategy, political commitments and military power, are inseparable; unless this be recognized, foreign policy will be bankrupt, as Mr. Walter Lippmann has shown in his persuasive book *U.S. Foreign Policy: Shield of the Republic*. The very existence of a nation depends upon its concept of the national interest and the means by which the national interest is promoted; therefore, it is imperative that its citizens understand the fundamentals of strategy. We do not have and do not wish to have a military class to whom these matters will be delegated with plenary powers. Our armed forces, including our officer corps, are recruited on a democratic basis. This is as it should be, since there is only one safe repository of the national security of a democratic state: the whole people.

Members of the armed forces, we venture to suggest, will be as much interested in the following chapters as the general reader and the student of international affairs. For example, a persistently debated question among tacticians and strategists deals with the relative merits of offensive and defensive warfare. It will be apparent to any reader of this book that this is not purely a military question, but is rather a complex of other questions, military, historical, social, and economic. If one were to generalize, one could say that the defensive usually enjoys a technical advantage over the offensive. But there are times when the offensive sweeps the defensive entirely aside, overwhelms it as a flood overwhelms all natural and artificial barriers in its path. This is true when a fundamental social revolution occurs, such as that in France in 1793 and in Germany with the advent of Hitler in 1933. The revolution not only adopts the Dantonian policy of audacity, again audacity, and always audacity, but it demoralizes the older order by a confusion of counsels and a conflict of ideologies. The offensive also takes precedence over the defense when there is some striking military invention such as gunpowder, the tank, and the airplane, or when older weapons are put to new uses in the manner in which the airplane is transformed into the Stuka and the long-range bomber. Throughout the nineteenth century most of the new weapons—notably the machine gun and the submarine—strengthened the defense. But the tank and the airplane reversed the former trend and have dominated the tactics and strategy of the present war. This close relation between social institutions and industrial technology, on the one hand, and the art of war, on the other, makes strange bedfellows—revolutionaries like Engels, Lenin, and Trotsky join conservatives like Ludendorff in the fascinating study of war and the means of waging it.

As society becomes more highly industrialized, the art of war becomes more complex. As an almost inevitable result, the logistical and tactical factors in military operations tend to condition the strategy of which they are theoretically but a servant. A strategist like Eisenhower or Wavell must have his work supplemented by a superb tactician like Montgomery, just as Lee had his Stonewall Jackson. The enormous technical preparations which precede a modern campaign preclude the possibility of rapid changes in strategy. Once a vast offensive gets under way, its momentum carries it on for a certain length of time regardless of whether political or other considerations require a change. When the war in Europe is at an end, the logistical and tactical factors of war in the Far East will do much to determine our strategy there.

Under modern conditions military questions are so interwoven with economic, political, social, and technological phenomena that it is doubtful if one can speak of a purely military strategy. Much of Hitler's success up to the invasion of Russia in 1941 was due to his remarkable understanding of this fundamental fact. His opponents in the field and in the chancelleries of Europe were still thinking, until the fall of France, in terms of the seventeenth century, when politics and war, strategy and tactics, could in some measure be put into separate categories. But in our day politics and strategy have become inseparable.

War now thrusts itself upon the attention of us all. And since it is the concern of all the people, all the people must realize that it *is* their concern. In wartime this involves a total effort; in time of peace, as in time of war, it demands wide understanding. It is to the cause of a broader comprehension of war and peace that this book is dedicated.

SECTION I

The Origins of Modern War: From the Sixteenth to the Eighteenth Century

CHAPTER 1. Machiavelli: The Renaissance of the Art of War

BY FELIX GILBERT

MANY are now of the opinion that no two things are more discordant and incongruous than a civil and military life. But if we consider the nature of government, we shall find a very strict and intimate relation betwixt these two conditions; and that they are not only compatible and consistent with each other, but necessarily connected and united together." These sentences, which open Machiavelli's book on the *Art of War*,[1] give the clue to an understanding of Machiavelli's interest in military questions. He did not approach them as a military technician; he had observed the decisive role of military power in politics, and he had drawn the conclusion that the existence and greatness of a state were assured only if military power received its appropriate place in the political order. He writes in the *Principe*[2] (and the importance which he attributes to this sentence can be seen from the fact that he repeats it in the *Discorsi*[3]) : "There cannot be good laws where there are not good arms, and where there are good arms there must be good laws." Accordingly, in the *Principe*, he admonishes the ruler to keep in mind that the preservation of his power depends on military strength : "A Prince should therefore have no other aim or thought, nor take up any other thing for his study, but war and its organization and discipline."[4] The *Discorsi* are concerned with the same problem; their grandiose theme is the interrelationship between the Roman military organization, the political constitution of the Roman republic and Rome's rise to world power, and from this lesson of Roman history he draws the conclusion: "The foundation of states is a good military organization."[5] And in turn, the *Art of War*, the great military textbook on which Machiavelli's fame as military thinker is mainly based, though dealing with the details of military organization and tactics, goes beyond them and discusses the political conditions and implications of a good military organization. Investigation of the role of military power in political life was the magnet to which all Machiavelli's political thinking was inevitably drawn.

I

How did it come about that Machiavelli focused his attention on the problem of the connection between political and military organization? The experiences of his own lifetime had given him an impressive lesson in observing the impact of the military factor on political life. He witnessed the loss of freedom of his

[1] Preface to Machiavelli's *Arte della Guerra, Tutte le opere storiche e letterarie di Niccolò Machiavelli*, ed. G. Mazzoni and M. Casella (Florence, 1929), p. 265.
[2] *Principe*, ch. 12 (*Opere*, p. 24).
[3] *Discorsi*, book 3, ch. 31 (*Opere*, p. 244).
[4] *Principe*, ch. 14 (*Opere*, p. 29).
[5] *Discorsi*, book 3, ch. 31 (*Opere*, p. 244).

native city because of the failure of its military machine; he saw Italy fall from independence into domination by foreign armies. Nevertheless Machiavelli's interest in this question was fundamentally the fruit of his unique political insight, an indication of his sensitive understanding of the political forces which shaped the political fate of his time. For the question of the true relation between political and military organization was at the bottom of the great revolutionary upheavals of the fourteenth and fifteenth centuries.

Only a discerning mind, however, would discover the link between the changes which occurred in military organization and the revolutionary developments which took place in the social and political sphere. To the ordinary observer, the connection between cause and effect in military developments seemed obvious. Because of the discovery of gunpowder and the invention of firearms and artillery it appeared that the armor of the knight was doomed and the collapse of the military organization of the Middle Ages, in which knights played the decisive role, had become inevitable. In his famous epic *Orlando Furioso* (1516), Ariosto, Machiavelli's contemporary and Italian compatriot, narrates[6] how Orlando, his hero and the embodiment of all knightly virtues, was forced to face an enemy with a firearm:

> "At once the lightning flashes, shakes the ground,
> The trembling bulwarks echo to the sound.
> The pest, that never spends in vain its force,
> But shatters all that dares oppose its course,
> Whizzing impetuous flies along the wind."

When the invincible Orlando succeeded in overcoming this redoubtable enemy and could choose from the rich booty:

> ". . . nothing would the champion bear away
> From all the spoils of that victorious day
> Save that device, whose unresisted force
> Resembled thunder in its rapid course."

Then he sailed out on the Ocean, plunging the weapon into the sea and exclaiming:

> "O! curs'd device! base implement of death!
> Fram'd in the black Tartarean realms beneath!
> By Beelzebub's malicious art design'd
> To ruin all the race of human kind. . . .
> That ne'er again a knight by thee may dare,
> Or dastard cowards, by thy help in war,
> With vantage base, assault a nobler foe,
> Here lie for ever in th' abyss below!"

In short, if firearms had not been invented or could now be banished, the world of the knights would live on forever in all its splendor.

[6] L. Ariosto, *Orlando Furioso*, libro 1, canto 9. English translation by John Hoole.

This dramatic explanation of the decline of the power of the knights hardly corresponds with reality. The history of military institutions cannot be separated from the general history of a period. The military organization of the Middle Ages formed an integral part of the medieval world and declined when the medieval social structure disintegrated. Spiritually as well as economically the knight was a characteristic product of the Middle Ages. In a society in which God was envisaged as the head of a hierarchy, each estate was assumed to fulfill a religious function and all secular activity had been given a religious meaning. The particular task of chivalry was to protect and defend the people of the country; in waging war the knight served God. He placed his military services at the disposal of his overlord, to whom the supervision of secular activities was confided by the church. Aside from its spiritual-religious side, however, the military bond between vassal and overlord also had its legal-economic side. His land, the fief, was given to the knight by the overlord and in accepting it he assumed the obligation of military service to the overlord in wartime. It was an exchange of goods against services as was fitting to the agricultural structure and manorial system of the Middle Ages.

A religious concept of war as an act of rendering justice, the restriction of military service to the class of land-holding knights and a moral-legal code which operated as the main bond holding the army together—these are the factors which determined the forms of military organization as well as the methods of war in the Middle Ages. The medieval army could be assembled only when a definite issue had arisen; it was ordered out for the purposes of a definite campaign and could be kept together only as long as this campaign lasted. The purely temporary character of the military service as well as the equality of standing of the fighters made strict discipline difficult if not impossible. A battle frequently developed into fights between individual knights and the outcome of such single combats between the leaders played a decisive role. Because warfare represented the fulfillment of a religious and moral duty, there was a strong inclination to conduct war and battles according to fixed rules and a settled code.

Since this military organization was a typical product of the whole social system of the Middle Ages, any change in the foundations of this system had inevitable repercussions in the military field. When rapid expansion of money economy shook the agricultural basis of medieval society the effects of this development on military institutions were immediate. It was primarily in the military field that those who were the protagonists of the new economic developments—the cities and the wealthy overlords—could make greatest use of the new opportunities: namely, to accept money payments instead of services, or to secure services by money rewards and salaries. The overlord could accept money payments from those who did not wish to fulfill their military obligations and, on the other hand, he could retain those knights who remained in his army beyond the period of war and for longer stretches of time by promises of regular payments. Thus he was able to lay the foundations of a permanent and professional army and to free himself from dependence on his vassals. This process—the transformation of the feudal army into a professional army, of the feudal state into the bureaucratic and absolutist state—was a very slow process and reached its climax only in the eighteenth century, but the true knightly spirit of the feudal armies died very quickly. We possess an illustration of this change in a fifteenth century ballad, describing life in the army of Charles the Bold of Burgundy.[7] Burgundy was a very recent political formation in the fifteenth century and the older powers considered her as a kind of parvenu; therefore, Charles the Bold was particularly eager to legitimatize the existence of his state by strict observance of all old traditions and customs and became in effect the leader of a kind of romantic revival of chivalry. It is the more revealing, therefore, that in this ballad, "knight, squire, sergeant and vassal" have only one thought, namely, "when will the paymaster come?" Here, behind the glittering facade of chivalry, is disclosed the prosaic reality of material interests.

In the armies of the greater powers, France, Aragon, or England, old and modern elements, feudal levy and professionalism, were mixed; but the great money powers of the period, the Italian cities, came to rely entirely on professional soldiers. Since the fourteenth century, Italy had been the "promised land" of all knights to whom war was chiefly a means of making money. The single groups, the "compagnie di ventura," were supplied and paid by their leaders, the "condottieri," who offered their services to every power willing to pay their price. Thus, in Italy soldiering became a profession of its own entirely separated from any other civilian activity.

The impact of capitalism and money economy also broadened the recruiting base of the armies. Money attracted into the military services new classes of men free from military traditions; and with this infiltration of new men, new weapons and new forms of fighting could be introduced and developed. Archers and infantry made their appearance in the French and English armies during the Hundred Years War. This tendency toward experimentation in new military methods received a further strong impetus from the defeats which the armies of Charles the Bold suffered at the hands of the Swiss near the end of the fifteenth century. In the battles of Morat and Nancy (1476), the knights of Charles the Bold, unable to break up the squares of Swiss foot soldiers and to penetrate into the forest of their pikes, were thoroughly defeated. This event was a European sensation. Infantry had won its place in the military organization of the period.

The importance of the invention of gunpowder has to be evaluated against the background of these general developments: first, the rise of a money economy; second, the attempts of the feudal overlord to free himself from dependence on his vassals and to establish a reliable foundation of power; and third, the trend toward experimentation in military organization resulting from the decline of feudalism.

[7] Ballad by E. Deschamps: "Quand viendra le trésorier?", printed in E. Deschamps, *Oeuvres complètes*, ed. Saint Hilaire, vol. 4 (Paris, 1884), p. 289.

Firearms and artillery were not the cause of these developments but they were an important contributory factor, accelerating the tempo of the evolution. First of all, they strengthened the position of the overlord in relation to his vassals. The employment of artillery in a campaign was a cumbersome task; many carriages were needed for transportation of the heavy cannon and their equipment, a specialized personnel of engineers was necessary, and the whole procedure was extremely expensive. The accounts of military expenditures for this period show that the expenses for artillery constituted a disproportionately large part of the whole sum.[8] Only the very wealthy were able to afford artillery. Also, the principal military effect of the invention of artillery worked in favor of the great power and against the smaller states and local centers of independence. In the Middle Ages, the final sanction of the position of the knight had been that, in his castle, he was relatively immune against attack. The art of fortification was much cultivated in this period.[9] Small states protected themselves by establishing at their frontiers a line of fortresses which enabled them to hold out even against superior forces. These medieval fortifications were vulnerable, however, to artillery fire. Thus, the military balance became heavily loaded in favor of the military offensive. Francesco di Giorgio Martini, one of the great Italian architects of the fifteenth century, who was in charge of the building of the fortresses for the Duke of Urbino, complained in his treatise on military architecture that "the man who would be able to balance defense against attack, would be more a god than a human being."[10]

These changes in the composition of armies and in military technique also transformed the spirit of military organization. The moral code, traditions, and customs, which feudalism had evolved, had lost control over the material from which the armies were now recruited. Adventurers and ruffians who wanted wealth and plunder, men who had nothing to lose and everything to gain through war, made up the main body of the armies. As a result of a situation in which war was no longer undertaken as a religious duty, the purpose of military service became financial gain. The moral problem arose whether it was a sin to follow a profession which aimed at the killing of other people. At the beginning of the fifteenth century, Christine de Pisan devoted a section of her military treatise to the problem of whether the acceptance of money for military services could be justified,[11] and one hundred years later, Martin Luther was faced with the necessity of answering the question whether a soldier could be a Christian.[12] In the most civilized parts of Europe, such as Italy, the people shunned all contact with the professional soldier.

Even among the statesmen, military virtues came into discredit: "Stability has increased so much," reported the ambassador of the Duke of Ferrara from

---

[8] Cf. for instance "Ordine dell'Esercito Ducale Sforzesco, 1472-1474," *Archivio Storico Lombardo*, III, ser. 1 (1876), 448-513.

[9] Cf. C. Oman, *A History of the Art of War in the Middle Ages* (London, 1924), I, 358.

[10] Francesco di Giorgio Martini, *Trattato di Architettura Civile e Militare*, ed. C. Promis (Torino, 1841), 131.

[11] Christine de Pisan, *Livre des faits d'armes et de cheualrie*, book 3, ch. 7.

[12] In his pamphlet of 1526: "Ob Kriegsleute auch ym seligen stande sein kuenden."

Florence in 1474,[13] "that, if nothing unexpected happens, we shall in future hear more about battles against birds and dogs than about battles between armies. And those who rule Italy in peace, will earn no less fame than those who kept her in war, because the true end of war is peace." The more elevated minds discussed the possibility of abolishing entirely the plague of war and soldiery. The composition and character of military organization, its place and importance in the social order had become problems which needed reexamination. The old classifications were no longer valid. A new era was beginning.

More than a lively interest in politics and a penetrating mind, however, are needed for the perception of the rise of new historical forces and of the development of a new political constellation. It is only when, through some shattering political event, the traditional prejudices and presuppositions have collapsed, that the inadequacy of existing political concepts becomes evident, and the way is cleared for a completely new appreciation and evaluation of the political situation. Such a situation occurred when, in the year 1494, a French army under Charles VIII, equipped with strong artillery and Swiss foot soldiers, invaded Italy and overthrew her entire political system. Guicciardini, Machiavelli's friend and the great historian of this epoch, called it "a year most unhappy for Italy and, indeed, the year which headed all the following years of misery, because it opened the door to an endless number of terrible calamities,"[14] and he gave a famous description of the far reaching revolutionary consequences of this French invasion:[15]

"The effects of the invasion spread over Italy like a wildfire or like a pestilence, overthrowing not only the ruling powers, but changing also the methods of government and the methods of war. Previously there had been five leading states in Italy, the church state, Naples, Venice, Milan and Florence; and the foremost interest of all these states had been to maintain the status quo. Each had tried to prevent the other from extending its territory and from becoming so powerful that it would be a threat to the others. They eagerly observed the slightest move on the political chess board and made a great fuss whenever the smallest castle changed its ruler. When war broke out, the forces were equal, the military organization slow and the artillery cumbersome, so that usually the entire summer was spent on the siege of one castle; the wars lasted very long and the battles ended with small losses or no losses at all. Through the invasion of the French, everything was thrown upside down, as though by a sudden hurricane; the bonds which held the rulers of Italy together were broken, their interest in the general welfare extinguished. In looking around and noticing how cities, dukedoms and kingdoms were shattered, each state became frightened and began to think only of its own security, forgetting that fire in the house of the neighbor could easily spread and bring ruin to himself. Now the wars became quick and violent, a kingdom was devastated and conquered more

---

[13] Cappelli, "Lettere e notizie di Lorenzo de' Medici," *Atti e Memorie della RR. Deputazione di Storia Patria per le Provincie Modenesi e Parmensi,* I (1863), 251.
[14] Francesco Guicciardini, *Storia d'Italia,* book 1, ch. 6. Ed., C. Panigada (Bari, 1929), I, 42.
[15] F. Guicciardini, *Storie Fiorentine,* ed. R. Palmarocchi (Bari, 1931), pp. 92-93.

quickly than previously a small village; the sieges of cities were very short and were successfully completed in days and hours instead of in months; the battles became embittered and bloody. Not subtle negotiations and the cleverness of diplomats, but military campaigns and the fist of the soldier decided over the fate of the states."

Guicciardini's words indicate how deeply the Italians felt the contrast between the conditions of the fifteenth and the sixteenth centuries. In the fifteenth century, conscious of their wealth and proud of the new life which their inventions and advances in art and scholarship had ushered in, they became used to looking down upon the other European powers whose social system and intellectual life was still in the shackles of superstition and prejudice. Now, in the sixteenth century, the fate of Italy lay in the hands of those very states which the Italians had believed they had a right to despise. Guicciardini's words indicate also the way in which the Italians explained their defeat. Because the superiority of Italian civilization in the economic and intellectual sphere was manifest, they blamed their neglect of the modern techniques of war. The girdle of castles and fortresses, with which each Italian state had protected the roads that led into its territory, had quickly fallen before the artillery of Charles VIII; the mounted Italian mercenaries had been unable to resist the impact of Charles's Swiss infantry and artillery. The modern military technique had won against the obsolete one. As Machiavelli says, it had been a "guerra corte e grosso"[16]— what we would now call a blitzkrieg. It was fought, he holds, "col gesso";[17] that is, the French were able to designate "with chalk" the houses in which they wished to billet their troops, without fear of resistance from the weak Italian forces. Since then, attracted by the easy French victories and the military impotence of the Italians, the Spaniards and the Germans had stretched out their hands for the same prize. To their dismay, the Italians were forced to become mere onlookers as their country became the battlefield of Europe and the center of attraction for all foreigners in search of military renown. All Italy admired and dreaded the names of the Gran Capitano, Gaetano di Consalvo, who, in his Neapolitan campaigns, had worked the miracle of transforming an indifferent crowd of Spanish mercenaries into a well-disciplined infantry, of Gaston de Foix, who, by quick movement of his troops and the innovation of night marches, outmaneuvered his more numerous enemies, and of Frundsberg, the organizer of the German "Landsknechte" and their leader in the future "sacco di Roma." Those who speculated about the fate of Italy arrived necessarily at the conclusion that the Italians had to reform their military institutions if they wanted to equal the foreign barbarians and become again master in their own house.

The general interest in military questions which this revolutionary period and the calamity of Italy were bound to evoke was intensified in Machiavelli's case by his practical experience in Florentine politics, where he had learned a special lesson in the field of military organization and its political implications.

[16] *Discorsi*, book 2, ch. 6 (*Opere*, p. 147).
[17] *Principe*, ch. 12 (*Opere*, p. 25).

It was the great tragedy of Machiavelli's life that he played an active political part only for a relatively short period, from 1498-1512, while Piero Soderini was the highest official, the Gonfaloniere, and directed the politics of the Florentine city-state. The coincidence of Machiavelli's political activity with Soderini's rule was no mere chance. After the expulsion of the Medici and a brief period of tumult and disorder, the aristocrats and the democrats, the two rival factions of Florence, had compromised upon the election of Soderini. Soderini, unable or unwilling to trust fully either of the two opposing factions, came to consider the permanent bureaucracy as the main prop of his regime. Machiavelli, the descendant of an impoverished noble family—neither an aristocrat nor a democrat—would never have had much chance of becoming a leading figure in either of the two Florentine parties. But as secretary in the chancellery, he belonged to the group which Soderini favored; there he had an opportunity to show his mettle. Soderini quickly recognized the talents of the ambitious young man; he drew him into his intimate circle and employed him in important diplomatic and administrative tasks. Machiavelli, therefore, came to grips with the great military problem which overshadowed Soderini's entire administration, the problem of the reconquest of Pisa.

Pisa, the great seaport at the mouth of the Arno, had made use of the confusion of the French invasion to free herself from the Florentine rule. The stabilization of Soderini's regime clearly depended on his success in reconquering Pisa. Year after year, the best Italian condottieri were taken into the services of Florence; the most daring and fantastic schemes, like that of depriving Pisa of her water supplies by diverting the course of the Arno, were tried. But every year when the approach of winter halted all military operations, Pisa remained unconquered. This failure was a continuous source of popular discontent against the Soderini regime. It resulted in a loss of prestige to Florence. Moreover, the necessity of maintaining perennially a troop of mercenary soldiers constituted a steady strain on the treasury and the purses of the taxpayers. Soderini and his friends desperately searched for a novel method by which the siege of Pisa could be brought to an end and the financial pressure eased. Among the ideas that were advanced, one suggested the employment of the man power of Tuscany for the formation of a popular militia. We do not know whether Machiavelli was the first to suggest this scheme but we do know that he composed the decisive memoranda on the basis of which was promulgated the Ordinanza (1506), i.e. the law which established obligatory military service for all men between 18 and 30.

The Ordinanza did not establish a rigorous or comprehensive system of conscription; it was hardly more than a first step in that direction. The obligation of military service did not apply to the citizens of Florence but was restricted to the people of the *contado*, to the population of the agricultural districts of Tuscany subjected to the rule of Florence. Even among them, only a small number were selected and special care was taken not to disturb the activities of civilian life. In peacetime, the training did not impose a very heavy burden on the conscripted; on Sundays and holidays the villagers were to 'practice the fundamentals of marching and of using their pikes. Twice a year the men from the various villages were to march to the central town of their district to be trained there for two days in larger formations. The Florentine politicians had not dared to accept more drastic proposals because they feared that the peasants of Tuscany, once armed, would revolt against the domination of Florence, or that, with the help of an efficient military organization, Soderini would make himself an absolute ruler.

That this half-hearted attempt to mobilize the man power of Florence led to any practical results, and that, from the year 1507, 2000 militiamen took part in the siege of Pisa, was mainly due to Machiavelli's industry. The work connected with conscription was transacted in his office; he rode about the country selecting the men who were to be taken into military service; and he supervised their training. He was also responsible for the choice of officers. When the militia was camping before Pisa, he was charged with their supply. Although the militia served only as a reinforcement of the mercenary troops, its participation in the siege was of great importance for the final success of the Florentines. The forces of the militia kept up the siege throughout the winter and, by preventing the Pisans from getting new provisions, starved them and forced them to surrender in 1509.

The good account which the militia had given of itself before Pisa had increased the confidence of the Florentines in this new institution. They relied heavily on their conscripted army, when, two years later, the army of the emperor approached Florence to reestablish the rule of the Medici. Against the seasoned troops of the imperial army, however, the militia failed disastrously. The militiamen had been massed in Prato, a small town, to guard the road to Florence. But the imperial troops succeeded in breaching the walls of Prato in their first attack. The militia was stricken by panic and fled without offering serious resistance. In the ensuing rout, more than 4000 people, mostly soldiers of the militia, met their death, a slaughter famous even in a period accustomed to cruelty and ruthlessness. The road to Florence was open and the Medici returned as victors to their native city.

The reestablishment of the Medici put an end to Machiavelli's political career. All his attempts to regain office were in vain. In the retirement which was forced upon him, he turned from action to thought, from political practice to political theory. In his reflections on his political experience the problems of war and military organization must have had a particularly bitter taste to him. Was not the militia, to a certain extent his brainchild and creation, one of the contributory factors in the fall of the republican regime and consequently in the unfortunate turn in his life? Yet Machiavelli's personal concern with this problem did not lead him to a querulous justification of his actions or to petty accusations of the mistakes of others. It is a distinguishing mark of Machiavelli's intellect that he envisaged the whole historical context behind the isolated political fact and that he was not content until he had discovered the general rule which explained single phenomena. Machiavelli's desire for justifying his actions was thus a stimulus to fruitful general reflections; this in turn revealed to him the impact which military factors had on the fate of Italy and on the development of his time. He deeply probed into the relations between political and military power. His own subjective experiences turned him to an objective examination of the military crisis of his time. He thus became the first military thinker of modern Europe.

II

As we have seen, Machiavelli became interested in military questions because, in his own experience, he had felt their impact on the general development of political events. An analysis of Machiavelli's military theories cannot be limited, therefore, to his military work, the *Art of War*, which is exclusively devoted to military matters. In his political and historical writings, war and military organization also play an important role, in the *Principe*, in the *Discorsi* and in the *Florentine History*. The differences which may be found between the military ideas of the *Art of War* and of Machiavelli's other historical and political writings spring from the different aims of these books. The *Art of War* gives a systematic, even largely technical, exposition of Machiavelli's military ideas; the *Principe* and the *Discorsi* present them as suggestions, in a somewhat aphoristic form. The *Art of War* is concerned with a positive program of military reform. The observations on military affairs in the *Principe* and the *Discorsi* are mainly of a negative nature; they were criticisms of the military institutions of the time.

Machiavelli's criticisms and negative remarks were directed against the military system which had flourished in Italy during the fifteenth century, prior to the French invasion. The condottieri and their mercenary horse companies were the butt of his contempt: "They are disunited, ambitious, without discipline, faithless, bold amongst friends, cowardly amongst enemies, they have no fear of God, and keep no faith with men."[18] His low opinion of their achievements finds satirical expression in his *Florentine History*, when he speaks of the battles which the armies of the condottieri had fought in the fifteenth century. In the battle of Zagonara, a victory "famous throughout all Italy, none were killed excepting Lodovico degli Obizzi, and he together with two of his men was thrown from his horse and suffocated in the mud."[19] In the battle of Anghiari "lasting from the twentieth to the twenty-fourth hour only one man was killed, and he was not wounded nor struck down by a valiant blow, but fell from his horse and was trampled to death."[20] In the battle of Molinella which lasted half a day "nobody was killed; only some horses were wounded and a few prisoners made on either side."[21] Machiavelli explained that the condottieri and their troops fought so badly because their interest in the war was purely mercenary. "They have no love or other motive to keep them in the field beyond

[18] *Principe*, ch. 12 (*Opere*, p. 25).
[19] *Istorie Fiorentine*, book 4, ch. 6 (*Opere*, p. 475).
[20] *Istorie Fiorentine*, book 5, ch. 33 (*Opere*, p. 528).
[21] *Istorie Fiorentine*, book 7, ch. 20 (*Opere*, p. 578).

a trifling wage, which is not enough to make them ready to die for you."[22] According to Machiavelli, prevalence of financial considerations determined the nature of the military organization as well as the conduct of war in Italy during the fifteenth century. Since soldiers were the working capital of the condottiere, he did not want to waste them. He shunned battles and preferred a war of maneuver. If a battle, however, could not be avoided, he tried to keep the losses down. This was a period of bloodless battles. On the other hand, short wars were not in the interest of the condottieri; they did not want to lose their employment; even when victory seemed assured, they drew the wars out over several campaigns. Machiavelli hinted that even the neglect of infantry in Italy had its cause in the financial interests of the condottieri. Foot soldiers could be much more cheaply equipped than mounted troops which were the working capital of the condottieri. Without riding troops, every power would have recourse to conscription as the cheapest way of outnumbering the army of its neighbor. Employment of foot soldiers would make the condottieri superfluous. And in general, he said, little improvement was possible under existing conditions. Since all the condottieri were guided by the same egoistic motives and considered war merely as a business proposition, it was in their common interest to observe the accepted rules of the business and to play the same game.

Machiavelli's characterization of the Italian methods of warfare in the fifteenth century should not be considered as historical truth. In depicting the condottieri of the fifteenth century, he, like the great portrait painters of his time, built up his characterization around a few dominating traits, suppressing everything which would confuse the preciseness of the outline and the clear conception of the type. It must be said against Machiavelli's description that, in the second half of the fifteenth century, the condottieri began to show interest in military innovations and, although on a relatively small scale, began to employ foot soldiers and artillery. Moreover, we must realize that there were intense rivalries between the condottieri; consequently, if only for reasons of their own personal ambition and prestige, they were most eager to defeat the enemy. If they frequently had to take recourse to a protracted war of maneuver, it was hardly due to conscious intention or actual ill will. Their strategy was dictated by the political situation of Italy in the fifteenth century. The relatively restricted means of the small states and their approximate equality of strength formed an insuperable obstacle to any concentration of power which could have made military reforms possible on a large scale.

Machiavelli was not unaware of this interrelationship between Italy's political system and the obsolescence of her military machine. He realized that more than measures of a military-technical character, more than the abolition of the mercenary system and the employment of infantry, were needed to enable Italy successfully to resist aggression by other European powers. Something was fundamentally wrong with the whole spirit in which the Italians conducted their wars. "That cannot be called war where men do not kill each

[22] *Principe*, ch. 12 (*Opere*, p. 25).

other, cities are not sacked, nor territories laid waste."[23] Complete destruction of the enemy state must be the chief aim in war; real war is a fight for existence and in such struggle everything is permitted. "Where the very safety of the country depends upon the resolution to be taken, no considerations of justice or injustice, humanity or cruelty, nor of glory or of shame, should be allowed to prevail."[24] Means of warfare should be judged solely with reference to their efficacy. Machiavelli admiringly says of Castruccio Castracani: "If he was able to win by fraud he never attempted to win by force, because he said that the victory, not the method of gaining it, brought glory to the victor."[25] Thus Machiavelli thought that a general's interest should not be restricted to purely military actions; he ought also to devise efficient methods of deceiving the enemy and employing ruses—like the spreading of false rumors—to discourage him.[26] Machiavelli was a great admirer of Frontinus (40-103), whose book on Stratagems was devoted to the subject of artifices in war, and he recommended many of Frontinus' devices. It would be farfetched to proclaim Machiavelli a forerunner of modern psychological warfare; his approach indicated, however, a new and radical conception of war. For in Machiavelli's own time—for instance, at the battle of Ravenna—the adversaries had still exchanged polite challenges and chivalrous compliments before the opening of hostilities; the view persisted, at least in theory, that wars were fought with clearly defined means and according to established rules. There were still traces of the medieval tradition which conceived of war as a trial by battle, in which the contestants had to meet under equal and fair conditions. This was the background of Machiavelli's revolutionary view that the employment of all possible forces was permitted in war: the whole state was conceived by him as a living being whose entire resources, whose strength, intelligence and courage, were put to test in wartime.

Machiavelli has often been criticized for his seeming misjudgment of the importance of the invention of artillery and for his underestimation of the role of money in war. Yet in the light of his radical concept of war, his views on these points appear quite logical and understandable. The famous chapter in the *Discorsi* on "the value of artillery to modern armies, and whether the general opinion respecting it is correct"[27] was not a dispassionate treatise on the effects of the invention of artillery upon the development of warfare, but rather was concerned with but one aspect of the problem: the importance of courage and initiative in relation to this new instrument of war. Machiavelli had heard people say "that hereafter wars will be made altogether with artillery."[28] All his arguments on this subject were meant to prove the incorrectness of this view. He did not deny that artillery increased the striking power but he impa-

[23] *Istorie Fiorentine*, book 5, ch. 1 (*Opere*, p. 499).
[24] *Discorsi*, book 3, ch. 41 (*Opere*, p. 256).
[25] *Vita di Castruccio Castracani* (*Opere*, p. 761).
[26] Cf. particularly *Discorsi*, book 3, ch. 14: "Of the Effects of New Stratagems and Unexpected Cries in the Midst of Battle."
[27] *Discorsi*, book 2, ch. 17 (*Opere*, pp. 162-166).
[28] *Opere*, p. 162.

tiently rejected the idea that artillery alone could be decisive. The result of the invention of artillery was not that war had become the specialty of technicians and engineers; the combination of all forces, military and spiritual, of a country still remained necessary; the ability of the commander and the courage of the soldier were always the decisive factors.

Machiavelli made the same point in his discussion of the role of money in war. In his chapter entitled: "Money is not the sinews of war although it is generally so considered"[29] he concluded that "It is not gold, but good soldiers that insure success in war, . . . for it is impossible that good soldiers should not be able to procure gold, as it is impossible for gold to procure good soldiers." Machiavelli's contemporaries, for instance Guicciardini,[30] concluded from these statements that Machiavelli was a mere theorist who had no acquaintance with practical affairs. But there are several passages in Machiavelli's letters[31] which show that, in weighing the possible chances of a political conjuncture, he took the financial factor into account. Machiavelli did not want to imply that financial resources are without importance for the conduct of war, but was thinking of the great Italian cities, like Florence and Milan, which, in spite of their wealth, had succumbed to foreigners. His contention was rather that the basis of political power is military power, and money constitutes political power only if it is in fact transformed into military strength.

Nevertheless, there is a wider meaning in Machiavelli's remarks on the relation between financial and military power. He had the feeling that the virtues which war demands are incompatible with the attitude resulting from commercial activities. He believed that the pacific inclinations of the Italians in the fifteenth century prevented the development of a true military spirit, and thought that there was an interrelationship between the spreading of this pacific atmosphere and the preoccupation with commercial interests. In one of his self-revealing letters to his friend Vettori, he said that "Fortune has decreed that since I cannot discuss silk making or wool manufacture, or profits and losses, I have to discuss politics. I must either make a vow of silence or talk about that subject."[32] Certainly he did not think it was entirely to his discredit that he felt misplaced in the society of the wool and silk merchants who ruled Florence. In his opinion, the great role which financial considerations played in the minds of the Florentines, was a reason for the political decline of Florence. In a state which aspired to political greatness, political interests must outweigh all others. Machiavelli's denunciation of the mercenaries, his criticism of the hesitant and ineffectual methods of Italian warfare found a positive expression in his demand for a new spirit of military heroism. Softness and a lack of fighting spirit, he thought, were the result of a preoccupation with personal well-being, inextricably connected with a society dominated by finance and commercial interest. Only a people, to whom the greatness of their country,

---

[29] *Discorsi*, book 2, ch. 10 (*Opere*, pp. 152-153).
[30] Francesco Guicciardini, "Considerazioni intorno ai Discorsi del Machiavelli," *Scritti politici e Ricordi*, ed. R. Palmarocchi (Bari, 1933), pp. 50-51.
[31] Cf. particularly the letter to Francesco Vettori, December 20, 1514.
[32] Letter to Francesco Vettori, April 9, 1513 (*Opere*, p. 882).

and not their personal fate, was their highest value, and who were willing to sacrifice everything for their political beliefs and to die for them, could provide the soldiers for an irresistible army.

The main principles of Machiavelli's ideas on military reform can be easily deduced from these criticisms of the previous system. He advocated an army of foot soldiers, formed on the basis of general conscription. Such military organization, however, would necessitate political reforms; it could be successful only if accompanied by a new spirit, which raised political values above all others. Moreover, people who rule themselves will be most willing to fight for themselves; there is an affinity between democracy and the idea of an army of conscripted foot soldiers.

Yet the reader will be disappointed who expects to find in Machiavelli's *Art of War*, his main work on military affairs, a detailed discussion of how these ideas can be applied to the conditions of the sixteenth century and a realistic description of warfare in this period. Machiavelli was a son of the Renaissance; his way of proving the validity of his ideas was to show that they had been valid in the ancient world. This approach determined the method and content of the entire book. He believed that the strongest point which he could make in favor of his theories was that the Roman army was an army of conscripted foot soldiers.

Machiavelli's book was to a large extent, therefore, an explanation of the Roman military institutions. The examples which he used were mostly taken from classical historians like Livy and Polybius; he followed widely—even slavishly—the classical authorities on military science; his treatise was, indeed, a kind of adaptation of classical military wisdom to his own time. His chief source of inspiration was Vegetius' *De Re Militari*. His own book was concerned with the same subjects: the selection of men, the equipment and training of soldiers, the character of battle, and the various incidents which might come up in the course of military action, the order of march and encampment, and the art of fortification. In the organization of these topics also, Machiavelli adhered to the scheme of Vegetius' book. One might perhaps say that one reason for the neglect of Machiavelli's *Art of War* in modern times is the fact that it is frequently considered as nothing but an Italianized and somewhat modernized Vegetius.

There were occasions when he deviated from his classical models. Since it is quite evident that Machiavelli considered the revival of ancient military science as his main task, they must have seemed particularly important to him and deserve particular attention. The most striking departure from Vegetius was the extensive treatment of the importance of battle in warfare. In contrast to Vegetius, who had dealt with this subject rather briefly, the battle formed the main topic of an entire book in Machiavelli's *Art of War*. This book, the third, containing the description of an imaginary battle, stands very distinctly in the center and heart of the whole work. The preceding two books, dealing with the selection of men and the training of the soldier, describe how an army fit for military action can be put into the field; they lead up to the battle. After the

climax of the battle the tension relaxes and the subjects of the succeeding books
—order of marching, camping, fortification—are taken up one after the other
in brief treatises only loosely connected with each other. The organization of
this part is much less rigid. Thus indirectly, by the form of composition, the
battle is set off as the chief part of the work. Moreover the importance of the
battle is impressed upon the reader in direct statements, scattered throughout
the whole book. "If [a general] wins a battle, it cancels all other errors and
miscarriages";[33] the battle is "the end for which all armies are raised, and
hence much care and pains are to be taken in disciplining them."[34] "The main
end and design of all the care and pains that are bestowed in keeping up good
order and discipline is to fit and prepare an army to engage an enemy in a
proper manner, because a complete victory commonly puts an end to war."[35]
Moreover, decision by battle cannot be avoided. If your adversary is decided
upon a battle, he can always force you into a position in which you have to
accept his challenge. "A general cannot avoid a battle when the enemy is
resolved upon it at all hazards."[36] Furthermore, since the invention of artillery,
castles and fortresses have been of no value in halting the advance of the
enemy. The battle is the necessary climax in every war.

Lastly the way in which the battle is treated in the third book accentuates the
importance which Machiavelli attributed to this section of the book; it is a
lengthy, carefully worked out description which makes full use of dramatic
possibilities.

He describes in detail what is now of mainly antiquarian interest, how
an army should be drawn up in battle line: the infantry, the main forces in the
middle; cavalry and light infantry on the sides to cover the flanks. After an
artillery volley, and after skirmishes by the cavalry and the light infantry, the
real battle opens when the field is cleared for the clash of the main forces, the
foot soldiers. Those who carry pikes form the first row and press against
the enemy; when the distance between the two enemies has become smaller, the
spearmen change their place with the swordsmen. This is the crucial moment.
The dexterity with which this maneuver is carried out will decide the issue:
"What carnage! What a number of wounded men! They begin to run away.
See, they are flying on the right and on the left. The battle is over. We have
gained a glorious victory!"[37] With these elated words, Machiavelli ends his
description of a battle. It is less an analysis of the possible maneuvers than a
narration of what a favorably placed observer would see; it is a battle painting
in words. Because the battle is seen from the outside, the actions of the battle
seem to be pre-determined; it seems to roll off according to plan, and the
elements of crisis, of suspense, and decision are lacking. The art of literary
analysis of a military action was still at its beginning.

Yet Machiavelli conceived of a battle as functioning like a well-oiled mech-

[33] *Arte della Guerra* (*Opere*, p. 275).
[34] *Arte della Guerra* (*Opere*, p. 303).
[35] *Arte della Guerra* (*Opere*, p. 352).
[36] Title of chapter 10 of *Discorsi*, book 3.
[37] *Arte della Guerra* (*Opere*, p. 309).

anism, because this concept corresponded to the course of a real battle. In this period, once a battle had started, there was little opportunity for initiative. When the squares of the foot soldiers pressed against each other, a maneuvering of these dense throngs was well-nigh impossible. The side which put more sustaining power behind its pressure, the side which had the strongest "push," won. The training which the soldier had received before the battle, therefore, played a decisive role; the outcome of the battle depended on the cohesion of the rows and the disciplined rhythm with which they executed their push. Frundsberg, for instance, was chiefly a drill master and owed his military fame to the discipline in which he had trained his renowned soldiers.

From the emphasis which Machiavelli puts on the battle, it can be seen that he was much concerned with the problem of military discipline. One might say that, next to the battle, this is the chief theme of the *Art of War*. The great importance of good discipline is stressed at every opportunity. It constitutes the foundation of a good army: "Good order makes men bold and confusion cowards."[38] Discipline is more effective than courage and can overcome mere power: "Few men are brave by nature, but good order and experience make many so. Good order and discipline in an army are more to be depended upon than courage alone."[39]

The problem of army discipline assumes for Machiavelli two different aspects. First of all, it is necessary to teach the individual soldiers the fundamentals of the use of arms and to accustom them to act in formation. "They must learn to keep their ranks, to obey words of command, and signals by drum or trumpet, and to observe good order, whether they halt, advance, retreat, are upon a march, or engaged with an enemy."[40] The second, more important, aspect of discipline—one that occupied a prominent place in military discussion during the following two centuries—was the problem of subdividing an army into smaller tactical units. In order to maintain discipline during action the throng of foot soldiers had to be broken up into smaller groups and so organized as to make possible a certain flexibility and maneuvering. In drawing up the battle line, Machiavelli recommended the formation of three groups, standing behind each other, so that the fight could be taken up again when the first push did not succeed. He advised taking the Roman legion as a model; the largest unit should be the battaglione, consisting, like the legion, of 6000 to 8000 men; like it, the battaglione should be divided into ten units, each commanded by its own officer. Even if so organized, Machiavelli considered a very large army unwieldy; he recommended as maximum size an army of 50,000 men "for more are apt to create discord and confusion, and not only become ungovernable themselves, but corrupt others that have been well disciplined."[41]

This is a striking example that, to Machiavelli's mind, there existed one definite norm which the military organization of each country should aim to

[38] *Arte della Guerra* (*Opere*, p. 392).
[39] *Arte della Guerra* (*Opere*, p. 362).
[40] *Arte della Guerra* (*Opere*, p. 292).
[41] *Arte della Guerra* (*Opere*, p. 347).

realize. That is the fundamental thesis of his entire military thought. He frequently spoke of the necessity of taking particular circumstances and special conditions into account but actually he was concerned only with the establishment of rules and precepts of general validity. This prevalence of general principles and the lack of realistic detail seem to make it doubtful that his military ideas could have had any relation to military practice.

Such doubts about the practicability of his work are further increased by the form in which the ideas are presented. The *Art of War* is conceived as a dialogue between three members of Florentine aristocratic families and the condottiere Fabrizio Colonna, taking place in the garden of the Rucellai family, a place famous for its philosophical meetings and discussion. The setting in itself is characteristic of the Platonic remoteness and the enthusiasm for ancient Rome which permeates the whole work, but this very fact tends to confuse the modern reader. He finds himself wondering whether the speakers are men of the Renaissance or of the ancient world, and whether the ideas they express refer to the present or to the past.

In short, the ideas expressed in the *Art of War* seem to bear a patina which makes the outlines of the work indistinct. Without a further elucidation and interpretation of its place in the history of military thought, it might be difficult to understand that this work represents an important step in the development of that thought and that it laid the foundation of a building which is still in the process of construction.

III

The *Art of War* became a military classic. It was printed in no less than seven editions in the sixteenth century and was translated in most European languages. Montaigne named Machiavelli next to Caesar, Polybius, and Commynes as an authority on military affairs.[42] Although, in the seventeenth century, changing military methods had brought other writers into the foreground, Machiavelli was still frequently quoted. In the eighteenth century, the Marshal de Saxe leaned heavily on him when he composed his *Reveries Upon the Art of War* (1757) and Algarotti—though without much basis—saw in Machiavelli the master who had taught Frederick the Great the tactics by which he astounded Europe.[43] Like most people concerned with military matters, Jefferson had Machiavelli's *Art of War* in his library,[44] and when the War of 1812 had increased American interest in problems of war, the *Art of War* was brought out in a special American edition.[45] Yet in the nineteenth century, when Machiavelli's fame as a political thinker reached a new height, his influence as a military authority began to pale. To many he remained the prophet of

---

[42] Montaigne, *Essais*, Livre 3, ch. 34: "Observations sur les moyens de faire la guerre de Julius Caesar."

[43] F. Algarotti, Lettres 8 and 9 of his work *Scienza Militare del Segretario Fiorentino*, printed in F. Algarotti, *Opere* (Venezia, 1791), vol. V.

[44] Catalogue of the Library of Congress 1815, i.e. Jefferson's Library.

[45] *The Art of War in Seven Books Written by Nicholas Machiavel . . . to Which Is Added Hints Relative to Warfare by a Gentleman of the State of New York* (Albany, 1815).

modern war, who had foreseen the great innovation of the French Revolution, general conscription. But others emphasized his failure to recognize the importance of artillery and saw in his enthusiastic recommendation of Roman military institutions an indication of his lack of realistic knowledge of military affairs.

It can hardly be said that, in this conflict of opinion, right or wrong is all on one side. For example, Machiavelli's recommendation of general conscription, startling as it is in the light of recent developments, is more a confirmation than a refutation of the view which ascribes to Machiavelli lack of insight into the real conditions of his time. For a penetrating analysis of the political forces of his period—the rise of money power, the growth of princely absolutism— would have shown that the immediate future belonged to the permanent professional army and that, under the existing conditions, a citizen army after the Roman pattern was purely a romantic dream.

We have seen, on the other hand, that Machiavelli's skepticism concerning the effects of artillery resulted from the tenable and sound thesis that, in spite of the impact of technical innovations, the fundamental elements of war remain the same. Furthermore, it should not be overlooked that Machiavelli's recommendation of the Roman example was not so utopian and impracticable as it seems to us today. The Roman legion was the pattern which stimulated the military reforms of the sixteenth century, first those of Francis I of France and then—still more important—those of Maurice of Nassau; the latter, on the basis of precise study of the Roman military techniques, created the regiment as the basic infantry unit and his methods of infantry organization and training were soon imitated all over Europe. Roman military methods had a direct and important influence upon modern developments. Thus the realism of Machiavelli's military ideas can be argued endlessly.

It is a mistake to evaluate Machiavelli's military theories from the standpoint of their immediate utility. The criterion of their value should be whether they represented a new and original approach to the problems of war and opened a new vista by which a clearer realization of the conditions and exigencies of warfare and military organization became possible. Not the usefulness of particular and specific measures, but the fruitfulness of his general methods and concepts is the yardstick with which to gauge Machiavelli's ranking in the history of military thought.

Machiavelli has been called the first modern military thinker. He was not, of course, the first to concern himself with military affairs; there were many before him and aside from him who had concerned themselves with these subjects. But Machiavelli raised military discussion to a new level and established the principles according to which intellectual comprehension and theoretical analysis of war and military affairs progressed. It is only by comparing his military ideas with those of earlier and contemporary military writers that we can understand clearly the peculiarities of his approach and the real nature of his achievement.

As part of the social activities of mankind, military affairs were included in

the social philosophy of the great medieval theologians who adapted the classical literature to medieval conditions and subordinated it to the ethical norms of
Christianity; by tradition, military problems remained a part of political
theory. Then the humanists put special emphasis on the military aspects of
political life. Because of their enthusiasm for Rome and Roman history,
military deeds and military history were seen as the essence of history and
political greatness. Finally, the rise of natural science and technology had given
birth to a new literature on technical topics; the technological parts of military
art—such as instructions in the use of certain weapons or the problems of
military architecture—were discussed in numerous treatises.

Unquestionably all three of these trends influenced the formation of
Machiavelli's military ideas. The establishment of the canonical authority of
Vegetius had been the work of the Middle Ages; the two best medieval
treatises on military art—those by Egidio Colonna[46] and Christine de Pisan[47]—
were chiefly revisions of Vegetius for the use of feudal times. Machiavelli, by
fitting Vegetius to the needs of his own period, followed in their steps. A
reflection of the technical discussions can be found in his interest in detailed
description of the handling of weapons and above all in the section on the art
of fortification,[48] which shows Machiavelli's familiarity with the problems of
military architecture.

The influence of the humanists, however, is most evident. "Nothing has
weakened the influence and power of Italy more than that the use of arms
became unfamiliar," wrote Platina in 1468.[49] Before Machiavelli, Florentine
statesmen, like Matteo Palmieri (1406-1475), had criticized the mercenary
soldier and advocated the arming of the citizens.[50] Humanists like Patricius
(1412-1494) had recommended obligatory military training for all young
men.[51] They had bewailed the softness of their time and the lack of military
heroism and considered the trading spirit as the cause of the political indifference of their period. At the end of the fourteenth century Salutati (1331-
1406) already had observed that Florence's policy was "not determined by
ambitions, which are typical of the nobility, but by the interests of trade; and
since nothing is more hostile and detrimental to merchants and artisans than
the disturbance and confusion of war, certainly the merchants and artisans who
rule us love peace and hate the waste of war."[52]

Because of these evident connections between Machiavelli's thought and
that of previous writers on military affairs, it is the more characteristic that he

[46] The third part of the third book of Egidio Colonna, De regimine principum, is concerned with military institutions and war: "Quomodo regenda sit civitas aut regnum
tempore belli."
[47] Cf. note 11 above.
[48] The seventh book of the Arte della Guerra.
[49] Platina, De vero Principe, book 3, ch. 6 (quoted in G. Gaida's Introduction to Platina's
Historici Liber de Vita Christi ac omnium pontificum, in Muratori, Rerum italicarum
scriptores [1913] III, Part 1, p. xiv).
[50] Matteo Palmieri, Libro della Vita Civile (printed 1529), p. 111 and p. 76.
[51] Franciscus Patricius, De institutione rei publicae (written about 1470, printed 1594),
p. 406.
[52] Coluccio Salutati, Invectiva . . . in Antonium Luscum de florentina republica male
sentientem (written 1400), ed. D. Moreni (Firenze, 1826), p. 181.

is not dependent on them for those ideas which form the gist of his military theory. Neither his recommendation of infantry, nor his emphasis on the decisiveness of battle, nor his general concept of war can be found in the previous military literature. On the contrary, Machiavelli stood in direct opposition to the tradition of military thought on all these points. Previously the dominating role of cavalry had not been doubted. For instance, Cornazzano in his *De Re Militaria* which was published in 1507 stated that "the modern age likes horsemen; they are most handy in the enterprises of war."[53] The captains were advised to shun battles because the outcome of a battle was always extremely uncertain. "Although intelligence, courage and knowledge of military science may help, Goddess Fortuna remains the decisive factor," wrote Patricius.[54] Moreover, the whole literature had still been dominated by the concept of the "bellum justum"; war was seen as an instrument of higher ethical purpose and the principles and methods of warfare remained subordinated to ethical standards and limited by rules.[55]

Yet the most important fact is not that Machiavelli deviated on these points from tradition, and enunciated views of a more realistic nature, but that all his views are connected and form complementary and necessary parts of one system of thought. They are all based on Machiavelli's concept of war and derive from it. To him political life was a struggle for survival between growing and expanding organisms. War was natural and necessary, it would establish which country would survive and determine between annihilation and expansion. War, therefore, must end in a decision, and a battle was the best method of reaching a quick decision, since it would place the defeated country at the mercy of the victor. Because of this central importance of the battle, however, its issue could not be left to mere chance, but had to be prepared so that, as far as possible, victory was assured. Thus efficient preparation for battle was the only criterion for the composition of an army. It necessitated a reexamination of the traditional methods of military organization. Moreover, political institutions, in their spirit as well as in their form, must be shaped in accordance with military needs. The humanists' indictment of their contemporaries for their lack of military heroism had been made without further discussion of its consequences; it was a moral appeal, unorganically attached to a completely different body of thought. But in Machiavelli's military theories, all his ideas reinforce each other and belong to an organic system; they are part and parcel of his general philosophy of politics and war.

Thus the originality and strength of Machiavelli's approach lies in his breadth of perception; he encompassed the whole complex of military prob-

[53] Cornazzano, *De Re Militaria* (Pescaro, 1507), Liber 2, ch. 1.
[54] Patricius, *op. cit.*, p. 374. Cf. also "Memoriale" of Diomede Carafa (written in 1479), published by P. Pieri in *Archivio Storico per le Provincie Napoletane*, LVIII (1933), p. 208.
[55] A trend toward Machiavelli's new and radical concept of war might be seen in the distinction which some of the humanists make between "wars for life and survival" and "wars for honor and fame," cf., e.g. M. Palmieri, *op. cit.*, p. 72 and L. Bruni Aretino, *Oratio in funere Nannis Strozzae* (1430), published in E. Baluze, *Miscellaneorum Liber Tertius* (Parisius, 1680), p. 246.

lems and realized that an inner connection exists between technical military detail and the general purpose of war, between military institutions and political organization. In shaping this new outlook, the historical situation, into which Machiavelli was born, and the practical experiences which he had undergone, certainly played a decisive role. Machiavelli's insight would not have been possible without the new vista which the political events of his time and the intellectual achievements of the Renaissance opened.

The military innovations of the period, the appearance of new weapons and, the rise of infantry, would under any circumstances have aroused reflection and comment. But in a period of social equilibrium and unquestioned acceptance of political values military thought would probably have been confined to the framework of single technical suggestions and to the discussion of details. The crisis of political institutions and values which signified the end of the Middle Ages provided the opportunity for a radical reexamination of their assumptions. Since the whole political world seemed to be in flux, it appeared possible to shape it according to new principles. This does not mean that such radical approach was customary among the thinkers of the period. A Machiavelli was needed to make use of the potentialities which the crisis of the period had opened, but the crisis of his time was likewise necessary to make Machiavelli's revolutionary grasp of political problems possible and fruitful.

For such examination of the political phenomena from the point of view of new principles, the spirit of the Renaissance was ideally suited. The presupposition on which the thinkers of the Renaissance proceeded and on which their achievements in philosophy and science were based, was the conviction that, behind the phenomena of social life and human activities, there were laws which reason could discover and through which events could be controlled. The view that laws ruled also over military events and determined them in their course was a fundamental assumption in Machiavelli's concern about military affairs; his military interests centered in the search for these laws. Thus he shared the belief of the Renaissance in man's reason and its optimism that, by the weapon of his reason, man was able to conquer and destroy the realm of chance and luck in life.

This was no shallow optimism. The men of the Renaissance did not think lightly of the power of Fortuna and regarded life as a dangerous fight between man's reason and this fickle goddess. Yet they had no doubt that, in the end, man's reason would win out. When Machiavelli set the battle into the center of his military theories, he could do so because the uncertainty of the outcome which previous writers had feared did not terrify him; by forming a military organization in accordance with the laws which reason prescribed for it, it would be possible to reduce the influence of chance and to make sure of success.

This belief in the supremacy of reason was also the true cause for Machiavelli's admiration for the Roman military institutions. It was not antiquarianism, not a personal bias, which led him to prefer this one historical example to all others. The invincibility of the Roman army and the course of Roman expansion were proof that the Romans had had the best organization reason could provide. Their military institutions were a realization of the eternal norm which should determine the military institutions of every country.

By establishing as the dominant purpose of war the complete subjection of the enemy, military thought was set up as an autonomous field, with its own logic and method; a discussion of military problems on a scientific basis was made possible. To be more precise, it was possible to gauge all military measures in relation to one supreme purpose and to have a rational criterion for them; moreover the successful outcome of a war was conceived as dependent on arranging one's measures in accordance with the rational laws determining the course of military affairs. In short, in Machiavelli's mind, it was the solution of an intellectual problem upon which success in war depended. Even if the term strategy did not exist then, this was the beginning of strategical thinking.

Military thought has proceeded ever since on the foundations which Machiavelli laid. This does not mean that Machiavelli's recommendations have been accepted as final truth. Yet further discussion did not develop in opposition to his view, but rather as an expansion and enlargement of his ideas. For instance, however important Machiavelli's idea of the decisiveness of battle was, it soon became clear that there was a real need for a much more thorough analysis of its consequences. Military theory could not stop with making rules for the formation of the correct battle order; it had also to scrutinize the course of events during the combat action. On the other hand, if a battle constituted the climax of war, it is clear that the whole campaign had to be planned and analyzed in respect to the decisive battle. Such considerations show that the role which theoretical preparation and planned direction of military action played in modern war was much greater than Machiavelli had envisaged. He had made a perfunctory acknowledgment of the importance of the role of the general, but in reality he had hardly said more than that a general should know history and geography. Later, the question of planning in military leadership and of the intellectual training of the general became central problems in military thought. In developing these problems, military thought advanced far beyond Machiavelli, yet these more modern conclusions were a logical continuation of the inquiry which he had started.

Nevertheless, there is one aspect in modern military thought which not only cannot be connected with Machiavelli's thought, but is in sharp contrast to it. Machiavelli was mainly concerned with a general norm, valid for the military organizations of all states and times; modern military thought emphasizes that actions under different historical circumstances must differ and that military institutions will be satisfactory only when they are fitted to the particular constitution and conditions of an individual state. Moreover, Machiavelli's emphasis on the establishment of military institutions and conduct of war according to rational and generally valid rules gave great weight to the rational factor in military matters. Although Machiavelli began as a vehement critic of the chesslike wars of the fifteenth century, eighteenth century generals returned to some extent to wars of maneuvering and this development is not entirely against the line of thought in military science which Machiavelli had started. When war is seen as determined by rational laws, it is only logical to leave nothing to chance and to expect that the adversary will throw his hand in when he has been brought into a position where the game is rationally lost. The result of considering war as a mere science or at least of overvaluing the rational element in military affairs leads easily to the view that war can be decided quite as well on paper as on the battlefield.

It has since been realized that war is not only a science but also an art. With the end of the eighteenth century and of the age of reason, there was a sudden recognition of the importance of other than rational factors. Not the general element, but the individual and unique feature of a phenomenon was considered as of supreme importance; the imponderables were seen as no less influential than the rational and calculable elements.

The introduction of these new intellectual trends—of the realization of the importance of uniqueness and individuality, of the recognition of the creative and intuitive element aside from the scientific—into military theory is connected with the name of Clausewitz. It is remarkable, however, that Clausewitz, who usually is extremely critical and contemptuous of other military writers, is not only very careful in examining suggestions made by Machiavelli but concedes that Machiavelli had "a very sound judgment in military matters."[56] This is an indication that, despite the new features which Clausewitz introduced into military theory and which are outside the framework of Machiavelli's thought, he agreed with Machiavelli in his basic point of departure. Like Machiavelli he was convinced that the validity of any special analysis of military problems depended on a general perception, on a correct concept of the nature of war. All doctrines of Clausewitz have their origin in an analysis of the general nature of war. Thus, even this great revolutionary among the military thinkers of the nineteenth century did not overthrow Machiavelli's fundamental thesis but incorporated it in his own.

[56] C. von Clausewitz, *Strategie*, ed. E. Kessel (Hamburg, 1937), p. 41.

CHAPTER 2. Vauban: The Impact of Science on War

BY HENRY GUERLAC

AN ALMOST uninterrupted state of war existed in Europe from the time of Machiavelli to the close of the War of the Spanish Succession. The French invasion of Italy which had so roused Machiavelli proved to be but a prelude to two centuries of bitter international rivalry, of Valois and Bourbon against Hapsburg. For a good part of this period epidemic civil wars cut across the dynastic struggle, never quite arresting it, and often fusing with it to produce conflicts of unbridled bitterness. Toward the end of the seventeenth century, when civil strife had abated and the chief states of Europe were at last consolidated, the old struggle was resumed as part of Louis XIV's bid for European supremacy, but with a difference: for now the newly risen merchant powers, Holland and England, which had aided France in bringing the Spanish dominion to an end, were arrayed against her. The Peace of Utrecht (1713) was an English peace. It set the stage for England's control of the seas, but by the same token it did not weaken France as much as her continental rivals had fervently desired. It left France's most important conquests virtually intact; it scarcely altered the instrument of Westphalia which was her charter of security; and above all it left her army—the first great national army of Europe—weakened but still formidable, and her prestige as the leading military power of the continent virtually undiminished.

The military progress of two hundred years was embodied in that army; and this progress had been considerable.[1] In the first place armies were larger. Impressed as we are by the first appearance of mass armies during the wars of the French Revolution, we are prone to forget the steady increase in size of European armies that took place during the sixteenth and seventeenth centuries. When Richelieu, for example, built up France's military establishment to about 100,000 men in 1635, he had a force nearly double that of the later Valois kings; yet this force was only a quarter as large as that which Louvois raised for Louis XIV.

This expansion of the military establishment was primarily due to the growing importance of the infantry arm, which was only twice as numerous as the cavalry in the army with which Charles VIII invaded Italy, but five times as great by the end of the seventeenth century. The customary explanation for this new importance of infantry is that it resulted from the improvement in firearms; and it is true that the invention of the musket, its evolution into the flintlock, and the invention of the bayonet, all led to a pronounced increase in infantry fire power, and hence to an extension of foot soldiery. But this is only

---

[1] In this and the following section I have relied heavily upon the works of Boutaric, Camille Rousset and Susane listed in the bibliography. Louis André's *Michel le Tellier* proved the most valuable single work concerned with army reform in the seventeenth century.

part of the story. The steadily mounting importance of siege warfare also had its effects, for here—both as a besieging force and in the defense of permanent fortifications—infantry performed functions impossible to cavalry.

European armies in the seventeenth century were bands of professionals, many of them foreigners, recruited by voluntary enlistment. Except for infrequent recourse to the *arrière-ban*, a feudal relic more often ridiculed than employed, and except for the experiment of a revived militia late in the reign of Louis XIV, there was nothing in France resembling universal service. In still another respect this "national" army seems, at first glance, hardly to have been representative of the nation. Whereas the nobility competed for admission into the elite corps of the cavalry and provided officers for the infantry, and whereas the common infantryman was drawn from the lowest level of society —though not always or preponderantly from the moral dregs as is sometimes implied—the prosperous peasant freeholder and the members of the bourgeoisie escaped ordinary military service whether by enlistment, which they avoided, or through the revived militia, from which they were exempt.

Did one whole segment of society, then, fail to contribute to the armed strength of the country? By no means. The bourgeoisie made important contributions to French military strength, even though they did not serve in the infantry or the cavalry. Their notable contributions fell into two main categories. First, they were important in the technical services, that is to say in artillery and engineering and in the application of science to warfare; and second, they were prominent in the civilian administration of the army which developed so strikingly during the seventeenth century, and to which many other advances and reforms are attributable. These technical and organizational developments are perhaps the most important aspects of the progress that has been noted above. In both, the French army led the way.

II

The army which Louis XIV passed on to his successors bore little resemblance to that of the Valois kings. The improvement in organization, discipline, and equipment was due chiefly to the development of the civilian administration at the hands of a succession of great planners—Richelieu, Le Tellier, Louvois, and Vauban—whose careers span the seventeenth century.

Until the seventeenth century army affairs were almost exclusively administered by the military themselves, and there was very little central control. The various infantry companies, which had at first been virtually independent under their respective captains, had, it is true, been coordinated to some extent by uniting them into regiments, each commanded by a *mestre de camp*, subject to the orders of a powerful officer, the *colonel général de l'infanterie*. But the prestige and independence of this high office was such as to weaken, rather than to strengthen, the hold of the crown over the newly regimented infantry. The cavalry, in the sixteenth century, had likewise been only imperfectly subjected to the royal will. By virtue of their prestige and tradition, the cavalry companies resisted incorporation into regiments until the seventeenth century. The elite corps of the gendarmerie, representing the oldest cavalry units, were controlled only by their captains and by a superior officer of the crown, the constable, who was more often than not virtually independent of the royal will. The light cavalry, after the reign of Henry II, was placed under a *colonel général* like that of the infantry. Only the artillery provided something of an exception. Here bourgeois influence was strong, a tradition dating back to the days of the Bureau brothers, and the effective direction was in the hands of a *commissaire général d'artillerie*, usually a man of the middle class. But even here the titular head was the grand master of artillery who, since the beginning of the sixteenth century, was invariably a person of high station. Thus, the army manifested a striking lack of integration. Other than the person of the king, there was no central authority. And except in the artillery there were no important civilian officials.

Richelieu laid the foundations of the civil administration of the army by extending to it his well-known policy of relying upon middle-class agents as the best means of strengthening the power of the crown. He created a number of *intendants d'armée*, who were usually provincial intendants selected for special duty in time of war, one to each field army. Responsible to the intendants were a number of *commissaires* who were to see to the payment of troops, the storage of equipment, and other similar matters. Finally it was under Richelieu that the important post of minister of war was to all intents and purposes created. Under two great ministers, Michel le Tellier (1643-1668) and his son, the Marquis de Louvois (1668-1691), the prestige of this office and the complexity of the civilian administration associated with it increased mightily. Around the person of the minister there grew up a genuine departmentalized government office complete with archives. By 1680 five separate bureaus had been created, each headed by a *chef de bureau* provided with numerous assistants. It was to these bureaus that the intendants, the commissioners, even commanding officers, sent their reports and their requests. From them emanated the orders of the minister of war; for only persons of great importance dealt directly with the minister, who had thus become, in all that pertained to important military decisions, the king's confidential adviser.

Judged by modern, or even Napoleonic, standards, the French army of Louis XIV was by no means symmetrically organized. There were gross defects of all sorts, anomalies of organization and administration, vices of recruitment and officering. But this army was no longer an anarchic collection of separate units, knowing no real master but the captain or colonel who recruited them. If it possessed a clearly defined military hierarchy with clearly defined powers, and if the royal authority could no longer with impunity be evaded by underlings or challenged by rebellious commanders—this was made possible by the painstaking work of the civilian administration during the seventeenth century. The great, semi-independent offices of the crown were abolished or brought to heel. Reforms were effected within the hierarchy of general officers to make powers more clear-cut and to eliminate vagueness of function and incessant rivalry among the numerous marshals and lieutenant generals. The principle of seniority was introduced. Unity of command was made possible by creating the temporary and exceptional rank of *maréchal général des armées*, held for the first time by Turenne in 1660. A host of minor reforms were also put through during this creative period, touching such diverse matters as the evil of plurality of office within the army, which was severely checked, venality of office, which proved ineradicable, the introduction of uniform dress and discipline, and improvements in the mode of recruiting, housing, and paying the troops.

Doubtless this sustained effort to systematize and order the structure of the army reflected what was taking place in other spheres. Throughout French political life traditional rights, and confusions sanctified by long usage, were being attacked in the interest of strengthening the central power. This cult of reason and order was not merely an authoritarian expedient, nor just an aesthetic ideal imposed by the prevailing classicism. Impatience with senseless disorder, wherever encountered, was one expression, and not the least significant expression, of the mathematical neorationalism of Descartes, of the *esprit géométrique* detected and recorded by Pascal. It was the form in which the scientific revolution, with its attendant mechanical philosophy, first manifested itself in France. And it resulted in the adoption of the machine—where each part fulfilled its prescribed function, with no waste motion and no supernumerary cogs—as the primordial analogy, the model not only of man's rational construction, but of God's universe. In this universe the cogs were Gassendi's atoms or Descartes' vortices, while the *primum mobile* was Fontenelle's divine watchmaker. We often speak as though the eighteenth or the nineteenth century discovered the worship of the machine, but this is a half-truth. It was the seventeenth century that discovered the machine, its intricate precision, its revelation—as for example in the calculating machines of Pascal and Leibnitz—of mathematical reason in action. The eighteenth century merely gave this notion a Newtonian twist, whereas the nineteenth century worshiped not the machine but power. So in the age of Richelieu and Louis XIV the reformers were guided by the spirit of the age, by the impact of scientific rationalism, in their efforts to modernize both the army and the civilian bureaucracy, and to give to the state and to the army some of the qualities of a well designed machine. Science, however, was exerting other and more direct effects upon military affairs, and to these we must now turn.

III

Science and warfare have always been intimately connected. In antiquity this alliance became strikingly evident in the Hellenistic and Roman periods. Archimedes' contribution to the defense of Syracuse immediately springs to mind as the classic illustration. The cultural and economic rebirth of western Europe after the twelfth century shows that this association was not fortuitous, for the revival of the ancient art of war was closely linked with the recovery and development of ancient scientific and technical knowledge.[2] Few of the early European scientists were soldiers, but many of them in this and later centuries served as consulting technicians or even as technical auxiliaries of the army. A number of military surgeons have their place in the annals of medical or anatomical science; while still more numerous were the engineers, literally the masters of the engines, whose combined skill in military architecture, in ancient and modern artillery, and in the use of a wide variety of machines, served equally to advance the art of war and to contribute to theoretical science. Leonardo da Vinci, the first great original mind encountered in the history of modern science, was neither the first nor the last of these versatile military engineers, although he is probably the greatest.

Throughout the sixteenth century and most of the seventeenth, before the technical corps of the army had really developed, a number of the greatest scientists of Italy, France, and England turned their attention to problems bearing upon the technical side of warfare. By the year 1600 it was generally realized that the service of outside specialists must be supplemented by some sort of technical training among the officers themselves. All the abortive projects for systematic military education, such as the early plans of Henry IV and of Richelieu, gave some place to elementary scientific training.[3] The great Galileo outlines in a little known document a rather formidable program of mathematical and physical studies for the future officer. Although organized military education, to say nothing of technical education, had to await the eighteenth century, nearly every officer of any merit by the time of Vauban had some smattering of technical knowledge, or regretted that he had not. The developments of science that brought this about are best described by a brief survey of the changes in military architecture and in artillery.

The art or science of military architecture suffered a violent revolution in the century following the Italian wars of Machiavelli's time. The French artillery—using the first really effective siege cannon—had battered down with ridiculous ease the high-walled medieval fortifications of the Italian towns. The Italians' reply was the invention of a new model enceinte—the main enclosure of a fortress—which, improved by a host of later modifications, was that which prevailed in Europe until the early nineteenth century. It was characterized primarily by its outline or trace: that of a polygon, usually regular, with bastions projecting from each angle, in such a manner as to subject the attacker to an effective cross-fire. As it was perfected by the later Italian engineers this enceinte consisted of three main divisions: a thick low rampart, with parapet; a broad ditch; and an outer rampart, the glacis, which sloped gently down to the level of the surrounding countryside.

Designing these fortresses became a learned art, involving a fair amount of mathematical and architectural knowledge. A number of scientists of the first rank were experts in this new field of applied science. The Italian mathema-

[2] In this section I have relied chiefly upon my own unpublished doctoral dissertation: *Science and War in the Old Régime* (Harvard, 1941).

[3] F. Artz, *Les débuts de l'éducation technique en France, 1500-1700* (Paris, 1938), pp. 38-43.

tician Tartaglia, and the great Dutch scientist, Simon Stevin, were as famous in their own day as engineers as they are in ours for their contributions to mathematics and mechanics. Even Galileo taught fortification at Padua.[4]

Francis I of France, aware of the skill of the Italian engineers, took a number of them into his service, using them in his pioneer efforts to fortify his northern and eastern frontiers against the threat of Charles V. This first burst of building activity lasted throughout the reign of Henry II, only to be brought to a halt by the civil wars. When the work was resumed under Henry IV and Sully, the Dutch were beginning to contest the primacy of the Italians in this field, and French engineers like Errard de Bar-le-duc were available to replace the foreigners.[5]

Errard is the titular founder of the French school of fortification, which may be said to date from the publication of his *Fortification réduicte en art* (1594). In the course of the seventeenth century there appeared a number of able engineers, some of them soldiers, others civilian scientists of considerable distinction. Among the men in the latter category can be mentioned Gerard Desargues, the great mathematician, Pierre Petit, a versatile scientist of the second rank, and Jean Richer, astronomer and physicist. In the development of the theory of fortification the great precursor of Vauban, one might almost say his master, was the Count de Pagan.

Blaise de Pagan (1604-1665) was a theorist, not a practical engineer. So far as is known he never actually directed any important construction. In engineering, as in science where he fancied himself more than the dilettante that he really was, his contributions were made from the armchair. He succeeded, however, in reforming in several important respects the type of fortresses built by the French in the later seventeenth century. Vauban's famous "first system" was in reality nothing but Pagan's style, executed with minor improvements and flexibly adapted to differences in terrain. Pagan's main ideas were embodied in his treatise *Les fortifications du comte de Pagan* (1645). They all sprang from a single primary consideration: the increased effectiveness of cannon, both for offense and in defense. To Pagan the bastions were the supremely important part of the outline, and their position and shape were determined by the help of simple geometrical rules that he formulated, with respect to the outside, rather than the inside, of the enceinte.

In the development of artillery there was the same interplay of scientific skill and military needs during the sixteenth and seventeenth centuries. Biringuccio's *De la pirotechnia* (1540), now recognized as one of the classics in the history of chemistry, was for long the authoritative handbook of military pyrotechnics, the preparation of gunpowder, and the metallurgy of cannon. The theory of exterior ballistics similarly was worked out by two of the founders of modern dynamics, Niccolo Tartaglia and Galileo. Perhaps it would not be too much to assert that the foundations of modern physics were a by-

---

[4] J. J. Fahie, "The Scientific works of Galileo," in *Studies in the History and Method of Science*, Charles Singer, ed. (1921), II, 217.

[5] Lieutenant Colonel Augoyat, *Aperçu historique sur les fortifications*, I, 13-21.

product of solving the fundamental ballistical problem. Tartaglia was led to his criticisms of Aristotelian dynamics by experiments—perhaps the earliest dynamical experiments ever performed—on the relation between the angle of fire and the range of a projectile. His results, embodying the discovery that the angle of maximum range is $45°$, brought about the widespread use of the artillerist's square or quadrant. But to Galileo is due the fundamental discovery that the trajectory of a projectile, for the ideal case that neglects such disturbing factors as air resistance, must be parabolic. This was made possible only by his three chief dynamical discoveries, the principle of inertia, the law of freely falling bodies, and the principle of the composition of velocities. Upon these discoveries, worked out as steps in his ballistic investigation, later hands erected the structure of classical physics.

By the end of the seventeenth century the progress of the "New Learning" had become compelling enough to bring about the first experiments in technical military education and the patronage of science by the governments of England and France. The Royal Society of London received its charter at the hands of Charles II in 1662, while four years later, with the encouragement of Colbert, the French *Académie Royale des sciences* was born. In both of these organizations, dedicated as they were at their foundation to "useful knowledge," many investigations were undertaken of immediate or potential value to the army and navy. Ballistic investigations, studies on impact phenomena and recoil, researches on improved gunpowder and the properties of saltpeter, the quest for a satisfactory means of determining longitude at sea: these, and many other subjects, preoccupied the members of both Academies. In both countries able navy and army men are found among the diligent members. In France especially the scientists were frequently called upon for their advice in technical matters pertaining to the armed forces. Under Colbert's supervision scientists of the *Académie des sciences* carried out a detailed coast and geodetic survey as part of Colbert's great program of naval expansion, and what is perhaps more important, they laid the foundations for modern scientific cartography so that in the following century, with the completion of the famous Cassini map of France, an army was for the first time equipped with an accurate topographic map of the country it was charged to defend.

IV

If we ask how these developments are reflected in the military literature of the sixteenth and seventeenth centuries, the answer is simple enough: the volume is, on the average, greater than the quality. Antiquity was still the great teacher in all that concerned the broader aspects of military theory and the secrets of military genius. Vegetius and Frontinus were deemed indispensable; and the most popular book of the century, Henri de Rohan's *Parfait Capitaine*, was an adaptation of Caesar's *Gallic Wars*. Without doubt the most important writing concerned with the art of war fell into two classes: the pioneer works in the field of international law; and the pioneer works of military technology.

Machiavelli had been the theorist for the age of unregulated warfare, but his influence was waning by the turn of the seventeenth century. Francis Bacon was perhaps his last illustrious disciple; for it is hard to find until our own day such unabashed advocacy of unrestricted war as can be found in certain of the *Essays*. But by Bacon's time the reaction had set in. Men like Grotius were leading the attack against international anarchy and against a war of unlimited destructiveness. These founding fathers of international law announced that they had found in the law of nature the precepts for a law of nations, and their central principle, as Talleyrand put it once in a strongly worded reminder to Napoleon, was that nations ought to do one another in peace, the most good, in war, the least possible evil.

It is easy to underestimate the influence of these generous theories upon the actual realities of warfare, and to cite Albert Sorel's black picture of international morals and conduct in the period of the Old Régime. Actually the axioms of international law exerted an undeniable influence on the mode and manner of warfare before the close of the seventeenth century.[6] If they did not put an end to political amoralism, they at least hedged in the conduct of war with a host of minor prescriptions and prohibitions which contributed to making eighteenth century warfare a relatively humane and well-regulated enterprise. These rules were known to contending commanders and were quite generally followed. Such, for example, were the instructions concerning the treatment and exchange of prisoners; the condemnation of certain means of destruction, like the use of poison; the rules for the treatment of noncombatants and for arranging parleys, truces and safe-conducts; or those concerned with despoiling or levying exactions upon conquered territory and with the mode of terminating sieges. The whole tendency was to protect private persons and private rights in time of war, and hence to mitigate its evils.

In the second class, that of books on military technology, no works had greater influence or enjoyed greater prestige than those of Sébastien le Prestre de Vauban, the great military engineer of the reign of Louis XIV. His authority in the eighteenth century was immense, nor had it appreciably dimmed after the time of Napoleon.[7] And yet Vauban's literary legacy to the eighteenth century was scanty and highly specialized, consisting almost solely of a treatise on siegecraft, a work on the defense of fortresses, and a short work on mines.[8]

[6] The notion has recently been stressed by Hoffman Nickerson, *The Armed Horde, 1793-1939* (New York, 1940), pp. 34-40.

[7] An eighteenth century writer on the education of the nobility suggests that the five most important authors a student should study are Rohan, Santa Cruz, Feuquières, Montecuccoli, and Vauban. Cf. Chevalier de Brucourt, *Essai sur l'éducation de la noblesse, nouvelle édition corrigée et augmentée* (Paris, 1748), II, 262-263.

[8] The works published in his lifetime were two: a work on administrative problems, called the *Directeur général des fortifications* (The Hague, 1685, reprinted in Paris, 1725), and his *Dixme Royale* (The Hague [?], 1707). A number of spurious works, however, had appeared before his death, purporting to expound his methods of fortification. His three treatises best known to the eighteenth century were printed for the first time in a slovenly combined edition titled: *Traité de l'attaque et de la défense des places suivi d'un traité des mines* (The Hague, 1737). This was reprinted in 1742 and again in 1771. The *Traité de la défense des places* was published separately by Jombert in Paris in 1769. No carefully prepared editions were published until 1795. Cf. bibliographical note.

He published nothing on military architecture, and made no systematic contribution to strategy or the art of war in general; yet his influence in all these departments is undeniable. It was exerted subtly and indirectly through the memory of his career and of his example, and by the exertions and writings of a number of his disciples. But by this process many of his contributions and ideas were misunderstood and perverted, and much that he accomplished was for a long time lost to view. Thanks to the work of scholars of the nineteenth and twentieth centuries, who have been able to publish an appreciable portion of Vauban's letters and manuscripts, and to peruse and analyze the rest, we have a clearer understanding of Vauban's career and of his ideas than was possible to his eighteenth century admirers. He has increased in stature, rather than diminished, in the light of modern studies. We have seen the Vauban legend clarified and documented; we have seen it emended in many important points; but we have not seen it exploded.

V

The Vauban legend requires some explanation. Why was a simple engineer, however skillful and devoted to his task, raised so swiftly to the rank of a national idol? Why were his specialized publications on siegecraft and the defense of fortresses sufficient to rank him as one of the most influential military writers?

The answers are not far to seek: these works of Vauban were the authoritative texts in what was to the eighteenth century a most important, if not the supremely important, aspect of warfare. In the late seventeenth century and throughout the eighteenth century, warfare often appears to us as nothing but an interminable succession of sieges. Almost always they were the focal operations of a campaign: when the reduction of an enemy fortress was not the principal objective, as it often was, a siege was the inevitable preliminary to an invasion of enemy territory. Sieges were far more frequent than pitched battles and were begun as readily as battles were avoided. When they did occur, battles were likely to be dictated by the need to bring about, or to ward off, the relief of a besieged fortress. The strategic imagination of all but a few exceptional commanders was walled in by the accepted axioms of a war of siege. In an age that accepted unconditionally this doctrine of the strategic primacy of the siege, Vauban's treatises were deemed indispensable and his name was necessarily a name to conjure with.

Yet only a part of the aura and prestige that surrounded Vauban's name arose from these technical writings. He has appealed to the imagination because of his personal character, his long career as an enlightened servant of the state, his manifold contributions to military progress outside of his chosen specialty, and his liberal and humanitarian interest in the public weal. From the beginning it was Vauban the public servant who aroused the greatest admiration. With his modest origin, his diligence and honesty, his personal courage, and his loyalty to the state, he seemed the reincarnation of some servitor of the Roman Republic. Indeed, Fontenelle, in his famous *éloge*, describes him as a "Roman, whom the century of Louis XIV seems almost to have stolen from the happiest days of the Republic." To Voltaire he was "the finest of citizens." Saint-Simon, not content with dubbing him a Roman, applied to him, for the first time with its modern meaning, the word *patriote*.[9] In Vauban, respected public servant, organizational genius, enlightened reformer, seemed to be embodied all the traits which had combined, through the efforts of countless lesser persons, to forge the new national state.

Still more felicitously did Vauban's technical knowledge, his skill in applied mathematics, his love of precision and order, and his membership in the *Académie des sciences*, symbolize the new importance of scientific knowledge for the welfare of the state. Cartesian reason, the role of applied science in society both for war and peace, the *esprit géométrique* of the age: all these were incarnated in the man, visible in the massive outline of the fortresses he designed.

VI

Vauban's career was both too long and too active for anything but a summary account in an essay of this sort. Scarcely any other of Louis XIV's ministers or warriors had as long an active career. He entered the royal service under Mazarin when he was in his early twenties and was still active in the field only a few months before his death at the age of seventy-three. During this half century of ceaseless effort he conducted nearly fifty sieges and drew the plans for well over a hundred fortresses and harbor installations.

He came from the indeterminate fringe between the bourgeoisie and the lower nobility, being the descendant of a prosperous notary of Bazoches in the Morvan who in the mid-sixteenth century had acquired a small neighborhood fief. He was born at Saint Léger in 1633, received his imperfect education—a smattering of history, mathematics and drawing—in near-by Semur-en-Auxois; and in 1651, at the age of 17, enlisted as a cadet with the troops of Condé, then in rebellion against the king. Sharing in Condé's pardon, he entered the royal service in 1653 where he served with distinction under the Chevalier de Clerville, a man of mediocre talents who was regarded as the leading military engineer of France. Two years later he earned the brevet of *ingénieur ordinaire du roi*; and soon after acquired as a sinecure the captaincy of an infantry company in the regiment of the Maréchal de la Ferté.

During the interval between the cessation of hostilities with Spain in 1659 and Louis XIV's first war of conquest in 1667, Vauban was hard at work repairing and improving the fortifications of the kingdom under the direction of Clerville.

In 1667 Louis XIV attacked the Low Countries. In this brief War of Devolution Vauban so distinguished himself as a master of siegecraft and the other branches of his trade that Louvois noticed his distinct superiority to Clerville and made him the virtual director, as *commissaire général*, of all the engineering work in his department. The acquisitions of the War of Devolution

9 Vauban, *Lettres intimes inédites adressées au Marquis de Puyzieulx (1699-1705). Introduction et notes de Hyrvoix de Landosle* (Paris, 1924), pp. 16-17.

launched Vauban on his great building program. Important towns in Hainaut and Flanders were acquired, the outposts of the great expansion: Bergues, Furnes, Tournai and Lille. These and many other important positions were fortified according to the so-called "first system" of Vauban, which will be discussed below.

This, then, was to be the ceaseless rhythm of Vauban's life in the service of Louis XIV: constant supervision, repairs, and new construction in time of peace; in time of war, renewed sieges and further acquisitions; then more feverish construction during the ensuing interval of peace. In the performance of these duties Vauban was constantly on the move until the year of his death, traveling from one end of France to the other on horseback or, later in life, in a famous sedan chair borne by horses. There seem to have been few intervals of leisure. He devoted little time to his wife and to the country estate he acquired in 1675, and he sedulously avoided the court, making his stays at Paris and Versailles as short as possible. The greatest number of his days and nights were spent in the inns of frontier villages and in the execution of his innumerable tasks, far from the centers of culture and excitement. Such free moments as he was able to snatch in the course of his engineering work he devoted to his official correspondence and to other writing. He kept in constant touch with Louvois, whom he peppered with letters and reports written in a pungent and undoctored prose. As though this were not enough Vauban interested himself in a host of diverse civil and military problems only indirectly related to his own specialty. Some of these subjects he discussed in his correspondence, while he dealt with others in long memoirs which make up the twelve manuscript volumes of his *Oisivetés*.

These memoirs treat the most diverse subjects. Some are technical, others are not. But in nearly all of them he answers to Voltaire's description of him as "un homme toujours occupé de sujets les uns utiles, les autres peu practicables et tous singuliers."[10] Besides discussing military and naval problems, or reporting on inland waterways and the interocean Canal of Languedoc, he writes on the need for a program of reforestation, the possible methods of improving the state of the French colonies in America, the evil consequences of the revocation of the Edict of Nantes, and—in a manner that foreshadowed Napoleon's creation of the Legion of Honor—the advantages of instituting an aristocracy of merit open to all classes, in place of the senseless and archaic nobility of birth and privilege.

The *Oisivetés* reveal their origin and belie their name. They were written at odd times, in strange places and at various dates. They are often little more than notes and observations collected in the course of his travels over the length and breadth of France; at other times they are extended treatises. What gives the writings a certain unity is the humanitarian interest that pervades them all and the scientific spirit which they reveal. The writings and the career of Vauban illustrate the thesis suggested earlier in this paper that in the seventeenth century scientific rationalism was the wellspring of reform. Vauban's

[10] Voltaire, *Le siècle de Louis XIV*, chapter XXI.

proposals were based on first-hand experience and observation. His incessant traveling in the performance of his professional duties gave him an unparalleled opportunity to know his own country and its needs. His wide curiosity and his alert mind led him to amass facts, with the pertinacity known only to collectors, about the economic and social conditions of the areas where he worked; and his scientific turn of mind led him to throw his observations, where possible, into quantitative form.

These considerations help us to answer the question whether Vauban deserves, in any fundamental sense, the label of scientist, or whether he was merely a soldier and builder with a smattering of mathematics and mechanical knowledge. Was membership in the *Académie des sciences* accorded him in 1699 solely to honor a public servant and was Fontenelle thus obliged to devote to him one of his immortal *éloges* of men of science?

Vauban's achievements are in applied science and simple applied mathematics. He was not a distinguished mathematician and physicist like the later French military engineer, Lazare Carnot. He made no great theoretical contributions to mechanical engineering, as did Carnot's contemporary, Coulomb. He invented no steam chariot like Cugnot. Aside from the design of fortresses, scarcely a matter of pure science, his only contribution to engineering was an empirical study of the proper proportions of retaining walls.[11] Vauban's chief claim to scientific originality is that he sought to extend the quantitative method into fields where, except for his English contemporaries, no one had yet seriously ventured. He is, in fact, one of the founders of systematic meteorology, an honor that he shares with Robert Hooke, and one of the pioneers in the field of statistics, where the only other contenders were John Graunt and Sir William Petty.[12] His statistical habit is evident in many of his military and engineering reports. Many of these are filled with apparently irrelevant detail about the wealth, population and resources of various regions of France.

From his harried underlings he exacted the same sort of painstaking survey. In a letter to Hue de Caligny, who was for a time director of fortifications for the northwest frontier from Dunkirk to Ypres, he expressed annoyance at the incomplete information he received in reports about that region. He urged Caligny to supply a map, to describe in detail the waterways, the wood supply with the date of cutting, and to provide him with detailed statistical information on population, broken down according to age, sex, profession, and rank. In addition Caligny was to give all the facts he could mass about the economic life of the region.[18] It was by information of this sort, painstakingly acquired as a by-product of his work as an army engineer, that Vauban sought to extend into civilian affairs the same spirit of critical appraisal, the same love of logic, order, and efficiency, which he brought to bear on military problems.

[11] A. Wolf, *History of Science, Technology and Philosophy in the Eighteenth Century* (1939), pp. 531-532; Bernard Forest de Bélidor, *La science des ingénieurs* (1739), Livre I, pp. 67-79.
[12] His right to pioneer status in meteorology rests upon a memoir on rainfall which he submitted to the *Académie des sciences*. Cf. Bélidor, *op cit.*, Livre IV, pp. 87-88.
[18] Georges Michel, *Histoire de Vauban* (1879), pp. 447-451.

VII

Vauban was one of the most persistent of the military reformers of the century. His letters and his *Oisivetés* are filled with his proposals. There were few aspects of military life, or of the burning problems of military organization and military technology, where Vauban did not intervene with fertile suggestions or projects for overall reorganization.[14]

The incorporation of his engineers into a regularly constituted arm of the service, possessed of its own officers and troops and its distinctive uniform, was something for which he struggled, though with little success, throughout his career.[15] His recommendations, however, bore fruit in the following century, as did also his efforts in the matter of scientific education for the technical corps. He enthusiastically praised the earliest artillery schools which were created toward the end of the reign of Louis XIV; and though he never succeeded in creating similar schools for the engineers, he established a system of regular examinations to test the preparation of candidates for the royal brevet, and took some steps to see that they were adequately prepared by special instructors.

Improvement of the artillery arm was a matter in which, as an expert on siegecraft, he was deeply interested. His studies and innovations in this field were numerous. He experimented with sledges for use in transporting heavy cannon. He found fault with the bronze cannon then in use, and tried to persuade the army to emulate the navy in the use of iron. He made numerous, but unsatisfactory, experiments on a new stone-throwing mortar. And finally he invented ricochet fire, first used at the siege of Philipsbourg, where the propelling charge was greatly reduced so that the ball would rebound this way and that after striking the target area, a peril to any man or machine in the near vicinity.

Vauban found space in his correspondence and in the *Oisivetés* to suggest numerous fundamental reforms for the infantry and for the army as a whole. He was one of the most tireless advocates of the flintlock musket for the infantry and was the inventor of the first satisfactory bayonet. As early as 1669 he wrote to Louvois strongly urging the general use of flintlocks and the abolition of the pike; and shortly thereafter he specifically proposed to substitute for the pike the familiar bayonet with a sleeve or socket that held the blade at the side of the barrel, permitting the piece to be fired with bayonet fixed.

He was preoccupied with the condition and welfare of the men as well as with their equipment. He sought to improve still further the mode of recruiting and paying the troops. To him is due in part the limitation of the practice of quartering soldiers on the civilian population which, after the peace of Aix-la-Chapelle, was supplemented by the creation of *casernes*.[16] These special

[14] P. Lazard, *Vauban, 1633-1707* (1937), pp. 445-500.
[15] H. Chotard, "Louis XIV, Louvois, Vauban et les fortifications du nord de la France, d'après les lettres inédites de Louvois adressées à M. de Chazerat, Gentilhomme d'Auvergne," *Annales du Comité Flamand de France*, XVIII (1889-1890), 16-20.
[16] Bélidor, *op. cit.*, Livre IV, p. 73.

barracks, many of them designed and built by Vauban, were chiefly used in frontier regions and recently conquered territory.

Vauban made no systematic study of naval construction, and what he knew seems to have been learned from Clerville who was skilled in this sort of work.[17] His first effort was at Toulon, where he improved the harbor installations, but his masterpiece was the port of Dunkirk. He devoted an interesting study to the naval role of galleys, in which he envisaged extending their use from the Mediterranean to the Atlantic coast, where they could serve as patrol vessels, as a mobile screen for heavier ships close to shore, or for swift harassing descents upon the Orkneys, or even upon the English coast. Closely related to these studies was his advocacy of the *guerre de course*, which he deemed the only feasible strategy after the collapse of the French naval power painstakingly built up by Colbert.

VIII

Vauban's most significant contributions to the art of war were made, as was to be expected, within his own specialties: siegecraft and the science of fortification. It was characteristic of Vauban's dislike of unnecessary bloodshed, as much as of the new spirit of moderation in warfare that was beginning to prevail in his day, that his innovations in siegecraft were designed to regularize the taking of fortresses and above all to cut down the losses of the besieging force. Before his perfection of the system of parallels, which he probably did not invent, attacks on well defended permanent fortifications took place only at a considerable cost to the attackers.[18] Trenches and gabions were employed without system, and as often as not the infantry was thrown against a presumed weak point in a manner that left them exposed to murderous fire.

Vauban's system of attack, which was followed with but little variation during the eighteenth century, was a highly formalized and leisurely procedure. The assailants gathered their men and stores at a point beyond the range of the defending fire and adequately concealed by natural or artificial cover. At this point the sappers would begin digging a trench that moved slowly toward the fortress. After this had progressed some distance, a deep trench paralleling the point of future attack was flung out at right angles to the trench of approach. This so-called "first parallel" was filled with men and equipment to constitute a *place d'armes*. From it, the trench of approach was moved forward again, zigzagging as it approached the fortress. After it had progressed the desired distance, the second parallel was constructed, and the trench was moved forward once more, until a third and usually final parallel was constructed only a short distance from the foot of the glacis. The trench was pushed ahead still further, the sappers timing their progress so as to reach the foot of the glacis just as the third parallel was occupied by the troops. The perilous task of

[17] Lazard, *op. cit.*, pp. 501-524; La Roncière, *Histoire de la marine française*, VI (1932), 164-169.
[18] For a description of early methods, cf. Gaston Zeller, *L'organization défensive des frontières du nord et de l'est au XVIIᵉ siècle* (1928), pp. 54-55.

advancing up the glacis, exposed to the enemy's raking fire from their covered way, was accomplished with the aid of temporary structures called *cavaliers de tranchées*, which were high earthworks, provided with a parapet, from which the besiegers could fire upon the defenders of the covered way. This outer line of defense could be cleared *par industrie*, that is, by subjecting the defenders to the effects of a ricochet bombardment, or by sending up grenadiers to take the position by assault under cover of a protecting fire from the *cavaliers*. Once the enemy's covered way was seized, siege batteries were erected and an effort was made to breach the main defenses.

The essential feature of Vauban's system of siegecraft, then, was the use he made of temporary fortifications, trenches and earthworks, in protecting the advancing troops. His parallels were first tried out at the siege of Maestricht in 1673, and the *cavaliers de tranchées* at the siege of Luxembourg in 1684. The perfected system is described at length in his *Traité des sièges*, written for the Duc de Bourgogne in 1705.

Vauban's work in military architecture has been the subject of considerable dispute, first as to whether the style of his fortresses showed great originality, second as to whether in placing them he was guided by any masterplan for the defense of France.

Until very recently even Vauban's most fervent admirers have agreed that he showed little originality as a military architect and added almost nothing to the design of fortresses he inherited from Pagan. Lazare Carnot admired Vauban in the manner characteristic of other eighteenth century engineers, yet he could find few signs of originality. "The fortification of Vauban reveals to the eye only a succession of works known before his time, whereas to the mind of the good observer it offers sublime results, brilliant combinations, and masterpieces of industry."[19] Allent echoes him: "A better cross section, a simpler outline, outworks that are bigger and better placed: these are the only modifications that he brought to the system then in use."[20] This judgment remained in vogue until very recent times. The most recent serious study, that of Lieutenant Colonel Lazard, has modified in Vauban's favor this somewhat unfavorable opinion.[21]

Lazard has made important changes in our interpretation of Vauban's methods of fortification. Whereas earlier writers have had the habit of referring to Vauban's *three systems*, Lazard points out that, strictly speaking, Vauban did not have sharply defined systems; rather, he had periods in which he favored distinctly different designs, all modifications of the bastioned trace discussed above. With this restriction in mind, it is convenient to retain the old classification.

Vauban's *first system*, according to which he built the great majority of his fortified places, consisted in using Pagan's trace almost without modification. The outlines of these forts were, whenever possible, regular polygons: oc-

---

[19] Didot-Hoefer, *Nouvelle Biographie Générale*, 1870.

[20] *Ibid.*, but cf. A. Allent, *Histoire du Corps Impériale du Génie* (1805), I (only one volume published), pp. 209-210.

[21] Lazard, *op. cit.*, pp. 377-394.

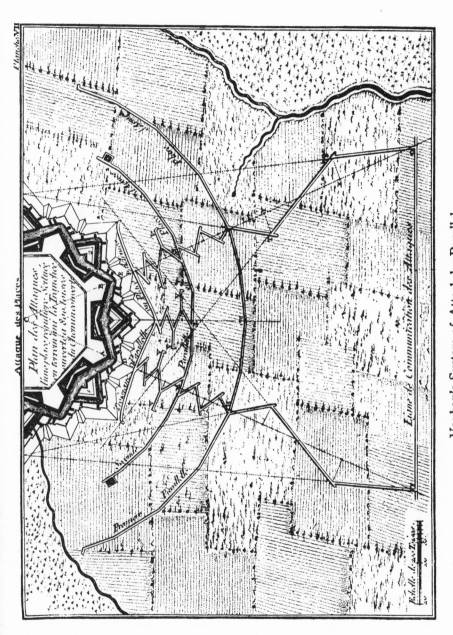

Vauban's System of Attack by Parallels

turn
next page

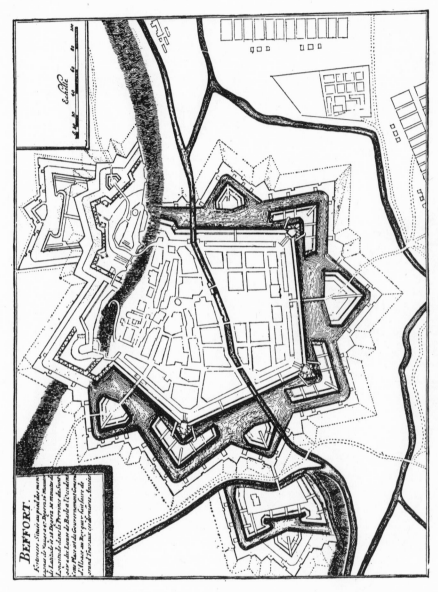

BEFFORT.

Vauban's Second System: The Fortress of Belfort

tagonal, quadrangular, even roughly triangular, as at La Kenoque. The bastions were still the key to the defensive system, though they tended to be smaller than those of Vauban's predecessors. Except for improvements of detail and the greater use of detached exterior defenses (such as the *tenailles* and the *demi-lune*, and other items in Uncle Toby's lexicon), little had altered since the days of Pagan. Since, therefore, most of Vauban's structures were built according to this conservative design, and since this was taken as characteristic of Vauban's work, it is not to be wondered at that later critics could find there little or no originality. The originality, according to Lazard, is evident rather in those other two styles which had little influence on Vauban's successors and which were exemplified in only a few samples of his work.

The *second system*, used for the first time at Belfort and Besançon, was an outgrowth of that previously used. The polygonal structure was retained, but the curtains (the region between the bastions) were lengthened, and the bastions themselves were replaced by a small work or tower at the angles, these being covered by so-called detached bastions constructed in the ditch.

The so-called *third system* is only a modification of the second. It was used for only a single work, the great masterpiece of Vauban at Neuf-Brisach. In this scheme the curtain is modified in shape to permit an increased use of cannon in defense, and the towers, the detached bastions and demi-lunes are all increased in size.

It is the second system which deserves our attention. Here, although his contemporaries could not see it, Vauban had made an important, even revolutionary improvement: he had freed himself from reliance on the main enceinte and taken the first steps toward a defense in depth. He had gained a new flexibility in adapting his design to the terrain without imperiling the main line of defense. In all previous cases adaptation had been through projecting crown works or horn works that were merely spectacular appendages to the primary enceinte; and when these were taken the main line was directly affected. The second system was rejected by Cormontaigne, and later by the staff of the Ecole de Mézières, whose ideas dominated the eighteenth century, and whose schemes of fortification were based squarely upon Vauban's first system. To them this second system seemed only a crude return to medieval methods. Only late in the eighteenth century do we find a revival of Vauban's second system: the revolt of Montalembert, which the Germans accepted long before the French, consisted chiefly in substituting small detached forts in place of the conventional projecting outworks, in reality part of the main enceinte.[22] Montalembert's great revolution, like the later advocacy of fortification in depth, was implicit in Vauban's second system, though whether Montalembert was inspired by it may well be doubted.

The confusion about his ideas that has existed until recently results from the fact that Vauban never wrote a treatise on the art of permanent fortification, never expounded it systematically as he did his theories of the art of

[22] Lazard, *op. cit.*, pp. 389-390; A. de Zastrow, *Histoire de la fortification permanente* (3rd ed., 1856), II, 62-208 (trans. from the German by Ed. de la Barre du Parcq).

attack and of defense. All the books which appeared in his own lifetime and thereafter, purporting to summarize his secrets, were the baldest counterfeits. Only the great work of Bélidor, which treated not of basic design or the problems of military disposition, but only of constructional problems and administrative detail, was directly inspired by Vauban.[23] There are, however, two treatises remaining in manuscript which deal with basic principles of fortification and which were directly inspired by him. One of these was written by Sauveur, the mathematician whom Vauban chose to instruct and to examine the engineer candidates; the other by his secretary, Thomassin. These are the best sources, aside from the forts themselves, for learning Vauban's general principles of fortification. It is possible to speak only of general principles, not of a dogmatic system, and these principles are exemplified equally well by all three of the Vauban styles. They are few enough and quite general. First of all, every part of the fort must be as secure as every other, with security provided both through sturdy construction of the exposed points (bastions) and by adequate coverage of the curtains. In general these conditions will be provided for if (1) there is no part of the enceinte not flanked by strong points, (2) these strong points are as large as possible, and if (3) they are separated by musket range or a little less. These strong points should be so designed that the parts which flank should always confront as directly as possible the parts they are protecting; conversely, the flanking parts should be visible only from the protected parts. A little thought will show that these basic principles are applicable to all of Vauban's schemes. The actual problem of building a permanent fortification consisted in so adapting the bastioned trace (or the polygonal trace with detached bastions) to the exigencies of a particular terrain that none of the basic principles was violated. Clearly this left the engineer a wide range of freedom and an admirable flexibility. It was by this method of work that the second style was developed, for Vauban himself tells us that it was not arrived at as a result of theoretical considerations but was forced on him by the terrain conditions at Belfort.[24]

IX

To what extent was the military building program of Louis XIV guided by some unifying strategic conception; and what is the evidence that this conception, if in truth there was such a thing, was due to the genius of Vauban? These are two of the most important questions, but they are not the easiest to answer.

The earlier biographers of Vauban, with characteristic impetuosity on behalf of their hero, leave us sometimes with the distinct impression that before Vauban France had no system of fortification worthy of the name, and that the ring of fortresses girding the kingdom by the end of his career represented the execution of some cleverly conceived master plan sprung from the mind of the great engineer. To these writers it was just as incredible that anyone

[23] Bélidor, *op. cit.*, Livre III, pp. 29-34, 35-43, 90-96.
[24] Letter to Louvois, October 7, 1687, cited by Zeller, *op. cit.*, p. 144.

besides Vauban could have had a hand in organizing this defensive system as it was that this system itself might have been the result of a slow historical growth.

Of late we have drifted perhaps too far in the other direction. Although, as we have seen, Vauban's technical reputation as a military architect has been enhanced by recent studies, there has been a simultaneous tendency on the part of certain writers to reduce him to the level of a great craftsman devoid of strategic imagination. He has been represented as a brilliant technician, executing blindly the tasks dictated by historical necessity or by the orders of superiors who alone did all the strategic thinking.

Who was there who was capable of challenging Vauban's authority in the field of his specialty? The answer is, the king himself. Louis XIV, it has been shown, was more than decently proficient in the art of fortification. He had studied it in his youth, and, during the early part of the reign, he had profited by the advice and instruction of Turenne, Villeroi, and Condé. Throughout his career he showed a constant interest in the most humble details connected with the art of fortification and on a number of occasions he resolutely opposed insistent recommendations of Vauban. Two important forts, Fort Louis and Mont-Royal, were created on the initiative of the king, and one at least of these was against the express advice of Vauban.[25] To one author, Louis the Diligent was in everything, even in these technical matters, the unquestioned master. Louvois was only an "excellent servant, not to say clerk," while Vauban in his turn "was never anything but the executor of his orders, albeit . . . an excellent one."[26] Another writer describes Vauban as "the chief workman of a great undertaking, the direction of which was never fully entrusted to him."[27] This interpretation is in fact inescapable. Vauban drew or corrected all plans for fortresses that had been decided upon; he submitted technical memoirs and recommendations; he gave his opinion on crucial matters when asked and sometimes when he was not asked. But his presence was not deemed necessary when the decisions were being debated. He was not a policy maker; his was only a consultative voice.

This should not lead us to underestimate his influence upon the royal decisions. Yet even if Vauban had had a master plan for the defense of France, it could only have been imperfectly executed. Many recommendations dear to Vauban's heart were rejected; many of his schemes were shattered by the realities of war and diplomacy. The peace of Ryswick in 1697, for example, marked Louis XIV's first withdrawal from the high watermark of conquest. To Vauban, who was not directly consulted about its terms, this treaty, though not as bad as he feared, was a great deception. Much work had to be done over to make up for the loss of Luxembourg—which he considered one of the strongest places in Europe—and of Brisach, Fribourg and Nancy.[28]

[25] Chotard, *op. cit.*, pp. 30-35; Zeller, *op. cit.*, pp. 96-117; Lazard, *op. cit.*, pp. 49-50, 202-204.
[26] Chotard, *op. cit.*, p. 36.
[27] Zeller, *op. cit.*, p. 118.
[28] Zeller, *op. cit.*, pp. 103-104; Th. Lavallée, *Les frontières de France* (Paris, 1864), pp. 83-85.

Did Vauban in reality have a master plan? On this question there is almost complete disagreement. The writers of the last century took it for granted that Vauban had a strategic pattern for his fortresses, though they were not altogether certain in what it consisted. One writer described it as "an assemblage of works sufficiently close to one another so that the intervals between them are not unprotected. Each of these works is strong enough and well provisioned enough to impose upon the enemy the obligation of a siege, yet small enough to demand only a small number of defenders."[29] With this interpretation Gaston Zeller is in categorical disagreement. He points out that Louis XIV and Vauban did not start work with a clean canvas, that neither of these men could have imposed a doctrinaire plan of defense without reference to the work that had gone before; and he indicates that many of the characteristics of the defense system were due to Francis I, Sully, Richelieu, and Mazarin, to their building programs and their treaties. Just as the actual frontier of the France of Louis XIV was the culmination of a long-sustained national policy, just so the disposition of the fortress towns was "the resultant of a long succession of efforts to adapt the defensive organization of the kingdom to the changing outline of the frontier."[30] In support of Zeller's contention that the fortress system was the work of historical evolution, not the work of a single man, is the evidence from the career of Vauban himself. The greatest number of strongholds that we associate with him were not *places neuves* but older fortresses, some dating back to Errard or his Italian predecessors, which Vauban modernized and strengthened. The fortresses did not in any sense constitute a system as Vauban found them; they were important only as separate units. There was no liaison between them and they were almost always too far apart. Each situation, moreover, had been chosen for its local importance: to guard a bridge, a crossroads, or the confluence of two rivers. Their total value depended not on their relative positions but rather upon their number.[31] Zeller and Lazard both agree that Vauban's general scheme resulted from a process of selection from among these fortresses. He made order out of prevailing chaos by choosing certain forts whose positions made them worth retaining and strengthening, and by suggesting that others be razed. His strategic vision could not work with complete freedom; he was limited— largely for reasons of public economy—to working with what France already possessed. It is easy to discover the principles which guided his process of selection and thus to find the key to his strategic thinking. To Zeller there is nothing outstanding about these principles; the "order" that Vauban effected fell far short of a great strategic conception. But Lazard is much more flattering. He takes the view that Vauban was the first man in history to have an overall notion of the strategic role of fortresses. He was not only an engineer but a *stratège*, and one with ideas far in advance of his own day.[32] Only

[29] Hennebert, cited by Chotard, *op. cit.*, p. 42.
[30] Zeller, *op. cit.*, p. 2.
[31] Zeller, *op. cit.*, p. 123.
[32] Lazard, *op. cit.*, pp. 408-421.

Vauban's own writings can allow the reader to decide between these two interpretations.

It should be remembered that as a result of the War of Devolution against Spain, his first war of conquest, Louis XIV extended his holdings along the northwest frontier deep into Spanish-held Flanders. The new positions—from Furnes near the coast eastward through Bergues and Courtrai to Charleroi—gave France a number of strong points scattered among the Spanish garrisons. Vauban's first great task was to strengthen and refortify these new acquisitions, and this occupied most of his time during the peaceful years from 1668 to 1672. In the spring of 1672, however, Louis launched his war against the Dutch. Vauban took the opportunity to raise for the first time the question of the general organization of the frontier. In a letter to Louvois, dated January 20, 1673, he wrote: "Really, my lord, the king should think seriously about rounding out his domain [*songer à faire son pré carré*]. This confusion of friendly and enemy fortresses mixed up pell-mell with one another does not please me at all. You are obliged to maintain three in the place of one."[33]

In 1675, a year which saw him busy consolidating French conquests in Franche Comté and elsewhere, Vauban made more specific suggestions. In September of that year he proposed the sieges of Condé, Bouchain, Valenciennes, and Cambrai. The capture and retention of these places would, he said, assure Louis' conquests and produce the *pré carré* that was so desirable. These towns were accordingly taken: Condé and Bouchain in 1676, Valenciennes and Cambrai in 1677. The Peace of Nimwegen, signed in August of the following year, gave France a frontier approximating the *pré carré*. She gave up some of her Flemish holdings but acquired instead Saint-Omer, Cassel, Aire, Ypres, and a half-dozen other important strongholds. To the eastward she gained Nancy in Lorraine and Fribourg across the Rhine. But Vauban was not satisfied with the western end of the frontier; he felt that the recent peace had disrupted it and left it open toward the Lowlands. In November 1678, three months after Nimwegen, he wrote the first of a series of important general statements on the organization of the northern frontier from the channel to the Meuse.[34]

Vauban opens by discussing the purposes of a fortified frontier: it should close to the enemy all the points of entry into the kingdom and at the same time facilitate an attack upon enemy territory. Vauban never thought that fortresses were important solely for defense; he was careful to stress their importance as bases for offensive operations against the enemy. The fortified places should be situated so as to command the means of communication within one's own territory and to provide access to enemy soil by controlling important roads or bridgeheads. They should be large enough to hold not only the supplies necessary for their defense, but the stores required to support and sustain an offensive based upon them. These ideas, enunciated tersely in this

[33] Lazard, *op. cit.*, p. 155; De Rochas, *op. cit.*, II, 89.
[34] Lazard, *op. cit.*, pp. 409-414; Zeller, *op. cit.*, pp. 96-98. This important memoir is printed *in extenso* in Rochas, I, 189 f.

memoir, were later elaborated and systematized by one of Vauban's eighteenth century disciples, the engineer and adventurer Maigret, whom Voltaire mentions in his *Charles XII*, and whose *Treatise on Preserving the Security of States by Means of Fortresses* became the standard work dealing with the strategic significance of fortifications. This book, all too little known, was used by the famous French school of military engineering, the Ecole de Mézières. In this work Maigret writes that "the best kind of fortresses are those that forbid access to one's country while at the same time giving an opportunity to attack the enemy in his own territory."[35] He lists the characteristics that give value and importance to fortresses: control of key routes into the kingdom, such as a mountain gorge or pass; control of the bridgeheads on great rivers, a condition eminently fulfilled by Strasbourg, for example; control of important communication lines within the state, as for example Luxembourg, which secured the emperor's communications with the Lowlands.

There were still other factors which might make a fort important. It might be a base of supplies for offensive action, or a refuge for the people of the surrounding countryside; perhaps it could dominate trade and commerce, exacting tolls from the foreigner; or perhaps it might be a fortified seaport with a good and safe harbor; a great frontier city with wealth, more than able to contribute the cost of fortification and sustaining the garrison; or a city capable of serving the king as a place to store his treasure against internal and external enemies.[36] The value of a fortress depends in large part, of course, upon the nature of its local situation. Art or science may make up for certain defects in the terrain but they can do little with respect to the matter of communication. Thus certain fortresses are advantageously situated because the defenders have the communications leading to it well under their control, whereas the enemy, in consequence, will have difficulty in bringing up the supplies necessary for a sustained siege.[37]

These criteria make it possible to select certain fortresses in preference to others but there still remains the question of their relation one to the other, of liaison. Vauban, in the memoir of 1678, concluded that the frontier would be adequately fortified if the strongholds were limited to two lines, each composed of about thirteen places, stretched across the northern frontier in imitation of infantry battle order.[38] This first line could be further strengthened and unified by the use of a waterline stretching from the sea to the Scheldt. Canals or canalized streams or rivers would link one fort with another, and the canals themselves would be protected at regular intervals by redoubts. This scheme was not original with Vauban; in fact it was in operation over part

---

[35] *Traité de la sureté et conservation des états, par le moyen des forteresses. Par M. Maigret, Ingénieur en Chef, Chevalier de l'ordre Royal et Militaire de Saint Louis* (Paris, 1725), p. 149.

[36] Maigret, *op. cit.*, pp. 129-148.

[37] Maigret, *op. cit.*, pp. 152 f., and 221-222.

[38] The first line: Dunkirk, Bergues, Furnes, Fort de la Kenoque, Ypres, Menin, Lille, Tournai, Fort de Mortagne, Condé, Valenciennes, Le Quesnoy, Maubeuge, Philippeville, and Dinant. The second line: Gravelines, Saint-Omer, Aire, Béthune, Arras, Douai, Bouchain, Cambrai, Landrecies, Avesnes, Marienbourg, Rocroi, and Charleville.

of the frontier even as he wrote. He was under no illusions as to the strength of the waterlines, for he saw that their chief purpose was to ward off the harassing raids by which small enemy detachments plagued the countryside. Should an enemy decide to attack the lines with an army, then the lines must be defended with an army.[39]

Such a project would of course necessitate new construction, but Vauban was careful to point out that it would also mean the elimination of numerous ancient strongholds, and he accordingly urged the razing of all fortresses remote from the frontier and not included in the two lines. This would not only be a saving for the treasury but, he urged, also a saving in man power: with the elimination of their garrisons, ten fewer strongholds would mean about 30,000 soldiers free for duty elsewhere.

This famous memoir of 1678 also embodied a consideration of possible future conquests and these indicate that, so far as the northern and eastern frontiers were concerned, Vauban was willing to pave the way for something more ambitious than a mere local rectification of a line. In the event of a future war, he said, certain enemy fortresses should be immediately seized. Dixmude, Courtrai, and Charlemont would open up the lowlands, while to the east, Strasbourg and Luxembourg were the supremely important cities to acquire. Not only did these fortresses have the most admirable features of size, wealth, and situation—in these matters they were the best in Europe—but they were the keys to France's expansion to its natural boundaries. Vauban would not have been Frenchman and patriot had he not accepted the familiar and tempting principle that France's natural frontier to the north and east was the Rhine. We know that he held this view and we can suspect that it was already clearly formulated in his mind early in his career. It certainly was later. Just before the peace of Ryswick, when he was terrified for fear France was about to lose both Strasbourg and Luxembourg, he wrote: "If we do not take them again we shall lose forever the chance of having the Rhine for our boundary."[40]

It is not easy to say with certainty whether this memoir of 1678 represents Vauban's mature and final view on the matter of permanent fortification. Vauban's later memoirs leave much to be desired as examples of strategic thinking about the role of fortresses. Except for a memoir on the fortification of Paris, in which he discusses at length the strategic importance of a nation's capital, most of the later studies are lacking in genuine strategic interest. They are concerned chiefly with detailed recommendations as to which fortresses should be condemned and which enlarged or rebuilt.

Despite these handicaps it is not hard to detect a series of changes in Vauban's opinions, due partly to a gradual evolution of his ideas, but chiefly to the changed conditions under which he was obliged to work in the later years of the reign. Increasing financial stringency and a growing drain on the man power supply encouraged Vauban to stress the razing of fortifications as much

---

[39] Lazard, *op. cit.*, pp. 282-284; Augoyat, *op. cit.*, I, 229.
[40] Lavallée, *op. cit.*, pp. 83-85.

if not more than new construction.[41] This led him to urge the destruction of many of the places that had been listed in his second line of defense in the memoir of 1678. At the same time the armies of Louis XIV were being thrown more and more on the defensive and Vauban adapted himself increasingly to defensive thinking. He followed the trend, that was becoming evident at the close of the century, toward still greater reliance upon a continuous waterline along the northern frontier. But he was aware of the peculiar weakness of this sort of defense. In 1696 he wrote a memoir in which he urged the creation of *camps retranchés*, fortified encampments to supplement the fortresses and to strengthen the waterline. The purpose of these encampments was either to guard the waterline in the interval between the fortresses or to strengthen the forts themselves by producing a veritable external defense. With a small army —smaller than the ordinary field army—camped beyond the outworks of a fortress and protected by elaborate earthworks it was possible either to interfere with any besieging forces unwise enough to tackle the fortress directly or to impose upon them a wider perimeter to be invested.

Taken together these two factors—first, the stress upon the continuous line supplemented by the fortified encampments; and second, the willingness to sacrifice the second line of forts he had favored in 1678—do not offer support to Lazard's assertion that Vauban was a pioneer advocate of the "fortified zone" which modern strategy has adopted. Quite the contrary, Vauban's thinking seems to have evolved in the direction of favoring a thinner and thinner line. He simplified that disorganized parody on a fortified zone which he had inherited from his predecessors. At first he reduced it to a double line of fortifications, a palpable imitation of the familiar infantry line, and then proceeded to simplify this still further into a single cordon, based on strong points linked by a continuous waterline and supported by troops. Perhaps it is not too farfetched to see in this a sign that the great engineer, toward the close of his career, was led gradually to lay more emphasis upon armies and less upon fortification. He seems almost to have come close to the idea of Guibert that the true defense of a country is its army, not its fortifications; that the fortified points are merely the bastions of that greater fortress of 'which the army forms a living and flexible curtain.

[41] Zeller, *op. cit.*, pp. 98-107.

CHAPTER 3. Frederick the Great, Guibert, Bülow: From Dynastic to National War

BY R. R. PALMER

THE period from 1740 to 1815, opening with the accession of Frederick the Great as king of Prussia, and closing with the dethronement of Napoleon as emperor of the French, saw both the perfection of the older style of warfare and the launching of a newer style which in many ways we still follow. The contrast between the two styles is the main subject of this chapter. Much of the old, however, was continued in the new. The underlying ideas sketched in the two preceding chapters were not outdated and they remain today essential to the theory of war. Machiavelli had made the study of war a social science. He had dissociated it from considerations of ethical purpose and closely related it to constitutional, economic, and political speculation. He had tried, in military matters, to enlarge the field of human planning and to reduce the field of chance. Vauban had opened up to military men the resources of natural science and technology. The government of Louis XIV, while enlarging armies beyond precedent, had advanced the principles of orderly administration and control. It had put a new emphasis on discipline, created a more complex hierarchy of tactical units, clarified the chains of command, turned army leaders into public officials, and made armed force into a servant of government. All these developments were accelerated and elaborated in the period of change with which this chapter deals.

The significant innovations concerned the constitution and the utilization of armies, i.e. man power and strategy. Citizen armies replaced professional armies. Aggressive, mobile, combative strategy replaced the slow strategy of siegecraft. Both had been anticipated by Machiavelli, but neither had been realized on a large scale since 1500. Together, after 1792, they revolutionized warfare, replacing the "limited" war of the Old Régime with the "unlimited" war of subsequent times. This transition came with the shift from the dynastic to the national form of state, and was a consequence of the French Revolution. War before the French Revolution was essentially a clash between rulers. Since that event it has become increasingly a clash between peoples, and hence has become increasingly "total."[1]

The dynastic form of state set definite limits to what was possible in the constitution of armies. The king, however absolute in theory, was in fact in a disadvantageous position. Every dynastic state stood by a precarious balance between the ruling house and the aristocracy. The privileges of the nobility limited the freedom of government action. These privileges included the right not to pay certain taxes and the right almost to monopolize the commissioned grades in the army. Governments, with their taxing power restricted, could not

[1] For the contemporary literature of the subject see M. Jähns, *Geschichte der Kriegs-wissenschaften vornehmlich in Deutschland* (3 vols.; Munich, 1891), Vol. III.

draw on the full material resources of their countries. Nor could they draw on their full human resources. Officers must come from a hereditary class which rarely exceeded two per cent of the population. Between populations as a whole and their governments little feeling existed. The tie between sovereign and subject was bureaucratic, administrative and fiscal, an external mechanical connection of ruler and ruled, strongly in contrast to the principle brought in by the Revolution, which, in its doctrine of responsible citizenship and sovereignty of the people, effected an almost religious fusion of the government with the governed. A good government of the Old Régime was one that demanded little of its subjects, which regarded them as useful, worthy, and productive assets to the state, and which in wartime interfered as little as possible with civilian life. A "good people" was one which obeyed the laws, paid its taxes, and was loyal to the reigning house; it need have no sense of its own identity as a people, or unity as a nation, or responsibility for public affairs, or obligation to put forth a supreme effort in war.

The army reflected the state. It was divided internally into classes without common spirit, into officers whose incentive was honor, class-consciousness, glory or ambition, and soldiers enlisted for long terms who fought as a business for a living, who were thought incapable of higher sentiments, and whose strongest attachment was usually a kind of naive pride in their regiments. The armies of Russia, Austria and Prussia were composed largely of serfs. Prussia and England used large numbers of foreigners. The Austrian forces were linguistically heterogeneous. In all countries the tendency was to recruit men who were economically the most useless, which is to say the most degraded elements in the population. Civilians everywhere kept soldiers at a distance. Even in France, which already had the most national of the large armies of Europe, cafes and other public places put up signs reading, "No dogs, lackeys, prostitutes or soldiers."[2]

To make armies of such motley hosts, of soldiers who were almost social outcasts and of officers who were often only youthful aristocrats, some kind of common purpose had to be created. For this end the troops had few moral or psychological resources in themselves. Governments believed, with good reason in the circumstances, that order could be imposed only from outside and from above. The horrors of an ungoverned soldiery were remembered, especially in Germany after the Thirty Years War. The enlightened monarchies of the eighteenth century tried to spare their civilian populations, both for humane reasons and as sources of revenue. To promote civil order, and to build morale among troops who could not be appealed to on a level of ideas, governments increasingly took good physical care of their men, quartered them in barracks, provided them with doctors and hospitals, fed them liberally, and established great fixed permanent magazines for their supply. It was feared that soldiers would desert if left to forage in small parties, or if not furnished with a tolerable standard of living, since to make a living, not to fight or die for a cause, was the chief aim of the professional soldier. And in truth, in the

---

[2] M. Weygand, *Histoire de l'armée française* (Paris, 1938), p. 173.

eighteenth century, both officers and men passed from one army to another, in war or in peace, with a facility inconceivable after the French Revolution.

Along with good care went a strict attention to discipline and training, also handed down from above. Only iron rule could make into a unified force men who had no cohesion in themselves. Rulers and aristocrats scarcely expected to find moral qualities in the lower classes who made up the soldiery—neither courage, nor loyalty, nor group spirit, nor sacrifice, nor self-reliance. Nor were these qualities in fact developed in the troops of the time, who, like the peoples in general of the dynastic states, felt little sense of participation in the issues of war. Soldiers could not be trusted as individuals, or in detached parties, or out of sight of their officers. Technical considerations also discouraged individuality. The poor state of communications and low quality of scouting (due in turn to the ignorance and unreliability of individual soldiers) made it more than ordinarily hazardous to divide an army in the field. The inaccuracy and short range of muskets made individual firing relatively harmless. As a result the ideal of military training was to shape a spiritless raw material into machinelike battalions. When engaged with the enemy each battalion stood close to the next in a solid line, the men being almost elbow to elbow, usually three ranks deep, and each battalion constituting a kind of firing machine, delivering a volley at the word of command. To achieve tactical alertness, long and intensive training was necessary. Two years were considered scarcely sufficient to turn a ragamuffin into a good professional soldier.

The constitution of armies strongly affected their utilization. For the governments of the Old Régime, with their limited resources, the professional armies were expensive. Each soldier represented a heavy investment in time and money. Trained troops lost in action could not easily be replaced. The great magazines of munitions and foodstuffs, which, in the poor state of transportation, had to be kept near the expected scenes of action, needed protection. In addition, in the latter part of the seventeenth century scientific progress improved the art of fortification, and a great revulsion spread through France and Germany against the chaotic and roving warfare of the so-called wars of religion, by which productive civilian life had been much impaired. The net result was to concentrate armies in chains of heavily fortified positions. Armies, and fragments of armies, were immobilized near their bases, from which they were not supposed to depart by more than five days' march. Even with magazines close behind them, they carried long baggage trains, so that a day's march was very short. Nor could the baggage trains be easily reduced: in most armies the aristocratic officers traveled in style, and the troops, fighting without political passion, would lose morale if their food supply became uncertain or if operations became distastefully strenuous.

A large-scale pitched battle between complete armies was in these circumstances a rare occurrence. It was not easy for a commander to establish contact with an unwilling enemy. Even with two armies face to face, to draw up a battle line took time, and if one side chose to depart while the other formed, no complete engagement would ensue. Battle was a tremendous risk. A margin of advantage gained on the battlefield could not easily be widened, because the technique of destructive pursuit was undeveloped. Military thinkers held that a state might suffer almost as much by victory as by defeat. Quick and decisive political results were in any case not expected from battle. Here the contrast between eighteenth-century and Napoleonic battles is especially clear. After Blenheim, Malplaquet, Fontenoy or Rossbach, the war dragged on for years. After Marengo, Austerlitz, Jena, Wagram or Leipzig, peace overtures began in a few months.

To sum up, many factors combined before the French Revolution to produce a limited warfare, fought with limited means for limited objectives. Wars were long, but not intense; battles were destructive (for the battalion volleys were deadly), but for that reason not eagerly sought. Operations turned by preference against fortresses, magazines, supply lines and key positions, producing a learned warfare in which ingenuity in maneuver was more prized than impetuosity in combat. War of position prevailed over war of movement, and a strategy of small successive advantages over a strategy of annihilation.

All this was changed in the upheaval which shook Europe after 1789. The "world war" of 1792-1815 was, except in the earliest years, and except for the struggle between France and Great Britain, a series of short wars each of which was promptly decided on the battlefield and concluded by the imposition of peace. Authorities agree that these wars marked a major turning point, closing a period which had begun about 1500, and opening a period from which we have not yet clearly emerged. Most writers attribute the change to the French Revolution, with the consequent nationalizing of public opinion and closer relations between governments and governed. This interpretation was established half a century ago by Jähns and Delbrück. There has been some evidence of a "revisionist" tendency, as in the writings of General Colin, who looked for a more material or at least technical explanation, and found it in the great improvements in the latter half of the eighteenth century in artillery, army organization, road building, and cartography. The burden of informed opinion, while recognizing the importance of technical progress, still considers the effects of the political revolution to have been more profound. As Delbrück said, the new *politisches Weltbild* of the French Revolution produced "a new constitution of the army, which first brought forth a new tactics, and from which a new strategy would then grow."[3]

The transition is evident in the works of the three writers treated below. Each of the three represents a significant stage in the history of military thinking. Frederick the Great embodied the utmost in military achievement that was possible in Europe in the conditions prevailing before the French Revolution. Guibert was a conscious disciple of Frederick, but he forecast more clearly than Frederick some of the transformations that were to come. Bülow, a contemporary of the revolutionary and Napoleonic wars, gradually perceived many of the lessons that they offered. Of the three, only Frederick was an

[3] H. Delbrück, *Geschichte der Kriegskunst* (7 vols.; Berlin, 1900-1936), IV, 363, 426; J. Colin, *Education militaire de Napoléon* (Paris, 1900).

experienced practical commander. His writings describe the actual warfare of the day. Guibert and Bülow, though army officers by training, commanded no armies; they were notable as critics, prophets and reformers. Frederick reveals a mind completely master of its subject. Guibert and Bülow, writing less from experience, aiming to go beyond existing conditions, were much less steady in their grasp. With their fluctuating and partial insights they may be taken to illustrate the difficulty, familiar in all ages, with which military theory adjusts itself to shifting realities in the world of fact.

II

Frederick the Great, invading Silesia without warning in 1740, gave Europe a taste of what later was to be called blitzkrieg. In three Silesian wars he managed to retain the coveted province, whose acquisition almost doubled the size of his small kingdom, and he proved himself, fighting at times against incredible odds, to be incomparably superior as a general to any of his opponents. His Prussia, in addition, possessed to the point of exaggeration the main features of the dynastic state. Of the chief states of Europe Prussia was the most mechanically put together, the most ruled from above, the least animated by the spirit in its people, and the poorest in both material and human resources. Frederick was also a voluminous and gifted writer. In the writings of such a king of such a kingdom, the generalities outlined in the section above take on definite and concrete form.

Frederick's first military work of importance was his *Principes généraux de la guerre*, written in 1746, and embodying the experience of the first two Silesian wars. It was circulated confidentially among his generals. The capture of one of these generals by the French in 1760 led to its publication. The king further developed his ideas in a *Testament politique* composed in 1752 for the private use of his successors to the throne. To this testament the *Principes généraux* was attached as an appendix. In 1768, when his wars were over and his ideas somewhat modified, he drew up a *Testament militaire* for his successors. To his generals in 1771 he issued his *Eléments de castramétrie et de tactique*. Continuously throughout his reign he composed special instructions for various branches of the army, which were brought together and published with his other writings in 1846. Among works which he made public are a didactic poem, *L'art de la guerre*, a number of political essays that touch on military questions, and the various histories and memoirs of his reign, together with their prefaces. In these writings contemporaries tried to discover the secrets of his generalship. His works were all written in French, except for the technical instructions which he wrote in German. His literary career reached over more than forty years. In general he adhered to the same ideas in army organization and tactics, but in the strategy and politics of war he moved from the sharp aggressiveness of 1740 to a philosophy of relative inactiveness.

The organization of the army was an old concern of the rulers of Prussia. In 1640, exactly a century before Frederick's accession, his great-grandfather, the Great Elector, came to the throne in the full fury of the Thirty Years War.

There was then no kingdom of Prussia, only parcels of territories along the flat north German plain, swarmed over and ravaged by the brutal mercenaries of every contending power. The Great Elector founded an army. To support this army he virtually founded a new polity and a new economy. With his reign began the distinctive features of Prussia. First, Prussia owed its existence and its very identity to its army. Second, military science, politics, and economics merged inseparably into a general science of statecraft. Third, Prussia, made by the Hohenzollern dynasty, was a triumph of careful planning. By the time of Frederick's father, Frederick William I, the king of Prussia was commonly considered one of the hardest working men in Europe. He directed the state in person, all threads came together in his hand, and the only center of unity was his own mind. Order, in Prussia, had not come from free discussion and collaboration. As Frederick the Great once observed, if Newton had had to consult with Descartes and Leibnitz, he would never have created his philosophical system.

A king of Prussia, in Frederick's view, must, to have an army, hold a firm balance between classes in the state, and between economic production and military power. He must preserve the nobility by prohibiting the sale of noble lands to peasants or townsmen. Peasants were clearly too ignorant to become officers;[4] to have bourgeois officers would be "the first step toward the decline and fall of the army."[5] Rigid class structure—with noble persons and inalienably "noble" land—was necessary to the army and to the state. A brave colonel, says Frederick, makes a brave battalion; and a colonel's decision in a moment of crisis may sway the destiny of the kingdom. But the king must make sure (so new, disjointed and artificial was the state) that these aristocrats have the desired spirit. In his first political testament Frederick confides to his successors that, during the first Silesian wars, he had made a special effort to impress upon his officers the idea of fighting for the kingdom of Prussia.[6]

For common soldiers Frederick often expressed a rough respect, as for men who risked their lives in his service, but his real interest in them was almost entirely on disciplinary and material questions. The peasant families (i.e. serfs, east of the Elbe) must be protected; their lands must not be absorbed by bourgeois or nobles; only those not indispensable in agriculture, such as younger sons, should be recruited. By and large, the peasants and townsmen are most useful as producers. "Useful hardworking people should be guarded as the apple of one's eye, and in wartime recruits should be levied in one's own country only when the bitterest necessity compels."[7] Half the army or more might be filled with non-Prussian professionals, with prisoners of war or with deserters from foreign armies. Frederick praises the Prussian canton system, by which, to equalize the burden of recruiting, specific districts were assigned to

---

[4] *Politisches Testament von 1752*, in *Werke Friedrichs des Grossen* (10 vols.; Berlin, 1912-1914), VII, 164.
[5] *Exposé du gouvernement prussien, des principes sur lequel il roule* (1775) in *Oeuvres de Frédéric le Grand* (30 vols.; Berlin, 1846-1856), IX, 186.
[6] *Pol. Test. 1752*, in *Werke*, VII, 146; *Oeuvres*, XXIX, 58.
[7] *Militärisches Testament von 1768*, in *Werke*, VI, 226-227.

specific regiments as sources of man power. By this system (and by the use of foreigners), he observed with satisfaction in 1768 that only 5000 natives of Prussia needed to be conscripted each year. Yet he was aware of the value of patriotic citizen forces, which he thought that the cantons produced by putting neighbors beside each other in war. Our troops, he wrote in 1746, recruited from "citizens," fight with honor and courage. "With such troops one would defeat the whole world, were victories not as fatal to them as to their enemies." Later on Frederick, like other *philosophes*, placed even higher theoretical value on patriotism. But he never did anything about it, nor could he, without revolutionizing his kingdom. In practice he assumed that common soldiers were without honor, and he died in the belief that to use foreigners to do one's fighting was only sensible statecraft.[8]

Frederick's soldiers felt no great inward attachment to him. Desertion was the nightmare of all eighteenth-century commanders, especially in disorganized Germany, where men of the same language could be found on both sides in every war. In 1744 Frederick had to stop his advance in Bohemia because his army began to melt away. He drew up elaborate rules to prevent desertion: the troops should not camp near large woods, their rear and flanks should be watched by hussars, they should avoid night marches except when rigorously necessary, they should be led in ranks by an officer when going to forage or to bathe.[9]

Working with untrustworthy material Frederick insisted on exact discipline, to which the Prussian armies had been habituated by his father. "The slightest loosening of discipline," he said, "would lead to barbarization."[10] Here again the army reflected the state. The aim of discipline was partly paternalistic, to make the soldier a rational being by authority, through preventing such offenses as drunkenness and theft. But the principal aim was to turn the army into an instrument of a single mind and will. Officers and men must understand that every act "is the work of a single man." Or again: "No one reasons, everyone executes"; i.e. the thinking is done centrally, in the mind of the king. All that can be done with soldiers, he said, is to give them *Korpsgeist*, to fuse their personalities into their regiments. As he grew older and more cynical, he observed that good will affected common men much less than intimidation. Officers must lead men into danger; "therefore (since honor has no effect on them) they must fear their officers more than any danger." But he added that humanity demanded good medical care.[11]

Made amenable by discipline the troops were to be put through careful training. Prussia was famous for its drillfields, where, to the admiration of foreign observers, battalions and squadrons performed intricate evolutions with

[8] *Principes généraux de la guerre* (1746) in *Oeuvres*, XXVIII, 7; *Lettres sur l'amour de la patrie* (1779), *ibid.*, IX, 211-244.

[9] *Prin. gén.*, *ibid.*, XXVIII, 5-6; *Ordres für die sämmtlichen Generale von der Infanterie und Cavalerie, wie auch Huzzaren, desgleichen fur die Stabsofficiere und Commandeurs der Bataillons* (1744), *ibid.*, XXX, 119-123; *Règles de ce qu'on exige d'un bon commandeur de bataillon en temps de guerre* (1773), *ibid.*, XXIX, 57-65.

[10] *Pol. Test. 1752*, in *Werke*, VII, 172.

[11] *Mil. Test. 1768*, *ibid.*, VI, 233, 237; *Oeuvres*, XXVIII, 5.

high precision. The aim was to achieve tactical mobility, skill in shifting from marching order to battle order, steadiness under fire, complete responsiveness to command. An army so trained, Frederick repeatedly said, allowed full scope to the art of generalship. The commander could form his conceptions in the knowledge that they would be realized. With all else shaped to his hand, his presiding intelligence would be free. Frederick therefore never tired of urging his generals to ceaseless vigilance over drill, in war and in peace. "Unless every man is trained beforehand in peacetime for that which he will have to accomplish in war, one has nothing but people who bear the name of a business without knowing how to practise it."[12]

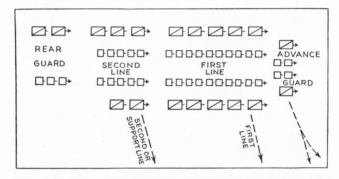

Frederick the Great's Order of March. By Turning His Column to the Right, He Could Execute "Left Wheel" by Platoons and Be in Battle Formation

Battle, with troops so spiritually mechanized, was a methodical affair. Opposing armies were arrayed according to pattern, almost as regularly as chessmen at the beginning of a game: on each wing cavalry, artillery fairly evenly distributed along the rear, infantry battalions drawn up in two parallel solid lines, one a few hundred yards behind the other, and each line, or at least the first, composed of three ranks, each rank firing at a single command while the other two reloaded. Frederick never departed from the essentials of this battle formation, though like all good generals he allowed himself liberty in adapting it to specific purposes. Battle order tended to determine marching order: troops should march, according to Frederick, in columns so arranged that by a quick turn the columns presented themselves as firing lines with cavalry on the flanks. Battle order was also the end object of severe discipline. It was not easy to hold men in the lines, standing in plain sight, elbow to elbow, against an enemy only a few hundred yards away. But orders were strict. "If a soldier during an action looks about as if to flee, or so much as sets foot outside the line, the non-commissioned officer standing behind him will run him through with his bayonet and kill him on the spot."[13] If the enemy fled, the victorious

[12] *Pol. Test. 1752*, in *Werke*, VII, 173-175; *Prin. gén.* in *Oeuvres*, XXVIII, 7.
[13] *Disposition, wie es bei vorgehender Bataille bei seiner königlichen Majestät in Preussen Armée unveränderlich soll gehalten werden* (1745), in *Oeuvres*, XXX, 146.

line must remain in position. Plundering the dead or wounded was forbidden on pain of death.

Frederick set a great value on cavalry, which constituted about a fourth of his army, but he used it in general only for shock action in solid tactical units. His scouting service was therefore poor; in 1744, with 20,000 cavalry, he could not locate the Austrians. Nor was he successful in the use of light infantry for skirmishing and patrolling. The Austrians had many light troops, mounted and foot, in their Croatians and Pandours; the French were to make use of light infantry in the untrained levies of the Revolution. Frederick hardly knew what to do with such troops, which, dispersed and individualistic, could not be extensions of his own mind.[14]

The middle years of the eighteenth century saw a more rapid increase in the use of artillery, in proportion to other arms, than any other period from the sixteenth century to the twentieth.[15] The Austrians, after their humiliating loss of Silesia, turned especially to artillery to meet the menace of Frederick's mobile columns. The French were the most progressive artillerists of Europe. Frederick often bemoaned this development, for Prussia of all major states could least afford an artillery race. The new vogue for artillery, observed the king in 1768, was a veritable abyss to the state's finances. Yet he joined the scramble; it was Frederick, with his appreciation of speedy movement, who introduced horse-drawn field artillery for shift of position during battle. He continued to insist that artillery was not an "arm" but only an "auxiliary," inferior to infantry and cavalry, but he gave increasing thought to its use, and one of his last writings, an *Instruction* of 1782, seems to show the influence of the French artillery theorists from whom Bonaparte was to learn. Frederick

[14] Delbrück, *op. cit.*, IV, 327-328.
[15] Cf. column II of the following table, which, compiled from data in G. Bodart, *Militär-historisches Kriegslexikon* (Vienna, 1908), pp. 612, 784-785, 816-817, shows the mounting intensity of war since 1600.

	I	II	III	IV
Thirty Years War	19,000	1.5	1	.24
Wars of Louis XIV	40,000	1.75	7	
Spanish Succession				.77
Wars of Frederick II	47,000	3.33	12	
Austrian Succession				.82
Seven Years War				1.40
Wars of French Revolution	45,000	—	12	
First Coalition				3.0
Second Coalition				4.4
Wars of Napoleon	84,000	3.5	37	
Third Coalition				7.0
War of 1809				11.0
War of 1812				5.2
American Civil War	54,000	3.0	18	1.0
War of 1870	70,000	3.3	12	9.0
Russo-Japanese War	110,000	3.75	3	1.0

Explanation of columns:

I. Average size of an army in battle, computed where possible from thirty battles in each war.
II. Number of cannon per 1000 combatants.
III. Number of battles in which the opposing armies together numbered over 100,000.
IV. Average number of battles per month.

here orders his artillery officers to avoid firing simply to satisfy the infantry or cavalry, to educate themselves in the discriminate use of ball and canister, and to concentrate their opening fire on the enemy's infantry in order to smash a hole in the enemy line and help their own infantry to break through.[16]

The use of the long unbroken battle array, since a frontal clash of two such solid lines would be butchery, caused Frederick to prize the flank attack, for which he designed his famous "oblique order," the advance of one wing by echelons with refusal of the other. Omitting tactical details, it may simply be said that Frederick's purpose in favoring this type of battle was, in case of success, to gain a quick victory by rolling up the enemy's line, and, in case of failure, to minimize losses, since the refused wing maneuvered to cover the withdrawal of the wing engaged. Frederick's superior mobility and coordination gave a special effectiveness to these flanking movements, which in themselves were of course among the oldest expedients of war.[17]

On these matters of army organization and tactics Frederick never seriously altered his opinions. He changed his mind on the larger issues of strategy. At first he seemed to introduce a new spirit, but in the end he accepted the limitations imposed by the political order, on questions of under what circumstances wars should be fought, and where and when battle should be joined.

His lightning attack on Silesia startled Europe. This first Silesian war (1740-1742) was a desperate gamble, played for what to a king of Prussia were very high stakes. In the second Silesian war (1744-1745, forming like the first a part of the War of the Austrian Succession) he aspired for a while even to the total destruction of the Hapsburg monarchy. The project failed, but Frederick retained Silesia. Thereafter his war policy became less ambitious. In the Seven Years War (1756-1763), after the battles of Rossbach and Leuthen, which probably saved Prussia from extinction, he was reduced to maintaining a brilliant defensive against the combined powers of France, Austria, and Russia, each of which had a population at least four times his own. Frederick's last war, that of the Bavarian Succession (1778-1779), dragged itself out in bloodless military demonstrations and promenades.

In the *Principes généraux de la guerre*, we find him calling for a strategy of blitzkrieg, though he did not use that term. The wars of Prussia, he says, should be "short and lively"; Prussian generals should seek a speedy decision.[18] These were in fact the principles on which he at first acted. It is notable, however, that the reasons given for these dashing operations were much the same as those which in later years made him increasingly cautious. A long war, he said, would exhaust the resources of Prussia and break down the "admirable discipline" of the Prussian troops. From preferring a short quick war it was no great distance to preferring either no war at all, or a longer war of low intensity in expenditure of men and material. In any case the governing condi-

---

[16] *Mil. Test. 1768*, in *Werke*, VI, 228 ff.; *Mémoires depuis la paix de Hubertsbourg*, in *Oeuvres*, VI, 97; *Eléments de castramétrie et de tactique* (1771), *ibid.*, XXIX, 42; *ibid.*, XXX, 139-141, 391-396.
[17] *Ibid.*, XXIX, 25; Delbrück, *op. cit.*, IV, 314-322.
[18] *Prin. gén.* (1746), in *Oeuvres*, XXVIII, 84.

tions were the same: the limited resources of the state, the dependence of armies on fixed magazines prepared beforehand, and the use of soldiers who, however well drilled, had no inward conviction to sustain them in times of trouble.

None of these conditions could Frederick overcome. He could not make Prussia a wealthy state; he could only economize its resources. He could not, like the governments of the French Revolution, let his armies live on occupied countries, although he recommended this procedure. His armies would melt away if dispersed to seek subsistence, and lose morale if they were not regularly supplied. Nor could he count on any welcome in occupied territories. His efforts to build a "fifth column" in Bohemia repeatedly failed. And he could not com-

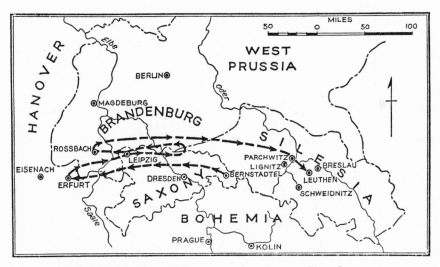

Frederick's Marches and Countermarches, Autumn Campaign, 1757

municate moral enthusiasm to his troops without changing his whole system and view of life.

In addition, when the Austrians strengthened their artillery and their fortifications after the loss of Silesia, they added technical hindrances to the development of aggressive strategy by Frederick. The old king, in his last years, repeatedly observed that conditions had changed since his youth—that henceforth Prussia could fight only a war of position. He himself, with his great permanent magazines and vulnerable frontiers, set a high value on fixed fortifications. Forts, he observed, were "mighty nails which hold a ruler's provinces together." To besiege and overwhelm such fortresses became a main object of warfare. The conduct of sieges had been a science since Vauban. Frederick carried on in this tradition. Even his concept of battle was colored by it. "We should draw our dispositions for battle from the rules of besieging positions." The two lines of infantry in battle order, he said in 1770, corresponded to the parallels formed by a besieging force. Even in occupying villages these principles should not be lost from mind. Nothing could be further from the direction in which military practice was to move. Napoleon was to conduct only two sieges in his whole career.[19]

Again unlike Napoleon, Frederick, though a successful battle general, was not fond of full-size battles, i.e. showdown clashes between the main forces of the belligerents. To his mind the outcome of battle depended too much upon chance and chance was the opposite of rational calculation. The supreme planning intelligence, the power of command to elicit obedience, which to Frederick were the first premises of scientific war, could not be relied on in the heat of a major engagement. "It is to be remarked in addition that most generals in love with battle resort to this expedient for want of other resources. Far from being considered a merit in them, this is usually thought a sign of the sterility of their talents."[20]

To annihilate the enemy's main combat force was thus not Frederick's usual strategic objective. He indeed realized that, if battle is fought, the winner should attempt a destructive pursuit of the enemy. But destructive pursuit was not easy to a Frederician army: the cavalry, trained for shock action in solid units, inclined to desert if scattered, fired neither by the half-barbaric ferocity of Croatian irregulars, nor by the political passion of more modern troops, was not suited to pursue a fugitive and broken army. Nothing like Napoleon's cavalry action after the battle of Jena would have been possible to Frederick. In effect for Frederick the purpose of battle was to force an enemy to move. "To win a battle means to compel your opponent to yield you his position."[21]

So Frederician war became increasingly a war of position, the war of complex maneuver and subtle accumulation of small gains; leisurely and slow in its main outlines (though never in tactics), and quite different from the short sharp warfare recommended in 1746. "To gain many small successes," he wrote in 1768, "means gradually to heap up a treasure." "All maneuvers in war," he added in 1770, "turn upon the positions which a general may occupy with advantage, and positions which he may attack with the least loss." He concluded also, from unfortunate experiences in Bohemia, that an army could not successfully operate far beyond its own frontiers. "I observe," he wrote in 1775, "that all wars carried far from the frontiers of those who undertake them have less success than those fought within reach of one's own country. Would this not be because of a natural sentiment in man, who feels it to be more just to defend himself than to despoil his neighbor? But perhaps the physical reason outweighs the moral, because of the difficulty in providing food supplies at points distant from the frontier, and in furnishing quickly enough the new recruits, new horses, clothing and munitions of war." Bonaparte, who could win battles in places as far from France as Austerlitz and Friedland, would

---

[19] *Mil. Test. 1768*, in *Werke*, VI, 247, 257; *Pol. Test. 1752*, ibid., VII, 176; *Eléments de castramétrie et de tactique* (1771) in *Oeuvres*, XXIX, 4, 21, 38.

[20] *Reflexions sur Charles XII* (1759), in *Oeuvres*, VII, 81; *Essai sur les formes du gouvernement* (1777), ibid., IX, 203.

[21] *Mil. Test. 1768*, in *Werke*, VI, 246-249; *Pol. Test. 1752*, ibid., VII, 174.

have smiled at such maxims of caution, though Borodino came to remind him of their force. For Frederick the rule held good.[22]

But although Frederick's strategic thinking remained within the old limits of the war of position, and although he remained disinclined to serious battle (it was his advisers who pressed for action in the year of Rossbach and Leuthen), he never favored passivity in operations. He continued to insist on the importance of surprise. He was prepared, in the years of peace after the Seven Years War, to spring at a moment's notice into Saxony or Bohemia, equipped with detailed maps and exact information, and with new ten-pound howitzers and new kinds of cavalry charges kept as a state secret. He favored offensive strategy in the field, as permitting more freedom of initiative; but would willingly fight on the defensive, as he often had to, when less strong than his enemy or when expecting to gain an advantage by time. It must however be an active and challenging defensive, which, while based on fixed fortifications, freely assaulted enemy positions and detachments. A commander, he said, "deceives himself who thinks he is conducting well a defensive war when he takes no initiative, and remains inactive during the whole campaign. Such a defensive would end with the whole army being driven from the country that the general meant to protect."[23]

Of the gains to be expected from war, under conditions then existing, he became increasingly dubious. Having made his debut by achieving the most successful revolution in the balance of power effected on the continent of Europe in his lifetime, he became with the acquisition of Silesia a man of peace, and ended by believing firmly in the value of the European balance now that Prussia was one of its main components. For Prussia he envisaged eventual expansion in Poland, Saxony and Swedish Pomerania; but (except for the first partition of Poland, which was accomplished without war and without disturbance to the balance of power, to the great satisfaction of diplomats) he was willing to leave this eventual expansion to his successors. He was a dynast, not a revolutionary or an adventurer; he could leave something to be done by others than himself. In 1775 he stood for the military status quo. "The ambitious," he wrote, "should consider above all that armaments and military discipline being much the same throughout Europe, and alliances as a rule producing an equality of force between belligerent parties, all that princes can expect from the greatest advantages at present is to acquire, by accumulation of successes, either some small city on the frontier, or some territory which will not pay interest on the expenses of the war, and whose population does not even approach the number of citizens who perished in the campaigns." Nor did he fear being crushed by his huge neighbors. "I perceive that small states [meaning Prussia, with its 5,000,000 inhabitants] can maintain themselves against the greatest monarchies [meaning France, Austria and Russia with their some 20,000,000 each], when these states put industry and a great deal

---

[22] *Ibid.*, VI, 248; *Oeuvres*, XXIX, 3; *Histoire de mon temps,* preface of 1775, *ibid.,* II, p. xxviii.
[23] *Mil. Test. 1768,* in *Werke,* VI, 253, 260-261; Jähns, *op. cit.,* III, 2027.

of order into their affairs. I find that the great empires are full of abuses and confusion; that they maintain themselves only by their vast resources and by the intrinsic force of their mass. The intrigues of these courts would ruin less powerful princes; they are always harmful, but do not prevent the keeping of numerous armies on foot." He seems never to have considered what would happen to the "equilibrium of Europe," should the greatest of the monarchies throw off its abuses and confusion, break down the limits set by the dynastic-aristocratic regime, and introduce into its affairs some of the attention to business already familiar in Prussia. He did not foresee the French Revolution.[24]

III

In France, however, the foundations of Napoleonic warfare were already being laid. The humiliating peace of 1763, by which France lost its empire overseas and its prestige in Europe, was followed by serious military thinking. Gribeauval revolutionized artillery by introducing the principle of interchangeable parts, improving the accuracy of fire, and heightening the mobility of guns through reducing weight. His reforms created the types that remained standard until the 1820's. The marshal de Broglie and the duke de Choiseul, in the 1760's, introduced a new and larger unit of army organization, the division. Developed gradually, the division came to be defined as a distinct, permanent, more or less equal part of an army, commanded by a general officer, and strong enough to engage the enemy successfully until other divisions reached the scene of action. Large armies ceased to be a single mass forming an unbroken front in battle; they became articulated wholes, with detachable and independently maneuverable members. Great new strategic and tactical possibilities were opened for a commander-in-chief, and at the same time, as divisional commanders, subordinate generals achieved an importance never enjoyed under Frederick. The revolutionary wars were the first in which the division was important. Napoleon and his marshals were the outcome.[25]

Along with practical innovations, after 1763, went a great deal of theoretical writing. Among the theorists was a young nobleman, the count de Guibert, who in 1772 published his *Essai général de tactique*. He was only twenty-nine, but his book made him a celebrity at once. He became a lion of the salons, fell in love with Mlle. de Lespinasse, wrote three tragedies in verse, served for a while in the War Office, and in 1789, at one of the district assemblies called to elect members to the Estates-General, he was liquidated from the incipient revolution by a combination of the reactionary, the disgruntled and the jealous. He died in 1790, crying on his deathbed: "I shall be known! I shall receive justice!"[26]

[24] *Pol. Test. 1752*, in *Werke*, VII, 158; *Histoire de mon temps*, preface of 1775, in *Oeuvres*, II, pp. xxviii-xxx.

[25] E. Picard, *L'artillerie française au XVIIIe siècle* (Paris, 1906); J. Campana, *L'artillerie de campagne, 1792-1901* (Paris, 1901); Weygand, *op. cit.*, p. 192; J. Colin, *op. cit.*, pp. 1-85.

[26] Editor's introduction, written in 1790, to Guibert, *Journal d'un voyage en Allemagne* (Paris, 1803); P. de Ségur, "Un grand homme des salons: le comte de Guibert, 1743-1790," in *Revue de Paris*, II (1902), 701-736; P. Vignié, "Un Montalbanias célèbre; le comte

Guibert was an unstable person, vain, unpredictable and brilliant, a *littérateur* and a *philosophe*, regarded by contemporaries as the embodiment of genius. He was inconsistent, overemphatic, swayed by the enthusiasm of the moment. When he wrote the *Essai* he had served as an officer in Germany and Corsica. Like other *philosophes* he warmly admired Frederick, who stood in their eyes for modernity and enlightenment. The great Frederick, according to rumor, was so annoyed to find his secrets divined by this impertinent youngster, that reading the *Essai* threw him into fits of rage. Whether the book divined old Fritz's secrets we cannot know; that it sometimes went beyond Frederician warfare is certain.

Two themes pervaded the *Essai général de tactique*. One demanded a patriot or citizen army. The other sounded the call for a war of movement. Both fell within Guibert's conception of *tactique*. The word at this time usually meant the maneuvering of troops, including under "grand tactics" what we call strategy, and under "elementary tactics" what we call tactics. This meaning Guibert rejected as too narrow. Tactics to him meant virtually all military science. It had two parts: first, the raising and training of armies; second, the art of the general, or what people then called tactics, and what we call tactics and strategy. Tactics, in his own enlarged sense, the young author wished to raise to the level of universal truth. "It becomes," he said, "the science of all times, all places and all arms . . . in a word the result of everything good which the military ages have thought, and of what our own age has been able to add."[27]

The theme of the citizen army was a common doctrine in *philosophe* circles. Montesquieu, Rousseau, Mably, and the host of lesser figures who by the 1770's made up liberal opinion maintained that, as a safeguard against tyranny, the citizens of a country must be trained to arms. A contributor to Diderot's *Encyclopedia*, J. Servan, who became war minister during the Revolution, published in 1781 a book on the citizen soldier. Guibert was riding the crest of a mighty wave. His *Essai*, dedicated "*à ma patrie*," proposing "to erect both a military and a political constitution" in which all Frenchmen, noble and commoner, king and subject, should glory in the title of "citizen," can be regarded as the leading *philosophe* work devoted to military science.

The present governments of Europe, Guibert begins, are all despotic machines. All peoples would overthrow them if they could. No people will fight for them. No government is really interested in military science. Even in Prussia discipline is purely external, the inhabitants are mostly unmilitary, and youth is not trained to warlike and Spartan habits. In France, where the king is not a soldier, conditions are even more relaxed. Peoples are indifferent to the fortunes of war, because prisoners are no longer slaughtered in cold blood,

de Guibert," *Bulletin archéologique de Tarn-et-Garonne*, LII (1924), 22-43; Guibert, *Précis de ce qui s'est passé à mon égard à l'Assemblée de Berry* (Paris, 1789); Jähns, *op. cit.*, III, 2059-2072.

[27] *Essai général de tactique* (1772) in *Oeuvres militaires du comte de Guibert* (5 vols.; Paris, 1803), I, 136-141. In his *Défense du système de guerre moderne* (1779), *ibid.*, Vols. III and IV, Guibert introduces the term *la Stratégique*.

and the civilians of a conquered province suffer no inconvenience except to pay a tribute often no heavier than their old taxes. In short, all the peoples of Europe are soft, and all the governments are weak. "But suppose," he says, "that a people should arise in Europe vigorous in spirit, in government, in the means at its disposal, a people who with hardy qualities should combine a national army and a settled plan of aggrandizement. We should see such a people subjugate its neighbors and overwhelm our weak constitutions like the north wind bending reeds."[28]

This remark has often been quoted out of context as a prophecy of the revolutionary and Napoleonic wars. It was no such thing. No such vigorous people, says Guibert, will arise. Russia under Peter might have become such at the beginning of the century, but even Russia is now too westernized, too habituated to "luxury" and the refinements of civilization. But though Guibert expects no change adequate to his theories, he observes that, in so effete a world, the country which reforms itself only slightly will have a great advantage over others. This much he hopes for France.

By introducing the vigor of its people into its army, France may develop a more decisive, swifter and more crushing kind of war. But even this much, though he hopes for it, he scarcely expects. The "vices" of modern warfare, he says, are incorrigible without political revolution. Revolution is out of the question—Guibert, like other *philosophes*, had little notion that revolutionary thinking might be followed by revolutionary behavior. What we must do, he says, "since we cannot have citizen troops and perfect troops, is to have our troops at least disciplined and trained." So, after the fanfare of general principles, as he works into his subject, Guibert arrives about where the great Frederick had started, at the idea, expressed by Frederick in 1746, that citizen soldiers were indeed the best, but that since most soldiers were not citizens they must be rigidly disciplined and trained.[29]

The second theme of the *Essai*, the demand for a war of movement, is accordingly far more developed than the theme of a citizen army. Through this second theme, as through the first, runs the same strain of primitivism, the same feeling that the culture of the eighteenth century is too complex and sophisticated, the same idealizing of rude and Spartan virtues. Guibert hopes to make war more mobile and decisive by simplification of its elements. He thinks the armies of his day too big, artillery overvalued, fortifications and magazines overgrown, the study of topography overdone. The European peoples, in his opinion, having no force of spirit, proliferate themselves in material objects and empty numbers. Lacking valor, they rely on money.

In his views on the size of armies and quantity of artillery, both of which were in the ascendant, reaching at Leipzig in 1813 the highest point attained in battle until the twentieth century, Guibert saw no further than his master Frederick, and remained within the school of limited war. However partial to citizen troops, he was no prophet of mass armies. Huge armies he regarded as

[28] *Essai général*, in *Oeuvres*, I, 1-23.
[29] *Ibid.*, pp. 1-151.

signs of the ineptitude of men in authority. A good general, he said, would be encumbered by an operating force of more than 70,000. On the contemporary artillery race he echoed Frederick's lamentations. Like Frederick, he regarded artillery only as an auxiliary, not as an "arm." The technical innovations of Gribeauval had, as usual, produced a wide split among experts. In a smaller way artillery was then in somewhat the position of aviation in our time. Guibert took a middle ground, favorable to Gribeauval, but he never fully appreciated the work of contemporary artillery theorists, such as du Teil, who were using the new mobility of guns to achieve heavy concentration of fire, and whose teaching shaped the mind of that most successful of all artillery officers, Napoleon Bonaparte.[30]

Guibert departed further from Frederick, and approached nearer to the practice of the world war which was soon to come, in his low opinion of fortifications and magazines. Armies, he thought, should live by requisitions on the countries they occupied. War must support war, as in the best days of Rome; troops should be frugal, have few needs, carry short baggage trains, endure scarcity and hardship without complaint. The present French system, he says, by which civilians accompany an army to supervise its provisioning, is ruinous, for military decisions come to depend on the consent of civil officials who care more about protecting supplies than about fighting the enemy. An army which travels light, living on the country, will gain new mobility, range of action and power of surprise.[31]

The art of fortification, Guibert thought, had been greatly overvalued since Vauban. Fortresses would become less necessary with the abolition of the large magazines which it was one of their functions to protect. Building chains of forts made war more costly than necessary. Dispersing the troops in garrisons made armies larger than necessary. The turning of military operations into a series of sieges made wars needlessly long. Nor would Guibert admit that fortified points had any real defensive value against a highly mobile army of the kind he envisaged. "As if," he wrote, "bastions alone could defend the cities which they surround, as if the destiny of these cities does not depend on the quality and vigor of the troops which defend and support them; as if, in short, fortresses poorly defended would not turn to the exhaustion, disgrace and certain enslavement of the conquered peoples who were their builders and masters." Forts, he concluded, should be few, very strong and entirely auxiliary to strategic movement.[32]

To accelerate movement Guibert had available the recent invention of the division. The divisional principle had not been carried very far in 1772, and Guibert failed to distinguish clearly between the new divisions in the French army and the temporary division of forces practiced by Frederick the Great. His doctrine, however, is clear, and marks an advance beyond Fredericks. Frederick's usual aim was to divide his army on the march in such a way that,

[30] *Ibid.*, I, 97, 445-472.
[31] *Ibid.*, II, 254-307.
[32] *Ibid.*, II, 208-220.

upon reaching the enemy, the parts would fall into place in a battle line planned in advance. The army marched as it intended to fight. Guibert emancipated marching order from this dependency on battle order. In marching, according to Guibert's conception, each division constitutes a column. These columns, in separating on the march, move more rapidly, cover a wider theater, and force the enemy to turn in a desired direction; for battle they concentrate, never having lost the higher unity which makes them a single army. The commander-in-chief, going ahead, surveys the field of prospective battle, determines his battle tactics in the light of what he sees, and arranges the placing of his divisions as they arrive upon the field. Battle becomes more flexible than before, more exactly adapted to terrain and circumstance, more susceptible to guidance by the commanding general after the armies are committed. Guibert credits Frederick with having used such a system at Hohenfriedberg, but in truth the idea was more Napoleonic than Frederician.[33]

The net message of the *Essai général de tactique*, in a sentence, was to call for a new kind of army, ideally a people's army, but in any case an army made more mobile by living on the country, more free to act because released from fortified points, more readily maneuverable because organized in divisions. With such an army the old war of position would yield to a war of movement. "In proportion as we fought more a war of movement, we should get away from the present routine, return to smaller and less overburdened armies, and seek less for what are called 'positions,' for positions should never be anything but a last resource for a mobile and well commanded army. When an army knows how to maneuver, and wants to fight, there are few positions that it cannot attack from the rear or cause to be evacuated by the enemy. Positions, in a word, are good to take only when one has reason not to try to act." And he sketches the lightning war which Bonaparte was to practice. A good general, he says, will ignore "positions" in the old-fashioned sense. "I say that a general who, in this matter, shakes off established prejudices will throw his enemy into consternation, stun him, give him no chance to breathe, force him to fight or to retreat continuously before him. But such a general would need an army differently constituted from our armies today, an army which, formed by himself, was prepared for the new kind of operations which he would require it to perform."[34] The Revolution was to produce this new kind of army.

Unfortunately for his reputation as a prophet, Guibert's only other completed work on military science, the *Défense du système de guerre moderne*, published in 1779, explicitly repudiated the main ideas of the *Essai*. "When I wrote that book," he said, "I was ten years younger. The vapors of modern philosophy heated my head and clouded my judgment."[35] In addition, after becoming famous by the *Essai*, he had met great Frederick, traveled through Germany, broken into society, been hailed as an expert, and become more contented with the world.

[33] *Ibid.*, II, 15-88.
[34] *Ibid.*, II, 249-254.
[35] *Défense du système de guerre moderne, ibid.*, IV, 212.

The "modern system" which the *Défense* tries to vindicate is simply the warfare of the day as contrasted with the warfare of classical antiquity. It is the conservative military technique of 1779. The body of the book deals with only one aspect of this "modern" war: the relative merits, debated for a generation, of column and line in the combat tactics of infantry. Guibert took the conservative side, defending the line, or principle of fire power, against the column, or principle of shock assault. To the body of this discussion Guibert added a final chapter, "The present system of war examined in relation to politics and administration." Here came the great recantation.

He will now have none of the idea of a citizen army. Citizen forces, while Guibert wrote, were fighting British and Hessian professionals in America. Many European officers watched the spectacle with interest; Lafayette, Berthier, Jourdan, Gneisenau were to bring back from America some favorable ideas on patriot soldiers and open fighting formations. Guibert insists that ex-civilians can never stand against professionals, and attributes the successes of the Americans entirely to the incompetence of the British. No modern state, he says, could possibly take the risk of using citizen levies, which were all very well for the ancients, among whom maneuvers were simple and firearms unknown, but which every nation of Europe has outgrown and discarded, except Turkey and Poland—and Poland is in ruins. In these contexts the word "citizen" meant hardly more than "inhabitant."[36]

Guibert also praises "modern," i.e. professional, war for the mild and even innocuous character which in the *Essai* was a main charge against it. Nowadays, he observes, a conquered country escapes the horrors of revenge and destruction, but "any country defended by its inhabitants must inevitably experience this kind of calamity." It is more humane for peoples to remain spectators to warlike violence. The emphasis on fortified positions, with all the subtleties of formalized maneuver, "may be an abuse . . . but certainly results advantageously for the tranquillity of nations and security of empires." The relative equality of training, discipline, resources, and talent among the military powers creates a salutary balance. So much the less, therefore, "will wars be decisive and consequently disastrous to the nations; the less possibility will there be of conquest, the fewer subjects of temptation for ambitious rulers, and the fewer revolutions of empires." Thus ends the thought of the *Défense*. It is scarcely distinguishable from that of Frederick the Great.[37]

Guibert, in both his books, glimpsed the difference between limited and unlimited war, or between the clashes of professional soldiers and the destructive struggles of peoples. He saw the close relation between warfare and the structure of government. His inconsistency was not logical but moral, an inconsistency of attitude, not of analysis. At twenty-nine, he looked upon the ideas of national armies and blitzkrieg strategy with favor. At thirty-five he looked upon these same ideas with disapproval. At neither time did he show much practical foresight, as distinguished from lucky predictions, or any sense that

---

[36] *Ibid.*, IV, 219-231.
[37] *Ibid.*, IV, 263-275.

the ideas which he favored in 1772, and rejected in 1779, would become realities for the generation then alive.

Before concluding the *Défense* Guibert took a parting shot at the *philosophes,* who sometimes showed pacifist inclinations, or at least objected to the wars fought by governments then existing. "To declaim against war," he said, ". . . is to beat the air with vain sounds, for ambitious, unjust or powerful rulers will certainly not be restrained by such means. But what may result, and what must necessarily result, is to extinguish little by little the military spirit, to make the government less interested in this important branch of administration, and some day to deliver up one's own nation, softened and disarmed—or, what amounts to the same thing, badly armed and not knowing how to use arms—to the yoke of warlike nations which may be less civilized but which have more judgment and prudence."[38] Here too was a prophecy for France. It was a warning not needed in the eighteenth century, however, for of the ideas of the *philosophes* it was not pacifism that was to prevail.

IV

In 1793 the revolutionary French Republic faced a coalition of Great Britain, Holland, Prussia, Austria, Sardinia and Spain. Of peoples living under one government the French were the most numerous and perhaps the most wealthy. A Committee of Public Safety, to meet the crisis, exploited their military potentialities in a way never possible under the Old Régime. Freed from the old special rights, local and class privileges, internal barriers and exclusive monopolies which had encumbered the monarchy, the Committee created a war economy by dictatorial methods, stimulated the national self-consciousness of the population, and introduced the principle of universal military service in the *levée en masse.* In this, the political side of warfare, the revolutionists were conscious of bringing about a new military order. They were less conscious of innovating in technical and strategic matters. Carnot's strategic ideas were rather old-fashioned.[39] Yet in leaving their armies to be supplied by requisitions rather than magazines the Republicans effected a revolution in logistics, and in throwing their half-trained troops into battle in rushing columns or in fanned out lines of *tirailleurs,* men who fought, fired and took cover as individuals (a practice suggested by the War of American Independence), they broke away from the Frederician system of solid battalions, and gave impetus to a revolution in tactics.

By 1794 the French took the offensive. In 1795 Prussia, Holland, and Spain withdrew from the war. In 1796 Bonaparte dropped into Italy out of the mountains. By 1797 the continent was at peace, and England negotiated. In 1798 war was resumed with the Second Coalition. In 1799 Bonaparte became autocrat of France. In 1800 he destroyed the Second Coalition, winning, again

[38] *Ibid.,* IV, 213.
[39] R. Warschauer, *Studien zur Entwicklung der Gedanken Lazare Carnots über Kriegführung* (Berlin, 1937).

by lightning operations in Italy, the first of his great, quick, decisive "Napoleonic" battles—Marengo.

A revolution had occurred in the art of war. Its significance dawned only gradually on observers. Certain civilians, Mallet du Pan and Gentz, for example, perceived some of the deeper causes sooner than professional soldiers. This is because the most fundamental change was in the political premises of military organization, in that new *Weltbild* whose coming, according to Delbrück, was necessary to the revolutionizing of warfare. In France the professional soldiers in these years were too busy in action to write treatises on what they were doing. In Germany Scharnhorst edited a journal and published piecemeal studies of events, and Gneisenau in a Silesian garrison town applied his American experiences to the training of troops; both were reeducating themselves in their profession, and both came forward after 1806 to rebuild the Prussian army. The military writers most in the public eye, in the years just before and just after 1800—Behrenhorst, Bülow, Hoyer, Venturini—seemed for a while to learn nothing from the facts before them. It is most instructive to dwell upon Bülow.[40]

Freiherr Dietrich von Bülow, like the count de Guibert, was a minor aristocrat with a modicum of experience in the army. To earn a living he wrote books on many subjects. He proved to be as erratic as Guibert, and even more pathologically egotistical. He repelled everyone by his claims to unrecognized wisdom, offended the Russians during the period of the Prusso-Russian alliance, was adjudged insane, and died in 1807 in confinement at Riga. He has since been called everything from a conceited crank to the founder of modern military science.[41]

His first military treatise, the *Geist des neueren Kriegssystems*, appeared in 1799, won great favor, and was soon translated into French and English. Geopoliticists today see in it a step in the development of their subject. Bülow concluded his book with reflections on political "space." He declared (contrary to Frederick) that, because of the modern military system, the age of small states was over. He held that state power tended to fill a certain area, and beyond that area to be ineffective; hence each power had natural frontiers; the attainment of these frontiers would produce a political balance and lasting peace, since each power would then have reached the natural limits of its action. There would be, he said, about a dozen states in Europe: the British Isles; France extending to the Meuse; a north Germany gathered around Prussia, reaching from the Meuse to Memel; a south Germany looking to Austria, which in turn would extend its borders down the Danube perhaps to the Black Sea; a united Italy; a united Iberian peninsula; Switzerland; Turkey; Russia; Sweden; and probably, though not necessarily, an independent Holland and an independent Denmark.[42]

[40] J. Mallet du Pan, *Considérations sur la nature de la révolution de France* (London, 1793) ; F. Gentz, *Von dem politischen Zustande van Europa vor und nach der französischen Revolution* (1801) ; and see Jähns, *op. cit.*, under the names cited.
[41] *Ibid.*, III, 2133-2145.
[42] R. Strausz-Hupé, *Geopolitics: the struggle for space and power* (New York, 1942), pp. 14-21 ; Bülow, *The spirit of the modern system of war* (London, 1806), pp. 187-285. The German original of this work seems to be unobtainable in the United States.

This was a surprisingly good anticipation of the map of Europe as it came to be by 1870. It was scarcely grounded on an accurate perception of the military situation in 1799. *Der Geist des neueren Kriegssystems* showed no real understanding of the wars of the Revolution. Only in the new open formation of *tirailleurs*, i.e. only in infantry tactics, did Bülow find any significant innovation.[43] He is credited with clarifying terminology, by giving currency, as words of distinct meaning, to the terms "strategy," "tactics" and "base of operations," though his definitions were not generally accepted. But the thesis of his book was a codification of obsolescent ideas.

Bülow's "modern system," like Guibert's, was simply the system developed since the seventeenth century. He claimed, however, to have discovered the true key to this system in the concept of the base of operations. He held also (as if they were new) to old notions of the geometry of war. The "base of operations" in his system must be a fortified line of prepared magazines; the two "lines of operations" projected from the ends of this base must converge upon the point under attack at an angle of at least ninety degrees. The attacking army must not move by more than three days' march from its magazines. The general should have as his principal objective, not attack on the enemy force, but the security of his own service of supply; and in offensive operations he should concentrate not against the enemy army, but against the enemy's supplies. Fighting should be avoided. A victorious general should refrain from pushing his advantage, "stopping judiciously in the midst of triumphs." Modern battles decide nothing; an enemy defeated on the battlefield can always attack again in a few days.[44]

The unreality of these conceptions had been shown as early as 1794, when the French cavalry rode into Amsterdam on the ice. The battles of Hohenlinden and Marengo, a few months after the publication of Bülow's book, came as an answer to his "system." This campaign opened his eyes. He wrote a book on it, perversely insisting that the French victories gave proof of his doctrine but in reality contradicting much of what he had said before. He learned, but he learned very reluctantly.

Marengo, said Bülow, in less than a month "has decided the destiny of the French Revolution and hence of humanity in Europe." Mobility is the secret of French success. Before a mobile army most fortifications are shown to be useless. Mobility and audacity are made possible by reduction of baggage trains and emancipation from magazines. Bonaparte, he observes, crossed the Alps with no food but biscuit, a compact, durable, portable nutriment that needs no cooking; and he arrived in Italy with a hungry army, planning to live on the country. How all this harmonized with the theory of the "base of operations" with its comfortable ninety-degree angle, Bülow failed to make clear, though he argued the matter at great length. He noted, as a source of the new boldness of action, the new type of personnel in the French army. The Austrian officers, he said, owe their positions to seniority. Their talents are

[43] *Ibid.*, pp. 109 ff.
[44] *Ibid.*, *passim*, but see pp. 1-25, 81-82, 108, 183-184.

average. "With the fermentation inseparable from revolution there have appeared in France men who in time of calm would not even have suspected what they were capable of. This sudden deployment of transcendent abilities is one of the first causes to which the marked superiority of the French in this war must be ascribed."[45]

Even with these explanations Bülow could not understand a blitzkrieg which astounded Europe. He called the French victory a portent, a miracle, a message from Providence. He became Bonapartist and pro-French. This made his position increasingly awkward as the national movement swept over Germany, and no doubt accentuated his paranoid inclinations.

Then came the campaign of 1805. In that year Austria and Russia joined with Great Britain in the Third Coalition. The two continental powers moved large armies westward. In these armies centered the highest hopes of aristocratic Europe. Seldom has disappointment been so swift. Bonaparte in a few days marched several army corps from coastal points to South Germany. There, at Ulm, he forced General Mack, reputed to be a master strategist, to surrender 30,000 men without serious fighting. Moving on to Vienna and into Moravia, he found the combined Austro-Russian forces eager to attack. He routed them at the village of Austerlitz.

Bülow immediately wrote a two-volume work on the campaign, published in the anxious months after Austerlitz, during which the Prussian state, having conducted a two-faced diplomacy, moved as if hypnotized toward the disaster of Jena. Bülow had to publish this work privately. It was too dangerous for anyone to touch but himself and it led to his own ruin. A strange and contradictory book, it reflected both his own mental unbalance and the general bewilderment of Europe. He wrote as one convinced that he alone saw the truth, that ignored though he was he must in duty give everyone advice, impelled by Kant's categorical imperative—metaphysics and military thought have gone together in Germany. He announced that he was destined to create a new theory of war, to be known as *Bülowisch*, by which all future officers would be formed. He berated Frederick the Great and the Frederician system, demanding the kind of regeneration which until Jena Prussia was not willing to undergo. Yet he said, too, that reform was hopeless, that Napoleon was about to unify Europe by war, and that the continental powers should accept his supremacy. Austerlitz, said Bülow, was the modern Actium.[46]

Bülow saw in the French victory of 1805 a proof of the doctrine of Guibert. He used a metaphor from business. The great art in war, he said, is to get the most out of one's capital, not to scatter an army in garrisons but to keep the whole of it constantly in circulation. Napoleon, more than others, "keeps his capital active." This was to recognize the obsolescence of the old war of position. At Ulm Mack had a strong army in a powerful position. Napoleon

---

[45] *Histoire de la campagne de 1800 en Allemagne et en Italie* (Paris, 1804), pp. 4-5, 16, 90, 92, 142 ff., 183. The German original, *Der Feldzug von 1800* (Berlin, 1801) is difficult to obtain.

[46] *Der Feldzug von 1805, militärisch-politisch betrachtet* (2 vols.; auf Kosten des Verfassers [Leipzig], 1806), I, pp. i-lxxvi, II, p. 158.

nevertheless forced him to surrender. He did it by applying Guibert's principles: skillful manipulation of the divisions (facilitated by the Napoleonic innovation of the army corps); physical dispersal of these divisions for speed in marching, and to cover a larger theater of action, without loss of unity of conception; simultaneous reconcentration at the objective with adoption of battle positions in the light of concrete local conditions. The result, according to Bülow, was "the most perfect manifestation of the superiority of strategy over tactics in modern war."[47]

As more depended on strategy and comparatively less on tactics, the problems of supreme command took on a hitherto unknown complexity and scope. Battle lost some of the element of pure chance which Frederick had feared in it, and which before the Revolution had served as a deterrent to aggressive operations. It became rather the test of elaborate preparations made long beforehand. Planning became more fruitful, prediction somewhat more possible, warfare more of a "science." Military command shaded into diplomatic relations on the one hand, and into domestic policy and constitutional practice on the other. On these matters Bülow had much to say.

Bülow, like Frederick, insisted on the need of a single unifying intelligence at the head of a state. He held that under modern conditions of strategy there could be no separation between politics and war—great soldiers must understand foreign affairs, as successful diplomats must understand military action. Of the advantage of uniting foreign policy and military responsibility in one mind Napoleon's career was an example and the fumbling of the Allied governments a kind of negative demonstration. A firm guiding intelligence also became more necessary with modern conditions of technology. The supreme command must rise above the specialists and the experts. The technique of fortification, the theory of artillery fire, military medicine, logistics, said Bülow, are only "preparatory sciences." "The science of employing all these things fittingly for the strengthening and defense of society is true military science." This is the real business of generalship. "Hear this plainly: when a chief of state is obliged to leave the guidance of the state's energies in war to a squad of mere specialists trained in the preparatory sciences, the inevitable outcome will be fragmentation and cross-purposes, of which the first result will be weakness—a stable full of calves and donkeys—and the end result dissolution; because the binding power of intelligence is missing, which unites the materials in one building, or in one purpose." Here again the lesson was driven home by the contrast between Napoleon and every other ruler of Europe.[48]

On man power, or the constitution of armies, Bülow had views not at all flattering to contemporary Prussia. He upbraided the Prussian government for blindly maintaining the Frederician system, of which he said even Frederick saw the weaknesses before his death—a system which left the common people demoralized and uneducated, subject to a discipline that violated the rights of

[47] *Ibid.*, I, pp. lviii-lix, II, pp. xxxiv, 109.
[48] *Ibid.*, I, 5-20.

man. He recommended the French system of universal conscription with its nationalistic effect on morale. "Even if we take a purely utilitarian view, an army could be regarded as the most general educational establishment for youth." Military science must face "a weighty matter of internal administration, the inspiring and rewarding of virtues and talents." Prussia, he observes, has produced few men of genius; yet resources are wasted unless able men control them. So Bülow calls for a policy of careers open to talent, and offers Napoleon's Legion of Honor as a model. He proposed a *Bund der Tugend*, in which men should be graded by intelligence, judgment, and utility to the state, and which, at least ideally, should efface the old aristocratic distinctions.[49]

All these ideas remained unassembled in Bülow's mind. He never attained that firmness of grasp and singleness of purpose which he recognized as essential to leadership. It is impossible to say what he felt his own aims to be. He seemed to favor the French Revolution, and spoke well of the rights of man; yet he was less a liberal than Gneisenau, to name another professional soldier for comparison. He called himself a Prussian patriot, but he despised Frederick, and said that Prussia by its very existence had ended the national existence of Germany. Sometimes he spoke as a German nationalist, but he remained stubbornly pro-French. Sometimes he favored a balance of power; again, he professed not to care whether the sovereigns of Europe maintained their independence. He certainly was a crusader, to what end is not clear. He was a vehement reformer but held reform to be a chimera. He was a kind of transcendental philosopher in military science, enjoying a sense of duty for its own sake without specifying its object. On the practical level, he advised Prussia, and all Europe, to come to terms with Napoleon after Austerlitz; he said that a Fourth Coalition would be useless and urged the continent to join with the French emperor for the humiliation of England. His attitude after Jena was simply, "I told you so."

Bülow by 1807 had given cause to the Prussian government to regard him as a madman, or at least as a nuisance in time of public disaster. He seemed to write for no purpose except to air his own views and the worst that can be said of the officials who sent him to prison, given the catastrophic conditions of 1807, is that in perceiving his faults they failed to recognize his merits. He was too irresponsible, vain, and vague to collaborate in the practical work of reconstruction. The world lost no Scharnhorst with his death.

As a theorist, he had the merit of sensing, though slowly and confusedly, the nature of the military revolution of his time. This revolution was not based on technology, despite important improvements in artillery; nor was it primarily a revolution of strategy in the strict sense, despite the heightened mobility and striking power of an army emancipated from magazines and organized in divisions. The military revolution was at bottom a political revolution. The driving force of the French was their new *politisches Weltbild*. This consisted in the fusion of government and people which the Revolution had effected. On the one hand the people, in a way not possible before 1789,

[49] *Ibid.*, II, pp. xviii-xxxii, 131-136; *Neue Taktik der Neuern* (Leipzig, 1805), p. 48.

felt that they participated in the state, that they derived great advantages from their government, and therefore should fight for it loyally and with passion. On the other hand the government, ruling by the authority of the nation and invoking its sovereign power, could draw upon human and material resources in a way not dreamed of by Frederick the Great. More temporary advantages of the French were revolutionary fanaticism and missionary zeal. The net result was that, after 1793, the wealth, man power and intelligence of France were hurled against Europe with irresistible effectiveness. During the nineteenth century the fundamental principle, the fusion of government and people, which may or may not be democratic, was built into the political system of most European states. The wars of kings were over; the wars of peoples had begun.

SECTION II

The Classics of the Nineteenth Century:
Interpreters of Napoleon

CHAPTER 4. Jomini

BY CRANE BRINTON, GORDON A. CRAIG AND FELIX GILBERT

THE World War between France and a series of coalitions (now generally known as the Napoleonic Wars) which lasted, with but few brief respites, from 1792 to 1815, brought to warfare one major innovation—the citizen army raised from a large population subject to universal service—and one very great military genius, Napoleon. Although the career and personality of Napoleon merit the extended and meticulous study they have been given, there can be no doubt that in the long history of warfare it is the innovation of the mass army which is the truly significant inheritance we have received from those troubled years.

The very eloquence of the famous decree of the Convention of August 23, 1793, calling for the *levée en masse* is still urgent upon us:

"ARTICLE I. From this moment until that in which our enemies shall have been driven from the territory of the Republic, all Frenchmen are permanently requisitioned for service in the armies.

"Young men will go forth to battle; married men will forge weapons and transport munitions; women will make tents and clothing, and serve in hospitals; children will make lint from old linen; and old men will be brought to the public squares to arouse the courage of the soldiers, while preaching the unity of the Republic and hatred against kings."[1]

The *levée en masse* was introduced at a moment when France seemed about to be crushed. Within a year of its adoption the French took the offensive. The powers of the coalition, suspicious of each other's aims, produced far less than their maximum strength at the critical moment. The revolutionary French republic, with a unity forced upon it by the government of the Terror, and with the tremendous man power provided by universal conscription, proved more than a match for its disorganized antagonists. In 1795 the coalition was broken. Spain, Holland and Prussia signed treaties of peace. In 1796 Bonaparte, one of the new generals who emerged from the period of the Terror, received his first independent command. He pierced into Italy over the Maritime Alps, divided the Sardinian and Austrian armies by the speed and precision of his movements, overwhelmed the Sardinians and quickly turned against the Austrians, upon whom in 1797 he forced the treaty of Campo Formio. France obtained its long coveted "natural frontier" of the Rhine, together with control over a new puppet republic in Lombardy. Though the old Venetian republic was wiped out and its territories ceded to Austria by way of compensation, the balance of power in Europe was revolutionized Peace lasted only a few months. Britain and Austria found a new ally in Russia. This Second Coalition was frustrated in a series of operations for which Bonaparte, without wholly deserving it, received the credit. The prestige

---

[1] *Réimpression de l'ancien Moniteur depuis la Réunion des États-Généraux jusqu'au Consulat*, August 25, 1793, XVII, 478 (Paris, 1840-1845).

of victory secured him in the dictatorship which he had arrogated a few months previously after the coup d'etat of Brumaire. The discouraged British signed at Amiens a treaty very unfavorable to themselves. For about a year in 1802-1803 there was no war.

Bonaparte meanwhile consolidated his authority in France, interfered in Italy, Switzerland, Holland, and Germany, provoked the fears and ambitions of the defeated powers, found himself again at war with England, and hence, through British diplomacy, menaced by a Third Coalition. To deal with Austria and Russia he suspended his plans for the invasion of England. In a few weeks he crushed the Austrians at Ulm and Austerlitz, and drove the Russians back over the Carpathians. Almost simultaneously the British naval victory at Trafalgar determined that there should be no invasion of England. France henceforth had to defeat England through land power if at all and Napoleon now had a new incentive for achieving domination of the continent. This he secured by meeting his remaining enemies separately and in quick succession, destroying Prussia as a great power by the battle of Jena in 1806 and persuading Russia to accept an alliance with himself after the battle of Friedland in 1807. After the treaty of Tilsit in that year Napoleon stood at the height of his power.

To subvert the economic structure of England he inaugurated the Continental System, prohibiting the entrance of British goods into the continent of Europe. This led to feverish efforts for the control of coastlines. Spain was occupied; the hold over Italy was tightened; Holland, the German North Sea littoral, and parts of Italy and Dalmatia were in turn annexed directly to France. Resistance developed in Spain, eventually supported by a British expeditionary force. The Austrian government, urged on by its patriots and encouraged by the Spanish rebellion, challenged the emperor of the French. Alone, and acting prematurely, it was defeated, for the fourth time, in the short but sanguinary War of 1809. For two years the continent enjoyed a troubled peace. Then the anti-French, pro-British forces in Russia won the ear of the czar. Napoleon, to hold Russia in line against England, plunged into the disastrous campaign of 1812. His army, numbering over six hundred thousand, drawn from increasingly younger French conscripts and from a dozen allied or subjugated nationalities, was probably the largest ever assembled in Europe, up to that time, for a specific campaign. Its fate is well known.

One by one, as the emperor's weaknesses revealed themselves, the still timid and half incredulous governments recently bound to Napoleon turned against him. To Britain and Russia were added Prussia, Austria, and the German states, together with Spanish rebels and Italian insurrectionists. Never before had the three great military monarchies of the continent acted together. Now that they did, the result, in October 1813, was the battle of Leipzig, the *Völkerschlacht* or "battle of the nations," which, with more than half a million men engaged, was to remain the greatest battle fought in Europe or America until 1914. It was a clash of national armies and it settled the fate of Napoleon. France, its man power depleted after twenty years, could not stand against a united Europe. British diplomacy joined the four chief powers in the strongest alliance yet achieved against France. Napoleon was sent to Elba in the spring of 1814. Except for the scare caused by his return, that was the end.

The scale of this warfare, the size of the armies involved, the speed with which Napoleon moved them, the completeness of his victories, the increasingly evident intention of the French to alter the whole European state system into a new continental order dominated by France—to contemporaries all this seemed something new, something for which history gave no precedent. In a sermon preached in Boston on April 5, 1810 William Ellery Channing said: "We live in times which have no parallel in past ages; in times when the human character has almost assumed a new form; in times of peculiar calamity, of thick darkness, and almost of despair. . . . Am I then asked, what there is so peculiar and tremendous in our times? I answer: in the very heart of Europe, in the centre of the civilized world, a new order has arisen, on the ruins of old institutions, peculiar in its character, and most ruinous in its influence. We here see a nation, which, from its situation, its fertility, and population, has always held a commanding rank in Europe, suddenly casting off the form of government, the laws, the habits, the spirit by which it was assimilated to surrounding nations, and by which it gave to surrounding nations the power of restraining it; and all at once assuming a new form, and erecting a new government, free in name and profession, but holding at its absolute disposal the property and life of every subject, and directing all its energies to the subjugation of foreign countries. . . . We see it dividing and corrupting by its arts, and then overwhelming by its arms, the nations which surround it."[2]

Not all contemporaries were as sure as Channing that what they were living through was absolutely unprecedented. To a distinguished Swiss officer in the French service, the career of Napoleon, novel though he knew it to be, in a sense seemed a perfectly explicable outgrowth of the warfare and politics of the eighteenth century. The more striking Napoleon's victories were, the more they seemed to this officer to be worth explaining in terms of those general truths, those "rules of old discovered, not devis'd," which the age of enlightenment was always seeking to make more precise and effective. In his attempt to explain Napoleon's career, General Jomini made his own contribution to the innovations of the age. He began, not indeed the study of war, but the characteristically modern, systematic study of the subject in the form it has retained ever since.

With Clausewitz, whom he antedates a bit, Jomini may be said to have done for the study of war something akin to that which Adam Smith did for the study of economics. Just as there were important books about economics before the *Wealth of Nations* appeared in 1776, so there were important works

---

[2] This sermon was reprinted in the *Christian Register*, CXX, August 1941, 248-249, exactly as given, save for the substitution of "Hitler" for "Napoleon" and of "Germany" for "France."

on war before the first volumes of the *Traité des grandes opérations militaires* in 1804. Most of these, and especially the writings of such immediate predecessors as Lloyd, Grimoard, Guibert, Bülow, Frederick the Great, were well known to Jomini, and he frequently acknowledged his debt to them.[3] Nevertheless, Jomini's systematic attempt to get at the principles of warfare entitles him to share with Clausewitz the position of co-founder of modern military thought.

Jomini's attitude toward his predecessors and toward his great contemporary throws an interesting sidelight upon his own position and aims. The two military writers whom Jomini treated with special attention were Bülow and Clausewitz. His work is studded with references to the former writer and criticisms of his theories.[4] Jomini accused Bülow of overemphasizing the scientific aspect of warfare. Bülow, he said, had regarded all of those who opposed the "trigonometric" concept of war as fools; yet his own doctrines, buttressed by scientific arguments, had been revealed as mere sophisms when tested in the wars of Napoleon.

But while Jomini opposed Bülow's tendency to place too much emphasis on the scientific aspect of war, he accused Clausewitz of making all military science impossible. "One cannot deny that General Clausewitz has great knowledge and a facile pen. But that pen, sometimes a little out of control, is above all too pretentious for pedagogical discussion, the greatest merit of which should be simplicity and clarity. Furthermore, the author displays too much skepticism in matters of military science. His first volume is merely a declamation against all theories of war, while the following two volumes, full of theoretical maxims, prove that the author believes in the efficacy of his own doctrines even though he doesn't believe in those of others."[5] After reading Clausewitz' treatise *On War* Jomini wrote that he found in this *savant labyrinthe* only a few striking ideas, and that the pervading skepticism of its author had convinced him of the necessity and the utility of "good theories."[6]

In his criticisms of Bülow and Clausewitz, we may find the key to Jomini's conception of the purpose of his own work. The excessive rationalism of Bülow taught him the necessity of revising eighteenth century concepts of war. But, as is evident from his strictures upon the Clausewitz treatise, he was never willing to abandon those concepts completely.

---

[3] The introductory chapter of the first complete edition of Jomini's *Précis de l'art de la guerre* (1838) is an interesting and very complete survey of military writers up to that time. In later editions of the book this chapter was brought up to date by Jomini's friend, the Swiss Colonel F. Lecomte.

[4] See *Traité des grandes opérations militaires* (4th ed., Paris, 1851), I, 415; II, 236, 273; III, 336; *Précis de l'art de la guerre* (Paris, 1838), I, 15, 234, 272-276.

[5] *Précis*, I, 20-22.

[6] *Ibid.* Later in the *Précis*, Jomini writes a criticism of Clausewitz which is typical of the vanity which often characterizes his work:

"Clausewitz' works have been incontestably useful, though frequently less for the author's ideas than for the opposing ones he provokes. They would have been even more useful if a pretentious style did not frequently render them unintelligible. But if, as a pedagogical writer, he has raised more doubts than he has revealed truths, as a critical historian he has been an imitator with few scruples. Those who have read my campaign of 1799, published ten years before his, will not deny my assertion, for there is not one of my reflections he has not repeated." (*Précis*, I, 32 note.)

II

Antoine Henri Jomini was born in 1779 in the canton of Vaud in French Switzerland of a good middle-class family which had emigrated from Italy several generations earlier. He received the conventional education of a young bourgeois destined to trade or banking, and was, indeed, in a banking house in Paris when he succeeded in getting himself an unpaid—and almost unofficial—staff position in the French army. For Jomini was seventeen when the glory of General Bonaparte's campaign in Italy flashed through the world, and made banking seem very dull indeed. The young Swiss, perhaps more ambitious, and certainly more curious, than adventurous, decided he too must be a soldier. Possessed of real administrative ability, he managed a bit unorthodoxly to work himself into the French army in the service of supply and to continue with minor staff work. During the delusive peace of Amiens he returned to commercial life but with the renewal of war found his niche as chief of staff for Marshal Ney, and started off on the great campaign that was to culminate in Austerlitz.

Jomini had in his half-dozen years of mingling with military men talked and thought a great deal about the art of war. It was the quickness and range of his mind applied to military matters that had impressed Ney, a brave soldier and a good tactician in action, but certainly not an inquiring student of the art of war. During the interlude of Amiens, Ney helped his protege to publish the first volumes of a great treatise on the campaigns of Frederick the Great, in which Jomini hazarded certain generalizations in military thought, and made certain comparisons between Frederick's generalship and Napoleon's. Jomini managed to get a presentation copy through to the emperor, and the emperor in the lull after Austerlitz managed to find a spare moment to have some of the book read to him. Impressed with the author's intuitive understanding of the Napoleonic touch, he had Jomini, his position at last regularized by an appointment as colonel in the French army, report to him at Mainz in September 1806.

The campaign of Jena was brewing in Napoleon's mind. Jomini, at the end of the conference, asked if he might rejoin the emperor four days later in Bamberg.

"Who told you that I am going to Bamberg?" asked the emperor—not, one assumes, without annoyance, for he supposed his destination a secret.

"The map of Germany, Sire, and your campaigns of Marengo and Ulm."

Our chief authority for this and other remarkable bits of divination by Jomini is unfortunately the diviner himself, and that mostly in his later years, when as a very old man in the Passy of the 1860's he brought back the past for distinguished guests like Sainte-Beuve. But if Jomini, like many another intellectual, was sometimes too right, there can be no doubt that he had an admirably clear understanding of Napoleon's strategical habit of mind, and that Napoleon appreciated the value of Jomini's writings.[7]

---

[7] For Jomini's often startlingly correct prophecies—a twenty-year old, he claims to have predicted Napoleon's march over the Alps in 1800—see Xavier de Courville, *Jomini, ou le devin de Napoléon* (Paris, 1935).

Though he rose to the position of *général de brigade* in the French army, and though he served as Ney's chief of staff in Prussia, in Spain, and again after the retreat from Moscow, though in the Russian campaign he served as governor of Vilna and later of Smolensk, Jomini never attained independent command, never came close to that marshal's baton men less intelligent than he discovered in their knapsacks. Jomini himself, and his biographers, have blamed the personal enmity of Berthier, the imperial chief of staff, and there seems no doubt that Berthier did dislike the self-confident Swiss. But Jomini was always being slighted and taking offense and resigning—he must have resigned and withdrawn his resignation half a dozen times in these years. It seems likely that his superiors, including the emperor himself, simply would not trust him with independent command of troops in the field.

Bitterly disappointed by lack of promotion, Jomini rode off to the Allied lines in August 1813 and offered his services to Alexander of Russia. Since Jomini was still a Swiss by citizenship, the act was something less than treason, and if it has kept Jomini from being a hero to the French, it has not made their historians bitter toward him, and has not kept their teachers from using his writings as textbooks. In the Russian service, where he held the rank of general until his death, he acted as military adviser, took a decisive part in the foundation of the Russian military academy, and found ample leisure to complete the historical and analytical studies he had begun after Marengo. His last years were spent now in Russia, now in France. At the time of the Crimean War, he was frequently consulted by the Russian emperor; in 1859, his advice was sought by Napoleon III before the latter embarked upon his Italian adventure. By the time of his death in Paris in 1869, Jomini's books were widely used in military education all over the world, and he had the satisfaction of knowing that he was regarded as something of an oracle.

Jomini's military career was certainly unusual. He rose neither through hard knocks in the ranks, nor through the formal conditioning of a cadet school. He slipped into an administrative position in the French army without previous military training. Inevitably, as a Swiss, something of an outsider, his peculiar professional position—and, it must be admitted, his temperament—prevented his ever attaining the full comradeship of arms. In one of his numerous conversations with distinguished visitors at Passy, Jomini is said to have remarked that, though he had seen a position taken at shoulder arms, he had never actually witnessed a charge with the bayonet, let alone taken part in one.[8] He had, moreover, that kind of jealous vanity peculiar to literary men and other intellectuals, display of which usually annoys the military, whose vanity is commonly under better discipline.

It is, however, quite wrong to think of Jomini as a "paper soldier," a pure theorizer, an intellectual who had no more direct concern with armies than the old-fashioned academic economist had with business. He did help move armies on the field. He had the grave responsibilities, as Ney's chief of staff, of getting things done in this imperfect world. He had to make important decisions,

---

[8] "General Jomini," *Every Saturday*, VII (1869), 567.

especially at Ulm and in Spain. He had first-hand experience of what it is now fashionable to call the "fog of war." His writings show it.

III

Jomini's writings on warfare may be divided into two groups, those mainly historical and those mainly theoretical or analytical. The division is not exclusive, for in military history Jomini is constantly seeking for the principles which explain why and how action was taken,[9] and in military theory he rarely goes far in abstract thinking without trying to buttress his theories with the facts of history. There are also a few pamphlets from his pen, mostly brief replies to his critics.

His histories, originally issued in twenty-seven volumes, cover the wars of Frederick the Great and the wars of the French Revolution and Napoleon, from 1792 to 1815. The Seven Years' War and the revolutionary wars are covered in detail. Napoleon's own military career after 1799 is treated rather more briefly in a four-volume work entitled *Vie politique et militaire de Napoléon*, which originally appeared in 1827. The book is based on a rather heavy and thoroughly eighteenth century literary device. It is written in the first person as if by Napoleon himself, who is introduced in a classical Other World—surely not the Christian Heaven—to justify before the shades of Alexander, Caesar, and Frederick the Great his conduct as a military leader. A separate volume deals with the campaign of Waterloo.

Jomini's military history is rather on the dull side, though perhaps no more than is usual in the genre, but it is clearly written and with a narrative flow that rarely bogs down in detail. He did a great deal of solid research, and his position in first the French and then the Russian armies opened to him material not accessible at the time to an outsider. But he wrote before professional historical writing had fully established its canons of investigation and presentation, and his works are without the modern *apparatus criticus* of footnotes and bibliographies. It is fair to say that his work as a military historian, though it broke some new ground, is definitely dated, and is now rarely read.

Jomini's theoretical writings have, however, survived, and have been a staple of military education for over a century. Jomini's first essay in military theory is to be found in the *Traité des grandes opérations militaires*, a work which is in the main a history of the Seven Years' War. It was the seventh and fourteenth chapters of this work which Napoleon had read to him after Austerlitz, and which so impressed him. They present, in their original form, the main principles of Jomini's military thought.

In the seventh chapter of the *Traité* Jomini presented his theory of "lines of operation" and drew the important distinction between exterior and interior

---

[9] The very titles of Jomini's chapters indicate this search for principles. Thus, in the *Traité* such chapter headings as the following occur: Chap. III. Observations on the operations of the first period; maxims on supply and sieges. Chap. V. Observations on Frederick's marching formations and on those of Guibert. Maxims on attacking an army on the march. Chap. VIII. Operations against the Russians and Swedes; battle of Jaegerndorf. Maxims on isolated attacks.

lines. Chapter fourteen of the *Traité* carried this discussion further, stressing the importance of the choice of a line of operations and showing how that choice must be influenced by geographical and even geometrical considerations. The *Traité* concluded with the famous thirty-fifth chapter in which Jomini rose above specific questions and attempted to generalize his experience and to formulate the fundamental principles inherent in all military operations.

Jomini's greatest theoretical work was the *Précis de l'art de la guerre*, which appeared in two volumes in 1838. It has had numerous later editions, and has been translated into the chief modern languages.[10] In this work, to a much greater extent than in the earlier one, the Swiss thinker is concerned with the problem of the validity of general ideas in military science.

Jomini took up the study of warfare, as he himself tells us, convinced that since it was a form of human activity here on earth, it must make some sense. He started out in definite reaction against such statements as the famous one of Marshal de Saxe, "War is a science covered with darkness, in the midst of which one does not walk with an assured step. . . . All the sciences have principles, but that of war as yet has none."[11] Against what he regarded as obscurantism like the foregoing, Jomini always maintained that the human mind is capable of discerning and stating in some systematic form methods which are likely to bring success in warfare, and methods which are not likely to bring success. "There have existed in all times," he said in the *Traité*, "fundamental principles, on which depend good results in warfare. . . . These principles are unchanging, independent of the kind of weapons, of historical time and of place."[12] In the *Précis*, Jomini stated that the essential object of the book was "to demonstrate that there is a fundamental principle in all operations of war, a principle which should preside over all measures adopted so that they may be successful."[13]

In his criticism of Bülow, Jomini had shown that he was opposed to "systems of war" which provide for all contingencies, which contain recipes, like cook books, and which present hard and fast rules for all matters of military organization.[14] Human intelligence, he felt, is incapable of inventing such a system, the more so because war is "an impassioned drama and in no way a mathematical operation."[15] To Jomini, the sphere of intelligence in warfare is restricted, but it is by no means excluded entirely. The training and discipline of soldiers is not essentially a matter of intelligence, nor will correct thinking alone win battles, where other qualities like courage and initiative are more important. But intelligence is supreme in its proper sphere and that sphere is strategy. In the field of strategy there are general rules and principles of eternal validity,

---

[10] An American translation by Major O. F. Winship and Lieutenant E. E. McLean, *Summary of the Art of War* (New York, 1854), is a model of what a translation ought not to be. American soldiers who had to read it must often have wondered what language they were reading.

[11] Quoted, with comments, by Jomini in the *Précis de l'art de la guerre*, in the preliminary chapter entitled "Notice sur la théorie actuelle de la guerre et sur son utilité."

[12] *Traité*, III, 333.

[13] *Précis*, I, 157-158.

[14] *Traité*, III, 335.

[15] *Traité*, III, 274-275.

which can be comprehended and formulated by the human mind. The main problem of military science is the establishment of these general principles. Jomini makes his position clear at the very beginning of the *Précis*:

"A general officer, after taking part in a dozen campaigns, ought to know that war is a great drama, in which a thousand moral or physical elements act more or less powerfully and which cannot be reduced to mathematical calculations.

"But I must equally well admit without qualification, that twenty years of experience have only fortified in me the following convictions:

"There are a small number of fundamental principles of war which may be disregarded only with the greatest danger and the application of which has, on the other hand, been crowned in nearly every case with success.

"The practical applications which derive from these principles are also few in number and, though they are modified sometimes by circumstances, they may nevertheless serve in general as a compass for the commander-in-chief of an army to guide him in the task, which is always difficult and complex, of directing operations in the midst of the noise and tumult of battle."[16]

Jomini, then, set out to make a kind of first approximation of these fundamental principles of the science of war. He was somewhat hesitant in face of the enormity of the task. "I have dared to undertake this difficult task without perhaps having the talent necessary to fulfil it. But it seemed to me to be important to lay the foundations, the development of which might have been delayed a long time if one had not profited by circumstances to settle them."[17] The work Jomini did was in effect scientific pioneering—not the first daring penetrations of an unknown country, but the first really good map making.

After experimenting with other formulations, Jomini decided that the fundamental principle of strategy consisted in:

"1. Bringing, by strategic measures, the major part of an army's forces successively to bear upon the decisive areas of a theater of war and as far as possible upon the enemy's communications, without compromising one's own;

"2. Maneuvering in such a manner as to engage one's major forces against parts only of those of the enemy;

"3. Furthermore, in battle, by tactical maneuvers, bringing one's major forces to bear on the decisive area of the battle-field or on that part of the enemy's lines which it is important to overwhelm;

"4. Arranging matters in such fashion that these masses of men be not only brought to bear at the decisive place but that they be put into action speedily and together, so that they may make a simultaneous effort."[18]

This very general and necessarily somewhat abstract formulation Jomini made more concrete by numerous specific instances from military history, pointing out that history proved that the most brilliant successes and the greatest defeats were the result of adherence to or violation of the fundamental principle.[19]

[16] *Précis*, I, 26-27.
[17] *Traité*, III, 336-337.
[18] *Précis*, I, 158. See also *Traité*, III, chapter xxv, *passim*.
[19] *Précis*, I, 161.

If the art of war consists of putting into action the greatest possible number of forces at the decisive point in the theater of operations, the means of accomplishing that is the choice of the correct line of operations. This, said Jomini, must be considered as the fundamental basis of a good plan of campaign,[20] and consequently the center and heart of all military theory.

Jomini's theory of lines of operation was first clearly stated in the seventh chapter of the *Traité*. A line of operations he defined as that part of the whole zone of operations which an army covered in carrying out its mission, whether it followed several routes or only one. The seventh chapter begins with a consideration of the campaign preceding the battle of Leuthen in the Seven Years' War. In that campaign, Frederick II had divided his army, leaving part of it in Silesia as he marched with the remainder into Saxony. In so dividing his forces, Jomini says, Frederick operated not on a single line of operation, but on a double line.

What are the relative advantages of the single and double lines of operation? The answer to this question depends upon which line, in a given situation, will succeed "in placing in action, at the most important point of a line of operations or of an offensive, more force than the enemy."[21] It is not the troops carried on the army lists, but those who are thrown into action who win battles.[22] Inherently, then, because it separates troops in the field, the double line of operations is extremely dangerous, unless the divided troops can be quickly reunited and the single line restored. Even when the double line is employed, therefore, it is necessary that all troops remain under one command.

Jomini holds that an army is safe in employing a double line of operations when it occupies the *interior lines*—that is, when the enemy is also using a double line of operations and when the enemy army can be less easily united than can the forces it opposes. "An army whose lines are interior and closer together than those of the enemy can by a strategic movement overwhelm the enemy forces one after the other, by reuniting alternately the mass of its forces."[23]

Throughout his work, Jomini places great emphasis upon the superiority of the interior position. For the army using a double line of operations, the interior line is essential unless it enjoys an overwhelming numerical superiority, and even in the latter case it is dangerous if the double lines are separated by several days' march. In conditions of numerical equality, wrote Jomini, the use of the double line of operations against an army whose forces are closer together (that is, on interior lines) "will always be disastrous, if the enemy profits from the advantage of his position."[24]

In the *Précis*, Jomini summarized his theories on this subject. Other things being equal, he wrote, a single line of operations on one frontier has a decided advantage over the double line of operations. At the same time, it should be

[20] *Précis*, I, 254.
[21] *Traité*, I, 417.
[22] *Traité*, I, 419.
[23] *Traité*, I, 413-415.
[24] *Ibid.*

noted that the double line often becomes necessary, because of the natural configuration of the theater of war or because the enemy has adopted the double
line and has made it expedient "to oppose a part of the army to each of the
great masses which he has formed." In the latter case, the advantage will rest
with the army which operates on interior lines.[25]

In view of these factors, the choice of a line of operations becomes important, since it may well decide the fate of a campaign. "It can repair the disasters
of a lost battle, make vain an invasion, extend the advantages of a victory,
assure the conquest of a country."[26] In the fourteenth chapter of the *Traité*,
Jomini outlined the factors which must influence this choice, and in the foreground of all such factors he placed those arising from the natural configuration of the zone of operations and from such things as existing roads and given
strategical points.

This leads naturally to a concept which has great significance in Jomini's
military theory. Each military operation will take place within a definite zone of
operations. Jomini, betraying the very mathematical tendency which he had
criticized in Bülow, regarded the zone of operations as consisting of a field
with four sides.[27] Two of these sides were occupied by the opposing forces. The
task of the commanding general was, in full consideration of the natural characteristics of the zone in which he was employed, to choose the line of operations which would be most effective in dominating three sides of the rectangular
zone. If he succeeded in doing this, the enemy would be crushed or would be
forced to abandon the zone of operations. It is hard to avoid the conclusion
that in his emphasis upon the necessity of dominating the zone of operations,
Jomini, like the theorists of the eighteenth century, regarded warfare largely
as a matter of winning territory.

It is apparent that, in Jomini's opinion, the task of the commanding general
is primarily an intellectual one. "It is," he says, "the combination of wise theory
with great character which will make the great captain."[28] A natural flair for
war, the ability to inspire troops, these things are also important; but the general, if he hopes to be successful, must have schooled himself in the fundamental principles of war. "Natural genius might, doubtless, by happy inspiration apply the principles as well as could the most well-versed theoretician. But
a simple theory, one free of all pedantry, going back to first causes but eschew-
ing absolute systems, based in short on a few fundamental maxims, may often
supplement genius and may serve to develop it by augmenting its confidence in
its own inspirations."[29]

The emphasis placed upon maxims is significant. Jomini believed that the
practice of warfare could be reduced to a set of general rules which could be

---

[25] *Précis*, I, 259. Jomini's writings on lines of operation had great influence upon the
theories of naval strategy developed by Alfred Thayer Mahan. See W. D. Puleston, *Mahan*
(New Haven, 1939), 79, and chapter 17 below.
[26] *Traité*, II, 272.
[27] *Traité*, II, 279.
[28] *Précis*, I, 130.
[29] *Précis*, I, 27.

learned and applied in all situations. In the thirty-fifth chapter of the *Traité* he undertook to formulate a set of such maxims. This formulation emphasizes, among other things, the importance of the "strategical initiative," the concentration upon one rather than several weak points in the enemy's lines, the importance of pursuing a beaten foe, and the supreme value of surprise.[30]

Jomini felt that the importance of the last-named element could not be over-emphasized. It is usually not enough to attack at a given point with a numerical superiority if the enemy is sure you are going to attack there at that time. He will get aid, he will entrench himself, he will be ready and you will not in fact have been carrying out Jomini's principle. You must as far as possible surprise the enemy. Naturally the campaigns of Frederick and of Napoleon gave Jomini ample illustrative material for this point. His favorite, perhaps because the surprise was achieved on such a huge strategic scale, both in time and in space, was Napoleon's campaign of 1800, when he moved an impossibly large army in an impossibly short time over impossible terrain—the great St. Bernard Pass—to score a strategic and not merely a tactical surprise over the Austrians.[31]

At several points Jomini comes very close to Clausewitz' famous doctrine that the object of war is the destruction of the enemy's armed forces. The great merit of Napoleon, he says, is that he went straight to the essential. "Rejecting old routine practices, by which one attempted the capture of one or two places or the occupation of a small frontier province, he seemed convinced that the first means of effecting great results was to concentrate above all on cutting up and destroying the enemy army, being certain that states or provinces fall of themselves when they no longer have organized forces to defend them."[32]

Nevertheless, Jomini stands on fundamentally different ground from Clausewitz. The central problem in warfare, in Jomini's opinion, is the choice of the correct lines of operation and the most important objective of the commanding general is the domination of the zone of operations in which he is engaged. Such domination is often impossible unless the enemy force is destroyed, but it should be remembered that when a commanding officer has chosen the correct line of operations he leaves two courses of action open to the enemy—either combat under unfavorable conditions or withdrawal from the zone of operations. Jomini's emphasis upon the choice of decisive maneuvering lines, his argument that the problem of the general is to effect the coincidence of the theoretically decisive lines with the actually existing roads, his constant use of diagrams, with its implication that each zone of operations can be reduced to a geometrical form—all of these things indicate that Jomini was thinking primarily not of the annihilation of the enemy but of the acquisition of territory.[33]

For that very reason, Jomini had a marked preference for the offensive.

[30] *Traité*, III, 338-353.

[31] It is usually admitted that Napoleon came near defeat at Marengo, which culminated this well conceived campaign, largely because he committed the error his opponents had so often committed. He separated his forces, sending Desaix almost out of reach. The return of Desaix to the battlefield saved the day. Jomini excuses Napoleon on the ground that he had been misled by an untrustworthy spy into thinking the enemy would not give battle at that point. Jomini, *Vie de Napoléon*, chapter vi.

[32] *Précis*, I, 201.

[33] See *Traité*, II, chapter xiv; *Précis*, I, chapter iii.

Even when a general is obliged by political or other considerations to assume the defensive, it should be what Jomini rather inelegantly calls an offensive-defensive, a position supported by actual forays against the enemy, by feints, attacks, and any other means necessary to keep from the mental and moral stagnation which so often destroys the conventional defensive position in war. No modern publicist has been more insistent on the weaknesses of what has been called the "psychology of the Maginot line," than Jomini. To await attack in a strong defensive position with no other purpose than to maintain oneself in this position is, he thinks, the worst of possible dispositions, a "vicious disposition." The fate of Daun at Torgau and of Marsin at Turin afford him adequate evidence of this truth.[34]

In contrast to Clausewitz, who bent his mind to the consideration of the nature and essential spirit of war, Jomini stands in the history of military thought as the theorist of strategy. He was not interested in the philosophical problems arising from the concepts of war-in-essence or war-in-being; he confined himself to what in his mind were the practical issues involved in warfare. In his theory the campaign occupies the central and decisive position. The purpose of warfare is to occupy all or part of the enemy's territory. Such occupation is accomplished by the progressive domination of zones of operation; and this domination is possible only if the campaign is planned carefully before the outbreak of hostilities. Wars are successful only when the lines of operation have been established beforehand and when available military means have been brought into relation with the geographical and strategical facts of the chosen zone of operations and its ideal mathematical configuration. The task of strategy is to make those preliminary plans.[35]

By defining the place of strategy in warfare, Jomini was able to distinguish clearly between strategy and such other fields of military activity as tactics and logistics. His *Précis* probably did more than any simple book to fix the great subdivisions of modern military science for good and all and to give them common currency.[36] On tactics and logistics, to which he consecrates the second

[34] *Traité*, III, chapter xxxv.
[35] "All those things which take in the general theater of the war are within the domain of strategy, which must thus include: 1. The definition of that theater and of the various plans it makes possible; 2. The determination of decisive areas which result from these plans and of the most favorable direction in which to operate; 3. The choice and establishment of the fixed base and zone of operation; 4. The determination of the objective proposed, whether offensive or defensive; 5. The operational and strategic fronts and line of defense; 6. The choice of lines of operations leading from the base to the objective or to the strategic front occupied by the enemy; 7. The choice of the best strategic lines to be taken for a given operation; the different maneuvers which will include these lines in the various possible plans of operation; 8. Possible operational bases, and strategic reserves; 9. The armies' marches considered as maneuvers; 10. Supply-depots considered in relation to the armies' marches; 11. Fortified places considered as strategic weapons, as refuges for an army, or as obstacles to its march; the making and withstanding of sieges; 12. Points at which fortified permanent camps, bridge-heads, etc. should be established; 13. Diversionary movements and detachments in force which may be useful or necessary." *Précis*, I, 154 f.
[36] Jomini defined tactics as the "maneuvers of an army on the battlefield, or combat maneuvers, and the various formations in which troops may be led to the attack." Logistics he defined as "the practical manner of moving armies, the material details of marching and formations, the layout of temporary camps and cantonments, in a word, the execution of the plans of strategy and tactics." *Précis*, I, 155.

volume of the *Précis*, Jomini is thoughtful, concrete, systematic and often suggestive. Here, as so often, he is not altogether original and certainly not profound, but he does do an admirable piece of elementary pedagogy which helps explain the great success of this manual in nineteenth century military education. But he was not interested in these lesser branches of war; his chosen field was strategy, and he stood in the forefront of the new strategical thinking of the nineteenth century.

But despite Jomini's position in nineteenth century military thought, it is apparent that he never completely dissociated his ideas from eighteenth century concepts. He criticized Bülow for an excessive rationalism, but his own thinking was strongly influenced by the prevailing rationalism of the preceding age. In his search for universally valid principles and infallible maxims, he tended to overlook the irrational factors in war which transcend the realm of calculability. It is true that he attempted to grapple with such questions. At the beginning of the *Précis*, he included a chapter on the *"politique de la guerre"* which was designed to treat unmilitary questions and a chapter on *"philosophie de la guerre"* which was intended to deal with the factor of irrationality.[37] Yet it is these very chapters which indicate the extent to which his thoughts were molded by the purely military, the purely rational.

In the first of these chapters, Jomini draws up a catalogue of the various types of war, distinguishing among them in accordance with their political purposes. He argues quite correctly that the political purpose of a war will play a great part in determining the nature of the war itself. But the most significant feature of his discussion is his failure to consider the possibility that war may have a dynamic tendency which drives it beyond its original limits and original purposes. In the second chapter, which is largely devoted to the question of national war and the influence of moral factors in war generally, there is a strong suggestion that Jomini had not completely learned the military lesson of the revolutionary period. He was by no means sure that the national war had come to stay; he was by no means convinced of the importance of the moral factor in war.

"Though it is permissible to believe that the support of political dogmas sometimes becomes an excellent aid, as has been seen in the article on wars of opinion, it must not be forgotten that the Koran itself would not take a province today. For that, cannon, bombs, bullets, powder, and muskets are necessary, and with such a load to be carried distances count for a great deal in strategic plans, so that nomadic adventures would no longer be in season."[38]

IV

Jomini's military thought is admirable evidence of a fact that many hopeful nineteenth century liberals refused to recognize: that is, that war is not an aberration of human life with a history all its own and alien to other kinds of history, but that it is an integral part of the history of civilization. For Jomini's

[37] *Précis*, I, chapters i, ii.
[38] *Précis*, I, 395-396.

thought is in many ways an almost perfect example of what Carl Becker has called the "climate of opinion" of the eighteenth century. Though all his writings are concerned with military matters, he is an unmistakable product of the age of enlightenment. A page or two of Jomini bears the stamp of that age as clearly as a letter of Horace Walpole, an epigram of Voltaire, a Louis XVI *mise en scène*, the Trianon, or Sans-Souci.

The eighteenth century is itself no simple matter, especially in the history of thought. There is nothing of Rousseau or Tom Paine in Jomini, nothing of the simpler rationalists like Holbach or La Mettrie. Jomini's is rather the eighteenth century of Montesquieu, whom, indeed, he resembles in many ways. It would not be amiss to say that Jomini's writings might be collected under the title *L'esprit de la guerre*. There is in both Montesquieu and Jomini the same love of generalization and system moderated by a respect for fact and a wide knowledge of pertinent facts, the same reasonableness of temper, the same desire for a quiet rather than a perfect world. Jomini is, indeed, chronologically one of the last of the genuine eighteenth century disciples of Montesquieu. He carried down into the age of Bismarck a habit of mind and feeling formed in the very last days of Frederick the Great and Voltaire.

It was largely upon the deeds of Napoleon and his revolutionary predecessors that Jomini exercised his reason and discernment. He has chiefly been known to later generations as the first great military thinker to comment upon Napoleon. Now, since Napoleon himself was in many important ways also a child of the eighteenth century and the enlightenment, Jomini had a task for which he was on the whole admirably fitted. Napoleon had learned a great deal from Frederick, Guibert, Gribeauval, and Bourcet; and Jomini was never so overcome by Napoleon's greatness and originality as to lose sight of a perfectly natural relation to his predecessors. Bonaparte's clear, mathematically trained reason cut through convention with incisiveness and fearlessness and he shared few of the sentiments the eighteenth century commonly labeled "prejudice." For these traits Jomini had a sympathetic comprehension. He understood admirably also what Napoleon had achieved as a military technician. Indeed, as has often been remarked, the broad outlines of Jomini's strategic principles are merely a generalized description of Napoleon's campaigns, especially those of 1796-1797 in Italy, of Marengo, of Austerlitz, and of Jena.

And yet Jomini misses part of Napoleon—or rather, he partly glimpsed and wholly disliked the romantic, the monstrous, the unearthly and impossible in Napoleon's career. The Napoleon who sought to impose his empire on all Europe, who helped make the mass state and the mass army, who marched on Moscow—it was this Napoleon of legend who moved the reasonable Jomini to wonder and apprehension. Napoleon had succeeded by breaking the stupid rules of custom and adhering to the sensible rules of nature and reason. But he had gone on to break a few of the rules of nature and reason also, and though he had no doubt been properly punished, his example might prove dangerous. Others might go on to fight wars with no rules whatever. War might become a bloody and most unreasonable struggle between great masses equipped with weapons of unimaginable power. We might see again wars of peoples like those of the fourth century; we might be forced to live again through the centuries of the Huns, the Vandals, and the Tartars.[39]

To contemporaries, Napoleon's ceaseless marches across Europe seemed without plan or system; his battles which he won by concentrating his striking power on one central point seemed inartistic and unnecessarily brutal. Jomini was the first to show that Napoleon's campaigns and battles were based upon the application of fundamental principles which had been valid in all times. He revealed the rational element in Napoleon's generalship. But, while Clausewitz regarded Napoleon as the "God of War," the lawgiver, the genius who made the rules, Jomini in his search for order, tended to make the rules all-important and Napoleon merely the instrument which brought them into play.

In the last analysis, the great wars for a reasonable man like Jomini were those of the eighteenth century, when men's lives were precious, if only because professional soldiers were expensive, when the great masters of one of the great branches of human thought could test out on the orderly chessboard of war their brilliant combinations, when officers were gentlemen, fighting gentlemen, all members of one great society. So Jomini came to feel. The great love, the real admiration, of this diviner of Napoleon, this writer whose name has been so inseparably linked with the emperor's, was not after all Napoleon: it was, quite logically and naturally, Frederick the Great.

What, then, is the importance of Jomini in the history of the development of modern military thought? With the passage of time much of his work has become obsolete. The progressive totalitarianism of warfare has effectually destroyed the validity of purely geographical campaigns and has made limited war impossible. The campaign of 1866 in Bohemia, which Jomini tried so hard to explain in terms of his own theories, proved that the progress of technical invention had cast serious doubt upon the superiority of interior lines of operation. Jomini's great service to military thought lay in another direction, in his clarification of the basic concepts of military science and in his definition of the sphere of strategy in warfare. In his emphasis upon the planning of operations, he made clear to his contemporaries the role which intelligence must play in war, and the establishment of general staffs and military academies throughout Europe showed that, in this respect at least, his influence would continue to be felt.[40]

---

[39] In this connection, Jomini's judgment on Napoleon's downfall is of some interest. Napoleon, he says, was well grounded in military science, "but his scorn of men made him forget its application. It was not ignorance of Cambyses' fate or of that of Varus' legions which caused his reverses, nor was it forgetfulness of Crassus' defeat, of the Emperor Julian's disaster or of the result of the Crusades. No, it was the fixed opinion he had that his genius assured him incalculably superior means while his enemies, on the other hand, had none. He fell from the topmost pinnacle of greatness through having forgotten that the mind and strength of man have their limits and that the more enormous the masses set in motion, the more the power of genius is subordinated to the unchangeable laws of nature and the less he can command events." *Traité*, III, 356-357.

[40] On the question of the development of the general staff, see Dallas D. Irvine, "The Origin of Capital Staffs," *Journal of Modern History*, X (1938), 161-179.

CHAPTER 5. Clausewitz

BY H. ROTHFELS

C LAUSEWITZ' military writings, particularly his book *On War*, hold a singular position in the history of military thought. The latter is reverently called a "classic," though one that seems to be more quoted than actually read. Although it contains large sections—particularly those dealing with tactics—the value of which has been weakened by the passage of time, it is nevertheless the first study of war that truly grapples with the fundamentals of its subject, and the first to evolve a pattern of thought adaptable to every stage of military history and practice. This achievement, to be sure, cannot be readily appreciated. Since Clausewitz' main work is unfinished—the author's early death in 1831 prevented him from completing a final revision—some inconsistencies have remained unsolved. The difficulties of interpretation are considerable, partly because of a philosophical terminology which seems "metaphysical." What harsher criticism could be made of a military writer! The Swiss Jomini, Clausewitz' contemporary, found his rival's pen "excessive and arrogant." And although French military theory of the late nineteenth century drew extensively on Clausewitz, a French author thirty years ago significantly complained that he was "le plus Allemand des Allemands . . . A tout instant chez lui on a la sensation d'être dans le brouillard métaphysique."[1]

More generally, however, it is in a different direction that Clausewitz' national characteristics and limitations are usually seen. He seems a foremost exponent of "Prussianism" and the "battle-mania" of the nineteenth century. His treatise *On War* is regarded almost as a textbook for Sadowa and Sedan. No less competent a man than Count Schlieffen has testified that Clausewitz "kept alive the conception of 'true War' within the Prussian officers' corps."[2] It is but natural that his critics should reverse this appraisal and hold Clausewitz responsible, to some extent, for the narrowing of the European military mind in the late nineteenth and the early twentieth centuries, for the "one-way" strategy that followed the Prussian victories. In the words of Captain Liddell Hart, the British military critic, the generals of the last half-century became "intoxicated with the blood-red wine of Clausewitzian growth."[3] More recently, an American author has complained that, from Clausewitz to Foch and Ludendorff, "military thinkers stubbornly identified the idea of war with that of the utmost violence."[4] Was not Clausewitz one of the "Mahdis of Mass" who gave theoretical justification to the "rage of numbers"? In his insistence that the enemy's field forces were the primary objective and battles the primary means of warfare, did he not neglect the intellectual achievement of the eight-

---

[1] Camon, *Clausewitz* (Paris, 1911), p. vii.

[2] Introduction to the 5th German edition of *On War* (1905). H. v. Moltke, when asked about his favorite books, listed the following: the *Bible*; Homer's *Iliad*; Littrow, *Wunder des Himmels*; Liebig, *Briefe über Agrochemie* and Clausewitz, *On War*.

[3] Liddell Hart, *The Ghost of Napoleon* (1933), p. 21.

[4] H. Nickerson, *The Armed Horde* (1940), p. 52.

eenth century theorists who had put emphasis on skill and refinement rather than on sheer force, on "rapier thrusts" rather than on "hammer blows," on indirect rather than on direct action? Did not the prevalence of Clausewitzian thought contribute also to a neglect of the experiences of the American Civil War, and to a sterile attitude which finally resulted in the deadlock of World War I? When this deadlock became apparent after the exhausting battles of Verdun, of the Somme, and of Flanders, an American author tried to revise Clausewitz, by falling back in a way upon the small, highly trained armies of the eighteenth century type.[5] Others went farther. It was in conscious opposition to Clausewitz that Captain Liddell Hart declared: "Strategy has to reduce the fighting to the slenderest possible proportions."[6] In support of this definition the same author outlined an analysis of what he called the "British way in warfare."[7]

When speaking of Clausewitz in the midst of a global war it is useful to keep this controversy in mind. It not only reflects a school of military thought characteristic of the period between the two world wars, it also reveals a fundamental problem of strategy and makes apparent an actual contrast between the "continental" tradition and the Anglo-Saxon or "insular" tradition, between those countries which possess a national mass army as part of their normal equipment and those which do not. This cleavage, however, stresses rather than diminishes the importance of a theory which certainly was not meant to be applicable only to certain national groups or for certain periods or regions. It is not the deductions which have been drawn from Clausewitz, some of them prejudiced or otherwise unwarranted; it is rather his real intention and achievement with which the present essay is concerned.

II

Clausewitz' written work on military affairs and the conduct of war was published in ten volumes after his death. That part of his writings which earned the praise of posterity and which gave Clausewitz his claim to fame was the treatise entitled *On War*. This study is divided into eight books. The first of these deals with "the nature of war"; the second, with "the theory of war." In the third book, Clausewitz discusses "strategy" and, in the fourth, "the combat." The fifth and sixth books are devoted to "military forces" and to "defense," while the concluding books contain Clausewitz' preliminary sketches for the discussion of "the attack" and "the plan of war."

What was the task which Clausewitz set himself in the study *On War*? Though rather given to understatement, he certainly wanted to do more than write merely for the next generation or for a Prussian military school. He was imbued with the spirit of search for the "absolute," for the very nature or the

[5] R. M. Johnston, *Clausewitz to Date* (Cambridge, Mass., 1917). In Germany at the same time Ludendorff began to be criticized in the light of eighteenth century strategy as well as of Clausewitz "correctly understood."

[6] On "Strategy," *Encyclopedia Britannica* (14th ed., 1929).

[7] In a more restricted and conservative sense Sir F. Maurice opposed "British Strategy" to continental warfare (1929).

"regulative idea" of things, a spirit which then dominated German philosophy. Significantly, Clausewitz, while specializing in the military field, also embarked upon broader studies of the methods of knowledge, of the validity of theoretical principles and their application to other "practical arts" besides that of warfare.[8] When he began to conceive his main work, in 1816 or 1817, he claimed that its scientific character would lie "in the endeavor to explore the nature of military phenomena, to show their affinity with the nature of the things of which they are composed. . . . Investigation and observation, philosophy and experience must neither despise nor exclude one another; they mutually guarantee each other."[9] Shortly before his death he pointed out that he held at least the "ruling principles" (*Hauptlineamente*) of his work to be correct. "They are the result of a very varied reflection permanently directed toward practical life."[10] And in his introduction of 1816-1817 he wrote: ". . . in the same way as many plants bear fruit only when they do not shoot too high, so in the practical arts the theoretical leaves and flowers must not be made to sprout too far but kept near the experience which is their proper soil."[11]

It is this close coordination of philosophy with experience which is the most significant and singular characteristic of Clausewitz' analysis of war. His is a position between two ages. While still belonging to the very German world of the eighteenth century *Dichter und Denker*, he proclaims the man of action, trained by history and experience. This intellectual position was favored by many circumstances of his career.[12] Born in 1780, Clausewitz first saw service at an early age in the Rhine campaign of 1793-1794. In the subsequent years of peace, by the hardest sort of work, he was admitted in 1801 to the Berlin Academy for young officers. Here he attracted the special attention of Scharnhorst, who was to become the reorganizer of the Prussian army.

[8] Evidence of this is to be found in unpublished manuscripts and in a study of architecture which the present writer has edited in *Deutsche Rundschau* (December 1917). See also Clausewitz, *On War*, Book II, especially Chs. II and III.

[9] Clausewitz, *On War*. Trans. by Colonel J. J. Graham. (First edition 1873.) New and revised edition with Introduction and Notes by Colonel F. N. Maude. Third Impression. 3 vols. (London 1918), I, p. xxix. In the present essay all quotations from Clausewitz' *On War* refer to the last named edition (cit. Graham-Maude). All quotations, however, include book and chapter in order to facilitate the use of other editions or the original text. For Colonel Graham's translation it must be said that unfortunately it is by no means free from misunderstandings and plain errors. To give one example: the last of the sentences quoted above runs in German: "Sie leisten einander gegenseitige Bürgschaft." Colonel Graham translates: "They mutually afford each other the right of citizenship!" In many cases the present author, therefore, has offered a translation of his own or drawn on translated quotations in books written in English. The Modern Library translation by O. J. M. Jolles of the University of Chicago was not available when this chapter was written.

[10] *Ibid.*, p. xxv.

[11] *Ibid.*, p. xxix. For the chronology of the various Introductions and Notices of Clausewitz and their bearing upon the interpretation of his work see H. Rosinski, *Historische Zeitschrift*, CLI, 278-293.

[12] For further biographical data see the article in the *Encyclopedia Americana*, VII (1941), 63. Briefer but more reliable in detail is the article by E. Kehr in *Encyclopedia for the Social Sciences*, III, 545. For a fuller account see the books of Karl Schwartz and the present writer in the bibliographical note.

During these years Clausewitz also came in contact with the Kantian philosophy, and doubtless received therefrom an important impulse.

In the campaign of 1806 Clausewitz served as a captain and aide-de-camp to a Prussian prince. Captured after the battle of Auerstädt, he had to spend more than a year in France and Switzerland. Upon his return he became an assistant to Scharnhorst and took an active part in the reform and moral regeneration of the Prussian army and the Prussian state. When in 1811 Prussia was forced into military "collaboration" with Napoleon, Clausewitz was, to borrow a phrase from the present day, one of the "Free Prussians." He took service with the Russians and at the beginning of the Wars of Liberation in 1813 he was a Russian colonel, serving first as liaison officer at Blücher's headquarters, then as chief of staff of the Russo-German legion. It was after the first peace of Paris that he was readmitted to the Prussian army. He became chief of staff of an army corps, which in 1815 took part in the battles of Ligny and Wawre. In the tactical sense, both battles were defeats, but strategically they paved the way to final victory. In that victory, in the crowning events of Waterloo, Clausewitz had no immediate share, but that seems in a way to be in keeping with his military career as a whole. Through a decade he had been very close to important and varied actions and yet always somewhat detached from them. In the midst of a passionate struggle, he maintained a striking clarity of mind and a thoroughly reflective attitude. When eventually peace was restored, his role became more and more that of a critical and synthesizing observer. From 1818 until 1830 he was managing director of the Military Academy at Berlin. His post was a merely administrative one which gave him no influence over the training of the Prussian officers' corps. Only a few friends knew of the scientific work in which Clausewitz was engaged. It was not behind his office desk but in his wife's drawing room that he set out to integrate into one conclusive conception the results of broad studies and the military experiences of his age.

III

The age of the French Revolution and of Napoleon was an era in which, in Clausewitz' own words, "War itself, as it were, had been lecturing."[13] War had reappeared as a terrible "act of violence," upsetting the territorial as well as the social order of Europe. The wars of this era were no longer fought over dynastic claims of limited scope; they involved the very existence of the nations concerned and, as in the religious wars of the sixteenth century, they involved opposing principles, opposing philosophies of life. These new tensions were intertwined with fundamental changes in the political and social structure of Europe, and this in turn, reacted upon the moral as well as the material means of warfare. The armies of the *ancien régime* had been composed of professional soldiers serving long enlistments, limited in numbers but highly trained. Each of them was a part of the invested capital of the state and had to be used

---

[13] "Da der Krieg selbst gewissermassen auf dem Katheder stand," *Ueber das Leben und den Charakter von Scharnhorst*, p. 23.

with caution. Moreover, a large percentage of these professional soldiers were foreigners or were drawn from the dregs of the population. An army so composed could not make effective appeal to individual military virtues or irrational forces, to national passions or the good will of citizens. It was kept together by a most rigid discipline; it was taught to march and fight in strict formations and under the closest supervision of its officers. It could send out neither skirmishing parties nor foraging detachments, the threat of desertion being obviously greater than the dangers of the enemy.

Armies, therefore, were largely dependent on magazines. Rapid marches, far-reaching thrusts, decisive pursuits were impossible or at least extremely dangerous. These limitations worked out in two ways. While a general could hardly allow his troops to be separated from his supply base by a march of more than two or three days, he found in the opponent's lines of communication a very promising objective. The average picture of the eighteenth century wars presents, therefore, a variety of more or less complicated maneuvers, of marches and countermarches. Fortresses in which magazines could be safely placed played an incommensurable part. Sieges and attempts at relief were more frequent than ordinary battles. Often the armies would oppose each other in fortified positions and remain immobile over long periods. In Clausewitz' words: "The army with its fortresses and some prepared positions constituted a state in a state within which the element of war slowly consumed itself."[14]

There were, of course, exceptions to this average picture. Gifted leadership or a clash of vital political interests would intensify war. But even a genius could not overreach the social and technical conditions of his time. There was also the beginning of a new appreciation of the imponderable factors in warfare, of the spirit of an army instead of its mechanical drill. New forms of organization, new tactical and strategical devices were outlined which would increase mobility. But progress was conditioned and delayed by the circumstances of the age.

It was the French Revolution which opened the way. While the revolutionary armies could not indulge in intricate maneuvers, they were free from conventional limitations; they could endure privations and fight wherever it seemed advantageous; they could attack regardless of cost in men because they could call upon the total resources of the nation. This change in social conditions made possible a highly mobile strategy. The divisional system developed; supply was largely provided by requisition. In the battles themselves the individual could be relied upon; deliberate fire, individually aimed, replaced or supplemented the rolling volleys; *tirailleur's* tactics were adopted in order to prepare for the mass attack.

Napoleon grasped these potentialities and added to them his personal genius of leadership. He first demonstrated what could be done with the new *levée en masse*. To contemporaries, the Italian campaign of 1796-1797 appeared as an eruption of elemental forces striking where they were not expected instead of where "good manners" would have indicated. Napoleon acted, in fact,

[14] Graham-Maude, III, 99 (Book VIII, Ch. III.B).

against all conventional rules; he placed his army on "the interior lines" between the Sardinians and the Austrians, without much regard for his own communications; he did not cover or conquer territory; and his sole aim was to battle and destroy the opposing armies. In Clausewitz' view Napoleon "hardly ever entered upon war without thinking of conquering his enemy at once in the first battle."[15] This was an "unmannerly" method of brutal directness. But this seemingly primitive boldness was combined with a particular care for technical details and a penetrating force of logic and calculation. The factor of surprise played a great part, whether Napoleon concentrated his loosely grouped divisions in a swift movement and fell like lightning upon the weakest point in the enemy's front, or whether he turned the flank with the bulk of his army and placed it astride the opponent's lines of retreat. Whenever possible his victory on the battlefield was pressed home and exploited with a relentless pursuit.

The ascendancy of the Napoleonic blitzkrieg was eventually checked when the growing size of the French armies was not matched by an increasing ability to handle them. Moreover, Napoleon's opponents learned the lesson. They adopted many of the new methods and aims, particularly the strategy of decision. Still more important was the fact that continental Europe in one way or another caught up with the social and moral conditions in which Napoleonic warfare was rooted. Whether in a more primitive or a more modern form, resistance to the French domination became an affair of the peoples themselves, in Spain and in Russia on the one hand, in Austria and Prussia on the other. The Prussian reformer Gneisenau wrote after the disaster of 1806: "One cause above all has raised France to this pinnacle of greatness, the revolution awakened all her powers and gave to every individual a suitable field for his activity. What infinite aptitudes slumber undeveloped in the bosom of a nation!" The awakening of these slumbering forces nationalized the armies throughout Europe and resulted in a theretofore unparalleled effort. In the campaigns of 1813 and 1814 about half a million Russians and Prussians were under arms, and in eight months the theater of war was transferred from eastern Germany to the center of France. While strategic conceptions still wavered, it followed from the very nature of the conflict that a solution could be reached only after the complete overthrow of the French armies.

Clausewitz, very naturally, was deeply and permanently impressed by these "lectures of war itself." Once war had revealed itself in its "absolute nature," he foresaw that the "push to the utmost" would not disappear again. "Everyone will agree with us," he later pointed out, "that bounds which to a certain extent existed only in an unconsciousness of what is possible, when once thrown down, are not easily built up again; and that, at least whenever great interests are in dispute, mutual hostility will discharge itself in the same manner as it has done in our times."[16] Clausewitz was certainly correct in connecting this push to the utmost with the fact, that, since the time of Bonaparte,

[15] *Ibid.*, I, 289 (Book IV, Ch. xi).
[16] *Ibid.*, III, 103 (Book VIII, Ch. iii.b).

war had become "an affair of the whole nation," and that the integration of new social forces resulted in war approaching "its absolute perfection." He was particularly anxious that this lesson should not be forgotten in his own country. Again and again he resorts in his writings to examples taken from the Napoleonic age. Even today the change in warfare which took place at the turn of the nineteenth century can be paraphrased best, perhaps, in Clausewitzian words. On occasion he went so far as to speak of Napoleon as the "God of War," and it is a familiar saying that Clausewitz codified Napoleonic warfare.

IV

A correct interpretation of Clausewitz cannot be confined to this short-range view. As indicated above he grappled with the fundamentals of war and was free from any tendency to dogmatize on the basis of recent events. This fact becomes apparent if Clausewitz is compared with some of the eighteenth century military theorists. The average picture of eighteenth century warfare was particularly congenial to the thought of an optimistic and rationalistic age. The *ancien régime* had not known the irrational atmosphere of deadly hostility or elemental hatred. The tensions between states were generally not powerful enough to drive war beyond its conventional limits. The "balance of power" implied a conservative trend. As there was a ceremonial of diplomacy, there was also something like a ceremonial of warfare, both being akin to the contemporary style of *rococo* art with its florid decorations. Society itself seemed to move in ornamental forms, until pastoral scenes and sentimental plays of shepherds and shepherdesses transformed the conflicts of life into an idyllic picture. Even warfare was praised because of its seemingly idyllic character, because peasants could plough and civil life could go on only a short distance from the battle front or the military camp. The brutal sword seemed replaced by the more elegant *rococo* rapier.[17]

Warfare in the *ancien régime* conformed, also, to the scientific spirit of the age. There was, of course, in the era of enlightenment, a very real opposition to war, in principle, which was based on both humanitarian and economic considerations. But at the same time many military thinkers found the contemporary warfare "ennobled" precisely by those limitations which resulted from the composition of the armies and other technical brakes. War after all had become scientific. What better evidence of progress could be thought of! Accordingly, much emphasis was laid upon the system of complicated movements which might save fighting altogether, upon geometrical relations and angles of operation, upon definite geographical points—watersheds, for example—the occupation of which would make victory almost mechanical. Mathematics and topography command the military leader. In the words of an English theorist (W. Lloyd) : "The general who knows these things, can direct war enterprises with geometrical precision and lead a continual war without

---

[17] Clausewitz significantly speaks of war spinning out time "with a number of small flourishes. . . . In these feints, parades, half and quarter thrusts . . . they find the aim of all theory, the supremacy of mind over matter." *Ibid.*, I, 229 (Book III, Ch. xvi).

ever getting into the necessity of giving battle." Another writer, the prince de Ligne, proclaimed that, war being scientific, it would only be natural to establish an international military academy.[18]

Clausewitz rejected both the optimism and the dogmatism of the eighteenth century theory. War, he held, was neither a scientific game nor an international sport, but an *act of violence*. In the nature of war there is nothing moderate or philanthropic as such. An often quoted sentence of the book *On War* reads: "We do not like to hear of generals who are victorious without the shedding of blood. If bloody battling is a dreadful spectacle, that should merely be the reason to appreciate war more and not to allow our swords to grow blunt by and by, through humanitarianism, until someone steps in with a sharp sword and cuts our arms off our body."[19] This statement again, of course, is rooted in painful experiences, but one should not lose sight of its specific implications. It implied, among other things, that science can neither moderate nor "ennoble" war, an opinion which in an unexpected sense has proved to be only too correct. In Clausewitz' view the scientific part of warfare, that is the one that can be measured and rationalized, is only of secondary importance. Clausewitz did not underrate the supply services[20] or the geographical nature of the theater of war. He admitted that mathematical and topographical factors are important in tactics, but, he pointed out that they are less important in strategy. "We therefore do not hesitate to regard as an established truth that in strategy more depends on the number and the magnitude of the victorious combats than on the form of the great lines by which they are connected."[21] Clausewitz liked to ridicule the "imposing" expressions, like "commanding ground," "sheltering position," "key to the country," which in his view were meant "to give a flavor to the seeming commonplace of military combinations. . . . The conditions have been mistaken for the thing itself, the instrument for the hand. . . . The occupation of such and such a position . . . is a mere sign of plus or minus which lacks substance. . . . This substance is a victorious battle."[22]

Clausewitz took up this issue in an early treatise of 1805.[23] While criticizing one of those predecessors who had attempted to make war scientific, he insisted upon the preeminence of *immaterial* and *moral* factors. From geometrical relations he turned to man and man's actions in the midst of those uncertainties

[18] For a broader analysis of these trends see the present author's *Clausewitz*, pp. 36-47 and A. Vagts, *A History of Militarism*, pp. 81-85.
[19] Graham-Maude, II, 288 (Book IV, Ch. xi).
[20] Clausewitz deals with the problems involved in a chapter on "Subsistence" (Book V, Ch. xiv). In the "modern" systems of exaction and requisition he sees a tendency to shorten the duration of wars. He assumes that under special conditions the magazine system may come up again. But he would not consider that system as an improvement in war on account of its being more humanitarian, "for war itself is nothing humanitarian." (Graham-Maude, II, 103.)
[21] *Ibid.*, I, 223 (Book III, Ch. xv).
[22] *Ibid.*, II, 130 (Book V, Ch. xviii).
[23] *Bemerkungen über die reine und angewandte Strategie des Herrn von Buelow* (1805). It may be noted that recently an even earlier prelude to Clausewitz' theory has been discovered, which does not touch, however, so much upon the principles. See Clausewitz, *Strategie aus dem Jahr 1804 mit Zusätzen von 1808 and 1809*. Ed. by E. Kessel (Hamburg, 1937).

which are the proper element of war. In a way, this was a Copernican revolution and, at the same time, a turn imbued with Kantian criticism. The very destruction of a dogmatic system makes true theory possible. In the final stage of the book *On War*[24] Clausewitz pointed out that theory does not mean a "scaffold" supporting man in action or a "positive direction for action." Theory rather means "an analytical investigation of the subject that leads to an exact knowledge; and if brought to bear on the results of experience, which in our case would be military history, to a thorough familiarity with it. The nearer theory attains the latter object, so much the more it passes over from the objective form of knowledge into the subjective one of skill in action." Theory, he added, "should educate the mind of the future leader in war, or rather guide him in his self-instruction, but not accompany him to the field of battle; just as a sensible tutor forms and enlightens the opening mind of a youth without therefore keeping him in leading strings all through his life." Hence, true theory cannot contradict or strangle creative practice as every dogmatizing of rational factors sooner or later does. In the treatise of 1805, we find, already clearly expressed, the argument which was to be repeated in the book *On War*: "What genius does must be the best of all rules, and theory cannot do better than to show how and why it is so."[25]

This point of view illuminates Clausewitz' true relationship to Napoleonic warfare. Contemporary events had widened the scope of analysis and brought out more clearly the structural element which constituted the concept of war. In Clausewitz' own words: "We might doubt whether the concept of the absolute character of war was founded in reality, if we had not seen real warfare make its appearance in our very times. . . . Without these warning examples of the destructive force of the element set free [theory] might have talked itself hoarse to no purpose; no one would have believed possible what all have now lived to see realized."[26] This very appraisal of the "genius" in warfare, together with Clausewitz' philosophical attitude, prevented him from dogmatizing on the most recent experience or any particular strategic or tactical device used by Napoleon.

V

In contrast to earlier theorists—and also to his contemporary Jomini[27]— Clausewitz' work is distinguished by the fact that it combines an analysis of the structural elements of war with an undogmatic elasticity and a great power of discrimination. Experience and philosophical thinking led him to the concep-

---

[24] Graham-Maude, I, 106-108 (Book II, Ch. 11).

[25] *Ibid.*, p. 100. In H. Cohen's view (*Von Kants Einfluss auf die deutsche Kultur*, p. 32) this statement reads as if "copied" from Kant's *Critique of Aesthetic Judgement*; it certainly testifies a Kantian method in describing the limits and the true objects of a theory of art. See also Clausewitz, *Werke*, VIII, 166.

[26] *Ibid.*, III, 82-83 (Book VIII, Ch. 11). In this quotation the term "real warfare" seems misleading, as Clausewitz generally contrasts "real" with "absolute" war. See below.

[27] Clausewitz knew of Jomini's *Traité des grandes opérations* when he wrote his additions to the Strategy of 1804. He found Jomini much more "solid" than Bülow, but lacking in the distinction between the incidental and the essential (Kessel, p. 72).

tion of what he called the "absolute war" or the "perfect war." This term is not free from ambiguity and needs some clarification. It is not identical with that of "total war,"[28] though in common usage both terms have more or less fused. In Clausewitz' view the conception of absolute war follows from the very nature of war itself. By definition war is *an act of violence intended to compel our opponent to fulfill our will.*[29] In another context Clausewitz defines war as belonging to the "province of social life. It is a conflict of great interests which is settled by bloodshed, and *only in that* is it different from others."[30] Physical force, therefore, is the specific means of war, and it would be absurd to introduce into the philosophy of war itself a "principle of moderation." Our opponent will comply with our will only if he is "either positively disarmed or placed in such a position that he is threatened with being disarmed."[31] From this it follows that the "disarming or overthrow of the enemy . . . must always be the aim of warfare." As both sides have the same aim, reciprocal action logically leads to an extreme. *"War is an act of violence pushed to its utmost bounds."*

Although somewhat simplified, this may be called Clausewitz' conception of "absolute war." He never fails to stress its theoretical importance. It is the duty of theory "to give the foremost place to the absolute form of war and to use that form as a general point of direction, so that whoever wishes to learn something from theory may accustom himself never to lose sight of it, to regard it as the natural measure of all his hopes and fears, in order to approach it *where he can* or *where he must.*"[32] Again, "a war directed toward great decisions is not only much simpler but also much more in accordance with nature, is more free from inconsistencies, more objective. . . ."[33] And again, "only through this kind of view [i.e. by looking at war in its absolute form] does war receive unity. Only by it can we see all wars as things of *one* kind; and it is only through it that judgment can obtain the true and perfect basis and point of view from which great plans may be traced and determined upon."[34] There seems little doubt that Clausewitz regarded and emphasized absolute war as "ideal" in the philosophical sense, as a "regulative idea" which gives "unity" and "objectivity" to very varied phenomena; an idea like that of perfect beauty in art which may never be attained but constantly approximated. He embraced the "push to the utmost" with the soldier's professional ardor and sense of responsibility; he saw in this form the "perfection of war." But there is also no doubt that absolute war was to him war in the abstract, or "war on paper" as he puts it on occasion.[35]

---

[28] Nor does the term fit in with the trinity of "absolute War," "instrumental War," and "agonistic fighting," which H. Speier has recently outlined (*American Journal of Sociology*, January 1941).
[29] Graham-Maude, I, 2 (Book I, Ch. I).
[30] *Ibid.*, p. 121 (Book II, Ch. III).
[31] *Ibid.*, pp. 3-5 (Book I, Ch. I).
[32] *Ibid.*, III, 82 (Book VIII, Ch. II).
[33] *Ibid.*, II, 409 (Book VI, Ch. XXX).
[34] *Ibid.*, III, 123 (Book VIII, Ch. VI.B).
[35] *Ibid.*, I, 78 (Book I, Ch. VII).

Clausewitz therefore follows up the logical definition of war with the remark: ". . . everything takes a different shape when we pass from abstraction to reality."[36] In the midst of his most philosophical chapter (I, 1) he lists a number of "modifications" which make war not an "ideal" but an "individual" process which is guided by laws of probability rather than rules of logic. War is not an isolated act; nor does it consist of one single action. Many factors, like new troops, the widening of the theater of war or of the system of alliances, may successively come into play: "Whatever one belligerent omits on account of weakness, becomes for the other a real objective ground for limiting his own efforts, and thus again, through this reciprocal action, extreme tendencies are reduced to efforts on a limited scale."

An important group of these modifications is discussed by Clausewitz in a number of chapters (I, IV-VII) which are characteristic of his realistic approach and can be appreciated even today by anyone who has served in time of war. They deal with "danger," with "bodily exertion," with "information in war" and with a number of other factors of uncertainty and chance which "separate conception from execution." Clausewitz sums up these factors under the heading of "friction," a term which has become an integral part of the military vocabulary. "Friction" is more than a merely mechanical process. The military machine, after all, is composed of individuals, each of whom has to pay his tribute to human frailty. "Friction," as Clausewitz puts it, "is the only conception which in a general way corresponds to whatever distinguishes real war from war on paper." An infinity of petty circumstances makes plans fall short of the mark. In this connection Clausewitz formulates a sentence which has become current in military manuals: "Everything is very simple in war, but the simplest thing is difficult. . . . Activity in war is movement in a resistant medium. Just as a man immersed in water is unable to perform with ease and regularity the most natural and simple movement, that of walking, so in war one cannot, with ordinary powers, keep even the line of mediocrity."[37]

The most important modification, however, results from the connection between war and politics. Before approaching this central problem of Clausewitz' theory a few more words must be said about the "main battle," the most specific means of warfare. It may be noted that the relationship between means and ends has a preeminent position in Clausewitz' thought. A good example of this is his definition of strategy and tactics: . . . *"Tactics is the theory of the use of military forces in combat; strategy is the theory of the use of combats for the object of the war."*[38] Clausewitz first formulated this definition in his early treatise of 1805,[39] in opposition to the view which merely distinguished between the conduct of movements within the enemy's visual field and those

[36] *Ibid.*, I, 6 (Book I, Ch. 1).
[37] *Ibid.*, pp. 77-79 (Book II, Ch. VII).
[38] *Ibid.*, I, 86 (Book II, Ch. 1).
[39] *Bemerkungen über die reine und angewandte Strategie des Herrn von Buelow.* The same definition underlies the "Principles of War" by which Clausewitz in 1812 supplemented the instruction of the Prussian crown prince (Graham-Maude, III, Appendix). See the recent and much improved reprint in English mentioned in the bibliographical note.

outside it. Whatever the technical value of his own definition,[40] it is characterized by his insistence upon a structural element, the cogent relationship between
means and ends. As Clausewitz puts it in the book *On War*: "Where troops
are assumed, there the idea of combat always must be present.[41] . . . Every
activity in war necessarily relates to combat, either directly or indirectly. The
soldier is levied, clothed, armed, trained, he sleeps, eats, drinks and marches,
all *merely to fight at the right time and place*."[42] This relationship, as it were,
repeats itself on a higher level. Combats are no more a means in themselves
than are troops. As troops are to be used for fighting, combats are to be used
for the objective of the war. This objective being the overthrow of the enemy's
will, it follows that the disarming of the opponent through a decisive battle
stands out as the most specific means of war. In many ringing sentences Clausewitz plays upon this conception: ". . . the destruction of the enemy's armed
force appears . . . always as the superior and more effectual means to which
all others must give way . . . the bloody solution of the crisis, the effort for the
destruction of the enemy's force, is the first-born son of war."[43]

Here again Clausewitz does not overlook the fact that, throughout history,
few wars have shown this cogent interplay of means and ends. Real war has
rarely culminated in one main battle; in many wars there has been no notable
combat at all. To solve this contrast between abstract and real war, Clausewitz
makes a highly interesting suggestion that further clarifies his concept of
absolute war. In his opinion, the "idea of a possible battle" serves as a "distant
focus" even in wars in which it does not materialize.[44] An army can avoid
fighting only if it is sure that the opponent will not appeal to the "Supreme
Court" of armed decision or that he will lose his case before it. One may say
that, in Clausewitz' thought, the main battle is something like the British
"fleet in being" which dominates events even if it does not actually appear.
Clausewitz himself draws another parallel: "The decision by arms is for all
operations in war, great and small, what cash settlement is in trade."[45] When
the German socialist Engels read this sentence, it struck him as particularly
suggestive. Even though cash settlement and battle may rarely occur, everything is directed toward them. If they occur they decide everything.[46]

The relationship of means and ends is also basic in Clausewitz' *political
interpretation* of war. Battles, wars and political transactions, he holds, form
a totality within which the whole dominates the parts, or the ends dominate the

[40] Recent military writers again define both terms rather from the angle of space and
time. F. Maurice, e.g., contrasts "the methods of employing troops in contact with the
enemy" with "the leading of troops up to the time of contact with the enemy" (*British
Strategy*, p. 51). Goltz defines strategy as the "science of directing armies," tactics as "the
art of leading troops" (*The Conduct of War*, p. 30).
[41] Graham-Maude, I, 88 (Book II, Ch. i).
[42] *Ibid.*, p. 37 (Book I, Ch. ii).
[43] *Ibid.*, pp. 41-45 (Book I, Ch. ii).
[44] *Ibid.*, p. 268 (Book IV, Ch. xi).
[45] *Ibid.*, p. 40 (Book I, Ch. ii).
[46] Marx-Engels, *Briefwechsel*, III, 235-236. See also the present author's essay on
"Marxismus und Auswärtige Politik" in *Deutscher Staat und deutsche Parteien* (1922),
p. 322. The quotation is a paraphrase taken from A. Vagts, *Militarism*, p. 192.

means. Sometimes it may appear that this order is reversed. The battle, by way of its decisive character, seems to overrule the purpose of war. In his disquisition on absolute war Clausewitz also points out that the military aim of overthrowing the enemy "replaces, as it were,"[47] the final object, the political aim. On the basis of this statement, it has been claimed that Clausewitz argued for the superiority and self-sufficiency of the military. To some extent this is true, for Clausewitz insisted that the general should be independent of political decisions and that, indeed, he should be in a position to influence them. "The political end," he said, "is . . . no despotic legislator. It must be adapted to the nature of the means and consequently may often be totally changed. . . . Strategy in general and the commander in chief in particular may demand that the political tendencies and aims shall not conflict with the peculiar nature of military means, and this demand is by no means a slight one. . . ."[48]

In formulating these statements, Clausewitz may have been thinking of the political whims of courtiers or of deliberating bodies which, so often in the eighteenth century, interfered with military operations. He may have been thinking also of the obvious fact that policy, as followed up in war, is dependent upon what is possible in the military sense. But he certainly had also in mind the vital character of military decisions, which, according to their nature, affect men in the most elemental sense, and cannot be "dictated" by policy. In this respect, no doubt, Clausewitz struck upon a fundamental verity which has proved true under every form of government. Even democracies have faced and will face situations in which military exigencies are bound to overrule political considerations.

It must be added, however, that the whole trend of Clausewitz' thought points rather to the opposite order of things. War is only part of a social totality; it differs from the whole only by its specific means. However strongly military needs may react "in certain cases upon political aims, they can only be regarded as modifying these aims. For political aims are the end and war is the means, and the means can never be conceived without the end."[49] This is the basic view which underlies one of the best known sentences of the book *On War*: "War," it reads, "is nothing else than the continuation of state policy by different means."[50] The superiority of the political aims in principle could not be stated more clearly. On other occasions Clausewitz returns to this point.[51] In its most elaborate and mature form, his statement runs as follows:

"War is nothing else than a continuation of political transactions intermingled with different means. We say intermingled with different means in order to state at the same time that these political transactions are not stopped by the war itself, are not changed into something totally different but substantially continue, whatever the means applied may be. . . . How could it be

---

[47] Graham-Maude, I, 2 (Book I, Ch. 1).
[48] *Ibid.*, I, 22-23 (Book I, Ch. 1).
[49] *Ibid.*
[50] *Ibid.*, I, xxiii.
[51] "War is not only a political act but a real political instrument, a continuation of political transactions, an accomplishment of these by different means. That which then remains peculiar to war, relates only to the peculiar nature of its means" (*Ibid.*, I, 23).

otherwise? Do the political relations between different peoples and governments ever cease when the exchange of diplomatic notes has ceased? Is not war only a different method of expressing their thoughts, different in writing and language? War admittedly has its own grammar but not its own logic."[52]

It has been stated with regret[53] that Clausewitz, while speculating on how to win wars, did not speculate on how to win peace. Policy, in his view, being an affair of governments, he certainly did not enter upon this field. But when he defines war as a continuation of political transactions "intermingled with different means" he stresses the fact that there is no definite interruption, no silence or political self-abdication *inter arma*. He would hardly have agreed with the opinion current in Germany during the last war that policy has to wait for the results that military operations may yield. In his view there was certainly nothing like "military isolationism."

This basic conception has an important bearing upon the theory of war itself. It reconciles, as it were, absolute with real war. State policy first is the "womb in which war develops."[54] Policy therefore determines the main lines along which war is to move. This is the correct order of things, provided that policy does not demand anything that is against the nature of war. In fact, it would be absurd to assume that generals can outline a plan of operation in the abstract. "Still more absurd is the demand of theorists that the available means of warfare should be laid before the general, so that he may draw up a purely military plan."[55] Obviously there is no plan of a merely military character. Every war is an individual progression of events. If the political tensions are of a very powerful character (and if adequate material means are given) the political aim may disappear behind, or rather coincide with, the military aim of disarming the enemy. In such a case real war approaches absolute war. Clausewitz was convinced, as has already been indicated, that this type of warfare would appear again and again in the age of nationalism. "The greater and more powerful the motives of a war, the more it affects the whole existence of the nations concerned, the more violent the tension which precedes the war,—by so much the nearer will the war approach its abstract form, so much the more purely military and less political will war appear to be."[56] It is a main task of theory to emphasize this fundamental trend of war which is—to repeat this sentence—"the natural measure of all hopes and fears." But theory has also to

---

[52] *Ibid.*, III, 121 (Book VIII, Ch. vi.b). It may be noted that Clausewitz does not speak of war being an "extension of policy," as P. Birdsall suggests in describing a militaristic character of policy derived from Clausewitz (*Versailles 20 Years After*, p. 116). It would not be against Clausewitz' view to assume that war as "a possible means" reacts upon the whole political system just as the battle reacts upon war. But his main concern is obviously to subordinate war to state policy.

[53] Liddell Hart, *The Ghost of Napoleon*, p. 121; A. Vagts, *Militarism*, p. 196; H. Nickerson, *The Armed Horde*, p. 143.

[54] Graham-Maude, I, 121 (Book II, Ch. iii).

[55] *Ibid.*, III, 126 (Book VIII, Ch. viii.b). Clausewitz applied this principle himself when in 1827 he was asked for his advice by an officer of the Prussian general staff. See *Zwei strategische Briefe von Clausewitz*, ed. by Hans Rothfels (*Wissen und Wehr* 1923, 3).

[56] *Ibid.*, I, 23 (Book I, Ch. i).

take into account that, with lesser tensions, war becomes more and more political. Its range extends through all degrees of importance and energy, from overthrowing the enemy at the one extreme, to mere demonstrations at the other. Thus, war is truly "chameleon-like, because it changes its colour in each particular case."[57]

It is in the light of this flexible interpretation that Clausewitz surveys the whole of military history.[58] No single event can be isolated from its socio-political preconditions and from the whole atmosphere of tension. When the monarchical powers invaded France in 1792, when principle met principle, the mere cannonade at Valmy was more decisive than a bloody battle in the Seven Years War.[59] A good many remarks of this individualizing character are still of current interest.

Clausewitz, for instance, is particularly interested in the problems arising from "wars of coalition." He points out that a state engaged in war against an alliance is confronted with the problem of deciding which of the allies, the stronger or the weaker, should first be overthrown. He points out further that, whatever the decision, the state must regard the bond uniting the enemy alliance as a legitimate military objective. The primary aim of overthrowing the enemy army may be modified by other circumstances as well. The conquest of territory, for example, is in itself a powerful weapon, since it destroys the enemy's ability to rebuild its army. Loss of territory coupled with military defeat will be effective in sapping the enemy's will. Thus the aim of disarming the enemy may be obviated by the psychological disarming which occurs when the enemy realizes that victory is either unlikely or too costly.

The basic problem confronting the strategist is therefore one of discerning the "center of gravity" against which the military push must be directed. According to a variety of circumstances this "center of gravity" may be differently placed. In most cases it lies in the armed force of the enemy. This was true not only of the Napoleonic Wars but also of those of Alexander, Gustavus Adolphus, Charles XII and Frederick the Great. If the enemy country is divided by civil dissension, however, the "center of gravity" may lie in the capital. In wars of coalition, the "center" lies in the army of the strongest of the allies or in the community of interest between the allies. In national wars "public opinion" is an important center of gravity, a vital military objective.[60] Considering this last point, Clausewitz seems almost to revive the eighteenth century concept of "unbloody" warfare. It would be more accurate to say, however, that he touches upon the most modern concept of a psychological warfare that precedes or accompanies or even replaces actual fighting.

[57] *Ibid.*, I, 25 (Book I, Ch. 1).

[58] He studied approximately 130 campaigns as is evidenced by his historical works (*Werke*, IV-X) and by unpublished manuscripts. An interesting summary is given in *On War*, Book VIII, Ch. III.B ("Of the magnitude of the object of the War and the efforts to be made").

[59] Graham-Maude, I, 223 (Book III, Ch. XVIII).

[60] The text is mainly a paraphrase of Book I, Ch. II and Book VIII, Ch. IV.

VI

Does not this highly elastic analysis of war blur the clean-cut lines of thought and confuse rather than enlighten the student of Clausewitz? In answering this question, two points should be made. First, it is the very avoidance of a universally binding theory that gives to Clausewitz' analysis a timeless quality and makes it important even today. It expressly appeals to the "tact," to the discriminating judgment of statesmen and generals. Only he who is familiar with the richness of possible solutions will "plunge like a fearless swimmer into the stream."[61] In the second place, this fullness of possibilities is no unordered chaos. Its backbone is the nature of things, its regulative idea is absolute war "in being." In Clausewitz' own words, no fault can be found with a commander who in a skillful way tries cautious methods, provided the premises on which he acts are well founded.[62] But he must be aware of the fact that "he only travels on side-tracks where the God of War may surprise him." The overthrow of the enemy is no supreme law, but, as has been indicated, a "general point of direction." Having grasped that fact, the commander must realize that "the best strategy is *always to be very strong*, first generally, then at the decisive point."[63] This sentence implies the distinction between decisive and secondary actions. It is at the decisive point that every available man must be assembled.

By further distinctions, Clausewitz tried to make his "open system" still more instructive. In a notice of 1827 he speaks of his intention of revising the book *On War* along two lines.[64] First, he wished to distinguish between *"two kinds of war,"* one in which the object is "the overthrow of the enemy," another, in which the object is merely "to make some conquests on the frontiers of his country, either for the purpose of retaining them permanently, or for turning them to account as a matter of exchange in the peace settlement." In the second place Clausewitz wished to stress the fact that war is but a continuation of policy and this point of view was intended to introduce "more unity" into the whole conception of war. This revision, Clausewitz thought, would "iron out some creases in the heads of strategists and statesmen."

As a matter of fact, Clausewitz revised parts of his main work along these lines.[65] In the eighth book, in which he deals with the *plan of war*, he carefully distinguishes between the "two kinds of war"—war to overthrow the enemy and limited war. A strategical operation, he points out, would have a very different meaning when applied in the one case from that which it would have when applied in the other. In the one case, only the final result counts; in the other, partial results may be piled up and the factor of time counted upon, until the enemy's will wears down. In the one, the conquest of territory is of no

---

[61] Graham-Maude, I, 21 (Book I, Ch. 1).
[62] *Ibid.*, I, 45 (Book I, Ch. 11).
[63] *Ibid.*, p. 207 (Book III, Ch. xi).
[64] *Ibid.*, pp. xxiii-xxiv.
[65] Clausewitz revised Book VIII and parts at least of Book I (probably Chs. 1-111) and of Book II (certainly Ch. 11). In his latest statement of 1830 he regarded only Book I, Ch. 1 as "complete." The text of the present essay refers as much as possible to the revised parts. But no useful purpose would be served by too strict an application of textual criticism.

avail unless the enemy forces are destroyed; in the other, actual possession may tip the scales: *beati sunt possidentes*. This distinction was not meant to be a historical one, as has been suggested. Clausewitz did not intend to contrast the warfare of the *ancien régime* with that of the nineteenth century, the one being a "strategy of attrition," the other, a "strategy of annihilation."[66] He does not use these terms; neither does his individualizing interpretation of historical conditions fit in with any dualism of this kind. He is rather bent upon a systematic orientation.[67] Limited warfare has occurred and will occur again in two cases: first, whenever the political tensions or the political aims involved are small; second, whenever the military means are of such a character that the overthrow of the enemy cannot be conceived of at all, or can only be approached in an indirect way.

With these views, Clausewitz at least touches upon the discussion with which the present essay started. His theory does not exclude the peculiar traditions of states without a national mass army and the peculiar means at the disposal of insular or oceanic powers. Small expeditionary forces and economic warfare cannot conceive of overthrowing the enemy in the specific military sense. Yet there remains Clausewitz' second thought, that of war as a continuation of policy, which is meant to introduce "more unity." In the first chapter of his work, the only one which Clausewitz finally regarded as complete, he again integrates the two kinds of warfare into one gradual development. The decisive sentence has been quoted before: "The greater and more powerful the motives of a war . . . by so much nearer will the war approach its abstract form. . . ."

It seems to the present writer that this conception is applicable to many military disputes of the recent past. The "two kinds of war" have remained distinct. The question as to whether military or political methods should be

---

[66] These terms were mainly proposed by Hans Delbrück in several treatises and finally in his *Geschichte der Kriegskunst*, Vol. IV. Consequently a long drawn struggle arose over the strategy of Frederick the Great. The most recent and constructive review is given by O. Hintze, *Delbrück, Clausewitz und die Strategie Friedrichs des Grossen.* (*Forschungen zur brandenburgischen und preussischen Geschichte*, XXXIII, 131-177.) For Clausewitz' intentions see also H. Rosinski, *Historische Zeitschrift*, CLI, 285-293.

[67] The distinction is apparently related to what Clausewitz calls the "philosophical-dynamic law which exists between the greatness and the certainty of success" (Graham-Maude, I, 34. Book I, Ch. 11). More expressly this "law" applies to attack and defense and a number of strategic and tactical operations: victories which are won near the frontier are easier, those won after a process of penetration are more decisive; operations on "exterior lines" ("concentric" operations) lead to more brilliant results; the advantages of operations on "interior lines" ("eccentric" operations) are more secure. (*Ibid.*, II, 152. Book VI, Ch. IV). "Flank and rear attacks have, as a rule, a more favourable effect on the consequences of the decision than on the decision itself" (*Ibid.*, I, 261. Book IV, Ch. VII). Clausewitz, like Jomini, has a distinct preference for the operation on interior lines and the tactical break-through. The advance on "separate converging lines" promises great successes but is very dangerous: "If, on account of the situation of the belligerents it must be resorted to, it can only be regarded as a necessary evil" (*Ibid.*, III, 146. Book VIII, Ch. IX). The task of "marching separately in order to battle united" appears to Clausewitz "a most risky one" (*Werke*, VI, 310). In this and other respects Clausewitz' opinions are outdated by the technical development since his time. An excellent introduction into his practical teaching is presented in the English reedition of his Instructions for the Prussian crown prince. The editor (see bibliographical note) follows the example of the last German edition by using different types for the outdated and the still valid sections of this treatise. The present essay intentionally concentrates on the structural elements of Clausewitz' theory which certainly represent his most enduring contribution to military thought.

emphasized was involved in the debates about eastern and western strategy, which divided opinion in Germany as well as opinion in England, during World War I. Reduced to its simplest form, the controversy raged over the question of whether the overthrow or the attrition of the enemy was aimed at. But the gigantic dimensions of modern warfare and the pressure of opposing principles integrated the two kinds into one "push to the utmost." Whether indirect action and the piling up of partial successes is advisable, whether the factor of time and attrition can be counted upon, is a matter of "tact," which does not affect the primary aim. What is "attrition" in one sense, sums up to "annihilation" in another, to say nothing about the weapons of blockade and counterblockade. The same point of view seems applicable to the recent debates about the small highly mechanized army or the mass army, about aerial warfare or continental fight-to-the-finish. In view of the decisive character of the present war the distinction is one of military means rather than of military ends. And there seems little hope for eighteenth century refinement or fighting of "the slenderest possible proportions."

The crucial problem comes up again, however, in connection with another of Clausewitz' distinctions, that of *defense* and *attack*. This, of course, is a stock distinction, politically, strategically, tactically. But Clausewitz interwove it into his analysis of the nature of war and gave it a new turn. His conception is very different from such views as a "high priest of Napoleonic warfare" might be expected to hold. For one thing, he puts a great deal of emphasis upon defense, a fact which is regarded by many military writers of the nineteenth century as a "dark stain" on Clausewitz' thought. Does not the assailant always impose the law of action, does he not enjoy all advantages of the initiative? Clausewitz is remarkably skeptical about these advantages and the "moral superiority" of the attack.[68] The element of surprise, of course, is important, particularly in tactics; it is less so, in Clausewitz' view, in strategy. While the assailant makes the first move, the defender holds all the advantages of the "last hand." Moreover, it is defense, which, in a way, first constitutes war. In a striking paradox, which refers to Napoleon but can easily be generalized, Clausewitz points out, that the (political) aggressor is "always peace loving,"[69] that is, he would like to invade his neighbors peacefully unless there is organized resistance.

Clausewitz' theory is, to a large extent, bent upon proving that the weak has at least a fair chance to resist a more powerful foe. He can do so, because defense is the "stronger form of warfare." Clausewitz did not, and could not, foresee how far his thesis would be supported, at any rate in tactics, by the development of quick-firing armament. The points which he stresses in favor of defense refer to tactics as well as to strategy and politics. The attacked, he holds, enjoys political sympathies and the moral advantages which are derived from defending his own country. He also enjoys the advantages of the theater of war, of fortresses, positions, and the use of the terrain. He profits from time and all unexpected events, from the wearing out of the enemy, his falling

---

[68] Graham-Maude, III, 31 (Book VII, Ch. xv).   [69] *Ibid.*, II, 155 (Book VI, Ch. v).

short of the mark, etc. In short, defense is the stronger form because of its very nature: "To preserve is easier than to acquire."[70] In a sentence, which seems particularly striking in the light of the experiences of 1942, Clausewitz suggests that everything that does *not* happen is to the defender's credit. "He reaps where he has not sowed."[71]

The advantages of defense, however, are counterbalanced by a "dialectic" relationship. The defense is the stronger form with the negative object; the attack is the weaker form with the positive object. In following up this positive object, it is the assailant who must reach a decision. If the object is great, he will have to strive for a decision in the sense of absolute war. Within the attack itself defensive action is only a "retarding weight," in fact, a "mortal sin,"[72] whereas defense necessarily includes the transition to offensive action. Absolute defense would contradict the nature of war. To borrow a famous phrase from the present day, wars are not won by "successful retreats and evacuations." Clausewitz, therefore, draws the conclusion that "a swift and vigorous assumption of the offensive, the flashing sword of vengeance, is the most brilliant point in the defensive."[73]

The dialectic relationship between attack and defense centers in one of the most instructive concepts of Clausewitz, in the concept of the *"culminating point."*[74] If the strategic offensive fails to reach a decision, the forward push inevitably exhausts itself. Some of the moral and material resources of the assailant increase as he advances; but generally, and for many reasons, he is bound to weaken himself. In view of the very obvious illustrations with which World Wars I and II have furnished every observer, it is hardly necessary to enumerate the several factors which "place a fresh load on an advancing army at every step of its progress." As Clausewitz wrote, he was thinking, of course, mainly of the experience of the campaign of 1812. But his interpretation strikes at a fundamental problem: "Beyond the culminating point the scale turns . . . and the violence of the reverse is commonly much greater than was that of the forward push." Here there is a true test of generalship. As Clausewitz points out, everything "depends on discovering the culminating point by the fine tact of judgment." As the progress of action continues, the assailant is "carried along by the stream . . . beyond the line of equilibrium." Like a horse drawing a load uphill, he may find it less difficult to advance than to stop. He may still be counting on the collapse of the enemy's will at the very moment when it is actually rising like the "fury of a wounded bull."

Clausewitz, no doubt, has a professional sympathy with the general who tries to reach the goal by "the last minimum of preponderance"; he gives, as it were, to audacity rather than to caution the benefit of the doubt. And yet while one general, through an excess of caution, fritters away his good fortune, another through recklessness tumbles into destruction. Useless expenditure is destructive expenditure. "Often all hangs on the silken thread of imagination."

[70] *Ibid.*, p. 134 (Book VI, Ch. i). [71] *Ibid.*
[72] *Ibid.*, III, 3 (Book VII, Ch. ii). [73] *Ibid.*, II, 154-155 (Book VI, Ch. v).
[74] The following according to Book VII, Ch. v and Appendix.

It is at this moment that the defender must prove his generalship by seizing the opportunity for the "flashing sword of vengeance." If, after a far-reaching advance, the assailant is forced back into the defensive, he lacks most of the advantages of the "stronger form." Moral and psychological factors turn against him. And yet he still holds one of the advantages of the defensive: he is in possession. It is here that the second kind of war appears again. While it is no longer possible to conceive of the overthrow of the enemy, there remains the chance of demonstrating that the opponent cannot reach this aim either. Such was the problem that Frederick the Great faced, and solved, in the second part of the Seven Years War. It would not be difficult to show that the same problem reoccurred in the most striking form in both World Wars. In fact, Clausewitz' concept and interpretation of the culminating point sheds a flood of light upon most recent events.

In this connection a last point must be stressed. In the discussion of the culminating point, as well as throughout Clausewitz' writings, the high evaluation of *moral* and *psychological* factors stands out as the most conspicuous of his permanent contributions to military thought. Some chapters of the book *On War* are specifically devoted to this subject (I, iii; II, iii; III, iii-viii). Clausewitz carefully analyzes the qualities which a commander in chief on the one hand, an ordinary general on the other, should possess. It is very significant that he ranks highest a harmonious combination of subjective qualities like audacity and other strong impulses of a soldierlike nature with the objective qualities of a steadfast character and a dispassionate intelligence. In his instruction for the Prussian crown prince Clausewitz postulated "heroic decisions based on reason."[75] As he says in the book *On War*, it is "to the cool rather than to the fiery heads" that we should prefer to trust the welfare of our brothers and children in time of war.[76] Or again: a strong mind is not one which is merely capable of strong emotions, but one that keeps its equilibrium amidst the most powerful emotions, so that in spite of the storm in the breast, perception and judgment can act with perfect freedom, like the needle of the compass in the storm-tossed ship.[77] It is strength of character which best overcomes the natural frictions, doubts, panics, and the line of mediocrity.

The military virtue of an army requires also more than mere bravery. It is not the "temper" but the "spirit" of an army that counts,[78] and it is certainly not numbers alone. Though Clausewitz stresses superiority in numbers, "first generally, then at the decisive point," he expressly fights against the "complete misconception"[79] of attaching exclusive value to numerical strength. On that point he wished no misunderstanding. Moreover, he insisted that the overthrow of the enemy should not be misunderstood as emphasis laid upon mere physical killing. The main battle involves the killing of the enemy's courage rather than of the enemy's soldiers.[80] This is Clausewitz' formula for the familiar military saying that a battle is never materially lost unless the commander's

---

[75] Graham-Maude, III, 183, Appendix.
[77] *Ibid.*, I, 60.
[79] *Ibid.*, I, 198 (Book III, Ch. viii).

[76] *Ibid.*, I, 71 (Book I, Ch. iii).
[78] *Ibid.*, I, 185 (Book III, Ch. v).
[80] *Ibid.*, I, 286 (Book IV, Ch. xi).

or the army's "spirit" is defeated. In the last analysis, it is the will which stands predominant and commanding in the center of the art of war, "like an obelisk toward which the principal streets of a town converge."[81]

Some of the remarks on army morale which Clausewitz deduces from his analysis of war may appear "romantic." Some imply a glorification of military virtues as such, which sounds strange to us. Clausewitz' basic view, however, his insistence upon the prevalence of the immaterial and imponderable amidst the most material and brutal facts of war, has certainly not become obsolete. It is no less applicable to the motorized and mechanized armies of today than it was to the foot-marching and horse-riding soldiers of the early nineteenth century. In the present conflict, the truth of Clausewitz' maxim is being daily proved. Physical forces are the "wooden hilt," but "moral forces" are the "shining blade" of the sword.[82]

[81] *Ibid.*, p. 78 (Book I, Ch. vii). [82] *Ibid.*, p. 178 (Book III, Ch. iii).

SECTION III

From the Nineteenth Century to the First World War

CHAPTER 6. Adam Smith, Alexander Hamilton, Friedrich List: The Economic Foundations of Military Power

BY EDWARD MEAD EARLE

O NLY in the most primitive societies, if at all, is it possible to separate economic power and political power. In modern times—with the rise of the national state, the expansion of European civilization throughout the world, the industrial revolution, and the steady advance of military technology—we have constantly been confronted with the interrelation of commercial, financial, and industrial strength on the one hand, and political and military strength on the other. This interrelationship is one of the most critical and absorbing problems of statesmanship. It involves the security of the nation and, in large measure, determines the extent to which the individual may enjoy life, liberty, property, and happiness.

When the guiding principle of statecraft is mercantilism or totalitarianism, the power of the state becomes an end in itself, and all considerations of national economy and individual welfare are subordinated to the single purpose of developing the potentialities of the nation to prepare for war and to wage war —what the Germans call *Wehrwirtschaft* and *Kriegswirtschaft*. Almost three hundred years ago Colbert epitomized the policy of the rising French monarchy of Louis XIV by saying that "trade is the source of finance and finance is the vital nerve of war." In our day, Goering has indicated that the political economy of the Nazi garrison state was aimed at the production of "guns, not butter." And a favorite device of Soviet preparation for total war was the slogan that it is better to have socialism without milk, than milk without socialism. Democratic peoples, on the other hand, dislike the restraints which are inherent in an economy based upon war and the preparation for war: *Wehrwirtschaft* is something alien to their way of life and beyond the bounds of what they consider necessary to their safety and prosperity. They prefer an economic system which is predicated upon individual welfare rather than upon the overweening power of the state. And they have a deep-rooted suspicion of coordinated military and economic power, as something which constitutes an inherent threat to their long-established liberties.

But whatever the political and economic philosophies which motivate a nation, it can ignore only at dire peril the requirements of military power and national security, which are fundamental to all other problems of government. Alexander Hamilton was enunciating a basic principle of statecraft when he said that safety from external danger is "the most powerful director of national conduct"; even liberty must, if necessary, give way to the dictates of security because, to be more safe, men are willing "to run the risk of being less free."[1] Adam Smith, who believed the material prosperity of the nation to be founded upon a minimum of governmental interference with the freedom of the individual, was willing to concede that this general principle must be compromised when national security is involved, for "defense is of much more importance than opulence."[2] Friedrich List, who disagreed with Smith on most subjects, found himself in perfect accord on this point: "Power is of more importance than wealth . . . because the reverse of power—namely, feebleness—leads to the relinquishment of all that we possess, not of acquired wealth alone, but of our powers of production, of our civilisation, of our freedom, nay, even of our national independence, into the hands of those who surpass us in might. . . ."[3]

For more than two centuries before Adam Smith published the *Wealth of Nations* western Europe was governed by beliefs and practices which, as a whole, are known as mercantilism. The mercantilist system was a system of power politics. In domestic affairs it sought to increase the power of the state against the particularist institutions which survived from the Middle Ages. In foreign affairs it sought to increase the power of the nation as against other nations. In short, the ends of mercantilism were unification of the national state and development of its industrial, commercial, financial, military, and naval resources. To achieve these ends the state intervened in economic affairs, so that the activities of its citizens or subjects might be effectively diverted into such channels as would enhance political and military power. The mercantilist state—like the totalitarian state of our time—was protectionist, autarkic, expansionist, and militaristic.

In modern terminology, we would say that the predominant purpose of mercantilist regulations was to develop the military potential, or war potential. To this end exports and imports were rigidly controlled; stocks of precious metals were built up and conserved; military and naval stores were produced or imported under a system of premiums and bounties; shipping and the fisheries were fostered as a source of naval power; colonies were settled and protected (as well as strictly regulated) as a complement to the wealth and self-sufficiency of the mother country; population growth was encouraged for the purpose of increasing military man power.[4] These and other measures were

---

[1] The *Federalist* (1787), No. VIII (New York, Modern Library edition, 1937, with an introduction by E. M. Earle), p. 42. All page references will be to this edition. The full text also is in Vols. XI and XII of Hamilton's collected *Works*, cited in footnote 30.

[2] *An Inquiry into the Nature and Causes of the Wealth of Nations.* Originally published in 1776. For convenience I have used the Modern Library edition (introduction by Max Lerner), which is a reprint of the edition of Edwin Cannan (London, 1904). The phrase here used is to be found in Book IV, Chap. II, p. 431.

[3] Friedrich List, *Das nationale System der politischen Ökonomie* (Stuttgart, 1841) in *Schriften, Reden, Briefe* (10 vols., Berlin, 1930-1935), VI (edited by Artur Sommer, Berlin, 1930), 99-100. This is the best edition of List's works, published in cooperation with the *Deutsche Akademie.* The quotation is from the English translation by Sampson S. Lloyd (London, 1885), pp. 37-38. Hereafter cited as *National System* from the English translation.

[4] A typical measure for encouraging population was prohibition of enclosure of pasture lands in favor of the extension of lands under cultivation of foodstuffs. A proclamation of 1548 in England, for example, stated that "the surety . . . of the Realm must be defended

designed with the major, if not the single, purpose of adding to the unity and strength of the nation.

War was inherent in the mercantilist system, as it is in any system in which power is an end in itself and economic life is mobilized primarily for political purposes. Representatives of a policy of power believe that their goals can be achieved "as well, if not better, by weakening the economic power of other countries instead of strengthening one's own. If wealth is considered as an aim, this is the height of absurdity, but from the point of view of political power it is quite logical. . . . Any attempt at economic advance by one's own efforts in one country must have appeared pointless, unless it consisted in robbing other countries of part of their possessions. Scarcely any other element in mercantilist philosophy contributed more to the shaping of economic policy, and even of foreign policy as a whole."[5] This logic was remorseless with the mercantilists and in large measure accounts for the almost continuous war—open or concealed—which raged in Europe from the middle of the seventeenth century to the early part of the nineteenth. Napoleon's Continental System and the retaliatory British Orders in Council were simply the culmination of a long series of similar measures.

From the mercantilist wars, England alone emerged triumphant. Achieving national unification earlier than any other European power, and enjoying the security which her insular position afforded, she was better able than the others to put "the might of her fleets and admiralty, the apparatus of customs and navigation laws, at the service of the economic interests of the nation and the state with rapidity, boldness, and clear purpose," and thereby to gain the lead in the struggle for commercial and political hegemony.[6] By 1763 England had crushed the commercial, colonial, and naval aspirations of Spain, Holland, and France. The resurgent France of the Revolution and Napoleon was crushed again at Waterloo. In 1815, despite the loss of her American colonies, Great Britain seemed to have arrived at world power in a manner and degree reminiscent of the great empires of antiquity. "In all ages there have been cities or countries which have been pre-eminent above all others in industry, commerce, and navigation; but a supremacy such as that [of Britain] which exists in our days, the world has never before witnessed. In all ages, nations and powers have striven to attain to the dominion of the world, but hitherto not one of them has erected its power on so broad a foundation. How vain do the efforts of those appear to us who have striven to found their universal dominion on military power, compared with the attempt of England to raise her entire territory into one immense manufacturing, commercial, and maritime city, and to become among the countries and kingdoms of the earth, that which a great

against the enemy with force of men, and the multitude of true subjects, not with flocks of sheep and droves of beasts." Cited by Eli Heckscher, *Mercantilism* (English translation by M. Shapiro, 2 vols., London, 1935), II, 44.

[5] *Ibid.*, II, 21, 24.

[6] This is a paraphrase, not a quotation, from Gustav Schmoller, *The Mercantile System and Its Historical Significance* (London and New York, 1896), p. 72. Translated from the German by W. J. Ashley. The German text is in *Das Merkantilsystem in seiner historischen Bedeutung*, first published in *Schmollers Jahrbuch* for 1884.

city is in relation to its surrounding territory; to comprise within herself all industries, arts, and sciences; all great commerce and wealth; all navigation and naval power—a world's metropolis. . . ." Thus wrote a German nationalist in 1841, in envy and in admiration.[7]

It was against the background of mercantilism and of a triumphant England that Smith the Briton, Hamilton the American, and List the German outlined economic and political policies for their respective countries. What they had to say concerning the economic foundations of military power can be understood only within the framework of their times and the spirit and special conditions of their respective countries.

II

When the *Wealth of Nations* was published in 1776, the time was ripe in Britain for critical reappraisal of the theories and practices of mercantilism. The revolt of the American colonies had focused attention upon the entire system of trade regulation which was involved in Britain's colonial policy. There was dissatisfaction with the wars which had been going on for over a century and with the mounting burden of war debts. Furthermore, after Britain's triumph over France in the Seven Years' War (1756-1763), there remained no serious rival to England in either commercial or naval power. Hence there was increasing skepticism concerning a political and economic philosophy by which "nations have been taught that their interest consisted in beggaring all their neighbors." The feeling began to grow, now that Britain's position as a world power seemed assured, that a more liberal policy might be initiated and that "the wealth of a neighboring nation, however dangerous in war and politics, is certainly advantageous in trade."[8] There was a growing conviction, too, that there had been abuses in the prevailing system, which enabled entrenched privilege to benefit from its association with the real or imagined interests of the nation. It was against these abuses that Smith struck out in attacking the merchant class in general and the chartered companies in particular for monopolistic practices, usurpation of governmental authority, and the fomenting of war.[9] "The capricious ambition of kings and ministers has not, during the present and the preceding century," he said, "been more fatal to the repose of Europe, than the impertinent jealousy of merchants and manufacturers. The violence and injustice of the rulers of mankind is an ancient evil. . . . But the mean rapacity, the monopolizing spirit of merchants and manufacturers, who neither are, nor ought to be, the rulers of mankind . . . may very easily be prevented from disturbing the tranquillity of any body but themselves."[10]

---

[7] List, *op. cit.*, p. 293.

[8] *Wealth of Nations*, pp. 460-461. Even before the Seven Years' War, David Hume in an essay on the *Jealousy of Trade* had gone counter to all mercantilist ideas in saying, "not only as a man, but as a British subject, I pray for the flourishing commerce of Germany, Spain, Italy, and even France itself," on the ground that all nations would flourish were their policies toward one another more "enlarged and benevolent." T. H. Green and T. H. Grose (eds.), Hume's *Essays Moral, Political and Literary* (London, 1898), I, 348.

[9] On the chartered companies, *op. cit.*, pp. 595-606.   [10] *Ibid.*, p. 460.

Smith's most trenchant criticisms of mercantilism were directed at its monetary theories, including the notion that the state must accumulate great stocks of bullion as a war chest. He admitted that Britain must be prepared to wage war, because "an industrious, and upon that account a wealthy nation, is of all nations the most likely to be attacked." Nor was he unaware that Britain's vast colonial and commercial commitments overseas required the maintenance of a substantial military and naval establishment. But he denied that war chests were essential or even useful to the effective defense of the nation, for "fleets and armies are maintained, not with gold and silver, but with consumable goods. The nation which, from the annual produce of its domestic industry, from the annual revenue arising out of its lands, labour, and consumable stocks, has wherewithal to purchase those consumable goods in distant countries, can maintain foreign wars there." This was proved by Britain's experience in defraying "the enormous expence" of the Seven Years' War from the profits of her expanded manufactures and her greatly increased foreign trade.[11] In other words, Smith believed that the ability of a nation to wage war is best measured in terms of its productive capacity, as was later to be argued so effectively by Friedrich List. Furthermore, he objected to war chests, as well as to war loans, as the principal means of financing wars. He favored heavy taxes instead. Wars currently paid for "would in general be more speedily concluded, and less wantonly undertaken" by governments, and "the heavy and unavoidable burdens of war would hinder the people from wantonly calling for it when there was no real or solid interest to fight for."[12]

Despite the fact that the *Wealth of Nations* became the bible, and Adam Smith the intellectual progenitor, of the laissez-faire school of nineteenth century British economic theorists, the truth is that Adam Smith did not really repudiate certain fundamentals of mercantilist doctrine. He rejected some of its means, but he accepted at least one of its ends—the necessity of state intervention in economic matters in so far as it might be essential to the military power of the nation. His followers were more doctrinaire free traders than Smith was himself, and they certainly were more ardent pacifists. "The first duty of the sovereign," he wrote, "that of protecting the society from the violence and invasion of other independent societies, can be performed only by means of a military force." But the methods of preparing this force in time of peace, and of employing it in time of war will vary according to the different states of society. War becomes more complicated and more expensive as societies advance in the mechanical arts; hence the character of the military establishment and the methods of supporting it will be different in a commercial and industrial state from that in a more primitive society.[13] In other words, as

[11] The discussion concerning war chests is in Book IV, Chap. I, especially pp. 398-415. The quotations here given are from pp. 399, 409, 679.

[12] *Ibid.*, pp. 878-879. The facts of history hardly support the thesis that governments or peoples carefully calculate the costs of war in advance of hostilities.

[13] *Ibid.*, Book V, Chap. I, Part I, pp. 653-669. Quotation on p. 653. Heckscher, *op. cit.*, understood fully the extent to which Smith accepted some of the basic tenets of mercantilism. Smith's admirer William Cunningham, in his monumental *Growth of English Industry and Commerce in Modern Times* (2 vols., Cambridge, England, 1882) seems to have

Marx and Engels later pointed out, the forms of economic organization in large measure determine what are to be the instruments of war and the character of military operations. It is inevitable, therefore, that military power be built upon economic foundations.

In so far as Great Britain was concerned, the heart of the mercantilist system—the ark of the covenant—was the Navigation Acts. Mercantilism in its other aspects may have been essential at an earlier period of her development, but by the end of the eighteenth century England was so far advanced industrially that protectionism was of much less importance to her than to France and the German states. She could have afforded, if necessary, to dispense with duties on most manufactures because she was without serious competition in her domestic and overseas markets. Indeed, she was later, in self interest, to abandon her earlier restrictive policies because she learned, as Bismarck said, that "free trade is the weapon of the strongest." But sea power was another matter, and anything related to it had to be judged by different criteria. The safety of the homeland and the empire demanded that Britain have virtually unchallenged control of the ocean highways; any power which thought otherwise was certain to earn implacable hostility. Furthermore, the entire superstructure of British industry, finance, and commerce was founded upon overseas markets and overseas sources of supply. Hence, her merchant marine was both an economic asset and an absolutely indispensable element in her military security, especially in an age when merchant vessels were readily converted into privateers or men-of-war. "Your fleet and your Trade," declared Lord Haversham in the House of Lords, "have so near a relation and such mutual influence on each other, they cannot well be separated: your trade is the mother and nurse of your seamen: your seamen are the life of your fleet: and your fleet is the security and protection of your trade: and both together are the wealth, strength, security and glory of Britain."[14]

For these reasons the real test of Adam Smith's views on mercantilism and power politics was his stand on the Navigation Acts and the fisheries. "The defence of Great Britain," he said, "depends very much upon the number of its sailors and shipping. The act of navigation, therefore, very properly endeavours to give the sailors and shipping of Great Britain the monopoly of the trade of their own country."

"When the act of navigation was made," Smith continued, "though England

missed the whole truth when he said that Smith treated "wealth without direct reference to power"; certainly Smith would not have subscribed to Cunningham's statement that "national rivalries and national power are mean things after all" and that the study of wealth had to be dissociated from these "lower aims." (Vol. I, pp. xxix, 593-594, especially note 2, p. 594.) Smith, writing shortly after the Seven Years' War and on the eve of the French and American revolutions, was keenly aware of the realities of power politics; Cunningham, writing almost midway in a century of peace, when war seemed remote, saw the situation differently. Smith's bitter opponent List missed the truth just as badly as Cunningham; he mistook the views of Smith's followers for those of Smith himself, as will presently be shown.

[14] Cited in G. S. Graham, *Sea Power and British North America* (Cambridge, Massachusetts, 1941), p. 15. This work should be consulted for an excellent discussion of the place of the navigation acts in British statecraft. See especially pp. 7-15.

and Holland were not actually at war, the most violent animosity subsisted between the two nations. It had begun during the government of the long parliament, which first framed this act, and it broke out soon after in the Dutch wars during that of the Protector and of Charles the Second. It is not impossible, therefore, that *some of the regulations of this famous act may have proceeded from national animosity. They are as wise, however, as if they had all been dictated by the most deliberate wisdom. National animosity at that particular time aimed at the very same object which the most deliberate wisdom would have recommended, the diminution of the naval power of Holland,* the only naval power which could endanger the security of England.

"The act of navigation is not favourable to foreign commerce, or to the growth of that opulence which can arise from it. . . . As defence, however, is of much more importance than opulence, the act of navigation is, perhaps, the wisest of all the commercial regulations of England."[15]

As regards the fisheries he took essentially the same point of view: "But though the tonnage bounties to those fisheries do not contribute to the opulence of the nation, it may perhaps be thought that they contribute to its defence, by augmenting the number of its sailors and shipping."[16] Smith likewise approved of the laws which authorized the payment of a bounty for the production of naval stores in the American colonies and prohibited their export from America to any country other than Great Britain. This typical mercantilist regulation was justified, in Smith's view, because it would make England independent of Sweden and the other northern countries for the supply of military necessities and thus contribute to the self-sufficiency of the empire.[17]

Furthermore, Smith was not averse to protective duties when they were required for reasons of military security. "It will generally be advantageous to lay some burden upon foreign, for the encouragement of domestic industry," he said, "when some particular industry is necessary for the defense of the country." Such protection was afforded the shipping industry by the Navigation Acts. But Smith was willing to pay bounties or to impose tariffs in the interest of other industries as well for the same public purpose: "It is of importance that the kingdom depend as little as possible upon its neighbours for the manufactures necessary for its defence; and if these cannot be maintained at home, it is reasonable that all other branches of industry be taxed in order to support them." With some reluctance he also approved of retaliatory duties and hence of what came to be called "tariff wars."[18]

Adam Smith was a free trader by sincere conviction. He completely demolished some of the theories which underlay mercantilism; and mercantilist practices, as they existed in the British Empire of his day, were repugnant to him. He was suspicious of state interference with private initiative, and he was no worshiper of state power for its own sake. But the critical question in determining his relationship to the mercantilist school is not whether its fiscal and

15 *Wealth of Nations*, Book IV, Chap. ii, pp. 430-431. Italics are mine.
16 *Ibid.*, Book IV, Chap. v, pp. 484-485.   17 *Ibid.*, pp. 545-546, 609-610, 484, note 39.
18 *Ibid.*, pp. 429, 434, 484-489 (especially note 39).

trade theories were sound or unsound but whether, when necessary, the economic power of the nation should be cultivated and used as an instrument of statecraft. The answer of Adam Smith to this question would clearly be "Yes" —that economic power should be so used.

This has not been altogether understood. Smith's followers, particularly in nineteenth century England, were responsible for presenting him as an uncompromising free trader. Some of his critics, particularly the Germans Schmoller and List, allowed cries of "free trade" to drown out the rest of Smith's teachings which would have been music to their ears. Thus in some quarters Smith has been considered a hypocrite—a British patriot who had seen his country outgrow the mercantilist strategy and tactics by which it rose to unchallenged power, and was then prepared to recommend the discarding of such strategy and tactics by other nations of lesser good fortune. That Smith was a British patriot need hardly be denied, but that he was a hypocrite is emphatically not true. He does not deserve the following withering indictment by List, who was more familiar with what he called "the school" of Smith's followers than with Smith himself :

"It is a very common clever device that when anyone has attained the summit of greatness, he kicks away the ladder by which he has climbed up, in order to deprive others of the means of climbing up after him. In this lies the secret of the cosmopolitical doctrine of Adam Smith, and of the cosmopolitical tendencies of his great contemporary William Pitt, and of all his successors in the British Government administrations.

"Any nation which by means of protective duties and restrictions on navigation has raised her manufacturing power and her navigation to such a degree of development that no other nation can sustain free competition with her, can do nothing wiser than to throw away these ladders of her greatness, to preach to other nations the benefits of free trade, and to declare in penitent tones that she has hitherto wandered in the paths of error, and has now for the first time succeeded in discovering the truth."[19]

III

More than three hundred years ago, Francis Bacon pointed out that the ability of a nation to defend itself depended less upon its material possessions than upon the spirit of the people, less upon its stocks of gold than upon the iron of determination in the body politic.[20] As a professor of moral philosophy, Adam Smith must have been acquainted with the works of Bacon. In any case, he believed that "The security of every society must always depend, more or less, upon the martial spirit of the great body of the people. . . . Martial spirit alone, and unsupported by a well-disciplined standing army, would not, per-

[19] *Op. cit.*, pp. 295-296. See a similar, but less vindictive, comment by Schmoller, *op. cit.*, pp. 79-80. A recent Nazi critic is also worth consulting in this same connection: P. F. Schröder, "Wehrwirtschaftliches in Adam Smiths Werk über den Volkwohlstand" in *Schmollers Jahrbuch*, Vol. 63 (1939), No. 3, pp. 1-16.

[20] *Essays Civil and Moral*, No. 19, "Of the True Greatness of Kingdoms and Estates." In the *Works of Francis Bacon*, edited by James Spedding, VII (Boston, 1840), 176 ff.

haps, be sufficient for the defence and security of any society. But where every citizen had the spirit of a soldier, a smaller standing army would surely be necessary." And Smith went even further in the belief that "even though the martial spirit of the people were of no use towards the defence of the society, yet to prevent that sort of mental mutilation, deformity, and wretchedness, which cowardice necessarily involves in it, from spreading themselves through the great body of the people, would still deserve the most serious attention of government; in the same manner as it would deserve its most serious attention to prevent a leprosy or any other loathsome and offensive disease, though neither mortal nor dangerous, from spreading itself among them. . . ." Only through "the practice of military exercises," supported by the government, could the martial spirit be effectively maintained.[21] During the nineteenth century many of Smith's followers, notably Cobden and Bright, were convinced pacifists, as well as ardent free traders, and would not have endorsed any such doctrine.

There is a long-standing and deeply rooted Anglo-American prejudice against "standing armies." The insular position of the British Isles made it possible for Parliament to "muddle through" in questions of national defense, and the long contest between Parliament and the Crown (in which the army was an instrument of the Stuarts) fostered the belief that a professional army was dangerous to civil liberty. On the continent of Europe the rivals of Great Britain had resorted to large standing armies as the bulwark of their strength, and under professional soldiers had made great progress in military organization and the art of war.[22] Nevertheless, Parliament continued during time of peace to maintain the army at inconsequential strength, persisted in the inefficient and demoralizing system of billeting of troops on the people, and continued its reliance on the militia, which Dryden had so effectively lampooned in *Cymon and Iphigenia:*

> "The country rings around with loud alarms,
> And raw in fields the rude militia swarms;
> Mouths without hands, maintained at vast expense,
> In peace a charge, in war a weak defence.
> Stout once a month they march, a blustering band,
> And ever, but in time of need, at hand."

At the end of the seventeenth century, Macaulay wrote, "there was scarcely a public man of note who had not often avowed his conviction that our policy and a standing army could not exist together. The Whigs had been in the constant habit of repeating that standing armies had destroyed the free institutions of the neighboring nations. The Tories had repeated as constantly that, in our own island, a standing army [under Cromwell] had subverted the

---

[21] *Wealth of Nations,* V, 1, 738-740.
[22] See Chapter II. For further material on Smith's convictions regarding the standing army, see a particularly valuable article by the late Professor Charles J. Bullock of Harvard, "Adam Smith's Views upon National Defense," *Military Historian and Economist,* I (1917), 249-257.

Church, oppressed the gentry, and murdered the King. No leader of either party could, without laying himself open to the charge of gross inconsistency, propose that such an army should henceforth be one of the permanent establishments of the realm."[23]

This was still the situation when Smith was professor of moral philosophy at Glasgow, 1752-1763, and delivered his famous lectures on justice, police, revenue, and arms.[24] In these lectures Smith broke with his famous teacher Francis Hutcheson, who had opposed a standing army on the ground that "the military arts and virtues are accomplishments highly becoming all honorable citizens" and that "warfare therefore should be no man's perpetual profession; but all should take their turns in such services."[25] This seemed to Smith an utterly impracticable program, and he took a categorical stand in favor of a professional army.

Smith admitted that a standing army might be a menace to liberty—after all, Cromwell had "turned the long parliament out of doors." But he believed that with proper precautions the army could be made to support, rather than undermine, the authority of the constitution. In any case, security demanded a well-trained and well-disciplined armed force; only then could the nation commit its fate to the god of battles. No militia, however trained and disciplined, could take the place of professional soldiers, especially in an age when the development of firearms put a greater premium on organization and order than on individual skill, bravery, and dexterity. The most elementary requirements of military precaution, therefore, demanded that the historic reliance upon the militia, and the traditional suspicion of the professional army, give way to the exigencies of the times. Furthermore, the sound economic principle of the division of labor demanded that war be made a vocation, not an avocation:

"The art of war," Smith wrote, "as it is certainly the noblest of all arts, so in the progress of improvement it necessarily becomes one of the most complicated among them. The state of the mechanical, as well as of some other arts, with which it is necessarily connected, determines the degree of perfection to which it is capable of being carried at any particular time. But in order to carry it to this degree of perfection, it is necessary that it should become the sole or principal occupation of a particular class of citizens, and the division of labour is as necessary for the improvement of this, as of every other art. Into other arts the division of labour is naturally introduced by the prudence of individuals, who find that they promote their private interest better by confining themselves to a particular trade, than by exercising a great number. But it is the wisdom of the state only which can render the trade of a soldier a particular trade separate and distinct from all others. A private citizen who, in time of profound peace, and without any particular encouragement from the public, should spend the greater part of his time in military exercises, might, no doubt,

[23] *History of England* (Riverside edition; Boston, n.d.), IV, 186-187.
[24] Edited by Edwin Cannan (Oxford, 1896) from notes taken by a student in 1763.
[25] Francis Hutcheson, *A Short Introduction to Moral Philosophy* (2 vols., Glasgow, 1764), II, 348-349.

both improve himself very much in them, and amuse himself very well; but he would certainly not promote his own interest. It is the wisdom of the state only which can render it for his interest to give up the greater part of his time to this peculiar occupation: and states have not always had this wisdom, even when their circumstances had become such, that the preservation of their existence required that they should have it."[26]

It is a coincidence, but a coincidence of significance to the English-speaking peoples, that 1776 was the date of publication of both the *Wealth of Nations* and the Declaration of Independence. Smith dealt at length with the relations of Great Britain with her American colonies, and what he had to say is of moment to any student of American or British history. For our present purposes, however, it is necessary to consider only Smith's attitude toward imperialism. He clearly believed that a colonial policy did not "pay" in the mercantilist sense. And although he thought that the Americans had not suffered, in fact, from the restrictions imposed by the mother country, such restrictions were nevertheless "a manifest violation of the most sacred rights of mankind," as well as "impertinent badges of slavery" imposed upon America by the official and mercantile classes of England. The value of colonies in an imperial system should be measured, in his judgment, by the military forces which they provided for imperial defense and by the revenue which they furnished for the general support of the empire. Judged by these criteria, the American colonies were a liability, not an asset, to Great Britain; they not only contributed nothing to imperial defense, but they required British forces to be dispatched to America and they had involved the homeland only recently in a costly war with France.[27] Stated in terms of a commercial and financial balance sheet, England would be better off without the colonies.

This is a parochial view of empire, which will be suggestive of Neville Chamberlain. But Smith did not propose that England accede to the American demand for independence; this would be "to propose such a measure as never was, and never will be adopted, by any nation in the world. No nation ever voluntarily gave up the dominion of any province, how troublesome soever it might be to govern it, and how small soever the revenue which it afforded might be in proportion to the expence which it occasioned. Such sacrifices, though they might frequently be agreeable to the interest, are always mortifying to the pride of every nation, and what is perhaps of still greater consequence, they are always contrary to the private interest of the governing part of it, who would thereby be deprived of the disposal of many places of trust and profit, of many opportunities of acquiring wealth and distinction, which the possession of the most turbulent, and, to the great body of the people, the most unprofitable province seldom fails to afford."[28]

[26] *Ibid.*, Book V, Chapter 1, pp. 658-659. In addition, see *Lectures*, Part IV, "Of Arms," of which the foregoing chapter is an elaboration.

[27] Smith was clearly wrong in saying that the "whole expence" of the Seven Years' War, as well as the cost of the wars which preceded it, should be charged to the colonies. The discussion on colonies is in Book IV, Chaps. 7 and 8.

[28] *Ibid.*, pp. 581-582. It is interesting to compare Smith's views on colonies with those of Jeremy Bentham, one of Smith's most faithful followers. Bentham agreed that the defense

Smith shrewdly foresaw that the War for American Independence would be a long and costly war. He even visualized a possible victory for the embattled colonists, who, from "shopkeepers, tradesmen, and attornies are become statesmen and legislators, and are employed in contriving a new form of government for an extensive empire, which, they flatter themselves, will become, and which, indeed, seems very likely to become, one of the greatest and most formidable that ever was in the world."[29] Smith was right, and among the attorneys who became statesmen was Alexander Hamilton, a giant among that remarkable galaxy of truly great men who brought into being the United States of America.

IV

With the exception of two years of travel on the Continent (1764-1766), Adam Smith's life was devoted entirely to academic pursuits. He was a student at Glasgow and Oxford, lectured at Edinburgh, and was successively professor of logic and professor of moral philosophy at Glasgow. After his return from Europe, he devoted himself to his great work, the *Wealth of Nations*, published fourteen years before his death.

Alexander Hamilton, on the other hand, was a man of action from his earliest youth. His life began inauspiciously on the tiny West Indian island of Nevis. His father was impecunious; and after the death of his mother in 1768, when he was only eleven years old, Hamilton had to make his own way in the world. He served as clerk in a general store, but soon went to New York, where he entered Kings College (now Columbia) in 1773. Within a year he became involved in the war of pamphlets which preceded the American Revolution and, while still in his 'teens, established a reputation as one of the most vigorous writers of his generation. He entered the army early in 1776, received a commission, fought with Washington on Long Island and at White Plains, Trenton, and Princeton. In March 1777, at the age of twenty, he was made military secretary to the Commander in Chief, with the rank of lieutenant colonel; as such, he was not only a confidant and adviser of Washington, but the author of a series of brilliant reports on army organization and administration.[30] Later he commanded an infantry regiment in Lafayette's corps, distinguishing himself by conspicuous bravery at Yorktown. He continued his military career long after the Revolution when, in 1798, he was commissioned major general and inspector general of the army, second in command to Washington, for the purpose of preparing for a threatened war with France.

Hamilton's role in bringing into being the Annapolis and Philadelphia conventions and, above all, his brilliant services in securing ratification of the

of colonies costs too much, but went farther and advocated the relinquishment by Britain of her existing colonies and the abandonment of all attempts to acquire new ones. *Principles of International Law*, in *Works*, edited by John Bowring, Vol. 2 (Edinburgh, 1843), essay IV, especially pp. 548-550.

[29] *Ibid.*, pp. 587-588.

[30] Hamilton's military papers are to be found in volumes 6 and 7 of his collected *Works*, edited by Henry Cabot Lodge, Federal Edition, 12 vols. (New York and London, 1904.)

Constitution, are too well known to need extensive comment. Quite aside from his other great state papers, his authorship of more than half of the *Federalist* would alone entitle him to high rank among political writers. He was the most influential single member of Washington's cabinet, roaming far afield from his own duties as Secretary of the Treasury. During the years 1789-1797 he probably did more than any other single person to formulate the early national policies of the United States, some of which came to have the binding force of tradition.[31] His tragic death in 1804, when he was only forty-seven, was a national disaster.

For the student of military affairs, Hamilton is a link between Adam Smith and Friedrich List. Hamilton was familiar with the *Wealth of Nations* and had it before him when, with the assistance of Tench Coxe, he wrote his famous "Report on Manufactures."[32] He agreed with Smith on the wisdom and necessity of a professional army, as well as on certain questions of economic policy related to national defense. Hamilton's influence on Friedrich List is evident in much of what the latter wrote. And in view of List's association with the protectionist groups in the United States, including the economist Mathew Carey, there can be little doubt that List considered the "Report on Manufactures" a textbook of political economy. Indeed, he invoked the support of Hamilton from time to time, and there is strong internal evidence throughout List's writings that Hamilton's ideas had a prominent place in his "national system."[33]

William Graham Sumner, an ardent free trader and hence an unsympathetic critic, said that Hamilton's concept of national policy was "the old system of mercantilism of the English school, turned around and adjusted to the situation of the United States."[34] There is some merit to the statement but not in the sense that Hamilton was a blind follower or admirer of mercantilist doctrines. As has been indicated above,[35] European mercantilists were concerned with two distinct but closely related things: national unification, as opposed to particularism; development of the resources of the nation, with special reference to its military potential. Hamilton was certainly a nationalist and he certainly believed in using economic policy as an instrument of both national unification and national power. Almost everything he said and believed can be related, in

[31] As there is no very satisfactory life of Hamilton, the article by Allan Nevins in the *Dictionary of National Biography* is probably the best source for the facts of his career. This article is done with the usual skill of one of America's foremost historians.

[32] This fact is established by W. S. Culbertson's admirable essay *Alexander Hamilton* (New Haven, 1911), pp. 90, 107-108, 127-129. See also Henry Cabot Lodge in *Works*, III, 417, and the article "Alexander Hamilton and Adam Smith," by Edward G. Bourne, *Quarterly Journal of Economics*, VIII (April 1894), 328-344. Concerning the role of Tench Coxe see note 61.

[33] William Notz, "Friedrich List in America," *American Economic Review*, XVI (June 1926), 249-265. Dr. Notz was one of the editors of the above-mentioned edition of the works of List. His admirable introductory essay to Vol. II (Berlin, 1931), pp. 3-61, is the best account of List's years in America and their significance to List's career as a whole. For estimates of Hamilton's influence on List see C. Meitzel, article on Hamilton in *Handwörterbuch der Staatswissenschaften*, fourth edition (1923), IV, 21, and M. E. Hirst, *Life of Friedrich List* (London, 1909), pp. 112-118.

[34] W. G. Sumner, *Alexander Hamilton* (New York, 1890), p. 175.　　[35] See Section I.

some manner, to this central theme. His advocacy of a well-rounded national economy which would include manufactures, his recommendations as regards the public debt (particularly the assumption of the debts of the states), his belief in a national bank, his concepts of foreign policy and security, his doctrine of the "implied powers" of the federal government, his conviction that the manufacture of munitions of war should be encouraged and if necessary controlled by the nation, his reports on military policy, his ardent espousal of the navy, even his attitude toward democratic government—all these can best be understood in relation first to his passion for national unity and second his jealous regard for the political and economic power of the nation.

On the other hand, it is doubtful if even Adam Smith could have written a fairer or more eloquent summary of the case for free trade than that which appears in Hamilton's "Report on Manufactures," submitted to the Congress, December 5, 1791.[36] Furthermore, if a system of industrial and commercial liberty, said Hamilton, "had governed the conduct of nations more generally than it has done, there is room to suppose that it might have carried them faster to prosperity and greatness than they have attained by the pursuit of maxims too widely opposite." There then would and could be a genuine international division of labor to the benefit of all. But liberty of trade and exchange have not prevailed; in fact, precisely the opposite is the case, and the nations of Europe, particularly those which had developed manufactures, "sacrifice the interests of a mutually beneficial intercourse to the vain project of selling everything and buying nothing." As a result, "the United States are, to a certain extent, in the situation of a country precluded from foreign commerce" and rendered impotent to trade with Europe on equal terms. This statement of the facts, continued Hamilton, is "not made in a spirit of complaint. It is for the nations whose regulations are alluded to, to judge for themselves, whether, by aiming at too much, they do not lose more than they gain. It is for the United States to consider by what means they can render themselves least dependent on the combinations, right or wrong, of foreign policy" of other states.[37]

The program set forth in his "Report on Manufactures" stamps Hamilton as an economic nationalist. His aim, he said, was to promote such manufactures "as will tend to render the United States independent of foreign nations for military and other essential supplies."[38] He believed that "not only the wealth but the independence and security of a country appear to be materially connected with the prosperity of manufactures. Every nation, with a view to those great objects, ought to endeavor to possess within itself, all the essentials of

[36] Works, IV, 70-198, especially 71-73, 100-101. The Report also is included in an admirably edited volume by Samuel McKee, Jr., Papers on Public Credit, Commerce, and Finance by Alexander Hamilton (New York, 1934).

[37] Ibid., pp. 73, 100-102.

[38] Ibid., p. 70. Compare with the statement in Washington's first annual message to Congress in 1790 that "the safety and interest [of a free people] require that they should promote such manufactories as tend to render them independent of others for essential, particularly military supplies."

national supply. These comprise the means of subsistence, habitation, clothing, and defence."

"The possession of these," Hamilton said, "is necessary to the perfection of the body politic; to the safety as well as to the welfare of the society. The want of either is the want of an important organ of political life and motion; and in the various crises which await a state, it must severely feel the effects of any such deficiency. The extreme embarrassments of the United States during the late war, from an incapacity of supplying themselves, are still matter of keen recollection; a future war might be expected again to exemplify the mischiefs and dangers of a situation to which that incapacity is still, in too great a degree, applicable, unless changed by timely and vigorous exertion. To effect this change, as fast as shall be prudent, merits all the attention and all the zeal of our public councils: 't is the next great work to be accomplished.

"The want of a navy, to protect our external commerce, as long as it shall continue, must render it a peculiarly precarious reliance for the supply of essential articles, and must serve to strengthen prodigiously the arguments in favor of manufactures."[39]

Hamilton believed that a young country like the United States could not compete with countries like Great Britain which had been long established in manufacturing. "To maintain, between the recent establishments of one country, and the long-matured establishments of another country, a competition upon equal terms . . . is in most cases, impracticable." Hence the industries of the newer country should enjoy the "extraordinary aid and protection of the government."[40] This aid and protection should be extended in the form of import duties (to the point of prohibition in some instances), restraints on export of raw materials, pecuniary bounties and premiums, drawbacks, exemption of certain essential raw materials from import tariffs, and other devices. This is the "infant industry" argument, but it also is the characteristic mercantilist case for autarky.

In determining the commodities on which duties are to be levied, and the amount of such duties, for the purpose of encouraging domestic manufactures, important and perhaps primary consideration should be given to "the great [factor] of national defense." Thus:

"Fire-arms and other military weapons may, it is conceived, be placed, without inconvenience, in the class of articles rated at fifteen per cent. There are already manufactories of these articles, which only require the stimulus of a certain demand to render them adequate to the supply of the United States.

"It would also be a material aid to manufactures of this nature, as well as a means of public security, if provision should be made for an annual purchase of military weapons, of home manufacture, to a certain determinate extent, in order to [assure] the formation of arsenals; and to replace, from time to time, such as should be drawn for use, so as always to have in store the quantity of each kind which should be deemed a competent supply.

"But it may, hereafter, deserve legislative consideration, whether manufac-

[39] *Ibid.*, pp. 135-136.   [40] *Ibid.*, pp. 105-106.

tories of all the necessary weapons of war ought not to be established on account of the government itself. Such establishments are agreeable to the usual practice of nations, and that practice seems founded on sufficient reason.

"There appears to be an improvidence in leaving these essential implements of national defence to the casual speculations of individual adventure—a resource which can less be relied upon, in this case, than in most others; the articles in question not being objects of ordinary and indispensable private consumption or use. As a general rule, manufactories on the immediate account of government are to be avoided; but this seems to be one of the few exceptions which that rule admits, depending on very special reasons."[41]

The "Report on Manufactures" also emphasizes the idea—to be developed at great length by Friedrich List—that a country with a diversified economy, including agriculture, manufactures, and commerce, will be more unified at home and stronger in its relations with other powers than it otherwise would be. But Hamilton made his best statement of this thesis in his first draft of Washington's "Farewell Address," which he wrote during the summer of 1796.[42] Hamilton visualized a nation in which sectional economies would interweave themselves into a common national economy and interest. The agricultural south would not merely contribute its own share to the national wealth but would share in the benefits of the industrial strength of the north. The west, especially after the development of adequate transportation, would offer a market for the manufactures and foreign commerce of the east and, in turn, would profit from the development of the "weight, influence, and maritime resources of the Atlantic States." Furthermore, "where every part finds a particular interest in the Union, all the parts of our Country will find greater independence from [i.e., by reason of] the superior abundance and variety of production incident to the diversity of soil and climate." The aggregate strength of a nation thus united by a common economic interest would be increased in every essential respect. The United States, by developing a diversified economy, would enjoy enhanced "security from external danger, less frequent interruption of their peace with foreign nations, and, what is more valuable, an exemption from those broils and wars between the [several] parts, if disunited, which their own rivalships, fomented by foreign intrigue . . . would inevitably produce." In consequence, the nation would profit from "exemption from the necessity of those military establishments upon a large scale which bear in every country so menacing an aspect towards Liberty." Thus did Hamilton link his economic system with national security.

[41] *Ibid.*, pp. 167-168. This is not the first occasion on which Hamilton made such a proposal as regards munitions. As chairman of a special committee of Congress he suggested in 1783 that "it ought to be made a serious object of policy, to be able to supply ourselves with all the articles of first necessity in war" and that to this end public manufactories of arms and munitions should be constructed. *Ibid.*, pp. 467, 475.

[42] For the text and all other details see Victor H. Paltsits, *Washington's Farewell Address* (New York, 1935), especially pp. 184-185. The extent to which Washington adopted Hamilton's argument in this respect will be evident by comparing the foregoing draft with the final manuscript, *ibid.*, pp. 143-144. For clarity, I have supplied punctuation in the text.

Hamilton's argument for an American navy and merchant marine was a similar amalgam of politics and economics. He was convinced that the United States was destined to become a great maritime power. The adventurous voyages of Americans to all quarters of the earth—"that unequalled spirit of enterprise . . . which is in itself an inexhaustible mine of national wealth"— had already "excited uneasy sensations" among Europeans, who "seem to be apprehensive of our too great interference in that carrying trade, which is the support of their navigation and the foundation of their naval strength." Some European states, by restrictive legislation, were resolved upon "clipping the wings by which we might soar to a dangerous greatness." But by a firm union, a flourishing merchant marine, prosperous fisheries (as a nursery of seamen), appropriate retaliatory navigation acts, and a navy "we might defy the little arts of the little politicians to control or vary the irresistible and unchangeable course of nature." The navy of the United States might not "vie with those of the great maritime powers," but it would at least "be of respectable weight if thrown into the scale of either of two contending parties," particularly in the West Indies. Our position, even with a few ships of the line, is therefore "a most commanding one," which would enable us to "bargain to great advantage for commercial privileges." Furthermore, "a price would be set on our neutrality and our friendship" in the event of a war between foreign powers. Hence, "by a steady adherence to the Union, we may hope, ere long, to become the arbiter of Europe in America, and to be able to incline the balance of power in this part of the world as our interest may dictate."[43] Surely, this is *Realpolitik* of a high order and shows that a strategy for America in world politics was evolved by the fathers of the republic.

It is imperative, Hamilton claimed, that the United States have an integrated national economy. To this great object, a navy would contribute, just as political and economic union would contribute to the growth of the navy:

"A navy of the United States, as it would embrace the resources of all, is an object far less remote than a navy of any single State or partial confederacy, which would only embrace the resources of a single part. It happens, indeed, that different portions of confederated America possess each some peculiar advantage for this essential establishment. The more southern States furnish in greater abundance certain kinds of naval stores—tar, pitch, and turpentine. Their wood for the construction of ships is also of a more solid and lasting texture. The difference in the duration of the ships of which the navy might be composed, if chiefly constructed of Southern wood, would be of signal importance, either in the view of naval strength or of national economy. Some of the Southern and of the Middle States yield a greater plenty of iron ore, and of better quality. Seamen must chiefly be drawn from the Northern hive. The necessity of naval protection to external or maritime commerce does not re-

---

[43] All quotations in the preceding paragraph and the one which follows are from the *Federalist*, No. 11. It should be noted that Hamilton did not wish us to pursue a balance of power policy in Europe. See, e.g., *Works*, IX, 327; X, 397.

quire a particular elucidation, no more than the conduciveness of that species of commerce to the prosperity of a navy."[44]

Hamilton's fiscal policy likewise had its political connotations. By funding the public debt, assuming the debts of the states, and founding a national bank, he hoped to link "the interest of the State in an intimate connection with those of the rich individuals belonging to it" and to turn "the wealth and influence of both into a commercial channel, for mutual benefit." Hence, a national debt might be a "national blessing" since it would be "a powerful cement to our Union."[45] He wanted the support of the merchant and propertied classes because he knew how they had been able to influence the government in England in the enactment of mercantilist legislation, and he believed that the economic motivation of politics was inherent in almost any society.[46] Furthermore, the establishment of the national credit on a firm basis was essential "as long as nations in general continue to use it as a resource in war. It is impossible for a country to contend, on equal terms, or to be secure against the enterprises of other nations, without being able equally with them to avail itself of this important resource; and to a very young country, with moderate pecuniary capital, and a not very various industry, it is still more necessary than to countries more advanced in both." One "cannot but conclude that war, without credit, would be more than a great calamity—would be ruin." Although admitting the legality of sequestration of private property in wartime, he opposed it on the grounds, among other valid reasons, that it would discourage foreign investment in American securities.[47] In short, he recommended that we "cherish credit as a means of strength and security."[48]

V

National security was a problem of absorbing interest to Hamilton, and he had a realistic appreciation of the factors which were pertinent to it. He understood that the distance of the United States from Europe and the vast extent of our territory were great assets to us, since they would make conquest by a foreign power difficult if not impossible. But he knew also that we were a young, undeveloped, and politically immature country, needing time to consolidate our position. Hence his reiterated emphasis upon national unity, his strictures against factionalism and sectionalism, his injunctions against "passionate attachment" or "rooted prejudice" as regards other nations, and his advice against political commitments abroad. Hence also his belief that "if we

[44] Compare this with the following statement which Theodore Roosevelt (who was a great admirer of Hamilton) made to a middle western audience in 1910: "Friends, the Navy is not an affair of the seacoast only. There is not a man who lives in the grass country, in the cattle country, or among the Great Lakes, or alongside the Missouri who is not just as keenly interested in the Navy as if he dwelt on the New England Coast, or on the Gulf Coast, or on Puget Sound." Speech at Omaha, September 2, reprinted in *The New Nationalism* (New York, 1910), p. 147.

[45] Letter to Robert Morris, 1780, in *Works*, III, 338, 387.

[46] On this point see also the *Federalist*, No. 10, written by Madison.

[47] "Second Report on the Public Credit" (December 1794), in *Works*, III, 199-300. Quotations are from pp. 295-296.

[48] Hamilton's draft for Washington's "Farewell Address." Paltsits, *op. cit.*, p. 193.

remain a united people under an efficient government the period is not distant when we may defy material injury from external annoyance."[49] But security is not possible without power, for "a nation, despicable by its weakness, forfeits even the privilege of being neutral."[50] Only if we are strong can we "choose peace or war as our interest guided by justice shall dictate."[51] But strength depends on union and, as Jay said, "on the government, the arms, and the resources of the country."[52]

Hamilton saw clearly, too, that we would never be altogether secure while European powers had substantial territories on this continent. He was opposed to transfers of American territory from one non-American power to another; consequently, he favored the purchase of Louisiana, even though it was effected by his opponent Jefferson. He even seems to have visualized the policy which came to be known as the Monroe Doctrine.[53] He was Anglophile, not only because he detested the radical principles of revolutionary France, but also because he believed that we were too weak for a definitive test of arms with Great Britain, as well as too dependent upon British toleration of our growing commercial strength.

Hamilton agreed with the preamble of the Constitution that a more perfect union, the common defense, the general welfare, and the preservation of liberty were inextricably interwoven. In No. 8 of the *Federalist* he wrote at length and with keen understanding on the delicate problem of reconciling military power with basic political liberties—a paper which shows striking resemblances to some of Adam Smith's ideas on the same subject. He pointed out also that it was not enough for a government to have authority to raise armies in time of war; it must maintain adequate forces in time of peace. Otherwise "we must expose our property and liberty to the mercy of foreign invaders . . . because we are afraid that rulers, created by our choice, dependent on our will, might endanger that liberty, by an abuse of the means necessary to its preservation."[54] In time of war, furthermore, the power of the executive must be adequate for "the direction of the common strength"[55] despite the traditional fear of Americans for centralized authority.

Like Adam Smith, Hamilton believed that the professional army should be the basis of national defense. As he wrote in the *Federalist*: "The steady operations of war against a regular and disciplined army can only be success-

[49] *Ibid.*, pp. 193-196.   [50] The *Federalist*, No. 11, p. 65.
[51] This famous phrase was Hamilton's, not Washington's. Paltsits, *op. cit.*, p. 196. Washington changed "dictate" to "counsel."
[52] The *Federalist*, No. 4, p. 65.
[53] For the nontransfer principle see "Answer to Questions Proposed by the President of the United States," September 15, 1790, in *Works*, IV, 338. Regarding the menace of European territories in America, see the *Federalist*, No. 24, pp. 150-151. The elimination of European influence on this continent is a fairly constant factor in American foreign policy; *cf.* E. M. Earle, "National Security and Foreign Policy," *Yale Review*, XXIX (1940), 444-460. The *Federalist*, No. 11, p. 69, indicates that, had he lived, Hamilton would have supported the Monroe Doctrine.
[54] The *Federalist*, No. 25, p. 156. On this same point see *ibid.*, No. 4 (by Jay), No. 23 (by Hamilton), and No. 41 (by Madison). The *Federalist*, in these and other numbers, is a textbook for students of military policy and national security.
[55] *Ibid.*, No. 74, p. 48.

fully conducted by a force of the same kind. Considerations of economy, not less than of stability and vigor, confirm this position. The American militia, in the course of the late war, have, by their valor on numerous occasions, erected eternal monuments to their fame; but the bravest of them feel and know that the liberty of their country could not have been established by their efforts alone, however great and valuable they were. War, like most other things, is a science to be acquired and perfected by diligence, by perseverance, by time, and by practice."[56]

During the latter part of the eighteenth century there was a widespread belief that parliamentary governments, especially those dominated by a commercial class, were less likely to be involved in war than monarchies. Hamilton thought any such opinion contrary to the dictates of common sense and the known facts of history. He was persuaded that popular assemblies were just as subject as other forms of government (perhaps more so) to "the impulses of rage, resentment, jealousy, avarice, and other irregular and violent propensities." He also disagreed with the view of the physiocrats that—to quote Montesquieu—"the natural result of commerce is to promote peace." On the contrary, in his judgment, commerce was more likely to be a cause of recurring wars. "Has commerce hitherto done anything more than change the objects of war? Is not the love of wealth as domineering and enterprising a passion as that of power or glory? Have there not been as many wars founded upon commercial motives since that has become the prevailing system of nations, as were before occasioned by the cupidity of territory or dominion? Has not the spirit of commerce, in many instances, administered new incentives to the appetite, both for the one and for the other?" He thought the answer to these questions clearly to be in the affirmative. War was too deeply rooted in human society, however changing its forms, to warrant belief in undisturbed peace and security.[57]

Surprisingly enough, Thomas Jefferson agreed with Hamilton that commerce was a potential cause of war. "Our people are decided in the opinion," he wrote John Jay from Paris in August 1785, "that it is necessary for us to take a share in the occupation of the ocean, and their established habits induce them to require that the sea be kept open to them, and that that line of policy

[56] No. 25, p. 157. Even earlier, Hamilton had given serious thought to a military policy for the United States. See a letter to James Duane in 1780 and Hamilton's report on behalf of a special committee of the Congress in 1783, in *Works*, I, 215-216; VI, 463-483. He believed that the army should be national in organization and loyalty; that a system of defenses should be built without reference to state lines; that the militia should be under national supervision as regards uniformity of service, training, and equipment; that there should be a national military academy; and that the manufacture of munitions should be encouraged and perhaps owned by the federal government. Hamilton also believed in the principle of universal liability to military service. *Ibid.*, VII, 47.

[57] The *Federalist*, No. 6, discusses the causes of war at length. Quotations are from p. 30. Concerning the view of the physiocrats and others that the influence of commerce was in the direction of promoting international peace see Edmond Silberner, *La guerre dans la pensée économique du xvie au xviie siècle* (Paris, 1939). In Nos. 3, 4, and 5 of the *Federalist* John Jay also discusses the causes of war and makes the remarkable forecast (in No. 4) that the growing trade with China would involve the United States in international conflict in the Far East.

be pursued, which will render the use of that element to them as great as possible. I think it a duty in those entrusted with the administration of their affairs, to conform themselves to the decided choice of their constituents; and that therefore, we should, in every instance, [even at the cost of almost certain war] preserve an equality of right to them in the transportation of commodities, in the right of fishing, and in the other uses of the sea."[58] And Jefferson gave practical effect to this belief when, as president, he waged war against the Barbary pirates, his pacifist convictions to the contrary notwithstanding.

Indeed, some measure of Hamilton's stature may be taken by observing further the extent to which Jefferson—his most bitter opponent—came to agree with him as regards economics and national defense. Jefferson was a free trader and an avowed enemy of manufactures. He detested Hamilton's protectionist program. But after his own experiences with the embargo and after observing the consequences of the war of 1812-1815 with Great Britain, he reluctantly came to the conclusion that the realities of power politics might require a change in the views which he had previously held. As he wrote the French economist and free trader Jean Batiste Say in March 1815:

". . . I had then [earlier] persuaded myself that a nation, distant as we are from the contentions of Europe, avoiding all offences to other powers, and not over-hasty in resenting offence from them, doing justice to all, faithfully fulfilling the duties of neutrality, performing all offices of amity, and administering to their interests by the benefits of our commerce, that such a nation, I say, might expect to live in peace, and consider itself merely as a member of the great family of mankind; that in such case it might devote itself to whatever it could best produce, secure of a peaceable exchange of surplus for what could be more advantageously furnished by others, as takes place between one county and another of France. But experience has shown that continued peace depends not merely on our own justice and prudence, but on that of others also; that when forced into war, the interception of exchanges which must be made across a wide ocean, becomes a powerful weapon in the hands of an enemy domineering over that element, and to the other distresses of war adds the want of all those necessaries for which we have permitted ourselves to be dependent on others, even arms and clothing. *This fact, therefore, solves the question by reducing it to its ultimate form, whether profit or preservation is the first interest of a State?* We are consequently become manufacturers to a degree incredible to those who do not see it, and who only consider the short period of time during which we have been driven to them by the suicidal policy of England. The prohibiting duties we lay on all articles of foreign manufacture which prudence requires us to establish at home, with the patriotic determination of every good citizen to use no foreign article which can be made within ourselves, without regard to difference of price, secures us against a relapse into foreign dependency."[59]

And although Jefferson never quite came to support Hamilton's views con-

---

[58] *Writings of Thomas Jefferson* (Memorial Edition), V, 94.
[59] *Op. cit.*, XIV (Washington, 1903), 258-260. The italics are mine.

cerning a standing army, he did come around to believing that much more thought must be given to the maintenance of a military establishment based upon universal liability to service. Commenting on a memoir of the Secretary of War, he wrote James Monroe in 1813: "It is more a subject of joy that we have so few of the desperate characters which compose modern regular armies. But it proves more forcibly the necessity of obliging every citizen to be a soldier; this was the case with the Greeks and Romans, and must be that of every free State. . . . We must train and classify the whole of our male citizens and make military instruction a regular part of collegiate education. We can not be safe till this is done."[60]

Alexander Hamilton can hardly be rated high as an economist, except, perhaps, in one respect—his effective statement of the "infant industry" argument for the protection of manufactures, in which he said with great effectiveness virtually all that can be said. In the formulation of this part of his famous report he had the active collaboration of Tench Coxe, his Assistant Secretary of the Treasury and one of the Philadelphia school of protectionists who had so marked an influence on Hamilton. But the historical significance of his plea for the development of American industry is greater than its inherent worth, for upon what he wrote was built the structure of American economic policy. As one who combines economics with politics and statecraft, however, Hamilton ranks with the great statesmen of modern times. He is, in fact, an American Colbert or Pitt or Bismarck. The power and effect of his ideas was indelibly impressed upon succeeding generations of Americans, so that in the realm of government and industry his influence is more marked than that of any of his contemporaries except Jefferson.[61]

VI

It is one of the ironies of history that Hamilton's political opponents Jefferson and Madison did more than Hamilton himself to give effect to his protectionist and nationalist views of economic policy. The Embargo, which Jefferson initiated in December 1807, the Non-Intercourse Act, and the succeeding war with Great Britain, upon which Madison reluctantly embarked, had the practical result of closing virtually all avenues of foreign trade and making the United States dependent upon its own resources for manufactures and munitions of war. The industries which were born under the stress and necessity of

[60] *Ibid.*, XIII, 261.

[61] Mr. Julian Boyd, librarian of Princeton University, has had the privilege of examining correspondence and manuscripts of Tench Coxe which indicate that the latter had an active part in the formulation and drafting of the "Report on Manufactures." The actual extent of Coxe's contribution to the final document must await release and publication of the Coxe papers by the Coxe family. For a very critical analysis of the report, pointing out certain inconsistencies and contradictions in the document, see Frank A. Fetter, in L. S. Lyon and V. Abramson, *Government and Economic Life* (2 vols., Washington, 1940), II, 536-540. A longer treatment of the same subject, less unfavorable to Hamilton, is E. C. Lunt, "Hamilton as a Political Economist," *Journal of Political Economy*, III (1895), 289-310. For the influence of the Philadelphia School see a paper by Professor Fetter, "The Early History of Political Economy in the United States," *Proceedings of the American Philosophical Society*, LXXXVII (1943), pp. 51-60.

the years 1808 to 1815 were the infants to which the nation gave protection in 1816 and in a succession of tariff acts thereafter.

While Americans were still smarting from the indignities inflicted upon the United States by Napoleonic France and Great Britain, there seemed to be substantial agreement upon governmental protection of manufactures. Madison and Jefferson, on the one hand, and the "war hawks" of 1812, Clay and Calhoun, on the other, found themselves in the same camp. Jefferson in January 1816 wrote an exceedingly bitter denunciation of those who cited his former free trade views as "a stalking horse, to cover their disloyal propensities to keep us in eternal vassalage to a foreign and unfriendly people [the British]." He called upon all Americans to "keep pace with me in purchasing nothing foreign where an equivalent of domestic fabric can be obtained, without regard to difference of price," for "experience has taught me that manufactures are now as necessary to our independence as to our comfort." For the sake of securing independence from others, "we must now place the manufacturer by the side of the agriculturist."[62] Hamilton himself could not have said more.

But as time went on, the old cleavages reappeared, and a bitter struggle over protectionism raged until the Walker Tariff of 1846 temporarily settled the issue. It was as a participant in this debate that Friedrich List made his appearance on the American scene and formulated the economic theories which were to have influence not only in the United States but, even more, in Germany. List was born in Württemberg in 1789, studied at the University of Tübingen (where he later served briefly as professor of politics), and entered public life as an ardent exponent of the *Zollverein*. His liberal and nationalist ideas kept him in constant hot water with the reactionary government of his native state, leading finally to his exile in 1825, when he came to America and settled among the Pennsylvania Germans at Reading. He became the editor of the Reading *Adler*, a German-American weekly with an influential voice in the affairs of Pennsylvania. His interest in commercial policy soon brought him into contact with the Pennsylvania Society for the Encouragement of Manufactures and the Mechanic Arts, which was under the vigorous and able leadership of Mathew Carey, Charles Jared Ingersoll, and Pierre du Ponceau, among others.[63] Although Mathew Carey was the more effective pamphleteer, List was able to write with a wider experience of economics and politics and became the foremost literary and scholarly propagandist of protectionism during his residence in America. He was lionized by Pennsylvania industrialists, met most of the prominent American statesmen of the day, was offered the presidency of Lafayette College, and, when he finally returned to Germany in 1832, did so as

[62] *Op. cit.*, XIV, 389-393. Letter to Benjamin Austin.

[63] This society seems to have been inspired by the earlier Philadelphia Society for Promotion of Domestic Industries, founded by Hamilton. The Pennsylvania Society published and distributed several editions of the "Report on Manufactures," as well as pamphlets by Mathew Carey, who did more than any other American except Hamilton to bring about the so-called American System. It sponsored the famous Harrisburg Convention of 1827, memorializing Congress in favor of higher tariffs (which materialized in the "Tariff of Abominations" of 1828), attracted nationwide attention by its effective propaganda, and in general served to put the state of Pennsylvania permanently in the protectionist camp in American politics.

a naturalized citizen and as a member of the consular service of the United States by appointment of Andrew Jackson. He was consul at Baden-Baden until 1834, at Leipzig (1834-1837), and at Stuttgart (1837-1845). He died by his own hand in 1846, after illness had terminated his public service.

List's intellectual history is fairly easy to trace. In his youth, "seeing to what a low ebb the well-being of Germany had sunk," he decided to study political economy and also to teach his fellow citizens the means, in terms of national policy, by which "the welfare, the culture, and the power of Germany might be promoted." He came to the conclusion that the key to the solution of Germany's problems was the principle of nationality. "I saw clearly that free competition between two nations which are highly civilized can only be mutually beneficial in case both of them are in a nearly equal position of industrial development, and that any nation which owing to misfortunes is behind others in industry, commerce, and navigation . . . must first of all strengthen her own individual powers, in order to fit herself to enter into free competition with more advanced nations. In a word, I perceived the distinction between *cosmopolitical*[64] and *political* economy. I felt that Germany must abolish her internal tariffs, and by the adoption of a common uniform commercial policy towards foreigners, strive to attain to the same degree of commercial and industrial development to which other nations have attained by means of their commercial policy."

The similarity of the foregoing views to the central themes of mercantilism —national unification and the development of national power through economic policy—is obvious.

"When afterwards I visited the United States," continued List, "I cast all books aside—they would only have tended to deceive me. The best work on political economy which one can read in that modern land is actual life. There one may see wildernesses grow into rich and mighty States; and progress which requires centuries in Europe, goes on there before one's eyes. . . . That book of actual life, I have earnestly and diligently studied, and compared with my previous studies, experience, and reflections. And the result has been (as I hope) the propounding of a system which . . . is not founded upon bottomless cosmopolitanism, but on the nature of things, on the lessons of history, and on the requirements of the nations."[65]

There is reason to believe that List formulated his views on politics and economics not, as he said, while a young man in Germany but only after his arrival in the United States. Certainly his *Outlines of American Political Economy* (a series of letters written to Charles Jared Ingersoll during the

[64] "Cosmopolitical" was the term by which List described the writings of Adam Smith, J. B. Say, and others of their "school." That he frequently misrepresented Smith's views must be apparent to any reader of the *Wealth of Nations* and *The National System of Political Economy*. List hopelessly confused *Smithianismus*—which was what anybody said Smith had said—with Smith's own ideas. On this point see the admirable Introduction by Professor J. S. Nicholson to the 1904 edition of Lloyd's translation of *The National System* (as cited in note 3 above).

[65] Author's preface to *The National System*, pp. xl, xlii. List always denied that he was a mercantilist, although he admitted that he had taken over "the valuable parts of that much-decried system." *Ibid.*, p. xliii.

summer of 1827, subsequently printed in pamphlet form and widely distributed by the Pennsylvania protectionists) contains all the essential ideas elaborated in *The National System of Political Economy*, which appeared fourteen years later. The *Outlines* so clearly show the influence of Hamilton and
Mathew Carey that there can be little reasonable doubt that American conditions and ideas were predominant, if not decisive, in the development of List's
economic theories.[66]

Nevertheless, List was first, last, and above all a German. He was always an
unhappy exile in America and acquired American nationality partly to avoid
the petty persecutions which had been his previous lot in his native land. He
admired and envied the vast undeveloped resources of the United States, the
youthful vigor of the country, its success in achieving political unification, the
Realpolitik of Hamilton, the lusty nationalism of Jackson, the American enthusiasm for railways and canals, and the seemingly unlimited possibilities for
the future of the United States as a world power.[67] But all of these things he
related to his hopes and aspirations for his own country, then so tragically
disunited. The Germany of his day might well have frustrated the determination of even a Colbert. Prussia, the dominant North German state, had more
than sixty-seven different tariffs within its own territories, with almost three
thousand articles subject to duties, to be collected by an army of customs
officials; it had boundaries meandering almost a thousand miles through the
rest of Germany, touching twenty-eight different states. Notwithstanding the
seemingly insuperable difficulties, List dreamed dreams and saw visions of a
new and greater Germany, unified by internal free trade, external protection,
and a national system of posts and railways; and, finally, rising to the stature
of a great European power. He lived to see only part of his program realized.
The *Zollverein*, which destroyed more obstacles to internal commerce and
political unity "than had been swept away by the political whirlwinds of the
American and French Revolutions," was partly the result of his untiring
efforts. His ceaseless propaganda for railways had some material results before
it wore him out and hastened his death. He did not live to see the Revolutions
of 1848, the successes of Bismarck, and the final creation of a German empire.
But that he is one of the makers of modern Germany has come to be more and
more appreciated with the passage of time. And he is also, alas, one of the
earlier exponents of that Greater Germany which has become the nightmare
of the civilized world.[68]

[66] This question has been debated with much heat. See Professor K. T. Eheberg's historical and critical introduction to the seventh edition of *The National System* (Stuttgart,
1853) for the viewpoint that Hamilton had little or no influence on List. *Contra* see Hirst,
op. cit., pp. 111-118 and, more especially, Ugo Rabbeno, *American Commercial Policy*
(London, 1893), an English translation of *Protezionismo Americano: Saggi Storizi di
Politico Commerciale* (Milan, 1893). Essay III, Chapters I (on Hamilton) and II (on
List), of Rabbeno's work is perhaps the fairest summary of the question.

[67] List firmly believed that the United States would, within a century, surpass Britain in
industry, wealth, commerce, and naval power. *The National System*, pp. 40, 77-86, 339.

[68] List has been adopted by the expansionists, the Pan-Germans, and even the Nazis as a
patron saint. For a characteristic pamphlet of the First World War see Karl Kumpmann,
Friedrich List als Prophet des neuen Deutschland (Tübingen, 1915). For the present day

VII

The primary concern of List's policies, both political and economic, was power, even though he linked power with welfare. In this respect, despite all his denials to the contrary, he was reverting to mercantilism. "A nation," he wrote, "is a separate society of individuals, who, possessing common government, common laws, rights, institutions, interests, common history, and glory, common defence and security of their rights, riches and lives, constitute one body, free and independent, following only the dictates of its interest, as regards other independent bodies, and possessing power to regulate the interests of the individuals, constituting that body, in order to create the greatest quantity of common welfare in the interior and the greatest quantity of security as regards other nations.

"The object of the economy of this body," he continued, "is not only wealth as in individual and cosmopolitical economy, but power and wealth, because national wealth is increased and secured by national power, as national power is increased and secured by national wealth. Its leading principles are therefore not only economical, but political too. The individuals may be very wealthy; but if the nation possesses no power to protect them, it and they may lose in one day the wealth they gathered during ages, and their rights, freedom, and independence too."

Furthermore, "as power secures wealth, and wealth increases power, so are power and wealth, in equal parts, benefited by a harmonious state of agriculture, commerce and manufactures within the limits of the country. In the absence of this harmony, a nation is never powerful or wealthy." Hence productive power is the key to national security. "Government, sir, has not only the right, but it is its duty, to promote every thing which may increase the wealth and power of the nation, if this object cannot be effected by individuals. So it is its duty to guard commerce by a navy, because the merchants cannot protect themselves; so it is its duty to protect the carrying trade by navigation laws, because carrying trade supports naval power, as naval power protects carrying trade; so the shipping interest and commerce must be supported by breakwaters—agriculture and every other industry by turnpikes, bridges, canals and rail-roads—new inventions by patent laws—so manufactures must be raised by protecting duties, if foreign capital and skill prevent individuals from undertaking them."[69]

Wealth is of no avail without the "unity and power of the nation." Thus modern Germany, failing to achieve either political unification or a "vigorous

see the best-selling novel *Ein Deutscher ohne Deutschland: ein Friedrich List Roman*, by Walter von Molo (Berlin, Vienna, Leipzig, 1931 and subsequent editions). This novel is valuable not as historical fiction but as an example of the Pan-German and Nazi mentality —bitterly hostile to Britain and France, patronizing toward the United States (whose independence is accredited to the military genius of Steuben), contemptuous of Austria. Von Molo makes many unsupported assertions, some of them inherently improbable, concerning the influence of List on Andrew Jackson, von Moltke, and others.

[69] *Outlines*, in *Schriften, Reden, Briefe* (hereafter cited as *Works*), II, 105-106. The similarity of the idea of harmonious interests to Hamilton's views on the same subject is obvious. See also *ibid.*, p. 374, note, in which the editor, Dr. Notz, relates List's doctrine not only to Hamilton but also to Daniel Raymond, Mathew Carey, and John C. Calhoun.

and united commercial policy," was for many generations unable to maintain the position among the nations to which her civilization entitled her and was "made a convenience of (like a colony)." Germany was several times "brought to the brink of ruin by free competition with foreigners, and thereby admonished of the fact that under the present conditions of the world every great nation must seek the guarantees of its continued prosperity and independence, before all other things, in the independent and uniform development of its own powers and resources."

Tariffs and other restrictive devices designed to develop such powers and resources "are not so much the inventions of mere speculative minds, as the natural consequence of the diversity of interests, and of the strivings of nations after independence or overpowering ascendancy"—in other words, the war system. "War or the very possibility of war makes the establishment of a manufacturing power an indispensable requirement for any nation of first rank." Just as it would be the height of folly for a state to "disband its armies, destroy its fleets, and demolish its fortresses" in the modern world, so it would be ruinous for a nation to base its economic policy on an unwarranted assumption of a state of perpetual peace and world federation which exists only in the minds of the free trade school.[70] The ability of a nation to wage war is measured in terms of its power to produce wealth, and it is the greatest possible development of productive power which is the goal of national unification and protectionism. Protectionist policies may for a time—but only for a time— result in a lower standard of living, because tariffs necessarily involve higher prices. But those who argue that cheapness of consumers' goods is a major consideration in weighing the advantages of foreign commerce "trouble themselves but little about the power, the honour, or the glory of the nation." They must realize that the protected industries are an organic part of the German people. "And who would be consoled for the loss of an arm by knowing that he had nevertheless bought his shirts forty per cent cheaper?"[71]

The greater the productive power, the greater the strength of the nation in its foreign relations and the greater its independence in time of war. Economic principles, therefore, cannot be divorced from their political implications:

"At a time where technical and mechanical science exercise such immense influence on the methods of warfare, where all warlike operations depend so much on the condition of the national revenue, where successful defence greatly depends on the questions, whether the mass of the nation is rich or poor, intelligent or stupid, energetic or sunk in apathy; whether its sympathies are given

---

[70] *Le système naturel d'économie politique* (1837), Chap. II, in *Works*, IV, 186. *The National System*, pp. 87, 91-92, 102-107. The reader need not be reminded that Adam Smith did not base his system upon any assumption of universal peace or a federation of the world. List himself, on some occasions, said that the ultimate goal of all society was a world state, although he was too much of a nationalist to be an evangelist for the idea.

[71] *Ibid.*, pp. 119, 140. Compare List's idea of productive power with Adam Smith's statement that the power to wage war is measured by "the annual produce of [a nation's] industry, from the annual revenue arising out of its lands, labour, and consumable goods." Above, Section II. See also Jefferson as regards price, Section V, and Hamilton as regards self-sufficiency in war time, Section IV above.

exclusively to the fatherland or partly to foreign countries; whether it can muster many or but few defenders of the country—at such a time, more than ever before, must the value of manufactures be estimated from a political point of view."[72]

List had a keen appreciation of the factors which enter into the military potential.

"The present state of the nations," he wrote, "is the result of the accumulation of all discoveries, inventions, improvements, perfections, and exertions of all generations which have lived before us; . . . and every separate nation is productive only in the proportion in which it has known how to appropriate these attainments of former generations and to increase them by its own acquirements, in which the natural capabilities of its territory, its extent and geographical position, its population and political power, have been able to develop as completely and symmetrically as possible all sources of wealth within its boundaries, and to extend its moral, intellectual, commercial, and political influence over less advanced nations and especially over the affairs of the world."[73]

From any such beliefs it is an easy step toward a policy of territorial expansion on the continent of Europe and colonial expansion overseas, and List did not hesitate to take the step. He wanted a unified Germany to hold sway from the Rhine to the Vistula and from the Balkans to the Baltic. He believed that "a large population and an extensive territory endowed with diversified natural resources are essential requirements of normal nationality; they are the fundamentals of the spiritual structure of a people, as well as of its material development and political power. . . . A nation restricted in population and territory, especially if it has its distinctive language, can possess only a crippled literature, only crippled institutions for promoting the arts and sciences. A small state can never bring to the fullest state of development its diversified productive resources." Hence small nations will maintain their independence with the greatest difficulty and can exist only by tolerance of larger states and by alliances which involve a fundamental sacrifice of national sovereignty.[74]

The foregoing is not very different from present-day German definitions of Lebensraum, as will be obvious from List's program for a Greater Germany. He advocated the inclusion in a unified Germany of Denmark, the Netherlands, Switzerland, and Belgium—the first three on grounds of race and language, as well as on grounds of economics and strategy. As regards Denmark, Belgium and the Netherlands, they were required because it was essential that Germany control the mouths of German rivers, plus the entire seacoast from the mouth of the Rhine to East Prussia, thus assuring the German nation "what it is now in need of, namely fisheries and naval power, maritime commerce and colonies." The acquisition of these three countries, together with Switzerland, furthermore, would assure Germany the natural boundaries of seas and mountains

---

[72] Ibid., pp. 168-169; also 118-119.   [73] Ibid., pp. 113-114.
[74] Ibid., p. 142. In this instance Lloyd's translation seems unsatisfactory and I have rephrased it in certain essential respects. For the German original see Works, VI, 210-211.

which are essential on both economic and military grounds.[75] Germany should likewise begin peaceful penetration of the Danubian territories and European Turkey. These areas were Germany's natural frontier, or *Hinterland*, and she had "an immeasurable interest that security and order should be firmly established" there.[76]

A nation should "possess the power of beneficially affecting the civilisation of less advanced nations, and by means of its own surplus population and of its mental and material capital to found colonies and beget new nations." When a nation cannot establish colonies, "all surplus population, mental and material means, which flows from such a nation to uncultivated countries, is lost to its own literature, civilisation, and industry, and goes to the benefit of other nationalities." This is notoriously true as regards German emigration to the United States. "What good is it if the emigrants to North America become ever so prosperous? In their personal relation they are lost for ever to German nationality, and also from their material production Germany can expect only unimportant fruits. It is a pure delusion if people think that the German language can be maintained by the Germans who live in the interior of the United States, or that after a time it may be possible to establish German states there." Hence the conclusion is inescapable that Germany must have colonies of its own, in southeastern Europe and in Central and South America. And such colonies should be supported by all the resources of the nation, including state-sponsored colonization companies and "a vigorous German consular and diplomatic system."[77]

List knew full well that his program for continental expansion and overseas colonies could not, in all probability, be realized without war. The advocates of a national system for Germany were aware, he wrote in a bitter polemic against *The Times* of London, that the future might bring national wars but they were therefore the more determined to mobilize the moral and material resources of the German nation in support of a national economy.[78]

It was England, of course, which stood in the way of German ambitions. She was the leading exponent of the balance-of-power policy which mobilized "the less powerful to impose a check on the encroachments of the more powerful." She stood virtually unchallenged in her position as an imperial power, which she had achieved by the development of her manufactures. Hence, "if the other European nations wish also to partake of the profitable business of cultivating waste territories and civilising barbarous nations, or nations once civilised but which are again sunk in barbarism, they must commence with the development of their own internal manufacturing powers, of their mercantile marine, and of their naval power. And should they be hindered in these endeavours by England's manufacturing, commercial, and naval supremacy, in the union of

[75] *Ibid.*, pp. 142-143, 216, 327, 332, 346-347. For some unexplained reason List was unimpressed by rivers as natural boundaries.
[76] *Ibid.*, p. 347. List said that it was better for Germans to emigrate to the Danube than to the shores of Lake Erie. For the frontier quotation see *Works*, V, 499-500.
[77] *Ibid.*, pp. 142, 216-217, 345-347.
[78] "Die Times und das deutsche Schutzsystem," *Zollvereinsblatt*, IV (1846), 693-694.

their powers lies the only means of reducing such unreasonable pretensions to reasonable ones."[79]

It was England, also, which stood like a colossus astride the sea lanes of the world, making it difficult for any other nation to achieve the sea power which was necessary to the fulfillment of its destiny. In a statement on British control of the seas which would do credit to Admiral Mahan, List wrote:

"England has got into her possession the keys of every sea, and placed a sentry over every nation: over the Germans, Heligoland; over the French, Guernsey and Jersey; over the inhabitants of North America, Nova Scotia and the Bermudas; over Central America, the island of Jamaica; over all countries bordering on the Mediterranean, Gibraltar, Malta, and the Ionian Islands. She possesses every important strategical position on both the routes to India with the exception of the Isthmus of Suez, which she is striving to acquire; she dominates the Mediterranean by means of Gibraltar, the Red Sea by Aden, and the Persian Gulf by Bushire and Karachi. She needs only the further acquisition of the Dardanelles, the Sound, and the Isthmuses of Suez and Panama, in order to be able to open and close at her pleasure every sea and every maritime highway."[80]

In view of Great Britain's overwhelming naval, commercial, and colonial strength, no single nation could successfully challenge her without powerful assistance from others. "The nations which are less powerful at sea can only match England at sea by uniting their own naval power"; hence every such nation "has an interest in the maintenance and prosperity of the naval power of all other nations"; and, together, they should "constitute themselves into one united naval power" for the purpose, among other things, of preventing undisputed control by Great Britain of the sea lanes of the world (especially those of the Mediterranean).[81] The part of wisdom would be for the continental nations to form a European bloc to check British power: "If we only consider the enormous interests which the nations of the Continent have in common, as opposed to the English maritime supremacy, we shall be led to the conviction that nothing is so necessary to these nations as union, and nothing is so ruinous to them as Continental wars. The history of the last century also teaches us that every war which the powers of the Continent have waged against one another has had for its invariable result to increase the industry, the wealth, the navigation, the colonial possessions, and the power of the insular supremacy [of Britain]."[82]

But List's strategical thinking never had parochial, or even continental, limits. Gazing far into the future, he saw the day when the Stars and Stripes, not the Union Jack, would wave over the seas, and when effective measures

---

[79] The National System, pp. 216-217, 330.

[80] Ibid., p. 38. As regards Panama, for the possession of which Britain was then contending with the United States, List proposed an internationalized waterway under German entrepreneurs: "Der Kanal durch die Landenge von Panama, ein Unternehmen für die Hansestädte," in Works, VII, 234-236.

[81] Ibid., pp. 332, 337.

[82] Ibid., p. 338.

would have to be taken by the other nations of the earth to curb the power of the United States:

"The same causes which have raised Great Britain to her present exalted position, will (probably in the course of the next century) raise the United States of America to a degree of industry, wealth, and power, which will surpass the position in which England stands, as far as at present England excels little Holland. In the natural course of things the United States will increase their population within that period to hundreds of millions of souls; they will diffuse their population, their institutions, their civilisation, and their spirit over the whole of Central and South America, just as they have recently diffused them over the neighbouring Mexican province. The Federal Union will comprise all these immense territories, a population of several hundred millions of people will develop the resources of a continent which infinitely exceeds the continent of Europe in extent and in natural wealth. The naval power of the western world will surpass that of Great Britain, as greatly as its coasts and rivers exceed those of Britain in extent and magnitude.

"Thus in a not very distant future the natural necessity which now imposes on the French and Germans the necessity of establishing a Continental alliance against the British supremacy, will impose on the British the necessity of establishing a European coalition against the supremacy of America. Then will Great Britain be compelled to seek and to find in the leadership of the united powers of Europe protection, security, and compensation against the predominance of America, and an equivalent for her lost supremacy.

"It is therefore good for England that she should practise resignation betimes, that she should by timely renunciations gain the friendship of European Continental powers, that she should accustom herself betimes to the idea of being only the first among equals."[83]

Friedrich List's views on England are an interesting study in psychology, perhaps more especially of German psychology. List enormously admired and envied Britain and British liberal institutions, and few men of any nationality have ever paid her more eloquent tributes. On the other hand, he feared and even hated Britain. He himself suffered from a persecution complex—arising out of petty ways in which he was harassed by official Germany—and it was therefore not surprising that he believed that Britain was actively engaged in frustrating the *Zollverein* and other steps toward German unification. Always cantankerous, he became involved in particularly vitriolic controversies with Englishmen—especially, of course, with the long-deceased Adam Smith and his living followers. At the very end of his life, on the other hand, he went to England in the vain hope of paving the way for an Anglo-German alliance. He prepared an elaborate memorandum on the subject which he submitted to Prince Albert, Sir Robert Peel (the prime minister), Lord Clarendon (the

---

[83] *Ibid.*, pp. 339-340. The same theme is developed at some length in a remarkable document written shortly before List's death in 1846: "Über den Wert und die Bedingungen einer Allianz zwischen Grossbritannien und Deutschland," *Works*, VII, 267-298. See also "Die vorige und die gegenwärtige Regierung von Nordamerika," *Staatslexikon* (1841), pp. 219 ff.

foreign secretary), and the King of Prussia. He had encouragement from de Bunsen, the Prussian ambassador in London, and from some British sources. But Peel could not accede to the plan, and List returned to Germany in the autumn broken in health and in spirit—on the verge of the suicide which occurred November 30, 1846.[84]

There are some fantasies in List's memorandum on the value and the conditions of an Anglo-German alliance, but it nevertheless reveals an acute appreciation of some of the strategic realities facing both countries in the middle of the nineteenth century. To begin with, List foresaw what Sir Halford Mackinder was to elucidate more than half a century later, that there was nothing eternal about British maritime supremacy. The development of steam railways and steam navigation, he thought, might give the continental powers advantages in relation to the British Isles which they did not then possess. The rising power of other nations, especially the United States, held the possibility that control of the seas might be threatened; without control of the seas, the unique advantages which Britain enjoyed from her insular position would become serious liabilities. List foresaw also the union of the Latin and Slavic races, through a Franco-Russian alliance, and believed that Britain and Germany should counterbalance any such combination by taking the lead of the Germanic peoples. He was convinced that Franco-Russian power would not only threaten Britain's interests in Europe and the east but would almost certainly crush Germany. Britain could use the help of a continental land power and Germany would welcome reinforcement from an insular sea power. All that Germany asked of Britain was sympathetic understanding and support for a moderate protective tariff in a unified Germany, which seemed to List a small price for Britain to pay for German friendship. Any such concession, List foresaw, would be resisted by the vested interests of British industry but, against these, Britain must set the fact that her position as a world power would be fortified and even extended.

List failed, as so many others have failed, to find a formula which would lead to Anglo-German solidarity because, for better or for worse, there has never been any agreement between the two nations on what constitutes a true community of interest and because so many moral and psychological factors have stood in the way of mutual understanding. He failed, also, because he could not undo in a few months the harm which he had done over the years by strident anti-British propaganda.

VIII

The greatest single contribution which List made to modern strategy was his elaborate discussion of the influence of railways upon the shifting balance of military power. He first became interested in railways during his residence in America, when he was one of the promoters of the Schuylkill Navigation, Railroad and Coal Company, a forerunner of the present Reading System.

[84] For the English mission see Hirst, *op. cit.*, pp. 97-106. For the memorandum on the proposed alliance, to be discussed in the next paragraph, see note 83.

Thereafter, railways were one of the passions of his life. His writings on railways fill two complete volumes and almost two pages of the index volume of his collected *Works*. During the years 1835 and 1836 he published *Das Eisenbahn Journal*, a magazine devoted to forwarding railway construction in Germany. To no other single cause did he give more devotion or more energy, for he saw, correctly, that a network of railways, ultimately incorporated in a truly national system, would be one of the forces which would cement German unification.

His interest in the economic effects of railways was to be expected, although he was much more foresighted than most of his contemporaries. But his understanding of the strategic implications to Germany of steam transportation is surprising and by any objective standards quite remarkable. Before the advent of the railway the strategic position of Germany was the weakest in Europe, with the result that she was the traditional battleground of the entire continent. List saw sooner than anyone else that the railway would make the geographical situation of Germany a source of great strength, instead of one of the primary causes of her military weakness. With political unification fortified by a nation-wide link of railway communications, Germany could be made into a defensive bastion in the very heart of Europe. Speed of mobilization, the rapidity with which troops could be moved from the center of the country to its periphery, and the other obvious advantages of "interior lines" of rail transport would be of greater relative advantage to Germany than to any other European country. In a word, List wrote, a perfect railway system would transform the whole territory of the nation into one great fortress, which could be readily defended by its entire combatant man power, with a minimum of expenditure and with the least disorganization of the economic life of the country. And after the conclusion of the war, the return of the troops to their homes could be brought about with equal facility and expedition. For all of these reasons, and others, List foresaw that the network of railway lines which he visualized for Germany in 1833—which is substantially that of the present *Reichsbahnen*—would enable the army of a unified Germany, in the event of invasion, to move troops from any point in the country to the frontiers in such a way as to multiply many fold its defensive potential and thus prevent the recurrent invasions which had been going on for over two hundred years. Ten times stronger on the defense, Germany also would be ten times stronger on the attack, should she undertake offensive war—which List thought unlikely.[85]

There was a note of urgency in List's pleas for railway construction in Germany. "Every mile of railway which a neighboring nation finishes sooner than we, each mile more of railway it possesses, gives it an advantage over us," he wrote. Hence "it is just as little left in our hands to determine whether we shall make use of the new defensive weapons given us by the march of progress, as it was left to our forefathers to determine whether they should shoulder

---

[85] For the 1833 plan see "Über ein sächsisches Eisenbahnsystem als Grundlage eines allgemeinen deutschen Eisenbahnsystems," in *Works*, Vol. III, Part 1, pp. 155-195. For the general strategic theory of railways see "Deutschlands Eisenbahnsystem in militärischer Beziehung," *ibid.*, pp. 260-270, the latter written 1834-1836.

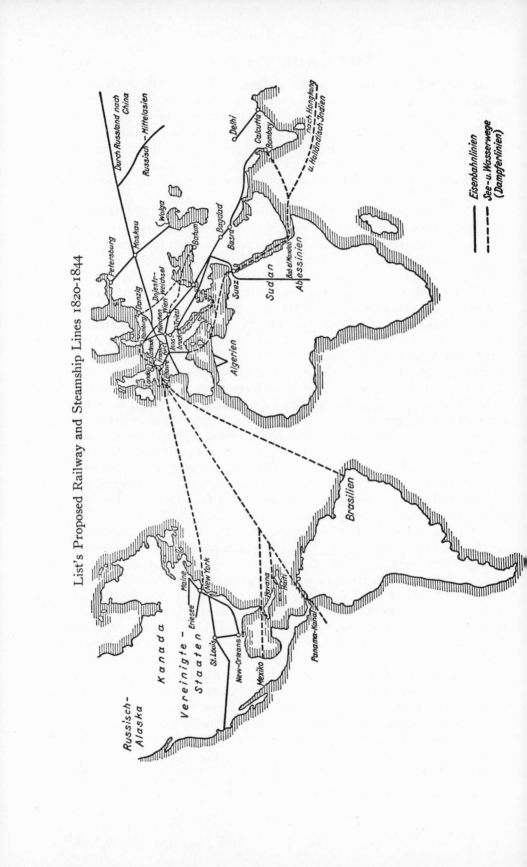

List's Proposed Railway and Steamship Lines 1820-1844

the rifle instead of the bow and arrow."[86] When it is considered that all of the foregoing was written before the American Civil War gave the first definitive proof of the military value of railways, it shows truly remarkable prescience.

List was wrong in thinking that railways would enable European states to reduce the size of their armies; on the contrary, as the Franco-German War subsequently showed, the railway simplified logistical problems and permitted the movement of larger armies, together with their astronomical quantities of munitions and supplies, than anyone had theretofore believed possible. List was also wrong in thinking that the construction of railways might render attack so costly to the attacker that the danger of war would be mitigated. But he was right in asserting that railway trackage and right of way were relatively less vulnerable to military destruction than many other permanent installations— a fact which has most recently been demonstrated in the German bombings of England and in Anglo-American aerial attacks on the continent.[87]

Even before Germany itself had a railway system, List's dreams went far beyond her borders into the rest of Europe and into Asia. In fact, he seems to have been the originator of the Bagdad Railway idea. In his project for an Anglo-German alliance he proposed that British communications with India and the Far East should be improved by railway lines extending from the English Channel to the Arabian Sea. The Nile and the Red Sea, he wrote, should be brought as close to the British Isles as the Rhine and Elbe were at the time of Napoleon; Bombay and Calcutta should be made as accessible as Lisbon and Cadiz. This could be accomplished by the extension of the projected Belgian-German railway systems to Venice, thence via the Balkans and Anatolia to the Euphrates Valley and the Persian Gulf and, finally, to Bombay. A Syrian spur would link the main line with Cairo and the Sudan. A telegraph line would parallel the railway, so that Downing Street would be in as easy touch with the East Indies as with Jersey and Guernsey. List also visualized a transcontinental line from Moscow to China.[88] None of these projects seemed to him any more ambitious or daring than the plans then being discussed in America for railways from the Atlantic to the Pacific.

To assure political security for the territories through which the proposed railways would pass, Germany and Great Britain should enter into an effective alliance defining their respective spheres of interest. The expansion of German rule over all of European Turkey would prevent interference by any power hostile to the British Empire—speaking in hyperbole, as he so often did, List said that "seventy or eighty millions" of Germans would constitute the guarantee which the situation required. Great Britain, on the other hand, should control all of Asia Minor, Egypt, Central Asia, and India—a vast territory which would more than compensate for the threat of a nascent American world power.[89]

[86] "Deutschlands Eisenbahnsystem . . . ," *op. cit.,* pp. 266-268.
[87] In addition to the foregoing see "Über ein allgemeines Eisenbahnsystem in Frankreich," *ibid.,* Vol. III, Part 2, pp. 564-573.   [88] See map, p. 150.
[89] For discussion of the railway to India see "Über . . . einer Allianz zwischen Grossbrittannien und Deutschland" cited in footnote 83. For details concerning the route of the

List's proposal concerning German control of European Turkey was, of course, closely connected with his desire to see large-scale emigration to the Danubian region and the Balkans. Indeed, all of his plans for railway construction were in some way linked with his passion for a unified and greater Germany. "A German railway system and the *Zollverein*," he wrote, "are Siamese twins. Born at the same time, physically knit together, of one spirit and one soul, they support each other and strive for the same great aim: the unification of the German tribes into one great, cultivated, wealthy, powerful, and inviolable German nation. Without the *Zollverein* no German railway system would ever have been even discussed, let alone constructed. Only with the aid of a German railway system is it possible for the social economy of the Germans to rise to national greatness, and only through such national greatness can a system of railways realize its full potentialities."[90]

IX

When List died in 1846, few of the causes to which he devoted his life were within reasonable hope of success. In 1846 Britain repealed the Corn Laws and the United States adopted the Walker Tariff, which seriously compromised the principles of autarky and protectionism and were, indeed, a step in the direction of free trade. Industrialization had proceeded but slowly in Germany and a German railway system existed only in blueprints. Conservatism and separatism continued to rule east of the Rhine, with the result that German national unification was not quite within reach. To be sure, List carried with him into another world the comfort of the *Zollverein*, a solid achievement for which he could justly claim a large share of credit. But it remained for historians to appreciate fully the importance of the *Zollverein* in the creation of the later German Empire.

Nevertheless, List's soul went marching on. Two years after his tragic death revolutionary movements swept Germany, giving birth to the hope that the German people would become a national state under liberal auspices—an event which List would have welcomed with all his heart, for he was an ardent believer in liberal, middle-class, constitutional government with adequate guarantees of individual liberty. But the liberal revolutions of 1848 failed and gave way to the policy of blood and iron. "German nationalists of conservative and traditionalist stamp could and did accept the economic teachings of List, while rejecting his political counsels [of liberalism and individual rights]; and an increasing number of German industrialists, regardless of nationalist or political bias, foresaw delightful solace for the woes of British competition in List's national programme. Even liberal nationalists of an ensuing generation, growing more in the grace of nationalism than in that of liberalism, came gradually

Constantinople-Bagdad-Basra-Bombay line see *Works*, Vol. III, Part 2, p. 679. The population of the German empire did not approach 70 millions until the eve of the first World War.

[90] "Das deutsche Eisenbahnsystem," in *Works*, Vol. III, Part 1, p. 347. Concerning railway expansion in the Danubian area: "Die Transportverbesserung in Ungarn," *ibid.*, pp. 434-460.

to agree with List's contentions. By 1880 the German national state, under Bismarck's nominal guidance, was actually treading the economic path which had been blazed by Friedrich List."[91]

In fact, Bismarck and his successors went even farther than List would have gone in the direction of economic nationalism and autarky. List had always opposed import duties upon foodstuffs. But the German tariff system as it developed under the empire was an all-inclusive plan giving protection both to the Junkers and to the industrialists, who were thus drawn together in support of economic nationalism, militarism, navalism, and colonialism. Whatever List might have thought of tariffs on grain, he could hardly have objected to the spirit and purposes of Chancellor Caprivi's statement to the Reichstag, December 10, 1891:

"The existence of the State is at stake when it is not in a position to depend upon its own sources of supply. It is my conviction that we cannot afford to dispense with such a production of corn as would be sufficient in an emergency to feed our increasing population . . . in the event of war. . . . I regard it as the better policy that Germany should rely upon its own agriculture than that it should trust to the uncertain calculation of help from a third party in the event of war. *It is my unshakable conviction that in a future war the feeding of the army and the country may play an absolutely decisive part.*"[92]

Much of the economic policy of the Second Reich was based upon the assumption that sooner or later Germany would be involved in a war to defend the realm and to win a recognized place in the sun. In preparation for such an eventuality German statesmen believed that they should depend upon Germany's inherent strength rather than upon the good will of her neighbors or the uncertainty of her overseas communications. The kaiser's statesmen may have been guilty of some distortion of List's ideas, but had List lived he would have understood full well the language which they spoke. And he also would have understood the autarkical motivation of the *Wehrwirtschaft* of the Nazis, however much he would have disapproved of Hitler's racial ideas and Himmler's disregard of individual rights.

List also, unhappily, laid the foundation for certain other basic concepts of Pan-Germanism and National Socialism, such as *Lebensraum*, the *Drang nach Osten*, naval and colonial expansion, the impermanency of frontiers, the permanent allegiance of the *Auslanddeutsche* to the fatherland, and the desirability of a continental bloc against Anglo-American power.

List, like Hamilton, was a leading figure in the revival of mercantilism in the modern world. Whatever may have been the virtues of mercantilism in the seventeenth and eighteenth centuries, its modern counterpart has been an incendiary force in a highly inflammable and explosive world. The new mercantilism is the more dangerous because it operates in our highly organized and closely integrated society. It is warp and woof with the war system. To a degree which

[91] C. J. H. Hayes, *The Historical Evolution of Modern Nationalism* (New York, 1931), pp. 272-273.
[92] Quoted by W. H. Dawson, *The Evolution of Modern Germany* (New York, 1908), p. 248. Italics are mine.

would have shamed the mercantilists of old, it has enlisted the power of the state for the further enhancement of state power. All of the old, familiar devices have been reinforced by a host of new ones in the form of quotas, boycotts, exchange controls, rationing, stockpiles, and subsidies. Out of the economic nationalism of the fifty years beginning in 1870 have come totalitarian economics, the totalitarian state, and totalitarian war, which are so inextricably interconnected that it has become impossible to tell which is cause and which is effect. In the name of national security, political authority has been extended into almost every domain of human activity.[93]

As an almost inescapable consequence of all this came the explosions of 1914 and 1939. One can understand them only with reference to the power concepts of nineteenth century Europe. The thinking of Adam Smith, Alexander Hamilton, and Friedrich List was conditioned by the fact that they were, respectively, British, American, and German. But in certain fundamentals of statecraft their views were surprisingly alike. They all understood that military power is built upon economic foundations and each of them advocated a national system of economics which would best meet the needs of his own country. That the world has come to grief as a result of neo-mercantilism is not necessarily their fault. For so long as nations continue to place their faith in unbridled nationalism and unrestricted sovereignty they will continue to rely upon whatever measures will, in their judgment, best guarantee independence and security.

[93] For further development of these ideas see E. M. Earle, "The New Mercantilism," *Political Science Quarterly*, XL (1925), 594-600. Also, with particular reference to totalitarian economics, A. T. Lauterbach, *Economics in Uniform: Military Economy and Social Structure* (Princeton, 1943), especially Chapters I-IV inclusive.

CHAPTER 7. Engels and Marx: Military Concepts of the Social Revolutionaries

BY SIGMUND NEUMANN

"T HE philosophers have only *interpreted* the world in various ways; the point, however, is to *change* it." This credo of Karl Marx in his *Theses on Feuerbach* (1845), at the beginning of his literary career, is a key to an understanding of the dynamics of the Marxian theory. It is primarily directed toward action. Theoretical analysis becomes nothing but spade work and preparation for the final revolutionary assault. To make the world revolution a reality, strategic considerations are therefore primary and fundamental. Hence Marx and Engels naturally gave unremitting attention to tactical problems and military considerations in all their writings.

Strangely enough, this crucial side of their teaching has been badly neglected in the literature on Marxism. Such an omission derives partly from the fact that the immense amount of material bearing upon strategic problems is scattered through all their writings and is not as easily available in one monumental work as is the case with *Capital*, the basic study of Marxian economic theory.

Of special importance for a comprehensive analysis of Marx and Engels as military strategists, apart from their pertinent historical sketches, are the rich Marx-Engels correspondence and their extensive journalistic writings. All these sources, still largely untapped, await systematic collection and elaboration; and they certainly deserve intensive study.

The complete disregard of Marxian strategy is, however, the result not only of such technical difficulties but even more of basic misconceptions in regard to their teachings. For the mind of the superficial observer concepts of military strategy and tactics seem to be alien to the spirit of these radical thinkers, whose declared policy was one of enmity toward the military machine, the military caste, and the military state; whose anticipated socialist order merged with the pacifist millennium; and whose position as "outsiders of the state" did not encourage a realistic consideration of military power and the planning of specific campaigns. And yet it would be utterly misleading to view these protagonists of the international class struggle as pacifists and unrealistic ideologists. A renewed emphasis on these military aspects of early Marxism may thus bring about a necessary correction of prevalent views.

Marxism superseded the earlier utopianism not only in a new "scientific" approach to social development but also in a realistic evaluation of political forces. The new teaching was meant to be eminently practical, an "applied science." While the succeeding generations were above all impressed by the theoretical edifice which Karl Marx and Friedrich Engels left behind them, the concrete analysis of historical problems seems to have been of equal interest to the two founders of Marxism. This "practical" side of the revolutionary teaching gains new significance at a time when the "Sixth Power," as Engels proudly called the Socialist International, has found a definite crystallization, however different from the original concept, in the U.S.S.R., controlling one-sixth of the world. Strategic considerations were the core of the political theory of the mid-nineteenth century revolutionaries.

It is not too much to state at the outset that the writings of Marx and Engels gain in significance and perspective while the twentieth century's pattern and problems of warfare become clear and fully developed. Marx and Engels can rightly be called the fathers of modern total war. What has long been recognized in the history of political organization and internal politics, namely, that the "totalitarian party" had its inception in the socialist movement, may be applied also to the field of military affairs. The proud discovery of Dr. Blau, a National Socialist strategist, that modern warfare is of a fourfold nature—diplomatic, economic, psychological, and only as a last resort military—was common knowledge to Engels and Marx. They were fully aware that military campaigns could be lost long before the first bullet was shot, that they would in fact be decided beforehand on the preliminary battlefronts of economic and psychological warfare. They certainly recognized that the many-fronted war was one and undivided and thus could be won or lost on the international battle line as well as by a nation's civil strife or within each citizen's faltering soul. War and revolution—unmistakably established as twin movements in our time—were at that early period seen in their fundamental and continuous interrelationship by these keen strategists of the world revolution.

Such a vista gave them a new insight into military affairs and into the character of modern revolutions such as none of their forerunners had fully grasped. What Marx and Engels called their dialectical approach to historical phenomena is nothing but this all-inclusive and dynamic view of the socio-political forces at work in the modern world. Such a comprehensive view, however, makes it impossible to single out clearly aspects of military strategy proper in the writings of Marx and Engels. To them war was fought with different means in different fields. In the words of the later militant syndicalist Georges Sorel, a general strike could become a "Napoleonic battle," just as the Crimean War could be regarded as a prelude to a great international civil strife. During the "promising" crises of 1857, Engels wrote to Marx: "A continuing economic depression could be used by astute revolutionary strategy as a useful weapon for a chronic pressure . . . in order to warm up the people . . . just as a cavalry attack has greater *élan* if the horses first trot five hundred paces before coming within charging distance of the enemy."

If one recognizes this essentially militant and activist nature of modern socialism, the roles of its leaders somewhat change in significance and Friedrich Engels deservedly gains in stature as compared with the master theorist, Karl Marx. Not only had Engels (as recent investigations have shown) actually written a good part of the historical studies hitherto attributed to Marx, but the "Carnot of the future revolution" also had a much greater insight into the actual forces at work in the world, thus foreseeing future trends and contributing, if only indirectly, to concepts and techniques of military strategy in decades to come.

In many respects the very opposite in character and temperament, Marx and Engels exemplify a friendship of an almost classic nature, thus refuting the frequent assertion of Marx's human coldness and aloofness. For a span of about forty years, the literary work of the one was complementary to that of the other. Theirs was a natural division of labor. Marx, revealing in his profound and searching work the stern intellectual tradition of his forebears, was clearly the better systematic thinker. Without him Engels' writings would have lacked direction and power of synthesis. Marx was probably also the better political strategist with a certain gift of sizing up a situation, especially in revolutionary moments—a quality which often kept his lifelong collaborator from hasty conclusions. But while the genuinely modest Engels freely consented to play second fiddle, his contribution was no less significant to their work as a whole. Since his early studies in England, and especially his groundbreaking book *The Condition of the Working Class in England*, he had helped to lay the foundations of the great socialist theory. All his life he untiringly brought together valuable material, selecting and combining it with a sure hand and a wealth of common sense. He had a feeling for what was in the air and for what promised results. His was a practical mind. Son of a Rhenish industrialist and for a good part of his life an *entrepreneur* in his own right (though against his own inclination) in mid-England's teeming city of Manchester, he knew from childhood the real nature of the rising factory system and was above all a man of action.

An enthusiastic rider and huntsman, he threw himself lightheartedly into his work even when it meant "leaping over the high fences of abstract thought." The somber Marx, who "struggled with the spirit of his time as Jacob wrestled with the angel and whose work came slowly to fruition," admired Engels' power. "He can work at any hour of the day, fed or fasting; he writes and composes with incomparable fluency." Engels himself said of his own style that, as with artillery, "each article struck and burst like a shell."

Such militant vocabulary was no mere play on words. Even in his most abstract writings, such as his *Anti-Dühring*, Engels made ample use of military terms and experiences, because he was by nature a soldier and warrior. Proud of his early military experiences in the service and above all as an active participant in the Baden insurrection during the German Revolution of 1848, he turned his attention chiefly to the study of military science during the many years of his exile in England, in order to be prepared for the coming revolution.

One may regret that the "general," as his friends jokingly nicknamed him, never had a chance to prove his mettle. Yet his influence on the tactics and strategy of the Russian Revolution are well known. Even his contemporary adversaries among the military experts respected his judgments. His articles on the Crimean War in the *New York Tribune* were attributed to the American General Scott (who, at the time, was running for the presidency). His pamphlet *Po and Rhine* was considered to be the work of the Prussian General von Pfuel.

Engels' writings in the field of military science are more numerous than the rest of his literary work. If these scattered publications had been brought together in one or more impressive volumes, this work would probably have been just as carefully studied by the military experts as Karl Marx's *Capital* has been by the learned opponents of Marxism. Engels wrote careful treatises on military campaigns and detailed studies on military techniques, thumbnail biographical sketches of military leaders and sharp reviews of books in the field of military science. Throughout all his work he shows an amazing familiarity with the deeds and writings of the great military strategists in history. At the same time his independent and original judgment is surprising. In his analyses of specific military events he was more farseeing than many of the better known military experts and even his journalistic treatments of military affairs are still of value.

One may say of his military writings what a critic once said about Clausewitz: "He is a genius in criticism. His judgments are as clear and weighty as gold. He shows how greatness in strategic thought consists in simplicity." Clausewitz, in fact, who so deeply influenced master minds of German military strategy, also impressed Engels greatly. Thus he wrote to Marx on September 25, 1857: "Among other things I am now reading Clausewitz' *On War*. A strange way of philosophizing but very good on his subject. To the question whether war should be called an art or a science, the answer given is that war is most like trade. Fighting is to war what cash payment is to trade, for however rarely it may be necessary for it actually to occur, everything is directed towards it, and eventually it must take place all the same and must be decisive."

The emphasis on decisive action and tactical offensive even on the defensive became the stock-in-trade of revolutionary strategy. Significantly enough, it was fully applied not only in the victorious campaigns of the Red Army during the long civil war but also in its heroic fight against the National Socialist assault in the renewed World War.

Militancy and preparedness for offensive action remained axiomatic for this soldier of the revolution and, under his influence, also for Marx throughout their lives. Beyond these fundamental concepts, however, there was a definite progress in the military thought of these social revolutionists—a development which led them to an increasingly more realistic, more circumspect, more dynamic interpretation of the military and political history of their time. Thus one may easily detect three stages in the unfolding of their thoughts. Beginning with the analysis of the *Civil War Tactics* of 1848 and their lessons in close investigations of the *military strategies* of the great powers in the 'fifties and 'sixties, they finally progressed to an original inquiry into the nature and concepts of the *revolutionary state*—studies which, in their combination of the experiences of revolutionary tactics in civil war and the military strategies of international conflagrations, approached the patterns of modern total war.

II

The Revolution of 1848, as is so often the case with lost causes in history, has been misjudged and underestimated in its spirit and performance. Different from the middle class pacifism of the post-Napoleonic era of reaction, which was in deadly fear of the "revolution," born and kept aflame on the battlefields, the radicalism of 1848 was eminently militant. It was an echo of the great tradition of 1793.

Marx did not mistake the essential militancy of mid-nineteenth century Europe even where it was dressed in peaceful attire. "It would be a great error," he wrote on January 28, 1853, in the *New York Tribune*, "[to think] that the gospel of peace as presented by the Manchester school has any philosophic significance. All it indicates is the substitution of a mercantile for a feudal method in the art of warfare—capital for cannons."

On the continent, to be sure, the movements of 1848 ended in utter defeat. After a successful beginning hopeless divisions soon split the revolutionary forces and a politically immature middle class succumbed to an experienced ruling caste. Thus the revolutionary dynamics faded away without visible result. And yet this civil war in Europe was a military event of great import. It was fought in Germany on the barricades and on the field of battle. The rebels were often led by trained officers, who had gone over to the revolutionaries, for the Prussian and Austrian armies were not free from what in the twentieth century would have been called "Bolshevik" influences.

Among the condottieri of the revolution there were colorful soldiers like the adventurous Otto von Corvin, some of whom later proved their military talents. George Weydemeyer, one of the first followers of Marx and Engels, had been a Prussian artillery officer and, after his emigration to the United States, distinguished himself as a colonel in the northern army during the Civil War. Wilhelm Rüstow, the Prussian staff officer who had turned revolutionary, won great military repute among the experts as a colonel, as an author and teacher in the military academy of the Swiss army, and afterward as chief of staff to Garibaldi in the conquest of Sicily and the march on Naples. In fact, official military circles, the contemporary military literature amazingly shows, looked upon the fighters of the barricades, however small their number, as a superior and insurmountable power, as puzzling to the professional military class as were the Riffs to the colonial armies of twentieth century Europe. Cavaignac, who first succeeded at Paris, in June 1848, in breaking the myth of the barricades, was celebrated as the great military genius of the century. The full military power of the Prussian army was required to defeat the Baden insurrectionists in a field campaign.

The movement of 1848, in spite or because of its failure, became the starting point of scientific socialism and the great heritage of its master theorists. The inquiry into its meaning—its military strategy and its historical background—represented the central theme of the writings of Marx and Engels during the first years of their exile. The lessons of defeat were to reveal the laws of a future strategy of insurrection. Such laws were first elaborated and clarified in the brilliant analyses of the revolutions of 1848-1849 in central Europe written by Engels and edited by Marx, in whose name this series of articles was published in the *New York Tribune* in 1851-1852. The most significant statement of this keen and realistic study on *Germany: Revolution and Counter-Revolution* was again brought into the historical limelight in Lenin's letter from Finland to the Bolsheviks in Petrograd on the eve of their revolution in 1917. "Insurrection is an art as much as war . . . and subject to certain rules of procedure. . . . Firstly, never play with insurrection unless you are fully prepared to face the consequences of your play. . . . Secondly, the insurrection-has given you. . . . In the words of Danton, the greatest master of revolutionary ary career once entered upon, act with the greatest determination and on the offensive. The defensive is the death of every armed rising. . . . Surprise your antagonist. . . . Keep up the moral ascendancy which the first successful rising policy yet known, 'De l'audace, de l'audace, encore de l'audace !' "

The full implications of the Marx-Engels concepts of revolutionary tactics can only be understood against the background of their complete philosophic system, based upon the "materialistic interpretation of history," and its emphasis on the prevailing economic conditions as a key to an understanding of the sociopolitical dynamics. In the *Communist Manifesto* this theory had been applied in rough outline to modern history in its entirety. It was also drawn on in the numerous essays dealing with contemporary affairs. In the light of such an interpretation, the rise and fall of the popular movements of 1848 were in the last analysis determined and conditioned by economic causes. In the words of Engels (in his introduction to a republication in London in 1895 of Marx's *The Class Struggles in France 1848-50*) : "The world commercial crisis of 1847 was the real cause of the February and March revolutions, and the industrial prosperity which arrived gradually in the middle of 1848, coming to full bloom in 1849 and 1850, was the vitalizing fact of the renascent European reaction. This was decisive." By the same token he stated, "A new revolution is possible only as a consequence of a new crisis, and it is also as certain as the latter."

The approach of a new economic crisis meant to Marx and Engels the clarion call of the revolution. Thus the depression of 1857 was judged by them as an indication of an impending revolutionary situation. Engels was delighted by the thought that he would perhaps soon be able to leave the exchange for the battlefield and his office stool for a horse. "Now our time is coming— this time it is coming in full measure : a life-and-death struggle. My military studies will at once become more practical. I am throwing myself immediately into the tactics and organization of the Prussian, Austrian, Bavarian and French armies. And apart from that I do nothing but ride, that is hunt; for hunting is a real cavalry school." But the "chronic crisis" did not lead to revolution and war.

While the master strategists no doubt failed in evaluating the real forces at work, they had introduced an important element into all future revolutionary planning. *Timing* became the yardstick of masterful strategy. The pupils of

Marx and Engels in the Soviet Revolution fully applied this lesson at home; the Third International obviously did not.

When the revolutionary situation had passed, Marx and Engels vigorously pointed out that any attempt at playing at revolution was futile and dangerous. Throughout all these years the workers were cautioned against an attempt at a *putsch* which would only play into the hands of the reactionary forces. Thus the master strategists fought against Schapper and Willich, who were agitating for such a premature enterprise in the early 'fifties. Instead of such a hopeless rising, they insisted on a strategy of long-range preparedness during this quasi-peaceful intermission. However impatiently Engels awaited the time when he could mount the horse again for "that great duel to the death between bourgeoisie and proletariat," he knew too well that the greatest danger for every such enterprise lay in its desire for action. Timing and patience became the main demand of sound strategy.

While Marx and Engels thus guarded themselves against the characteristic pitfalls of an *emigré* existence, they turned their exile into a challenging and productive experience. The first decade of their London exile became a period of *weltpolitische Lehrjahre* of Marxism. Here Marx and Engels entered the world-wide events of nineteenth century middle class civilization. Separated now from the local, particularistic, and limited aspects of the German small states and French petty politics, the two penetrating thinkers could gaze upon a broader vista. "Nothing but an objective account of the totality of all mutual relationships of all the classes of a given society . . . can serve as the basis for the correct tactics of the advanced class." Such an "objective account" in regard to social forces is made in Marx's masterful study *The Eighteenth Brumaire*. The tactical lesson of this great defeat of the French Revolution at the hands of Napoleon the Little is seen by Marx in the need for developing the "democratic energy of the peasantry." Engels comes to the same conclusion in his contemporary study of the *German Peasant War*. "The whole thing in Germany will depend on the possibility of backing the proletarian revolution by some second edition of the Peasant War," Marx stated in his correspondence with Engels. From now on the peasantry as a possible ally or driving force in the coming social revolution played a major part in their considerations. Especially the prospects in the Russian social scene were almost exclusively measured in terms of the fate of the peasanty. The emancipation of the serfs was hailed as a turning point in political history and a new line-up of revolutionary forces. "At the next revolution," Marx wrote, as commander in chief of the world revolution issuing Napoleonic commands from his wretched home in London, "Russia will kindly join the rebels." Henceforth a Russian revolution became a permanent factor in their political speculations. From this realization a direct line led to the Soviet upheaval of 1917. The armies, recruited primarily from the peasantry, had defeated the revolutions of 1848 everywhere; an alliance with the revolutionary peasants saved the Soviet Russia of Civil War days. This was the lesson of the victorious revolution and of its intellectual pathfinders.

Even more significant for the development of revolutionary thought was the turning of the fathers of socialism toward the study of international affairs in general. They soon began to realize that the revolution of 1848 had failed to a large extent because of its international implications. In fact, from the early days of the *Neue Rheinische Zeitung*, to which Marx was called as editor "to produce the most radical, the most spirited, and the most individual journalistic enterprise of the first German revolution," the two friends had realized how closely foreign policy and internal affairs were connected; they also saw that the future of the European revolution would not be determined by the efforts of one country alone. Such a realization necessitated close attention to the problems of socialism and foreign politics and a serious consideration of military affairs for a realistic revolutionary strategy. It is a major contribution of Marx and Engels, often overlooked by their interpreters, that they raised the social dynamics of their time beyond the insurrectionary stage of the isolated *putsch* to the plane of world politics.

III

Marxist strategy reaches its second stage during the early 'fifties. Strange as it may seem, it is in these years of their exile that the two expatriates discover their own national ties. Engels, no doubt, is more outspoken in his expression of deeper loyalties and of sincere patriotism; but even Marx, often unconsciously, reveals definite national biases in his attacks on his political adversaries. What is more important, the socialist leaders now begin to take full stock of national individuality and its growing importance in international affairs. They carefully note the awakening nationalism in central and eastern Europe and, in fact, expect from these movements of independence a renewal of revolutionary impulses which will destroy the political apathy of the reaction in mid-nineteenth century Europe.

Typical of such hopes were Engels' great expectations for the Hungarian revolution under the leadership of Kossuth, whom at that time (in contrast to his later opinion) he regarded as "a combination of Danton and Carnot." It has been suggested that the daily reports of the military campaign in Hungary, which Engels wrote for the *Neue Rheinische Zeitung*, awakened in him his lifelong interest in the problems of the general staff officer.

Internationalists as they claimed to be, the socialist leaders began to think in terms of international power politics long before the spokesmen of the middle class parties emancipated themselves from narrow provincialism. Thus every political action in whatever country it might occur was viewed in terms of the larger European issues. Such international orientation, to be sure, was at first utterly dogmatic and only a rough approach to reality. Political divisions were simply drawn according to the formula of the two Europes: reaction versus revolution, czarism versus the progressive west. For a long time France was regarded as the revolutionary homeland. An alliance of the western powers to fight Russia, a war between Jacobin France and the Holy Alliance—

that was the international policy which Marx and Engels had strongly recommended in 1848. When the expected clash between east and west finally came in Crimea, it was, however, a conflict between the Czar and the usurper Napoleon, with Britain supporting France. Still, the Marxist strategists were hopeful that in time the war would release the forces of revolution.

The Crimean War was the first occasion for Engels to analyze in detail the military problems of the time. Now he even tried to make the study of military science his life profession; but he failed to find a desired position with the London *Daily News*. The only outlet for his extraordinary knowledge became the articles which Karl Marx regularly contributed to the *New York Tribune*. They showed mastery of technical material and keen strategic judgment, and they were accordingly praised by the American public as stirring contributions from expert hands.

At the beginning of the war, Engels had great hopes for quick and energetic action on the part of the allied forces in the Black Sea and (in combination with Sweden and Denmark) in the Baltic, which would lead to the destruction of the Russian navy and the conquest of all naval fortifications. "The giant without eyes" would thus be forced to his knees by a great pincer movement, and an impending internal revolution would soon end the hothouse development of Czarist Russia. But the undecided attitude of Prussia and Austria created extraordinary difficulties in military strategy. Their neutrality prevented great land battles; Austria's mobilization neutralized a substantial part of the Russian army; and the hope for active Hapsburg participation delayed allied action for five months. Engels regarded such a delay as a tactical blunder but, with Marx, also suspected in Palmerston a secret ally of "his friend Czar Nicholas," following in this respect the lead of the much talked about Scottish monomaniac Urquhart.

A careful analysis of the organization and tactical qualities of the combatant armies, however, never left any doubt in the mind of Engels about the superiority of the allied nations. Up to the battle of Inkerman, the supremacy of their artillery and cavalry had been proved; now the Russian infantry, victorious though it had been as a mass army, showed its inability to cope with modern military techniques and the tactical movements of small detachments. Many years later, Engels characterized the Crimean War to the Russian economist Danielson as "a hopeless struggle between a nation with a primitive technique of production and others which were up-to-date." Confidence in an allied victory did not prevent Engels from sharp criticism of the organization of the English army. The scandalous lack of food, clothing, and medical care had also aroused the anger of the British public. True to the Marxian interpretation of history, Engels put the chief blame on the ruling classes.

An important feature of the Crimean War was its emphasis on fortifications and siege warfare. To a superficial observer this fact might have indicated a change in the art of war, a "slipping back" from the age of Napoleon to the age of Frederick the Great. Engels, however, did not come to any such conclusion. "Nothing could be less true," he stated after the fall of Sebastopol. "Today fortifications have no other importance than to be concentrated points in support of the movements of a field army. Their value is relative. They are no longer independent factors in military campaigns, but valuable positions which it might or might not be wise to defend to the last." For this reason, he concluded, the Russians had been equally right in avoiding an open battle and in considering the safety of their army more important than the abstract value of a fortress. After all, on the eve of the Crimean War, Engels had not only read the writings of the great military strategists since Napoleon—Jomini, Willisen, Clausewitz and others—but had also followed closely Napoleon's campaign in Russia. Thus he knew what difficulties the Allied forces had to face after the conquest of Crimea in order to come to grips with Russia. The problems of logistics in the land of the wide spaces seemed to be insurmountable, and the Allied desire for an early termination of the war was therefore understandable.

To such an impasse Engels' answer, however, was an appeal to revolutionary strategy. "A war of principle" seemed to him the solution for both the Allies and Russia, the one appealing to the revolutionary forces of rising nationalism in Germany, Poland, Finland, Hungary, Italy; the other to Pan-Slavism. These possibilities of an ideological warfare were a correct estimate of the situation, and Napoleon III himself later confessed to Queen Victoria that a continuation of the war would necessarily have led him to call to arms the peoples striving for independence. Much as Engels would have welcomed such a turn, neither Nicholas nor Napoleon was prepared to unloose dynamics of warfare which became decisive in the patterns of twentieth century militarism.

The end of the Crimean War in 1856 shattered Engels' hopes for greater revolutionary upheavals. It also hardened the opinions of the revolutionary strategists in respect to the danger of Bonapartism. Bonapartism and Pan-Slavism now became the major themes of Engels' strategic considerations. The fear of the rising expansion and national ambitions of Russia was, of course, inextricably mixed with the undiminished hate of her reactionary absolutism, whose military intervention had destroyed the revolution of 1848. But the bitter and most personal controversy between Marx and Vogt showed to what extent strategic thoughts of military security were at the base of the Marx-Engels fight against this "Russian Pan-Slavist" who would not care if "Bohemia, right in the heart of Germany, should become a Russian province." According to Engels, a German renunciation of Bohemia would mean the end of German national existence; the direct way from Berlin to Vienna would thus lead via Russia. Strategic, cultural, and economic considerations now convinced Engels that all parts in eastern and southeastern Europe which in the past had been won by Germany should remain German. He strongly rejected the dissolution of the great cultural nations and the creation of splinter states incapable of an independent national existence—and all that in the name of national self-determination. Dangers of twentieth century developments were thus well foreseen.

No less timely for modern discussion was Engels' fight against Bonapartism.

Its real strength and danger he rightly recognized in its demagogic appeal to the latent economic expansionism of a dissatisfied middle class and to the "patriotism" of the revolutionary masses. Engels carefully scrutinized the military implications of the Napoleonic ambitions in two excellent pamphlets *Po and Rhine* (1859) and *Savoy, Nice and the Rhine* (1860). In the first essay, he attacked the popular thesis of his day as it prevailed among the military experts (such as General von Willisen in his *Italian Campaign of the Year 1848*), namely, that the Rhine should be defended on the Po, which was regarded as an integral part of Germany. In an authoritative analysis of the courses of the upper Italian rivers and of the strategic position of the Italian fortifications (a study which would inflate the pride of contemporary geopoliticians) Engels proved that control of the Po Valley was not a military necessity for the defense of Germany's southern frontier. Moreover, he showed that, hidden behind so-called military arguments, the real motivation for such strategies were political ambitions for a renewal of the Holy Roman Empire and a German claim to become the arbiter of Europe. He specifically warned against an annexationist policy of a greater Germany whose "liberation" of weak neighbors would make her the most hated nation in Europe. The Rome-Berlin Axis and the New Order of a Hitler-Germany were thus rejected eighty years before their inauguration.

Even more astonishing was Engels' discussion of the possible strategy of a western campaign. Here he tried to prove that France, having fortified Paris, could now abandon her traditional claim to the left bank of the Rhine. Again, as he had done in the case of Germany-Austria's claims in northern Italy, Engels disproved, exclusively in terms of military evidence, the validity of the French plea for "a natural frontier." The strategy of French campaigns was directed primarily toward the defense of Paris, and justifiably so because the centralization of France made Paris the key of the country. The surrender of the capital spelled the doom of the empire. With the recent fortification of Paris, however, Vauban's threefold ring of fortifications was superfluous and meant only a useless diversion of military forces. Engels saw the real danger to French security in her weak Belgian frontier, because in spite of European treaties, "history has yet to show that in case of war Belgium's neutrality is more than a scrap of paper." On the basis of such a realistic evaluation Engels elaborated his plan for a successful military campaign. Based upon fortified Paris, France could defend herself offensively on the Belgian frontier. "If this offensive is repulsed the army must make a final stand on the Oise-Aisne line; it would be useless for the enemy to advance farther, since the army invading from Belgium would be too weak to act against Paris alone. Behind the Aisne, in unchallengeable communication with Paris—or at the worst behind the Marne with its left wing on Paris—the French northern army could take the offensive and wait for the arrival of the other forces." Fifty-five years later French *poilus* arrived in taxicabs, fulfilling Engels' prophetic prediction of the miracle of the Marne.

A decade later, during the Franco-Prussian War, Engels gave a similar test of his mastery of campaign strategy. In a series of articles written for the London *Pall Mall Gazette* he suggested in a detailed analysis the sudden shift of the Prussian army marching on Châlons toward the Belgian frontier, and thus was the only European observer to predict the strategy which led to Moltke's decisive victory at Sedan.

Savoy, Nice and the Rhine pointed at another element of military strategy the full meaning of which was not realized until the First World War: the specter of a two-front war resulting from a Franco-Russian alliance. "Has the Rhineland no other calling," he disgustedly exclaimed, "but to be cursed by a war in order to give Russia a free hand on the Vistula and the Danube?" Russia remained the main threat to European liberty, though Engels now harbored the vain hope that this danger would soon be checked by a new ally of the revolution, the liberated serfs. "The struggle which has now broken out in Russia between the ruling classes of the rural population and the ruled is already undermining the whole system of Russian foreign policy. The system was possible only so long as Russia had no internal political development; but that time is past."

Greater attention, on the other hand, was to be given to the plans of Napoleon III. Engels seriously studied the prospects of a French invasion of England and the defense of the British Isles. In this connection he published in two technical journals of military science (the *Darmstädter Allgemeine Militär Zeitung* and the *Volunteer Journal of Lancashire and Cheshire*) a number of articles dealing especially with the volunteer riflemen. Some of these articles were brought out in 1861 in pamphlet form: *Essays Addressed to Volunteers.* Despite his great sympathy for the riflemen and their less rigid system of drill, Engels came to the definite conclusion that they were no match for the newly increased French army, which he called the "best military organization in Europe."

The great event in the military history of the following years was the American Civil War. Contrary to most of the official military authorities, who, at the time, showed no interest in this long bitter struggle—Moltke is said to have stated that he did not care to study the "movements of armed mobs"—Engels regarded it as a "drama without parallel in the annals of military history." It was a revolutionary war not only in its first strategic use of railways and armored ships over a vast area of military operations but also in its "world-transforming abolition of slavery." In the preface of the first edition of *Capital* (1867) Marx wrote: "As in the eighteenth century the American War of Independence sounded the tocsin for the European middle class, so in the nineteenth century the American Civil War sounded it for the European working class."

While Engels' sympathies were most definitely on the side of the north, he was appalled by its "slack management" as contrasted with the deadly earnestness of the south. In a letter to Marx, on November 5, 1862, he said that he could not "work up any enthusiasm for a people which on such a colossal issue allows itself to be continuously beaten by a fourth of its own population." He was even doubtful as to the outcome of the war. It was Marx who warned him, rightly, not to be prejudiced by a one-sided attention to military aspects. Only when Lee, whose superior strategy he had admired, was surrounded and Grant, like Napoleon, delivered his battle of "Jena" capturing the whole of the enemy's army, did Engels recognize the remarkable discipline and morale of the northerners who had entered the war "sleepily and reluctantly."

With the rise of Prussia under Bismarck's leadership, the military thoughts of the revolutionaries turned again to European battle grounds. The short Danish War proved to Engels that the German infantry was superior to the Danes and that "the Prussian firearms, both rifles and artillery, were the best in the world." Still, he underestimated the military striking power of Prussia. Indeed, in an article in the *Manchester Guardian* written on the eve of the battle of Sadowa, he went so far as to predict her defeat in the pending war. He sharply attacked Moltke's plan for the campaign, only to admit the following day that the Prussians, "in spite of their sins against the higher laws of warfare, had not done badly." Engels' surprising miscalculation was largely derived from his erroneous appraisal of Prussia's internal situation. The bitter constitutional struggle over the army reforms in the early 'sixties, had been mistaken by him, as by so many among the popular opposition, for a disintegration of the army and a prelude to revolution. "If this chance passes without being used . . . then we can pack up our revolutionary bags and turn to studying pure theory," he confessed. Indeed, another revolutionary situation had passed and the day after Sadowa Engels was quick to recognize the fact. With his unqualified respect for the Prussian army he also accepted the political consequences of its victory. "The simple fact is this," he wrote to Marx, "Prussia has five hundred thousand needle guns and the rest of the world has not five hundred. No army can be equipped with breech loaders in less than two or three, or perhaps five, years. Until then Prussia is on top. Do you suppose that Bismarck will not use this moment? Of course he will."

While the great social revolutionary now recognized in Bismarck the real Bonapartist, more dangerous than Napoleon III, and while he regretted a German unification "temporarily flooded with Prussianism," he equally rejected the unrealistic refusal of socialist leaders like Wilhelm Liebknecht "to look at the facts." Instead of their shadowy opposition, Engels renewed the struggle with the Prussian Junker upon the very basis created by the Prussian successes.

The grandeur of the dialectical outlook of Marx and Engels now faced a test. In the hard school of their exile they had learned to see the particular developments of classes and nations in their great European context and to unfold their own revolutionary generalship on the basis of an "account of the objective state of social development." Thus they reached the third stage of their military strategy, the approach to the revolutionary state.

IV

The strategy of the revolutionary state as visualized by Marx, and more especially by Engels, remained only fragmentary, to be sure. The status of the socialist movement, whose spokesmen they were, denied them a full materialization of their ideas and certainly their practical application in a socialist order. One might add that even the leadership of the socialist parties at that time showed a definite reserve, if not opposition, to Engels' concepts. Still, the direction of his policies became clear at this stage, crowning a lifelong study of military strategy and shaping the future development of democratic radicalism in Europe.

At the base of Engels' positive military policy was the doctrine of the democratic army, the nation in arms, and the belief in its progressive realization. Indeed, in Engels' pamphlet *The Military Question and the German Working Class* (1865) this vision had already appeared. It became his guiding principle during the next thirty years.

The study of the Prussian military question, published at the height of the constitutional conflict between the feudal ruling class and the rising liberal bourgeoisie, was primarily written as a primer for the workers' party. Engels' advice to the proletariat, fighting for its own political emancipation, was to support the bourgeoisie against the forces of reaction (now fashioned in its new type of the Bonapartist state in which every vestige of political power was withdrawn from both workers and capitalists alike). What gave this essay its special significance was not only the shrewd appraisal of the strength and weaknesses of the middle class opposition and the amazing command of technical details concerning the history of Prussian army organization since the Napoleonic Wars, but also its realistic support of the army reforms in view of Prussia's increase in population and wealth and especially in regard to her neighbors' military potential. In fact, Engels' attack was directed in large part against the bourgeoisie, which had lost its strategic advantage and had failed to win over the army during these critical years. This fundamental failure, Engels claimed, was above all responsible for the stagnation of democratic development in Germany after 1870. The development of the army, in his judgment, was an integral part of social growth.

In earlier studies, such as the articles written for the *New American Cyclopaedia* (edited by George Ripley and A. Dana, New York, 1860-1862), Marx and Engels had emphasized the social basis and preconditions for military organization, past and present. Now they realized that the army itself could serve as a social agency of the first order; in fact, it could serve as the major channel through which a democratic society might emerge. The formula was simple and it obviously followed the historical trends introduced by the French Revolution. The emancipation of the bourgeoisie and peasantry had opened the way for the modern mass army. General conscription, if practiced consistently, guaranteed the strongest and most efficient army for defense of the nation against the outside world. By the same token, it necessarily transformed the inner character of the armed forces, changing them into a people's army simply by the weight of the increasing number of the lower classes.

Proudly Engels could exclaim in 1891, "Contrary to appearance, compulsory military service surpasses general franchise as a democratic agency. The real strength of the German social democracy does not rest in the number of its voters but in its soldiers. A voter one becomes at twenty-five, a soldier at twenty; and it is youth above all from which the party recruits its followers. By 1900, the army, once the most Prussian, the most reactionary element of the country, will be socialist in its majority as inescapably as fate."

Obviously, Engels miscalculated the staying power and inner dynamics of established institutions; no less did he mistake the tempo of great historical transformations. Yet his view was part and parcel of his optimistic belief in the final confluence of democracy and the socialist state. Such conviction, however, did not mislead Engels in underestimating the military needs of the present state, especially in view of a constantly threatening world war of "unexampled violence and universality." The final decision in such a general European war, he surmised, would rest with England since she could blockade either France or Germany and so starve one country or the other into submission.

"We cannot demand that the existing military organization should be completely altered while the danger of war exists," he wrote to Bebel in October 1891. In a series of articles entitled *Can Europe Disarm?* (1893), he suggested, as a means of preventing a war the "gradual diminution of the term of military service by international agreement," such service at first to be for two years. Yet consistent with his basic conviction, he stated that "I limit myself to such proposals as any existing government can accept without endangering the security of its country"; and while he regarded the *Miliz* system as a final goal, he hastened to say to Marx "only a Communist society could get really near the full *Miliz* and even that approach would only be asymptotic."

Whether Engels' last development contradicted the revolutionary policies of his early days is an open question. Both evolutionary and revolutionary socialism, twin brothers in conflict, can claim him as their master. To Engels these contradictions were almost meaningless. Fighter and soldier that he was, he would not be satisfied with slow and tedious reforms. At the same time, the great strategist of socialism was too astute not to recognize that every fight was dependent on the weapons available and that every people and every period would demand a different method.

Toward the end of his life, Engels gave testimony to such necessary changes of revolutionary strategy in the introduction of the newly edited *Class Struggles in France 1848-1850* (London, 1895; reprinted under the title *The Revolutionary Act*, New York, 1922). "The fighting methods of 1848," he stated, "are obsolete today in every respect." Gone was the day of the barricades, of street corner revolutions. In fact, Engels was right in pointing out that "even during the classic period of street battles, the barricade had a moral rather than a material effect." If it held until it had shaken the self-confidence of the military, the victory was won; if not, it meant defeat. Even in 1849 the chances of success were rather poor. "The barricade had lost its charm; the soldier saw behind it no longer the people but rebels . . . the officer in the course of time had become familiar with the tactical forms of street battles. No longer did he march in direct line and without cover upon improvised breastworks, but outflanked them through gardens, courts, and houses." Since then much more had been changed, all in favor of the military; on the side of the insurgents all the conditions had become worse. New armaments and incomparably more effective ammunition, products of large industry, could no longer be improvised by the insurrectionists. The newly built quarters of the large cities erected since 1848 had been laid out in long, straight, and wide streets as though made to order for the effective use of the new cannon and rifles.[1] The ruling classes should not expect the revolutionary to select these new working class districts for a barricade battle. "They might as well ask of their enemies in the next war to face them in the line formation of Frederick the Second or in the columns of whole divisions à la Wagram and Waterloo. The time is past for revolutions carried through by small minorities at the head of unconscious masses. When it gets to be a matter of the complete transformation of the social organization, the masses themselves must participate, must understand what is at stake; that much the history of the last fifty years has taught us."

The legal conquest of the state was the order of the day. There was but one means whereby the steadily swelling growth of the militant socialist forces could for the moment be stemmed—a collision on a large scale with the military, a blood-letting like that of 1871 in the short-lived Paris Commune. This first attempt at a "Socialist Republic" has often been praised in the literature as the great object lesson for the European revolutionaries of the following decades. Marx had analyzed it carefully in his *Civil War in France* (1871); no doubt Lenin's ideas had been tremendously affected by this event; and its influence was certainly visible in the early development of the Soviet regime. Yet the Commune suggested hardly anything in respect to the military strategy of the social revolutionaries. In fact, a renewal of a Paris Commune, though it might be provoked by a threatening *coup d'état* of reactionary forces, was not in line with the idea of the master strategist Engels. In this last stage of his revolutionary career, he saw the triumph of socialism through the democratic processes of the franchise, as he visualized the victory of democracy through the channels of universal military service.

"The nation in arms" was the declared ideal of Engels as military strategist. To aim at the destruction of militarism in the existing state of society, Engels regarded as futile ideology. Instead of that, to eradicate its feudal traditions and to awaken the democratic tendencies inherent in universal compulsory military service—this seemed to him the only promising policy. His followers in the twentieth century were numerous and his influence reached far beyond

[1] Cf. the very different structure of the working class districts, which were erected in Vienna during the days of the republic, and the strategic significance of these fort-like blocks during the upheaval of 1934.

party lines. The "nation in arms" has materialized fully in modern Soviet Russia, and it has become the guiding principle of the militant democracies in the present war. No doubt, Engels would have agreed fully with one of his outstanding disciples, the French socialist Jean Jaurès, who in his *Armée Nouvelle* stated: "Governments will be far less ready to dream of adventurous policies if the mobilization of the army is the mobilization of the nation itself. . . . If a nation which wants peace is assailed by predatory and adventurous governments in quest of some colossal plunder or some startling diversion from their domestic difficulties, then we shall have a truly national war . . . the 'nation in arms' represents the system best calculated to realize national defense in its supreme and fullest form. The nation in arms is necessarily a nation motivated by justice. It will bring to Europe a new era, it will bring hopes of justice and peace."

CHAPTER 8. Moltke and Schlieffen: The Prussian-German School

BY HAJO HOLBORN

FOR half a century after the peace of Vienna, Prussia abstained from active participation in European wars. When in the 'sixties, the Prussian army emerged as the most powerful force on the continent, it had for almost two generations no practical experiences of war. It had undertaken some insignificant campaigns during the revolution of 1848-1849 and had been mobilized repeatedly between 1830-1859 in anticipation of conflicts which did not materialize. In the same period the Russian, Austrian, French, and British armies had been fighting wars. The superiority of the Prussian army in the 'sixties was made possible only by its organization, by its peacetime training, and by the theoretical study of war which had been brought to perfection in the half-century before Sadowa and Sedan.

The Prussian army of the nineteenth century was created by four men: Frederick the Great, Napoleon, Scharnhorst, and Gneisenau. Frederick bequeathed precious memories of victory and endurance in adversity, which are so essential for the pride and self-reliance of an army. In addition, he impressed upon his military successors the knowledge that even the peacetime life of an army consists of hard labor and that battles are won first on the training ground. There was undoubtedly in the Prussian army an overemphasis on the minutiae of military life, which was originally counterbalanced by the strategic genius of the king. He did not train younger strategists, however, and it was a foreign conqueror who reminded the Prussians of the role which strategy plays in warfare, and two young officers, both non-Prussian by birth, had to remold the Prussian army, which they did largely along the modern French pattern. Thus Napoleon became the second taskmaster of the Prussian army, and—after Jena—Scharnhorst and Gneisenau adapted the Prussian army to the new type of warfare.

The Prussian military reformers knew that new methods of war were an expression of the profound social and political changes which the French Revolution had produced. The army of Frederick the Great had been a force of mercenaries isolated from civilian society. Only the noble-born officer's sense of honor and loyalty was glorified while the rank and file were kept together by a brutal discipline. The Prussian military reformers undertook to transform the army of the age of despotism into a national army. To this end they introduced universal conscription of a more radical type than had ever been attempted before. Napoleon's Treaty of Tilsit hampered the immediate realization of Scharnhorst's ideas, but in the Prussian military law of 1814, drafted by his pupil, Boyen, his plan became the permanent order of Prussia's military system.

Conscription became the rule in practically all countries on the continent, but outside of Prussia it amounted merely to the conscription of the poor, since the well-to-do were allowed to make money payments or purchase substitutes. In Prussia, all groups of the population actually served. In this respect, the Prussian army was more clearly a citizens' army than that of any other country. Unfortunately, the Prussians were not democratic citizens, but remained subjects of a bureaucratic absolutism. There was also a recrudescence of the privileged position of the Prussian gentry in government and army, and the junker class continued to monopolize the officers' positions. National service, the logical outcome of national and liberal thought in America and France, became in Prussia a device for strengthening the power of an absolutist state.

The dream of the Prussian military reformers of creating a true citizens' army was frustrated by the political reaction after 1815. The legacy of their strategic and tactical knowledge fared better, though even here the old school scored certain successes. The Prussian field service regulations of 1847 tried to revive Frederician tactics which Scharnhorst's order of 1812 had wisely excluded. Still Scharnhorst's and Gneisenau's strategic ideas were not forgotten in the Prussian army.

Among the contemporaries, these two officers from Hanoverian and Austrian families were the only equals to Napoleon in the art of war. An early death in the summer of 1813 kept Scharnhorst from ever assuming high command in the field. Gneisenau, as the chief of staff of the Prussian army from the fall of 1813 to the summer of 1815, was destined to prove that the new Prussian school of military thought could produce not merely a new philosophy, but also men able to translate their insight into action.

There has been much controversy about which of the two was the greater general. Clausewitz, friend and pupil of both, gave the crown to Scharnhorst because he combined a profound contemplative mind with a deep passion for action. Schlieffen found Gneisenau superior because he seemed to have higher perspicacity and determination on the battlefield. From an historical point of view, however, it is most important to remember that both officers, the calm and self-possessed Scharnhorst and the impetuous and generous Gneisenau, represented a new type of general. Both were born leaders of men, the one possibly greater in educating them for war, the other in directing them on the battlefield, but both these children of Germany's philosophical age, of the epoch of Kant and Goethe, believed that thought should lend wings to action.

The new Prussian strategy sprang from an original interpretation of Napoleon's art of war. To most nineteenth century students of war before Sadowa and Sedan, Jomini's writings seemed the last word on Napoleonic strategy. Had not Napoleon himself said that this man from Switzerland had betrayed the innermost secrets of his strategy? Napoleon, however, though admiring Jomini, had also remarked that he set down chiefly principles, whereas genius worked according to intuition.[1] Jomini's cold rationalism was not capable of doing justice to the spontaneity which was the hidden strength of Napoleon's

---

[1] General Baron Gourgaud, *Sainte Hélène, Journal inédit, 1815 à 1818* (Paris, 1899), II, 20.

actions. The interpretation of Napoleon's strategy, which Scharnhorst found and which animated Gneisenau's conduct of the campaigns of 1813-1815, was based on an historical and inductive method which gave full credit to the creative imagination of the commander and the moral energy of his troops. In Clausewitz' work *On War*, the new philosophy found its classic literary expression.

The new Prussian school of strategy created its own organ in the Prussian general staff, which became the brains and nerve center of the army. The origins of the general staff go back to the decade before 1806, but not before Scharnhorst's time did it receive its characteristic position. When, in 1806, Scharnhorst reorganized the ministry of war, he created a special division which was charged with the plans for organization and mobilization and with the peacetime training and education of the army. Under the jurisdiction of this section came also the preparation of military operations by intelligence and topographical studies, and finally the preparation and direction of tactics and strategy. As minister of war, Scharnhorst retained the direction of this section and exercised a strong influence on the tactical and strategical thought of the officers in it by training them in war games and staff maneuvers. It became customary to assign these officers as adjutants to the various army units, which went far to extend the control of the chief of staff over all generals. The young men with the purple-striped trousers carried strategic thought into all sections of the army.

Under Scharnhorst, the general staff was still a section of the war ministry, under which it would have remained if Prussia had received a parliament. The absolutistic structure of the Prussian government, however, made it possible to divide military responsibility under the supreme command of the king. In 1821, the chief of the general staff was made the highest adviser of the king in matters of warfare, while the ministry of war was restricted to the political and administrative control of the army. This decision was of far-reaching consequence, since it enabled the general staff to take a leading hand in military affairs, not merely after the outbreak of war, but also in the preparation and initial phase of a war.

II

Moltke was destined to take full advantage of the traditional ideas and institutions which were created during the wars of liberation. Like Scharnhorst and Gneisenau, he was not a Prussian by birth, but came from the neighboring Mecklenburg. His father was an officer of the king of Denmark, who, as the Duke of Schleswig and Holstein, was then still a German prince. Moltke was brought up as a Danish cadet, becoming a lieutenant in 1819. His experiences at school had been unhappy, however; his relations with his father were not close; nor did service in the Danish army hold out great prospects. In 1822, Moltke applied for a commission in the Prussian army in which his father had started his military career before transferring to the Danish army.

The Prussians put the young lieutenant through a stiff examination and made him begin at the very bottom of the military ladder again. After a year, however, he was favored by admission to the war college which was under Clausewitz' direction. Clausewitz gave no lectures, however, and Moltke did not come under his spell before 1831, when Clausewitz' work was posthumously published. From his studies at the war college, Moltke gained his lasting interest in geography, physics, and military history which were well presented at the school. In 1826, Moltke returned to his regiment for two years, but most of this time was again given to theoretical work, this time to the teaching of the officers of his division. In 1828, he was assigned to the general staff to which he belonged for more than sixty years.

With the exception of five years as a lieutenant in the Danish and Prussian armies, Moltke never served with the troops. He had never commanded a company or any larger unit when, at the age of sixty-five, he took virtual command of the Prussian armies in the war against Austria. The years from 1835-1839, which he spent in Turkey as a military adviser of the Sublime Porte, gave him some actual war experiences in the futile campaign against Mehemet Ali of Egypt. The Turkish commander threw the good advice of the young captain to the winds, and Moltke saw war at its worst among defeated troops.

When he returned to Berlin, the hardest period of his life was over. As a lieutenant, he never had a penny to spend. Dire need dictated his writing a short story for a popular magazine, or historical essays. In order to purchase mounts, without which he could not accept a commission on the general staff, he translated six volumes of Gibbon's history only to discover that his publisher was insolvent. It is impressive to see how the young Moltke wrestled with such materialistic problems and yet acquired an Attic education in such a Spartan setting. His chief work in his early years was concerned with topography, but he went beyond into all the other aspects of geography and penetrated deep into history as well. His learning and education were remarkably well rounded, and with them grew his power of expression. Moltke became one of the foremost writers of German prose.

He did not become, however, a statesman or original political thinker. Scharnhorst and Gneisenau had been statesmen as much as generals and their military reforms aimed directly at a reform of the whole life of the nation. This had made them suspect in the conservative atmosphere of the Prussian or, for that matter, of the Austrian and Russian courts. As soon as the French Revolution and Napoleon seemed defeated, they were called Jacobins, and Gneisenau and the younger reformers were retired. Moltke was conscious of the natural interrelation of generalship and statesmanship, and took a lively personal interest in politics. He abstained from active participation in political affairs, however, and never questioned the powers that be. He was convinced of the superiority of monarchical government and found its special justification in the fact that it allowed the officers to manage army affairs without interference from nonprofessional elements. The defeats of German liberalism in the revolution of 1848-1849, and again in the 'sixties, were highly gratifying to him.

An officer of his quiet manner, conforming political views, and wide learning was well received at court. In 1855, Frederick William IV made him aide-de-camp to his nephew, Prince Frederick William, the future emperor Frederick III. This appointment brought Moltke into contact with the prince's father, known as the soldier-prince, and William I apparently discovered in Moltke talents which seemed to recommend him for the position of chief of the general staff.

One of William's first actions when in 1857 he became regent of Prussia was to appoint Moltke to the post. Still William I was immediately more interested in the political and technical reorganization of the army, and the figure of the minister of war, Roon, overshadowed the silent chief of staff in the councils of state. What Roon and William proposed was a decided improvement in the efficiency of the army, but it meant at the same time the ultimate abolition of those militialike sections of the army in which a more liberal spirit had survived. The popular *Landwehr* (territorials or national guard) was curtailed in favor of a greatly expanded standing army. This gave the professional royalist officer corps unchallenged control over all military establishments of the nation. The Prussian parliament fought this measure, but the reorganization became effective under Bismarck even without parliamentary consent. The ensuing constitutional conflict was still raging when the battle of Sadowa was fought. The parliamentary opposition, however, broke down when the Bismarckian policy and Moltke's victories fulfilled the longing for German national unity. Moltke's successful strategy, therefore, decided two issues: first, the rise of a unified Germany among and over the nations of Europe; second, the victory of the Prussian crown over the liberal and democratic opposition in Germany through the maintenance of the authoritarian structure of the Prussian army.

The role which Roon, as minister of war, played in the years of political conflict made him the most influential figure in the army before 1866. William I was so used to taking military advice from him that the chief of the general staff was almost forgotten. The unpretentious Moltke was little known in the army, and even during the battle of Sadowa, when an officer brought an order from him to the commander of a division, the latter replied, "This is all very well, but who is General Moltke?" Moltke's rise to prominence among the advisers of the king was sudden and unexpected, though it was the logical outcome of Prussian military history following the days of Scharnhorst and Gneisenau.

His aloofness from the political scene in the years from 1857-1866 allowed him to give his undivided attention to the preparation of future military operations. The revolutions of 1848-1849, the rise of the Second Empire in France, and the Crimean War had already shown that a new epoch of European history had opened in which military power was freely used. Moltke began at once to overhaul the plans which the Prussian general staff had drawn up. His predecessor, General Reyher, incidentally one of the few Prussian generals who had come up from the ranks, had been a man of great vision and a remarkable teacher of strategy. Moltke could count on the ability of the Prussian officer to find original solutions for the tactical problems of war. In fact, the officers silently dropped the official service regulations of 1847 as soon as they crossed the Bohemian frontier in 1866 and followed largely their own ideas.

The peacetime formation of the Prussian army was a more highly developed system than that of any country. With the exception of the guard troops, the regiments drew their recruits and reservists from their local districts. The Hapsburg empire with its nationality problems could not use such a system. Moreover, after 1815, the Prussian army had retained an organization in army corps which Napoleon had created during his campaigns, but which had been given up by France under the Bourbons. With the exception of Prussia, army corps were formed on the eve of war, which again acted as a brake upon rapid mobilization and upon the capacity of troops and leaders in the performance of large-scale operations.

Rapid as the mobilization of the Prussian army was, comparatively, Moltke accelerated it still further. The unhappy geographical structure of the Prussian monarchy of this period, with its far-flung east-west extension from Aix-la-Chapelle to Tilsit severed by Hanover, aggravated Prussia's military problems. The railroad age offered a remedy which Moltke exploited to the full. Moltke had begun to study railroads before a single line had been built in Germany. He apparently believed in their future, for when in the early 'forties, railroad building got under way, he even risked his savings by investing in the Berlin-Hamburg railroad. His speculative interest was enhanced by his matrimonial concern, namely to cut down the distance which separated him from his young bride in Holstein! But his military thinking was always awake. In 1847-1850, troops of various nations were for the first time moved by rail. In 1859, when Prussian mobilization was pending during the Italian war, Moltke could test the facilities for the rail transportation of the whole army and could introduce important improvements.

The railroads offered new strategic opportunities. Troops could be transported six times as fast as the armies of Napoleon had marched, and the fundamentals of all strategy—time and space—appeared in a new light. A country which had a highly developed system of rail communications gained important and possibly decisive advantages in warfare. The speed of the mobilization and of the concentration of armies became an essential factor in strategic calculations. In fact, the timetable of mobilization and assemblage, together with the first marching orders, formed in the future the very core of the strategic plans drawn up by the military staffs in expectation of war.

In addition to making use of the modern railroads, Moltke proposed to employ the dense road system which had come into being in the course of the industrial revolution. Napoleon had already pointed the way by dividing his army on marches and had set, in the campaign of 1805 which led to the surrender of the Austrian army at Ulm, a classic example for the strategic use of separate marching orders. An army column is, however, not ready for battle, and it takes a full day to deploy a corps of 30,000. The changeover from marching to battle formation was accordingly a time-consuming process, and armies had, therefore, to be massed days before the battle. After 1815, road conditions improved greatly and new tactics became possible. In 1865, Moltke wrote: "The difficulties in mobility grow with the size of military units; one cannot transport more than one army corps on one road on the same day. They also grow, however, the closer one gets to the goal since this limits the number of available roads. It follows that the normal state of an army is its separation into corps and that the massing together of these corps without a very definite aim is a mistake. A continuous massing becomes, if merely on account of provisioning, embarrassing and often impossible. It makes a battle imperative and consequently should not take place if the moment for such a decision has not arrived. A massed army can no longer march, it can only be moved over the fields. In order to march, the army has first to be broken up, which is dangerous in the face of the enemy. Since, however, the concentration of all troops is absolutely necessary for battle, the essence of strategy consists in the organization of separate marches, but so as to provide for concentration at the right moment."

It is probable that Moltke already envisaged operations in which the concentration of the army would take place on the battlefield itself, thus discarding the Napoleonic principle that the army should be concentrated well before the start of a battle. Still Moltke's direction of operations in the weeks before Sadowa did not disregard the Napoleonic rule from the very beginning. He could have drawn the armies together before the battle but he decided at a late date to continue their separation and to achieve their union on the battlefield. After Sadowa, he summed up his ideas thus: "It is even better if the forces can be moved on the day of battle from separate points against the battlefield itself. In other words, if the operations can be directed in such a manner that a last brief march from different directions leads to the front and into the flank of the enemy, then the strategy has achieved the best that it is able to achieve, and great results must follow. No foresight can guarantee such a final result of operations with separate armies. This depends, not merely on calculable factors, space and time, but also often on the outcome of previous minor battles, on the weather, on false news; in brief, on all that is called chance and luck in human life. Great successes in war are not achieved, however, without great risks."

The last remarks permit a glimpse at Moltke's philosophy of war. As a loyal student of Clausewitz, Moltke was anxious to extend the control of reason over warfare as far as possible. He knew, however, only too well that the problems of war cannot be exhausted by calculation. War is an instrument of policy and, though Moltke maintained that a commander should be free in the actual direction of military operations, he admitted that fluctuating political aims and circumstances were bound to modify strategy at all times.

While the impact of politics on strategy confronted a general with an element of uncertainty, Moltke felt that the mobilization and initial concentration of the army was calculable since it could be prepared a long time before the outbreak. "An error," he said, "in the original concentration of armies can hardly be corrected during the whole course of a campaign." The necessary orders, however, can be deliberated long before and, assuming that the troops are ready for war and transportation is properly organized, they will inevitably lead to the desired results.

Beyond this stage, war becomes a combination of daring and calculation. After actual operations have begun, "our will soon meets the independent will of the enemy. To be sure, we can limit the enemy's will if we are ready and determined to take the initiative, but we cannot break it by any other means than tactics, in other words, through battle. The material and moral consequences of any larger encounter are, however, so far-reaching that through them a completely different situation is created, which then becomes the basis for new measures. No plan of operations can look with any certainty beyond the first meeting with the major forces of the enemy. . . . The commander is compelled during the whole campaign to reach decisions on the basis of situations which cannot be predicted. All consecutive acts of war are, therefore, not executions of a premeditated plan, but spontaneous actions, directed by military tact. The problem is to grasp in innumerable special cases the actual situation which is covered by the mist of uncertainty, to appraise the facts correctly and to guess the unknown elements, to reach a decision quickly and then to carry it out forcefully and relentlessly. . . . It is obvious that theoretical knowledge will not suffice, but that here the qualities of mind and character come to a free, practical and artistic expression, although schooled by military training and led by experiences from military history or from life itself."

Moltke denied that strategy was a science and that general principles could be established from which plans of operations could be logically derived. Even such rules as the advantages of the inner line of operation or of flank protection seemed to him merely of relative validity. Each situation called for a definition in terms of its own circumstances, and for a solution in which training and knowledge were combined with vision and courage. In Moltke's opinion, this was the chief lesson to be derived from history. Historical study was also of the greatest usefulness in acquainting a future commander with the complexity of the circumstances under which military actions could take place. He believed that no staff or army maneuvers, indispensable as they were for the training of staff officers, could put before their eyes as realistic a picture of the significant aspects of war as history was able to do.

The study of military history was made one of the central responsibilities of the Prussian general staff and not left to a subordinate section. Moltke set the style by his classic monograph on the Italian war of 1859, first published in 1862, which aimed at an objective description of the events in order to draw from them valid practical conclusions. The histories of the wars of 1866 and 1870-1871 were later written in a similar manner under his direction.

Moltke took the view that strategy could benefit greatly from history, provided it was studied with the right sense of perspective. His own practice exemplifies the benefits which he derived from historical study. He knew, of course, of Napoleon's occasional use of detached corps for attacks against the flank or rear of the enemy. These operations with detailed units, however, had not affected Napoleon's general principle of strong concentration of the gross of the army, and his belief in the irresistible power of central attack. The advantages of such a strategy had been great in the age of Napoleon, but they had not shielded him against ultimate defeat. The battle of Leipzig had shown the possibilities of concentric movements of individual armies which Scharnhorst had predicted in his advice that one should never keep an army aimlessly massed, but always fight with concentrated forces. In Moltke's opinion, the progress of technology and transportation made it possible to plan concentric operations on a much larger scale than had been used half a century before.

Important as history was for the officer, Moltke pointed out that it was not identical with strategy. "Strategy is a system of *ad hoc* expedients; it is more than knowledge, it is the application of knowledge to practical life, the development of an original idea in accordance with continually changing circumstances. It is the art of action under the pressure of the most difficult conditions."

Accordingly, the organization of command held a prominent place in Moltke's ideas on war. He treated the subject with great clarity in his history of the Italian campaign. No war council could direct an army, and the chief of staff should be the only adviser of the commander with regard to the plan of operations. Even a faulty plan, provided it was executed firmly, was preferable to a synthetic product. On the other hand, not even the best plan of operations could anticipate the vicissitudes of war, and individual tactical decisions which must be made on the spot. In Moltke's view, a dogmatic enforcement of the plan of operations was a deadly sin and great care was taken to encourage initiative on the part of all commanders, high or low. Much in contrast to the vaunted Prussian discipline, a premium was placed upon independent judgment of all officers.

Moltke refrained from issuing any but the most essential orders. "An order shall contain everything that a commander cannot do by himself, but nothing else." This meant that the commander in chief should hardly ever interfere with tactical arrangements. But Moltke went beyond this. He was ready to condone deviations from his plan of operations if the subordinate general could gain important tactical successes, for, as he expressed it, "in the case of a tactical victory, strategy submits." He remained unmoved when certain generals in the first weeks of the Franco-Prussian War by foolhardy, though gainful enterprises, wrecked his whole plan of operations.

Moltke did not wish to paralyze the fighting spirit of the army or to cripple the spontaneity of action and reaction on the part of subordinate commanders. The modern developments had placed a greater responsibility upon them than was the case in former ages. One of the chief reasons why Napoleon kept his army close together was his wish to keep all troops within the reach of his direct orders. Moltke's system of disposition in breadth made the central direction of the battle itself extremely difficult, although the marches prior to the battle could be easily arranged by telegraph. Moltke directed most movements in the war of 1866 from his office in Berlin, and arrived on the theater of war just four days before the battle of Sadowa. He confined himself very wisely to general strategic orders. To ensure an adequate, and this meant free, execution of strategic ideas, army commands were created while the authority in tactical questions rested with the commanders of corps and divisions.

Moltke's strategic thought and practice met its first and greatest test in the Austrian campaign of 1866. His role in the war which Austria and Prussia conducted against Denmark in 1864 had been modest. In the latest phase of the war he had quickly stopped the bungling which characterized the regime of the old Field Marshal Wrangel, and his critical counsel established him in the eyes of William as a prudent strategist. In the discussion of war plans against Austria he became increasingly prominent so that William I, on June 2, 1866, directed that all orders to the army should be issued through him. Since the king henceforth accepted Moltke's advice almost unconditionally, the sixty-five-year old general, who had thought of retirement, found himself the virtual commander in chief of the Prussian army.

The first test of his generalship was at the same time the greatest one in his career. The forces were more evenly matched than later in the Franco-Prussian war, and Moltke had to overcome more obstinate geographical and political problems. The war of 1866 and particularly the Bohemian campaign also illustrate the strategic side of war in a much clearer form than the Franco-Prussian or for that matter most other wars.

William I wished to avoid the war with Austria into which Bismarck ultimately pushed him. The Prussians thus began their mobilization much later than the Austrians and even then it remained doubtful whether the king could be persuaded to declare war, thereby enabling the army to take the offensive. The original strategic problems were accordingly very delicate. From Bohemia and Moravia the Austrians could have operated against either Upper or Central Silesia or marched into Saxony to threaten Berlin, possibly after effecting a union with the Bavarian army in Northern Bohemia or Saxony. Whether one or the other of these possibilities could be realized depended entirely upon the date of the actual opening of war. Naturally enough, Moltke supported Bismarck in urging the king to act soon, but he avoided prejudicing the political issue by military measures—in contrast to his nephew, who as chief of staff had to inform William II in August 1914, that the strategic plans of the general staff had deprived the government of its freedom of action.

The elder Moltke's moves were aimed in the first place at making up for the delay caused by the belated start of the Prussian mobilization. In addition, he wished to cope with a possible Austrian advance against Saxony and Berlin or against Breslau in Central Silesia while Upper Silesia remained originally unprotected. Whereas the Austrians could employ only one railroad line for their mobilization in Moravia, Moltke used five to transport the Prussian troops from all over Prussia to the neighborhood of the theater of war. As a conse-

KEY

   I   First Silesian Army

  II Second Silesian Army

  E Army of the Elbe

Bohemian Battle Area, 1866, Showing Main Railroads and the Advance of the Prussian Armies

quence, on June 5, 1866, the Prussian armies were spread over a half circle of 275 miles from Halle and Torgau to Goerlitz and Landeshut. The original placement of the Prussian troops was safe as long as the Austrian forces were far to the south. In point of fact, they were not even in Bohemia, as Moltke assumed, but still in Moravia.

Moltke, of course, never planned to leave his troops at their points of disembarkation but began at once to draw them closer toward the center around Goerlitz. At all times he refused, however, to order a full concentration in a small area as was advocated by most Prussian generals and even by members of his own staff. On the other hand, he too felt somewhat worried when he ultimately learned that the main Austrian forces were assembling in Moravia and not in Bohemia, a fact which seemed to point to a contemplated Austrian offensive toward Upper Silesia. Reluctantly he allowed the left wing to extend toward the Neisse River, thus again spreading the Prussian armies over a distance of more than 270 miles from Torgau to Neisse. His hesitation was chiefly caused by uncertainty about the policy of William I and not by military considerations. In Moltke's opinion, everything would be well if he did not miss the opportunity of achieving the ultimate concentration of the Prussian armies along the shortest route, which meant by a forward move into Bohemia.

Moltke had chosen Gitschin as the point for such a concentration—not because it offered important strategic advantages of itself, but merely on account of distances. It was about equally close to the two main Prussian armies, the Second Army under the Crown Prince, Friedrich Wilhelm, which formed the left wing in Silesia, and the First Army under Prince Friedrich Karl, which had its base around Goerlitz. At the same time, Gitschin was equally distant from Torgau and Olmuetz, that is, from the Prussian Elbe Army and from the Austrian main army. Provided the Prussian armies could begin marching on the same day on which the Austrian army left Moravia their concentration should have been completed before the Austrians arrived at Gitschin.

It was not before the twenty-second of June that officers of the Prussian vanguard handed Austrian officers notification of the Prussian declaration of war, but Prussia had opened hostilities against other German states on June 16. Thus the Elbe Army began to occupy Saxony on the same day on which the Austrian army started its march from Olmuetz to Josephstadt at the upper Elbe.

The Austrian army was worthy of the best traditions of Austrian military history. Its morale and enthusiasm were high; its officers, among them some of the best generals of the period, had great ability and practical experience. Certain branches of the services, namely cavalry and artillery, were definitely superior to those of the Prussian army. The strength of the latter was in its infantry which excelled both in tactics and arms. The Prussian needle-gun by itself, however, could not have achieved victory as was proved in the war against France where the Prussians fought against an infantry armed with superior rifles. It was the outmoded shock tactics of the Austrian infantry together with its old-fashioned guns which put the Austrians at a decided disadvantage.

The scales were turned, however, by the lesser strategic ability of the Austrian high command. Benedek was a fine soldier with a distinguished record of war service to the Hapsburg empire. He was at his best in battle; fearlessly and correctly he directed even the retreat of his beaten army on the battlefield of Sadowa. But he had grown up in the classic school of strategic thought and his chief strategic adviser, General Krismanic, whom he had not selected, lived largely in the operational thought of the eighteenth century. These elements determined the strategic conduct of the war by the Austrian high command. They meant formation in depth and emphasis upon the maintenance of naturally strong positions. Moltke, on his part, showed that space could be conquered by time.

The Austrian army moved from Moravia in three parallel columns. Though the strain of such marching arrangements was considerable, the Austrians reached their goal quickly and in good order. But after the arrival of the vanguards in Josephstadt on June 26, at least three days were needed to mass the army again. This loss of time probably saved the Prussian armies.

In spite of Moltke's continuous warnings, the First Army had made slow progress, since Prince Friedrich Karl wanted to wait for the Elbe Army, which, after occupying Saxony, was to be joined to his command. This gave Benedek an opportunity to use the inner line of operations. Which of the two about equally strong Prussian armies Benedek should have attacked has been an interesting controversy among students of military history. Probably Benedek's judgment was right when he considered chiefly an attack on the First Army. He failed, however, to recognize in time that he had only one or possibly two days in which he could have taken the offensive against one of the Prussian armies without having to fear the other in his rear. Since the Austrian high command believed rather in the tactical advantage of strong positions than in the priceless value of time, and since the early concentration of the army hindered its mobility, the opportunity slipped by. When Benedek discovered the mistake, it was even too late to retreat behind the Elbe at Josephstadt and Koeniggraetz, and he had to accept battle with the river at his rear.

The danger of an Austrian attack against one of the two Prussian armies having passed, Moltke began to delay the concentration of the armies, keeping them at one day's distance from each other in order to achieve their union on the battlefield. During the night of July 2, the last orders were given. They were actually bolder than their execution made them appear. According to Moltke, the left wing of the Second and the right wing of the First Army were supposed to operate not merely against the flanks but also against the rear of the enemy. Moltke conceived of Sadowa as a battle of encirclement. But the Prussian generals did not follow him and the Austrian army got away—though losing a fourth of its strength. An immediate pursuit was impossible since the troops of the Second Army had run into the front of the First, thus causing a mix-up of all army units which could not be easily disentangled. Four years later, the battle of Sedan proved that the Prussians had learned their lesson, although they fought a smaller French army on that occasion.

Moltke had shown by his strategy that the much-vaunted inner line of operations was merely of relative significance. He summed up his experiences in these words: "The unquestionable advantages of the inner line of operations are valid only as long as you retain enough space to advance against one enemy by a number of marches, thus gaining time to beat and to pursue him, and then to turn against the other who is in the meantime merely watched. If this space, however, is narrowed down to the extent that you cannot attack one enemy without running the risk of meeting the other who attacks you from the flank or rear, then the strategic advantage of the inner line of operations turns into the tactical disadvantage of encirclement during the battle."

These sentences have often been interpreted as a definite condemnation of operations along the inner line and a recommendation of concentric maneuvers. This was not Moltke's opinion. During the Franco-Prussian war of 1870-1871, he used both concepts freely and successfully, depending chiefly upon the actions of the enemy. Moltke's strategy was characterized by his openness of mind and by the elastic changes from one device to the other.

It has been suggested that Moltke's strategy reflected the superior military strength which Prussia enjoyed at that time, but such a statement is true only within certain limitations. In 1866, Moltke had to create the slight superior strength of the Prussian armies in Bohemia, which, incidentally, was not to be found in manpower. He took the risk of denuding all Prussian provinces of troops and of leaving only an extremely small army to deal with Austria's German allies. If the Bohemian campaign had dragged on or turned into a deadlock, Napoleon III could have used the chance to take the Rhineland and to settle the fate of the continent. Nor were possibilities of this sort entirely lacking during the war of 1870-1871.

After the treaty of Frankfort, Prussia-Germany could breathe more freely, provided the government succeeded in preventing the military cooperation of her foremost neighbors, France and Russia. Moltke had considered this eventuality for the first time in 1859, but it had been a passing cloud on the political horizon. From 1879 on, the possibility of a Franco-Russian coalition loomed larger and larger in the thoughts of the general staff. With the conclusion of the Franco-Russian alliance in the early 'nineties, it became the major strategic consideration.

Moltke's plans in this situation were in line with his strategy in the past, namely to fight one enemy with as little as possible in order to make available superior forces with which to crush the other. His advice was to stay on the defensive in the west and to take the offensive against Russia. Germany, in possession of Alsace-Lorraine, could defend her western frontier with small forces whereas she could not hope to achieve rapid decisions against the rising line of French fortifications. Greater results could be hoped for in Russia. Moltke's second successor as chief of the general staff, Count Schlieffen, reversed the sequence in 1894; from that time on, the German plans for a two-front war envisaged making the first offensive in the west.

III

Count Alfred Schlieffen, born in 1833, was the descendant of a noble family which had given many outstanding civil servants and officers to the Prussian monarchy. His reticence, his limited eyesight, and his genuine interest in studies seemed to predestine him for civil rather than military service, and it was not until he was serving his year in the army that he decided to become an officer. From 1858 to 1861 he attended the war college, and his subsequent appointments indicate clearly that his superiors had earmarked him for high staff positions. At due intervals he changed from posts in the Great General Staff to staff work with the troops until in 1876 he became for seven years commander of the First Guard Uhlan regiment at Potsdam. From 1883 to his retirement in 1906, he was again with the general staff, first as head of various sections and after 1891 as its chief.

Schlieffen's career prior to his last fifteen years in the general staff brought him somewhat more into contact with the life of the troops than had been the case with Moltke. He had also acquired more practical war experience than Moltke had gained before 1864. During the war of 1866, Schlieffen was a member of the staff of a cavalry corps. In this capacity he saw the battle of Sadowa which made a great impression upon him. To his regret, he did not participate in the frontier battles of the Franco-Prussian war. But he had opportunities to demonstrate his talent when he served on the staff of an army during the Loire campaign and collected variegated impressions and ideas about war and generalship.

Compared to Moltke's early struggles, Schlieffen's rise to prominence was fairly easy. Nor did he have to fear that his advice as chief of the general staff would not be accepted. Before 1866 the influence of the general staff had not been unchallenged, but the supreme authority which the office inspired after Sadowa and Sedan fell to Schlieffen with Moltke's toga. Schlieffen was thus able to concentrate even more than Moltke on military problems and to ignore their political implications. The increasing professionalization which was a characteristic of life in the late nineteenth century was reflected in the history of the German army command. When Schlieffen studied at the war college in the 'fifties, specialists and technicians were already beginning to overshadow that older philosophical and historical universalism which had been Moltke's spiritual food and which explained to a large extent his Olympian calm and serenity. Schlieffen had a Promethean nature, which drove him by an unending zeal to achieve the impossible, but his efforts were restricted to his professional sphere. On the political causes and consequences of war, he did not speculate very much. Moltke, too, had kept from meddling in politics, but he had been very conscious of political forces and tried to adapt his strategy to them. Schlieffen's whole life and thought were devoted to strictly military problems.

The sudden death of his young wife after a brief marriage left Schlieffen oblivious to everything but his duties as chief of staff. Every phenomenon of life was immediately examined for its war potentialities. There was something inhuman in his ascetic devotion to his military tasks although it seemed superhuman to his loyal students. To those anxious to pierce the secrets of modern generalship his mind exercised a compelling fascination.

Though the membership and sections of the Prussian general staff expanded after 1891, it did not alter as an institution. Its functions continued to be the education of the army for war and the preparation of operations. Schlieffen's chief technical contributions were the further development of railroad transportation, the creation of mobile heavy artillery—carried through against the stanch opposition of the army conservatives—and the introduction of certain new branches such as the army railroad engineers and the air corps. He was keenly interested in the progress of modern technology, but was not too successful in convincing the semifeudal officer corps of the necessity of making full use of new discoveries. The officers remained extremely suspicious of modern technology and were not willing to assign to scientists and engineers a major role in the management of military affairs. During the first World War only the heavy artillery and the military railroad engineers were equal to their task. The German air force and signal corps were inadequate and, what was worse, the army command was unprepared to employ the new technical possibilities. The tank battles of 1918 produced a change of heart, the results of which can be seen in the coordination of military and technical services in Hitler's army.

Schlieffen gave so much thought to modern technology, not because he believed that technology would dethrone strategy but because he saw in it a new challenge to military leadership. In his sarcastic manner he once remarked of the progress of modern technology: "Dividing its precious gifts among all evenly and impartially, it created the greatest difficulties and considerable disadvantages for all."[2] Of the two chief results of the industrial age, speed and mass production, only the latter seemed to affect warfare. Modern mass armies with their tremendously increased firing power apparently eliminated the strategy of mobile warfare as it had been developed by Napoleon and Moltke. But Schlieffen criticized those officers who meekly accepted this loss of mobility and maneuver and could not think of other solutions for it than defensive positions or frontal attacks. "The Russo-Japanese war," he wrote, "has proved that mere frontal attacks can still be successful in spite of all difficulties. Their success is, however, even in the best case only small. To be sure, the enemy is forced back, but after a little while he renews his temporarily abandoned resistance. The war drags on. Such wars are, however, impossible at a time when the existence of a nation is founded upon the uninterrupted progress of commerce and industry. . . . A strategy of attrition will not do if the maintenance of millions of people requires billions."[3] In Schlieffen's opinion only a strategy of annihilation could preserve the existing social order.

[2] "Der Krieg in der Gegenwart" in Schlieffen, *Cannae*, p. 274.
[3] *Ibid.*, pp. 279-280. Cf. *Dienstschriften*, I, 86-87.

Schlieffen was not a prophet of total war, although he feared that fundamental social changes would become certain in the course of a long war. The anxiety that the contemporary strategists might fail added a somber shade to all his thought and teaching. It was as a thinker and teacher of strategy that he gained his place in history. Whether he would also have proved himself a great commander in war cannot be said. He possessed qualities which seemed to qualify him singularly well for the high command in war. His thought was never confused by minor considerations nor perturbed by mere incidents, and in spite of his cool remoteness of manner his personality radiated a sovereign strength which electrified all around him. His students were convinced that he would have become as great a master on the battlefield as at the map table, and that none of the generals of the first World War measured up to his greatness. Like Scharnhorst, however, Schlieffen died on the eve of great decisions, occupied to his last moments in January 1913, with the solution of Germany's military problems. His shadow hung over the first, as it is still over the second World War, and his influence on German military thought can hardly be overestimated. As late as 1938 General Beck, the chief of the German general staff in the period of German rearmament, called him the "first among the classic teachers of strategy," though General von Fritsch added to his own praise the warning that "the enormous progress of technology since Schlieffen's death may make some of his rules appear as no longer strictly valid."[4]

The Franco-Russian alliance of 1893 made it certain that in case of a European conflict Germany would have to fight a two-front war. It seemed hopeless to compete with the Franco-Russian bloc in numbers. For political reasons Austria-Hungary was unable to expand her armaments considerably. Italy could not be counted on as an ally, while as time progressed Britain became more clearly a potential military adversary. On the other hand, Germany still held the advantage of the central position on the continent, and, if she took the risk of an uneven distribution of her troops, of greater striking power in one theater of war during its initial phase. In Schlieffen's opinion this temporary German superiority had to be employed where it was most likely to result not merely in victorious battles but in a speedy decision of the war. Moltke's recommendation of defensive warfare against France and an offensive against Russia did not seem to promise this success. Operations in the east would be time-consuming, since the vast eastern plains permitted the Russians to employ evasive tactics. A stalemate in the west and protracted warfare in the east would make Britain the arbiter of Europe. Even without expecting British intervention Moltke had warned that in contrast to the wars of 1859, 1864, or 1870-1871, future wars were likely to last for many years. To overcome the danger of protracted warfare, Schlieffen decided in 1894 that, in the event of war, offensive action should be undertaken first against France.

Schlieffen's reasoning was based on the fact that France was the more powerful military enemy of Germany and the one whose concentrated strength could be met and disposed of at an earlier date of the war. Control of France

[4] In the introductions to Vol. I and II of Schlieffen, *Dienstschriften.*

would make a British intervention improbable or ineffective. However, in order
to achieve a supreme decision of the European war through a campaign in
France it would not be enough to force the French army to retreat into the
interior or even to capture Paris. It would be necessary to annihilate the total
armed strength of Germany's western enemy.

The first Schlieffen plan for an offensive against France, drawn up in 1894,
was hardly adequate for the achievement of such a super-Sedan; it still rested
almost completely upon frontal attack from Lorraine, and the growing strength
of French armaments made it a costly and precarious scheme. Between 1897
and 1905 Schlieffen evolved his grandiose plan of a German offensive which
was to gain its irresistible momentum from the weight of a powerful German
right wing wheeling through Luxembourg, Belgium, and southern Holland.
The famous memorandum of 1905 gave these strategical ideas for a western
campaign their classic form, though Schlieffen continued to his death to reexamine and reformulate its problems. The decision to open a two-front war
with a lightning offensive against France was accepted by both the German
general staff and the German government in spite of the great political risks
which the violation of Belgian and Dutch neutrality imposed. The east was to
be guarded only by small forces, and up to eight-ninths of the German army was
to be used to eliminate the armed might of France. However, the German army
of the east, according to Schlieffen's hopes, was not to fall back at once behind
the fortresses of the Vistula, but should first attempt to attack the Russian
armies which would be forced by the Masurian Lakes to divide their forces in
invading Eastern Prussia. The inner line coupled with a strategy of encirclement might enable the numerically inferior army to achieve a victory.

This dream of an eastern victory was realized in the battle of Tannenberg
of August 28, 1914, in which the Samsonov army was annihilated. The plan
for such a battle was originally conceived by Schlieffen and was often tried out
with his officers. In 1914 Hoffmann and Ludendorff executed an oft-posed
Schlieffen war game problem in a manner which would have delighted their
military teacher. Schlieffen had never expected that an eastern battle of this
type would be decisive in terms of the strategy of the war. He had only hoped
that such a victory would gain time for the completion of the great operations
in the west.

However, as the strategy of a battle the eastern or Tannenberg type seemed
to Schlieffen the highest achievement of generalship. In the first volume of
Hans Delbrück's *History of the Art of War*, published in 1900, which dealt
with the strategy of antiquity, he found the prototype of this form of battle in
the Carthaginian victory at Cannae. In 216 B.C. Hannibal had annihilated a
vastly superior Roman army by boldly accepting temporary defeat in the
center in order to be strong enough to crush the enemy's wings and to encircle
his legions. Schlieffen judged that all the great commanders of history had
aimed at the Cannae scheme. Frederick the Great did not have forces powerful
enough to perfect such annihilating blows, but in Schlieffen's opinion Frederick's important victories were incomplete Cannaes. Napoleon at the height of

his career had shown the Hannibal touch, as for example in the great campaign of 1805 which culminated in the capture of Mack's army at Ulm. The defeat of Napoleon, in turn, was the result of a Cannae strategy, particularly in the battles of Leipzig and Waterloo. The same was true in the case of Sadowa, which was well conceived but not too brilliantly executed. In recent times Sedan had constituted a true Cannae.

Schlieffen was prone to simplify military history. He had formed his strategical concepts in studying modern tactics and too easily projected modern ideas into the historical past. It is questionable whether Napoleon's strategy of central break-throughs illustrated the lesser form of his military genius. It was correct to emphasize, however, that as a consequence of modern firing power frontal attack had become exceedingly costly and less effective. This had already been the case in the epoch of Moltke, but the progress of military technology in the half century after Sadowa made tactical defense even more powerful. On the other hand, the lines of communication of modern armies grew more vulnerable since ammunition was a vital necessity in maintaining the fighting quality of modern mass armies. A thrust into the enemy's rear was, therefore, the real aim of a flanking attack. It was not enough to roll the enemy's wings toward the center of his position. This latter method would only lead to what Schlieffen, using an expression of Napoleon's, called an "ordinary" victory; the thrust into the rear would mean a battle of annihilation. Schlieffen was convinced that the encirclement battle, preferably conducted as an attack against both wings of the enemy, was the highest achievement of strategy. To master the problems of such a strategy was imperative for a numerically weaker army, since in it was to be found the only hope for victory. Frederick the Great's statement that there was no need to despair in the face of a superior enemy, provided the dispositions of the general made up for numerical shortcomings, expressed Schlieffen's own confident hope which he tried to infuse into the German officer corps.

It was difficult enough to use a Cannae strategy in a situation where space for maneuver was available, as in Eastern Prussia where single armies were expected to fight in the initial stage of the German-Russian war and the railroad could be used for surprise operations of relatively small armies. But the growth of mass armies and the space which they needed as a result of modern firearms made the scheme inapplicable to the western European theater of war. Armies of millions would cover all available space along the Franco-German frontier and possibly extend from the Channel to Switzerland. A simultaneous assault against the two wings of the French army was impossible since the French right wing was protected by Belfort and the Swiss Jura fortifications. Only an offensive through Belgium against the left wing of the French army held out the chance for a thrust into the enemy's rear. The Schlieffen plan has been compared to Frederick's oblique battle order of Leuthen in 1757, when 70,000 Austrians were defeated by an army of 35,000. However, Frederick's forces were too weak to allow him the full exploitation of his flanking tactics for a strategy of encirclement. Schlieffen could assemble sufficient strength for a battle of annihilation by temporarily ignoring the Russian threat to Germany's eastern provinces.

In his memorandum of 1905 Schlieffen counted on using against France eight armies comprising altogether 72 divisions, 11 cavalry divisions, and 26½ *Landwehr* brigades. In addition he intended to use eight *Ersatz* corps as soon as they could be mobilized. The overwhelming mass of these armies was to be concentrated between Metz and Aix-la-Chapelle with the greatest power again assigned to the right wing. An army of only nine divisions, three cavalry divisions, and one *Landwehr* brigade was to be placed between Metz and Strasbourg while the southern Alsace was to be left unprotected with merely three and a half *Landwehr* brigades covering the right bank of the upper Rhine. The ratio of strength between the right and left wing of the German army was to be about 7 :1.

The offensive, which in its first stage was to reach a line from Verdun to Dunkirk, revolved around Metz. On the thirty-first day after mobilization the

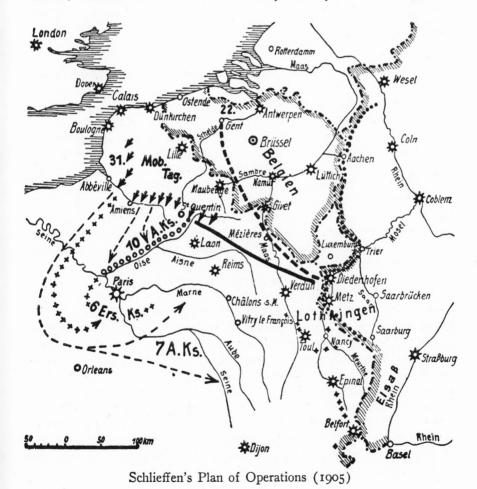

Schlieffen's Plan of Operations (1905)

Somme should have been reached and Abbéville and Amiens passed. The next and decisive phase was to be taken up by operations against the lower Seine, the crossing of which would lead to the final stage of the battle. At this moment the German right wing was to turn toward the east, and operate south of Paris against the upper Seine, thus throwing the French armies against their own fortresses and the Swiss frontier.

The boldness of the Schlieffen plan lay in the risks which he was willing to take in order to assemble the vast superiority of the German right wing. This right wing was not only to be strong enough to smash any opposition while marching through Belgium, but to maintain its forward drive over a period of five to seven weeks, while continuously unfolding toward the north and west. The French and German armies were expected to be of roughly equal strength, and only by denuding Alsace and even offering the French the opportunity of crossing the upper Rhine was it possible to gather sufficient power for his ambitious scheme. Schlieffen assumed that the French would not leave their fortresses in force and that French troops invading Alsace or southern Germany would soon be drawn back with magnetic power by the threat of the wheeling German right wing. If so, they would probably be too late to affect the decision of the campaign.

In reviewing war games in 1901 Schlieffen pointed to the contrast between his and Moltke's strategical concepts in the following terms: "In 1870 we were able to attack the front of the enemy. Our numerical superiority then enabled us after making contact with the enemy to wheel around his extended wing and strike his flank. Now, we can never count on numerical superiority, only at best on equal numbers. Ordinarily, we shall have to be satisfied with being considerably inferior in numbers. Necessity compels us to think of a way in which to conquer with numerically weaker forces. There is no panacea, not just *one* scheme, but one idea seems to be well founded: if one is too weak to attack the whole, one should attack a section. There are many variations of this. One section of the enemy's army is its wing, and consequently one should attack a wing. This is difficult in the case of a company, a battalion, or a detachment, but it grows simpler the stronger the enemy is, the farther his lines extend, the more time it takes to support the attacked wing by the other one. How is the enemy's wing to be attacked? Not with one or two corps, but with one or more armies, and the march of these armies should be directed, not against the flank, but against the enemy's line of retreat, in emulation of what was demonstrated at Ulm, in the winter campaign of 1807, and at Sedan. This leads immediately to a disturbance of the enemy's line of retreat and through it to disorder and confusion which gives an opportunity for a battle with inverted front, a battle of annihilation, a battle with an obstacle in the rear of the enemy."[5]

These words contain the essence of Schlieffen's strategic thought which wrestled with the problem of how to fight brief and decisive wars against a stronger enemy. He did urge the expansion of the army, and this took place at various times between 1891 and 1906, though at the time of his resignation

[5] Schlieffen, *Dienstschriften*, I, 86-87.

Germany trained only 54 per cent of her registered young men compared to France's 78 per cent. But he never advocated the full mobilization of German manhood, only the provision of enough troops for operations of the contemplated character. Strategical thought and ability meant to him more than sheer superiority in mass and numbers.

It was in this spirit that Schlieffen tried to train the members of the general staff, aiming throughout the years for ever greater perfection. The first two volumes of his official military writings, published by the German general staff in 1937 and 1938, make it possible to study the progress of his strategical concepts and teachings in the fifteen years during which he held Scharnhorst's and Moltke's position in the German army. Most of Schlieffen's ideas were the logical continuation of the classic tradition of Frederick, Napoleon, and Moltke, and their adaptation to modern war conditions. Schlieffen, however, went beyond his great predecessors by planning beforehand not merely the mobilization, transportation, first concentration, and the direction of the offensive, but also the decisive battle itself. This gives the so-called "Schlieffen plan" of 1905 its place in the history of strategy. It should be noted that it is not quite correct to speak of a Schlieffen plan of 1905 since the document was only the sketch of a plan counting on forces which were not yet at the disposal of the German high command.

Napoleon occasionally boasted of having worked out beforehand the total course of his campaigns. But opposing statements and the actual practice of his generalship contradict this. On the whole, he would have agreed with Moltke's statement that the "independent will of the enemy" made it impossible to predetermine the course of war. In contrast, Schlieffen believed that the will of the enemy could be immobilized by forcing him from the outset into the defensive. This required the highest speed in all operations, though this was hardly new in military history. Higher mobility has been at all times one of the fundamental prerequisites of military success. Schlieffen merely applied the old principle to countries with highly developed systems of transportation and communication.

In addition, however, Schlieffen insisted that the success of such strategic operations depended on the full control of all available space in order to keep the enemy from launching strategic operations of his own. His plan for an invasion of northern France through Belgium and southern Holland was largely dictated by the desire to hold the French army within the predictable and inescapable line between the channel and the Swiss Alps. He warned again and again that it was essential for the German army to reach the channel and Abbéville, since otherwise outflanking moves by the enemy could hardly be avoided. Faced by a German army with its right wing safely protected by the French coast the French army would find it extremely difficult to extend its left wing in time and in sufficient strength to halt the German offensive, particularly if the center of the French army was at the same time locked in battle.

The Schlieffen plan left the French hardly any strategical choice except one which was bound to increase their troubles. The French could take the offensive against Alsace-Lorraine but they could not gain decisive results from this; on the contrary, they would have removed troops even farther from the crucial area of northern France. Schlieffen was not willing to count on such errors which he liked to call a "favor" (*Liebesdienst*) of the enemy. Naturally he hoped that mistakes of the enemy would facilitate successful operations, but he expected them to occur as the result of hasty countermoves to the unexpected German dispositions. He was bold and ready to run risks, but he was no gambler and therefore refused to anticipate a faulty French plan in drawing his own strategical blueprint.

If Schlieffen was to turn a whole war into a single battle, preconceived and prepared simultaneously with the plans for the mobilization, transportation, and first concentration of the army, he needed a fully integrated organization of the army command. Early, he envisaged the commander in chief as the first World War saw him: "The modern commander-in-chief is no Napoleon who stands with his brilliant suite on a hill. Even with the best binoculars he would be unlikely to see much, and his white horse would be an easy target for innumerable batteries. The commander is farther to the rear in a house with roomy offices, where telegraph and wireless, telephone and signalling instruments are at hand, while a fleet of automobiles and motorcycles, ready for the longest trips, wait for orders. Here, in a comfortable chair before a large table, the modern Alexander overlooks the whole battlefield on a map. From here he telephones inspiring words, and here he receives the reports from army and corps commanders and from balloons and dirigibles which observe the enemy's movements and detect his positions."[6]

He explained, however, to his officers: "It will be impossible always to issue a specific order of battle. In any case, the order of attack has been replaced by the marching order, and by this is meant not the marching order which leads directly to the battlefield, but the marching order which starts the army moving after its first concentration has been achieved, and which leads only in the end to an encounter with the enemy. Assuming a smooth and normal course of events, army corps meeting the enemy will simply have to form a battle line and attack. Out of the direction of their marching orders follows the envelopment, the break-through, etc., in brief, the form of the battle which the commander-in-chief has planned. But it is unlikely that everything can be anticipated so that such a simple course can be assumed. Various incidents may happen which may necessitate a certain deviation from the original plan here and there. It will not be possible to ask the commander for orders in this case, since telegraph and other communications may not work. The corps commander will be faced with the necessity of arriving at a decision of his own. In order that this decision should meet the ideas of the commander-in-chief he must keep the corps commanders sufficiently informed, while, on the other hand, the latter must continuously strive to keep in mind the basic ideas of all the operations and to enter into the mind of the commander-in-chief."[7]

In contrast to Moltke, whose flexible plans of operations had made allowance

---

[6] Schlieffen, *Cannae*, p. 278.   [7] Schlieffen, *Dienstschriften*, II, 49.

for considerable mistakes in execution, Schlieffen's strategy called for a high degree of exactness. To be sure, Moltke was not in a position to impose his will to the same extent as his fortunate heirs. In any event, however, modern mass armies needed stricter coordination if they were to remain manageable in mobile operations and particularly in circumstances where maneuverability was supposed to make up for the relative weakness in numbers.

Schlieffen, therefore, laid an even sharper emphasis on strategy in training the members of the general staff. Many officers were surprised to see how even the youngest officers of the general staff were permitted to direct maneuvers of large army units. This did not lead to a disregard of tactical knowledge in the German army since its mastery was wide-spread, but often enough it produced a lack of respect for the older front-line commanders who were far removed from the source of strategic wisdom. The fateful role which Lieutenant Colonel Hentsch played during the battle of the Marne by ordering the retreat of the German armies on the right wing against the better judgment of the commander and chief of staff of the First Army illustrates the point. However, the whole state of the German high command under Schlieffen's successor, the younger Moltke, had changed so completely that a direct comparison is impossible. Schlieffen, in speaking about the future command, had always assumed that the army would have a real leader as its chief, not a man who would vacillate in his own strategical concepts and would, at a crucial moment, let a junior member of his staff make historic decisions in his place.

Nor was the French campaign of August 1914 a test of the Schlieffen plan. The situation of 1914 was different from the one existing in 1905, which Schlieffen tried to solve in his memorandum. Since 1905 Russia had gained fresh strength, while in France, under the influence of Colonel Grandmaison's teaching the idea of offensive warfare had been firmly established in the general staff. The younger Moltke followed Schlieffen in planning for a weak defense in the east and strong offensive operations in the west. In the latter, however, Schlieffen's conception of the wheeling right wing was greatly changed. Whereas Schlieffen had postulated a ratio of 7:1 between the right and left wing of the German army, Moltke accepted 3:1. Moltke was worried about the expected French drive into Alsace-Lorraine and rightly so, but there was no justification for strengthening the southern wing of the German army excessively, since it enjoyed in the defensive the advantage of strong fortifications. In fact, Moltke himself would have made different dispositions if he had not believed that a French offensive would afford the opportunity for a decisive battle in Alsace-Lorraine. If the French left their fortresses and sent half of their army into Lorraine, their southern wing could be driven against the Vosges and the Rhine and its annihilation would have shattered all hopes for further French resistance. Moreover, the Lorraine battle could be fought three to four weeks before a decision could be expected under the Schlieffen plan of 1905.

Thus, the right wing of the German army assumed a different meaning in Moltke's view; its chief task was to induce the French to launch their offensive into Lorraine. In this case the German armies had to make their way through Belgium, but the continuation of their march toward Paris might become less important if the German left wing was meanwhile able to strike. In vague emulation of his uncle, Moltke chose an open system of strategy. While Schlieffen wanted to give the French no chance of determining the course of

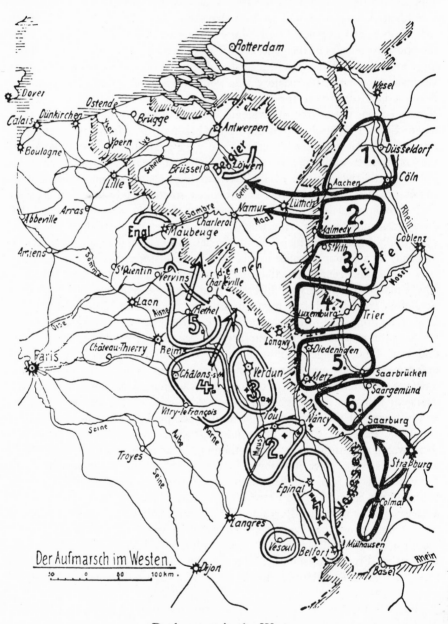

Deployment in the West

German strategy by their actions, Moltke made the conduct of operations partly dependent upon "the enemy's independent will." This attitude introduced an element of uncertainty about the ultimate strategical aims of the German high command. To ensure coordination and to make the final decision about the course of operations Moltke ought to have kept in closest contact with the troops and molded their actions according to his system of military thought.

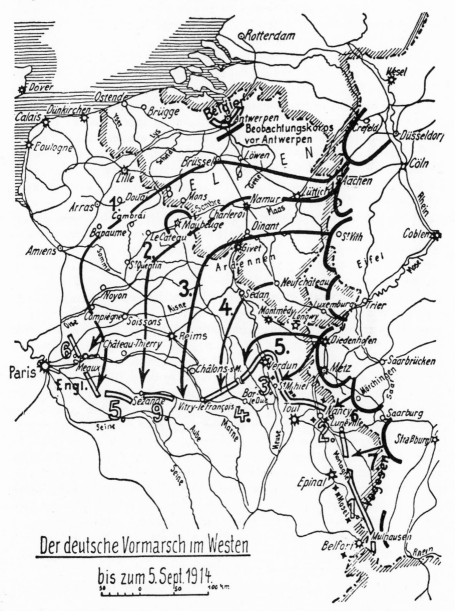

German Advance in the West to September 5, 1914

It may well be that Moltke could have realized his dream of a decisive battle in Lorraine if he had insisted that the Sixth and Seventh Army act in accordance with the original plan under which they were supposed to fall back, in order to draw the French away from their fortresses, and if he had simultaneously taken steps to slow up the rapid progress of the German right wing and to transfer strong forces to Lorraine. Ludendorff's chief of operations in 1917-1918, General Wetzell, presented a well argued case in 1939 for the success of such a scheme.[8] Instead, Moltke tolerated a premature offensive of the Sixth Army by frontal attack which forced the French back, but merely into the safety zone of their own fortresses. Thereafter he did not even prohibit the direct assault of the French lines which brought the Sixth Army into serious difficulties and made it impossible to withdraw troops for use on the right wing. Probably he found in the memory of the elder Moltke a justification for his leniency toward the individual army commanders, though his great personal modesty and his lack of resoluteness may have been the real reason.

The German right wing of 1914 was from the outset too weak to accomplish all the aims which Schlieffen had assigned to it. It was very unlikely that the German armies could reach the channel and operate west and south of Paris. The German right wing was further weakened by the transfer of two army corps from Belgium to Eastern Prussia on August 25. They were missing during the battle of the Marne and were still rolling toward the east when the battle of Tannenberg was fought. Undoubtedly troops to strengthen the German Eighth Army in the east should have been taken from the German left wing where reserves were available.

Moltke acted under the impression of the Lorraine battles of August 20-23, which seemed to have opened the opportunity for great decisions. On the other hand, the progress of the German armies through Belgium had been steady and rapid, and they had just passed the line Brussels-Namur, which was, next to Liége, the most dreaded bottleneck. Still their real task had just begun since they had not succeeded in breaking the morale of the Allied armies or in encircling individual enemy units, like the British Expeditionary Force or the French Fifth Army. The need for fresh troops was bound to grow in new weeks of forced marches. In spite of the transport of the two German army corps to the east and unnecessary employment of an additional corps for the siege of Maubeuge a little while later, the German high command had the means at its disposal to strengthen the German right wing by bringing up troops from the left by railroad or at least by turning the operational direction of the German center more toward the right. Moltke continued to believe, however, that the German left wing, though perhaps unable to perform its original strategical task, was keeping strong French forces busy, thus making their transfer to the Paris region impossible, and enabling the German right wing to achieve the aim of the Schlieffen plan.

This faith in the Schlieffen plan, which he himself had weakened and

[8] "Das Bild des modernen Feldherrn," *Militär-Wochenblatt*, pp. 2257-2264; 2329-2338; 2406-2409.

changed, was now a desperate one. It is apparent that a strong offensive against the western flank was at this stage the last remaining chance for German success, and consequently that everything should have been done to give the First, Second, and Third German armies the overwhelming strength that they needed for delivering a decisive blow. Moltke could even have given those armies a brief period of badly needed rest, provided he had taken all measures to ensure the superior strength for the coming battle. In the meantime, however, the French high command was maintaining a firm direction of the operations of all its armies even in the general retreat, and the railroad system of France was being used to the best advantage to equalize the odds and to prepare an offensive for the right moment. Moltke was out of touch with the actual situation along the front, communications were poor, railroad transportation was limited and neglected, but the chief cause of trouble was the uncertainty about the ultimate strategical plan of the German chief of staff. The commanders of individual German armies committed errors of both omission and commission, but they could not be blamed severely since the commander in chief left them largely in the dark about the meaning of the operations as a whole.

Finally the First Army succeeded in achieving the impossible by extricating itself from supreme danger by sheer boldness. It seemed just about ready to harvest a substantial part of the gains expected under the old Schlieffen plan, when Lieutenant Colonel Hentsch, whom Moltke sent to investigate the situation of the western flank, gave the order to retreat. Moltke, who had tried to follow an open system of strategy and to avoid the rigid premeditation and control of the Schlieffen plan, had generally accepted the initiative of the various army commanders during the early strategy of the war. Now, at the height of the crisis, he attempted to restore coordination by vesting authority in a younger member of his staff. Hentsch was an able officer who distinguished himself later as chief of staff in the Serbian campaign of 1915. His decisions of September 8, 1914 were largely the result of his surprise at finding the real battle conditions of the German right wing much more serious than the high command stationed in Luxembourg had visualized.

The Schlieffen plan, even in its emasculated form and in spite of the irresolute and ambiguous direction of German strategy, still gave the German offensive of 1914 a deadly force. It would not have fallen short of ultimate success if the German high command had from the beginning placed full faith in it instead of being diverted by the Eastern Prussian war and by the notion of a decision in Lorraine. Schlieffen had been right when he predicted that it would be extremely difficult to achieve a more than ordinary victory in Lorraine since the French would be able to retire to their fortifications. He did not deny that in war new situations might arise and make fresh dispositions on the part of the chief commander necessary. However, the concentration upon the right wing would have unified all operations, and the commander in chief could have contented himself with impressing the ultimate meaning of the common operations upon the individual army chiefs. Even in the event of a breakdown of communications, the army commanders would have been likely to act in accord-ance with the underlying general conception of the operations.

Schlieffen's plan of 1905 was not his last answer to the problems of a future war. As has been stated before, he was inclined to assume in 1905 that the French would not leave their frontier fortifications. This was entirely correct, the French plan of mobilization No. 15[9] envisaging a strategic defensive. In the years of his retirement he noticed, however, the growing influence in France of the "neo-Napoleonic" school of offensive warfare. He was afraid that the French general staff would try to meet the German offensive through Belgium by an early occupation of the line Namur-Brussels-Antwerp. Schlieffen's fear was not wholly justified. It was true that in 1911 General Michel, who was then considered for appointment as chief commander by the French Supreme War Council, presented a plan which anticipated almost exactly the German opera-tions of 1914. In its actual execution the Michel plan came close to the Allied operations in 1940, and one may even suspect that Gamelin was influenced by it. In 1911 the plan was not accepted and Joffre was designated as commander in chief. Plan No. 17 was adopted, which was built on the assumption that the Germans were likely to invade Belgium but would not have sufficient troops to pass beyond the Meuse. No French officer or historian has ever tried to explain this gravest blunder of French military intelligence.

It is interesting to see that Schlieffen even in his years of retirement did not become a victim of wishful thinking by counting on favors of the enemy. He continued to keep abreast of new developments in the field of tactics and tech-nology. Fearing a French occupation of the Antwerp-Namur line and a subse-quent offensive in Lorraine, he argued that this would necessitate the creation of strong reserves behind the German front. As the best countermove he pro-posed to use these reserves from the very beginning to strengthen the German battle forces and to take the initiative along the entire front from Belfort to Liége. He still maintained his belief that only the progress of the German right wing would bring great strategic results. At the same time he thought he discovered new strategical opportunities in the field of railroad transportation. The dictum of the older Moltke that an error in the original concentration of the army could not be rectified during a campaign, began to lose its validity under the conditions existing in western and central Europe. The density of railroad lines would enable the commander to switch troops from one flank to the other, and in the last war games of the general staff which Schlieffen directed such maneuvers were practiced on a large scale. The German high command of 1914 showed, however, that these ideas had not yet affected the thought of Schlieffen's successor. The French proved far superior in using this new source of mobility, as they were also the first to discover the possibili-ties of motor transportation when Galliéni requisitioned the taxicabs of Paris

[9] The plan was drawn up in 1903 and slightly modified in 1906 and 1907. The chief source for the French plans of mobilization from 1871-1914 is the official French work on the first World War: *Les Armées Françaises dans la Grande Guerre*, ed. by Ministère de Guerre, Etat Major de l'Armée, Service Historique Vol. I, 1 and Vol. I, Annexes (Paris, 1922).

to move troops to the embattered Marne positions, and when Foch used motor transport to move 60,000 men to Flanders later in September 1914.

The battle of the Marne did not ruin Schlieffen's fame among German officers. On the contrary, the long and hopeless war of position in the west, with all its consequences for Germany's social and economic order, appeared to be the dire outcome of the disregard of his military genius. In the east Schlieffen's teachings had led to many striking victories, like Tannenberg and the winter battle of the Masurian Lakes in 1914, or Herrmannstadt at the opening of the Rumanian campaign of 1916. Such victories enabled the German army to fight a world-wide coalition for four years and brought triumph almost within reach. Thus even the ultimate defeat of Germany in 1918 did not weaken the belief in Schlieffen's mastery of modern war which became steadily greater as Germany launched her rearmament.

Although there existed an orthodox Schlieffen school in the German army during the interwar period of 1920-1939 the stategists of the *Reichswehr* and of Hitler's army—men like Groener, Seeckt, Fritsch, and Beck—were by no means uncritical in their adaptation of Schlieffen's strategical concepts. The new German army was imbued with his belief in the power of the strategic initiative, of mobility, and of encircling maneuver. But at the same time the new lessons of the World War had been driven home by defeat. After the German offensive against France had come to a standstill in September 1914, the problem of defensive warfare had assumed great significance. Most of the fighting along the endless eastern front had also been in fact defensive fighting. After 1933 while the German army was still weak, the problem of defense ranked uppermost in the thought of the general staff. In the intellectual arsenal of the present German general staff there are probably more ideas about defensive warfare than were put into practice in the early years of the second World War. Field Marshal von Leeb's studies on defense of 1937-1938 shed some light on these efforts.[10] It is also clear that the German officers are now inclined to put greater reliance on fortifications than formerly.

Of much greater importance was the experience gained in the use of frontal attacks. The trench warfare of 1914-1918 compelled the German high command to seek a new strategic approach to a decisive battle. The tactical frontal assault merely confirmed Schlieffen's prediction that the enemy would be forced to retreat but would be free to resume the fight in new positions. For decisive results the lines of the enemy had to be pierced to such an extent as to endanger his rear communications and to destroy his freedom of action. The first example of a successful strategy of break-through was the battle of Gorlice-Tarnow in May 1915 which had been planned by Seeckt. The results were considerable and would have been greater if the operations had been more fully prepared. A week later the Allies launched their assault against the German positions at Arras and Labassée. Thereafter both the Allied and the German

---

[10] First published in the *Militärwissenschaftliche Rundschau*, 1937. Thereafter as separate volume: Wilhelm Ritter von Leeb, *Die Abwehr* (Berlin, 1938). English translation by S. T. Possony and D. Vilfroy, *Defense* (Harrisburg, Pa., 1943).

generals worked on the problem of a break-through with subsequent strategic exploitation. The most ambitious attempt was Ludendorff's offensive in France during the spring of 1918, undertaken with the aim of forcing a decision of the war before Germany's strength gave out. No strategic success was obtained, however. A deep bulge was driven into the Allied front, but no deadly rupture occurred.

The failure of the spring offensive of 1918, together with the breakdown of the Schlieffen plan in 1914, were the chief subjects of critical military discussion in Germany after 1920. Two years before the second World War the chief artillery expert of the Ludendorff era, Krafft von Dellmensingen, summed up the debate in his comprehensive book *The Break-through* for the use of Hitler's army. In a way he still reiterated Schlieffen's doctrine that "the break-through always remains the most difficult form of a decision," that it is "only a preparatory move, which may inflict heavy penalty upon the enemy," but that "the final victory can only be achieved by subsequent operations of encirclement."[11] However, the author added that the necessity of attempting a break-through could hardly be avoided in future. "No army will consequently be as one-sided as to cultivate exclusively theories of encirclement."[12]

The solution seen here, as in similar German studies, was the restoration of surprise and mobility through mechanized and motorized forces. It was in this field more than in any other that the German army learned from its enemies. It was not forgotten that those "black days" of July 18 and August 8, 1918, when Allied tanks rolled forward at Soissons and Amiens, had sealed the defeat of Germany. These tank assaults were great tactical achievements even though their strategic scope was limited. The German general staff of the 1930's endeavored to expand such tactical possibilities, in particular by combining them with the use of aerial weapons. But its major aim was to develop tactics which would lead to the resurrection of Schlieffen's strategy of encirclement and annihilation which seemed to have lost its edge in the war of position from September 1914 to 1918.

In this sense the German offensive against France in 1940 was still animated by Schlieffen's ideas. The Schlieffen plan itself was not revived, however. The plan of 1940 called for a break-through of the center of the Allied front, which was achieved at Sedan on May 14, 1940. Two battles of encirclement were to follow, the first one to be fought against the French and British armies of the north which were to be driven to the Belgian coast and the English channel. The second battle of encirclement was to be launched from the Sedan-Abbéville line and was aimed at forcing the southern French armies against the Maginot fortifications and the Swiss frontier. Hitler was right, therefore, in stressing before the Reichstag the distinction between the campaign of 1940 and that of 1914. The second phase of the French war offered certain parallels to the Schlieffen plan, while the first phase, the battle of Belgium, resembled in some measure the strategy of the elder Moltke against MacMahon at Sedan. How-

11 K. Krafft von Dellmensingen, *Der Durchbruch*, Hamburg, 1937, p. 405.
12 *Ibid.*, p. 407.

ever, both phases of the battle of France were dependent upon the initial break-through, which was made possible by the tremendous superiority of the offensive over the defense at that particular moment of history.

This rare opportunity affected the general character of German strategy profoundly, but it was Schlieffen's teachings which had helped to guide German military thought toward a new faith in mobile warfare. His influence on the last half-century of German military history was unique. Still, though the military tradition of Germany reached a new height through his activities, there were already signs that the German school of strategy had lost the idealistic energy and realistic strength of earlier times. Scharnhorst and Gneisenau had been military as well as national reformers. They had wished to reform the Prussian army not merely for the war of liberation from Napoleon but to build a more liberal Prussia as well. These two reformers considered the problems of war with a view to the peace to follow, knowing that the social implications of any military organization were far-reaching. In full harmony with their ideas, Clausewitz taught that war was a political act and that politics and warfare had "the same logic although they were using a different grammar."

The period of reaction after the Congress of Vienna frustrated all attempts to maintain a close contact between the army and the new social and political forces. The Prussian gentry regained full control over the army, which became the chief bulwark of monarchical conservatism in Germany. Outside of the preservation of the existing monarchical institutions and particularly the royal prerogatives in army affairs, the officers took no part in political matters and kept aloof from the new ideas of the century.

Moltke the Elder still had a great deal of Clausewitz's universal intellectual interest, and the stormy decades of European history between 1848 and 1871 stirred his thoughts deeply. His own political answers to the political problems of his time were all strictly on the conservative side, however, even more so than was the rather opportunistic policy of Bismarck. Moltke strongly asserted his authority against the chancellor in military affairs, but he accepted Bismarck's political leadership. Bismarck's successes at home and abroad enabled the army to fall back into its reactionary lethargy, for they freed it from all fear of parliamentary or popular control. Consequently the army was willing to follow the imperial government blindly even after 1890 when William II took over the functions which William I and Bismarck had held.

Schlieffen, as the chief military adviser of the crown, should have raised his voice against the dangers to German security which William's policies created. The naval program of William II and Tirpitz drove Britain into the opposite camp, but Schlieffen did not warn the government, although the state of German armaments and war plans made it impossible to disregard the threatening character of the international situation. Schlieffen's plan of operations, as has been seen, was based upon the expectation that the complete defeat of France would induce Britain to make peace. This was, however, not more than a hope since the Prussian general staff never considered an invasion of England. If the German-Russian war had continued after a defeat of France, Britain could at least have crippled German commerce and industry and thus forced that complete change of the German economic and social system which Schlieffen dreaded so much.

It was even more surprising that Schlieffen was unconcerned about the role of the German navy in a national program of defense. It could have no part in the type of war Schlieffen had planned for, and the building of a navy of such a size was, therefore, a waste of money and manpower. The army felt this all the time, for it was unable to procure enough funds and officer candidates for the formation of the new divisions which were needed for the implementation of the Schlieffen plan. Schlieffen did not complain, however, nor did he seem worried about the international aspects of the naval building program whose ultimate effect would be to bring a British army to the continent. He pondered even less the problems of the existing system of German government, in particular whether the army might not need a closer contact with the new social forces in order to reach the peak of its effectiveness in days of national emergency.

Schlieffen never questioned the autocratic rule of William II. Even in his own military domain he abstained from intimating that the emperor's display of military ineptitude might be ruinous to the monarchy of Frederick the Great. William II liked to gain maneuver victories by colossal cavalry charges, which he would lead himself, and his subsequent critiques of staff performances betrayed a similar lack of dexterity in military affairs. Such practices caused fear and resentment among the officers, which Schlieffen would parry with the statement that criticism of the emperor would undermine the monarchical authority upon which the morale of the Prussian army rested. His blind faith in monarchy was not even shaken by the appointment of the younger Moltke as chief of staff, though it led Schlieffen to voice grave admonitions that the serious character of Germany's strategical situation would allow no false military steps.

Schlieffen's blind faith in monarchy kept him from recognizing that the deepest problems of war transcend the realm of mere military proficiency. No modern general can hope to emulate Marlborough, Prince Eugene, Frederick the Great, or Napoleon in combining political with military command. Military and political affairs have grown too complex, and their mastery requires long professional experience in either field. Still, the fact that war is an act of politics has not changed. The highest form of strategy is the outcome of military excellence enlightened by critical and constructive political judgment. This truth, of which the founders of the Prussian school of strategy were well aware, was forgotten by Schlieffen and his students.

The failure of the war of movement in the west succeeding the battle of the Marne confronted the military leaders of the second German empire with the full impact of politics, international and national, upon the conduct of war. It became necessary to improvise a total war mobilization and war economy. The outcome of the war became dependent upon an amphibious rather than a land strategy. The pupils of Schlieffen were ill prepared for the reality of a world-wide war, which had to be fought as much with political, economic, and psychological weapons as with infantry and artillery. Ludendorff thought that he could direct those forces as easily as the army, and by subordinating the government to the orders of the high command turned Germany into a virtual military dictatorship. But the army failed to hold the home front.

The German revolution of 1918 made the officers the logical target of popular indignation and, though the army never forgot its humiliation and remained a reactionary and subversive power in the short lifetime of the German Republic, the best brains of the German army had become convinced that Ludendorff's error should not be repeated. This did not mean in their opinion that the army should ally itself with the new popular forces. It meant that they should cooperate with any political movement that would permit them to concentrate, as in Schlieffen's day, on strategy as a professional, or as they liked to call it, "unpolitical" task. This theory led even the most intelligent members of the German general staff to tolerate Hitler's rise to power in Germany. The German generals received from Hitler all the tools for renewing the war of 1914 to 1918, but they are learning again in our day that "war is an act of politics." This time, however, the logic of Nazi politics is dragging them to defeat in a war for which they had worked out the plans of a brilliant Cannae strategy. Schlieffen and his followers always disregarded the historical fact that Carthage was defeated in spite of Hannibal's victories.

CHAPTER 9. Du Picq and Foch: The French School

BY STEFAN T. POSSONY AND ETIENNE MANTOUX*

THE long period of virtually uninterrupted peace which intervened between the fall of Napoleon and the Italian and German wars of unification led quite naturally to a stagnation in the development of military theory. Between Waterloo and Solferino, the French army had no experience of full scale combat—the Crimean War was largely a war of sieges —and it was content to rest on the laurels it had won in the period of the Revolution and the Empire. The conquest of Algiers stimulated a certain amount of interest in new theories of colonial warfare,[1] but as far as the European continent was concerned all possible wisdom seemed to be contained in Napoleon's campaigns and seemed to be definitely formulated in Jomini's handbooks. These works served as a sort of vade mecum for all French officers, and all other theoretical knowledge and study seemed superfluous.[2]

It was not until after the Austro-Prussian war of 1866 and the Prussian triumph at Sadowa, which far overshadowed the French success of 1859, that French military authorities began to suspect that all was not well with their army organization and staff work. Reforms were instituted in great haste, and there developed a lively literary discussion on the problems of military organization. The most important and most lasting contribution to this debate is represented by the work of Ardant du Picq.

I

Of the life of Ardant du Picq very little is known, except that he was born on October 19, 1831 at Périgueux (Dordogne)—a region where many great Frenchmen have seen the light of day; that he was a graduate of Saint-Cyr and served in the Crimean War until taken prisoner at Sebastopol; and that he resumed his military service in Syria and Algeria and was killed while leading his regiment near Metz in the first days of the Franco-Prussian War (August 15, 1870), in much the same manner in which Turenne lost his life at Sassbach. When he died, he had just published his study on *Combat in Antiquity*, in an edition intended only for the private use of a few officers. A new edition was printed in a military magazine, the *Bulletin de la Réunion des Officiers*, in 1876/77. In 1880, some of his writings were published for the first time in book form, but not until in 1902 were his *Etudes sur le Combat* published in their entirety. This edition served as a basis for all subsequent editions; it purports to reproduce du Picq's complete works, in so far as they can be complete, as he left most of his manuscripts in an unfinished form. However,

---

* Dr. Possony wrote the section on Du Picq, Dr. Mantoux the section on Foch.

[1] See Chapter 10.

[2] See the two basic studies by Dallas D. Irvine, "The French and Prussian Staff Systems before 1870," *Journal of the American Military History Foundation*, II (1938), 192-203 and "The French Discovery of Clausewitz and Napoleon," *Journal of the American Military Institute*, IV (1941), 143-161.

there is reason to doubt that all the papers he left are really found in this edition, for it seems incredible that such a master of the pen should have written only a few comments on what is more or less one and the same subject, and that he should never have written anything before he was far in his forties. At any rate, his real biography is yet to be written; and notwithstanding the fact that his *Battle Studies* were, with the exception of Tolstoi's *War and Peace,* the most widely read book in the French trenches during the First World War, Ardant du Picq is still consistently ignored by the great international encyclopedias. At the present moment when almost all connections with France are severed, it is naturally difficult to trace the line of Ardant du Picq's intellectual ancestry and to place his writings in their due historical perspective. Nevertheless, it is possible to determine some of the thinkers and experiences which had probably the greatest influence upon his mind.

Ardant du Picq himself acknowledges that his military thinking was much influenced by Marshal Bugeaud, the foremost French soldier at the time when du Picq began his military studies. The marshal was a compatriot of his and was deputy from his native Périgueux. It can be presumed that Marshal Bugeaud, in some way or other, protected and advised the student officer from his home district in his military career. Bugeaud, in turn, was less influenced by Napoleon than by Suchet, who was one of the most brilliant of Napoleon's marshals and, at the same time, the most original and scientific among them. Bugeaud's soberness and objectivity permeate du Picq's whole work.

Whatever the direct influence of Bugeaud on du Picq may have been, he undoubtedly had an intimate knowledge of General Trochu's book *L'Armée Française en 1867* in which an important chapter is devoted to the problems of combat and panics, and military psychology in general. Trochu had been Bugeaud's trusted and preferred aide-de-camp, and his book—a product of the self criticism which permeated the French army after Sadowa—was one of the best sellers at the very time when Ardant du Picq began to write his study. In fact, it may be assumed that the reading of Trochu's book suggested to Ardant du Picq the composition of his own. If we add that du Picq has drawn some of his arguments from the Marshal de Saxe, Guibert, the Prince de Ligne, and Marshal Marmont, we may hope to have named his most important military ancestors.

Another source of his inspiration certainly was his own military service. Two experiences were outstanding in his military career: the incompetence of military leadership during the Crimean War which, among other things, led to the paradoxical situation that a numerically strong army was actually weak on the battlefield, because it comprised too large a percentage of nonfighting personnel; and his services in Syria and Africa where he learned that "the theory of the big battalions is a despicable theory," since it is simply not true that victory must go to the big battalions. If anywhere, it is in Africa that one must learn "to distrust mathematics and material dynamics as applied to battle principles." The highly trained French army had no trouble in repeatedly defeating superior numbers of Arabs, proving time and again the truth of Napoleon's dictum: "Two Mamelukes held three Frenchmen; but one hundred French cavalry did not fear the same number of Mamelukes; three hundred vanquished the same number; one thousand French beat fifteen hundred Mamelukes." "Such," adds du Picq, "was the influence of tactics, order and maneuver."

And he goes on to challenge Napoleon's mass theory: "Let us take Wagram, where his mass was not repulsed. Out of twenty-two thousand men, from three thousand to fifteen hundred reached the position. Certainly the position was not carried by them, but by the material and moral effect of a battery of one hundred pieces, cavalry, etc., etc. Were the nineteen thousand missing men disabled? No. Seven out of twenty-two, a third, an enormous proportion may have been hit. What became of the twelve thousand unaccounted for? They had lain down on the road, had played dummy in order not to go on to the end."

As a last important military fact which greatly influenced Ardant du Picq's thought, we must mention the heated discussion about the reform of the French army which developed after the startling Prussian victory of 1866. In 1867, Marshal Niel took over the Ministry of War not only in order to equip the French army with new and improved weapons but also to strengthen it by building up a large and powerful reserve and, in fact, to introduce universal conscription. It is well known that Niel with his military reforms could not prevail over the spirit of party politics. The Bonapartist party was afraid of putting additional military burdens upon the voters and the opposition was afraid lest the army reform should stabilize the regime of Napoleon III. It is less well known that the French army itself was in part strongly opposed to Niel's ideas. Trochu's book was intended to show possible compromises between the urgency of military reforms and the abhorrence of the French officers for conscription. Ardant du Picq was a convinced opponent of Marshal Niel and his book can be interpreted, at least to a certain extent, as a refutation of the marshal's ideas on universal military service. In the discussion of the relative attributes of the small professional army and the large conscript force, du Picq declared for the former, enlisting thus on what history was to prove the wrong side of the question.

Ardant du Picq approaches the military problem in a scientific way. War, after all, is a matter of combat and battle; its atom, so to speak, is the combat of hostile military units. Surprisingly enough, however, the theorists of war did not bother to determine the structure of this atom. They were satisfied to deal with more abstract subjects and with what they considered the general principles of war. But in things military, it is as impossible as in other fields of science and art to know anything unless the fundamental facts are clearly ascertained and understood. When Ardant du Picq tried to assemble the basic facts of battle, he discovered that those facts were unknown. In fact, after perusing the greatest military authorities one could hardly suspect that war was anything more than a game of chess.

Du Picq considered his first task to be one of establishing these facts. Since his own experience was limited, he resorted to the then novel expedient of drawing up a questionnaire, which he circulated among his fellow officers. At this time and in the traditionalist background of the French army, this method must have been something of a shock to his colleagues; and though many of them probably did not suspect it to be a sort of subversive maneuver, most of them, certainly, were inclined to consider Ardant du Picq a bore. No doubt the correct answering of du Picq's questions would have required much work, perhaps even the writing of whole books, if indeed his questions were not far beyond the grasp of the then average French staff officer. And yet, du Picq's questionnaire is a unique document of lasting value which even today could serve a very useful purpose. Let us give a brief outline of his questions:

What—he asked—became of the disposition of your troops and their marching order under the influence either of terrain or of approaching danger, or of both? If the order had been changed, could the rearranged disposition be maintained at further approach? What happened when the troops came within the range of the enemy's cannon and rifle bullets? At what moment and at what distance was a new disposition taken either spontaneously or on command to react with fire or charge, or with both at the same time? How did the firing begin, how was it continued, and how did the soldiers adjust themselves to firing conditions—how many shots were fired, how many men were lying down? How was the charge made: at what distance did the enemy flee or at what distance did your attack break down either under the enemy fire, or because of his firmness and resolution, or under the impression of some enemy movements? What was the attitude and the behavior (that is, order, disorder, cries, silence, excitement, or composure) of the officers and the soldiers on your side and on the enemy side before, during, and after the charge? Were the soldiers always manageable, had they always acted upon orders, or did they, at some moments, show a tendency to leave the ranks, either lagging behind or plunging forward? If discipline could not be maintained throughout the engagement, at what precise moment did effective control slip from the chief of battalion, at what moment from the captain, from the lieutenant, and from what moment on (if such a thing did happen) was there a sort of disorderly impulse which carried away both officers and soldiers? Where and when did the successfully charging troops come to a halt? When and where could the officers regain control? Such details which throw light on either the material or the moral elements of military action and which enable us to understand it as closely as possible are, according to Ardant du Picq, much more instructive for soldiers than the best discussions of war plans and general strategy by the most famous military leaders.

The results of du Picq's use of the questionnaire were not gratifying. Consequently, realizing its limitations, he tried to supplement it by reconstructing the reality of battle from the writings of the ancients. He turned to antiquity not only because the military experiences of the Greeks, and more especially of the Romans, were rated very highly, but also because the ancient authors were much more outspoken about fundamental military facts than the moderns. The secret of the Roman legion, the invention of which was ascribed to divine inspiration, had haunted military thinkers ever since the time of Machiavelli. How is it, Ardant du Picq asked, that the Romans, who were not essentially a brave people, were regularly victorious and defeated the most courageous nations with inferior numbers, though "it is acknowledged that the valorous impetuosity of the barbarians, Gauls, Cimbri, Teutons, made [the Romans] tremble?" Du Picq's answer, based upon that of Polybius, was that "to the glorious courage of the Greeks, to the natural bravery of the Gauls, [the Roman] opposed a strict sense of duty, strengthened by a terrible discipline of the masses. . . . The Roman, a politician above all, . . . had no illusions. He took into account human weakness and he discovered the legion."

The study of the battles of the ancients led du Picq to discover two facts which he considered of fundamental importance. The first is described in the following phrase of Marshal de Saxe: "The human heart is the starting point in all matters pertaining to war." The second is that in all battles of antiquity there is a sharp discrepancy between the losses of the victor and the defeated respectively, those of the defeated being immensely higher. At the beginning of human conflict, individuals battled against individuals, each man for himself. It is therefore often assumed by uncritical minds that a battle is nothing but the sum of numerous single combats. Two armies clash head on; each soldier begins duels with such of the enemy as are next to him; that side which, as a result of these single combats, has suffered the most serious losses, is thus confronted with a numerical superiority and has to give ground. Thus a battle is considered to be nothing but a magnified duel, the outcome of which depends upon the ability of individual soldiers, so that the army which is composed of the best bayonet fighters and sharpshooters must inevitably win. However, since the individual Gaul was undoubtedly an excellent fighter and a soldier much superior to the individual Roman, the Gauls, if this theory were true, should always have defeated the Romans and not been defeated by them. Furthermore, the tremendous discrepancy of the losses would remain unexplained.

The reality of combat, therefore, must be quite different. It is not the total number of combats which decides the issue. As a matter of fact, "success in battle is a matter of morale." "In battle, two moral forces, even more than two material forces, are in conflict. The stronger conquers. The victor has often lost . . . more men than the vanquished. . . . With equal or even inferior power of destruction, he will win who is determined to advance, who . . . has the moral ascendancy. Moral effect inspires fear. Fear must be changed into terror in order to conquer. . . . The moral impulse lies in the perception by the enemy of the resolution which animates you. . . . Maneuvers . . . are threats. He who appears most threatening wins."

In other words, du Picq, like certain writers before him such as Guibert and the Prince de Ligne, denies the existence of the shock and the physical impulse. According to him, there is a shock neither between two opposing cavalry units nor between two opposing infantry units; nor can the cavalry smash an infantry line by the mere impulse of a shock. A battle cannot, therefore, be compared to a duel in which the physically stronger and the materially better equipped party must win. It is not the physical destruction produced by weapons which brings about the decisions, nor is that side bound to be defeated which has been physically annihilated in battle. The fact is that defeat threatens the party whose moral cohesion has broken down. Weapons are only effective in so far as they influence the enemy's morale. A battle is a contest between two opposing wills, a clash of two moral powers, and not, or only to a certain extent, a collision between physical forces.

A charge is not successful because its material power exceeds the violence the enemy can bring to bear against it. A charge succeeds against an enemy who falls back, or breaks down because its opponent stands firm. The issue is decided *before* the two parties come to grips in so-called hand-to-hand fighting. This alone explains why well-equipped armies have been so often defeated by armies having only mediocre weapons and why entrenched troops or garrisons of strong fortifications have so frequently given way or surrendered prematurely. "When confidence is placed in superiority of material means, . . . it may be betrayed by the action of the enemy. If he closes in upon you in spite of your superiority in means of destruction, the morale of the enemy mounts with the loss of your confidence. His morale dominates yours. You flee." The *abordage* or head-on clash never takes place because one of the two parties evades, avoids, or eludes the enemy—if it does not flee or surrender. A fair combat between two battle lines both of which are striking with the utmost energy and resolution therefore never occurs. *"Le choc est un mot. . . . Les ouragans de cavalerie qui se rencontrent, c'est la poésie, jamais la realité."* "Never do two equal resolutions meet each other in battle. . . . The *abordage* is never reciprocal. . . . The enemy never holds his position, because, if he is holding, you flee. . . . If there were a mêlée, there would be mutual extermination, but never a victor. . . . By instinct, man prefers always the fight at a distance to the fight at close quarters." This, in effect, is the explanation for the discrepancy in the losses: losses do not occur during a head-on fight of two battle lines. In reality one army charges, the other gives way, and *then* the slaughter occurs. Casualties are inflicted on the enemy not during the shock but during the pursuit.

The starting point of productive military thinking therefore cannot be military virtue, let alone heroism; it should be fear. The skeptical, illusionless Roman "took into account human weakness and discovered the legion." Nothing can be changed in the heart of men, but discipline may make the soldier suppress his fear for a few minutes more—the very minutes which are necessary to achieve victory. The Roman generals raised the morale of their troops not by enthusiasm but by anger. "A Roman general . . . made the life of his soldiers miserable by excessive work and privations. He stretched the force of discipline to the point where, at a critical instant, it must break or expend itself on the enemy." "There is a point beyond which men cannot bear the anxiety of combat in the front lines without being engaged." Du Picq agreed with the picturesque and profound saying of General Bourbaki that an attack is *au fond*, nothing but an "escape by advance."

These, then, were the most important lessons to be gained from a study of ancient battles and according to the contemporary material and experience gathered by du Picq, they had not been disproved by modern war. But there were, he pointed out, many legends still current about successful charges and military feats which blinded everybody, "the generals as well as the *bon bourgeois* and which were the everlasting cause of the recurrence of the same kind of blunders." The famous bridge of Arcole was never taken by frontal attack, nor was the battle of Solferino in which, according to Moltke, the French had used the bayonet "with true enthusiasm," initiating a new period of bayonet worship. The Austrians, in particular, remembered their outmoded shock tactics and based their whole army reform upon the tactical idea of decisive bayonet charges, entirely oblivious of the fact that the French had used their bayonets only *after* the collapse of the Austrian lines. Moltke was a much more perspicacious observer of the true facts and from this battle drew the lesson that the Prussian fire had to be improved to the utmost. The battle of Sadowa in 1866 was the result of the respective conclusions which he and the Austrians had drawn from Solferino.

Shock tactics which never had the efficiency usually ascribed to them could not, evidently, be rendered more effective under modern conditions. "It is strange but true that the nearer we approach the enemy, the less we are close up. Good-bye to the theory of pressure. If the front rank is stopped, those behind fall down rather than push it. . . . Today more than ever flight begins in the rear, which is affected quite as much as the front. . . . This shows the error of the theory of physical impulse." The main intellectual task of the modern soldier is to get rid of such antiquated prejudices and to develop tactics that will make men fight with their maximum energy. Tactical methods must, of course, change continuously. As Napoleon rightly pointed out, an army is not of good quality unless it changes its tactics every ten years. Yet discipline and confidence are "absolutely at the basis of tactics" and their essentials are much less subject to continuous change. They form one of du Picq's most important problems.

Discipline and confidence are dependent partly upon military organization and the quality of leaders, partly upon what we might call military sociology. It is necessary that the officers be sufficiently trained to foresee the enemy's action. But even more important is resolution. "The formula is R (resolution) and R and always R and R is greater than all the MV^2 in the world." But it is not sufficient that the superior officers alone should be animated by resolution; all the grades of the military hierarchy must have a strong resolution, particularly those officers who actually lead their troops into battle. Du Picq is strongly opposed to the "tendency to oppress subordinates; to impose upon them, in all things, the views of the superior; not to admit honest mistakes, and to reprove them as faults; to make everybody, down to the private, feel that there is only one infallible authority. A colonel, for instance, sets himself up as the sole authority with judgment and intelligence. He takes all initiative away from subordinate officers, reducing them to a state of inertia resulting from their lack of confidence in themselves and from their fear of being severely reproved." Let us assume that "this firm hand, which directs so many things, is absent for a moment. All subordinate officers up to this moment have been held with too strong a hand, which has kept them in a position not natural to them. They now act like a horse always kept on a tight rein, whose rein is suddenly loosened or missing. They cannot in an instant recover that confidence in themselves which had been painstakingly taken away from them very much against their will." Finally, in addition to resolute officers, every company must have resolute soldiers and non-commissioned officers who, so to speak, form the moral skeleton of the unit; who in a crisis may serve as rallying points; and who are able to inspire weakening soldiers with new strength and moral support.

All these different kinds of resolution must be integrated into one single purpose, into a unity which "alone produces fighters." The integrating factor which "produces the greatest unity from top to bottom . . . between the commanding officers, between the commanding officers and men, between the soldiers . . . and which permits no one to escape action is . . . iron discipline." "With the Romans, discipline was severest and most rigidly enforced in the presence of the enemy. It was enforced by the soldiers themselves. Today, why should not the men in our companies watch over discipline and punish themselves?" For this seems to be "the only way to preserve discipline, which has a tendency to go to pieces . . . at the moment of greatest need."

As has already been pointed out, the Romans based discipline almost exclusively upon anger, fear, and punishment. But "Draconian discipline is not in harmony with our customs." What, therefore, are the elements of discipline in a modern army? First of all, the officers must have a good deal of self-confidence. They have to be trained to adhere strictly to the following basic rules (which were later espoused by Foch): "To verify: observe better. To demonstrate: try out and describe better. To organize: distribute better, bearing in mind that cohesion means discipline." Thus, du Picq replaces the Roman "anger" by "cohesion." This is due to the fact that in ancient times to retire from action was both a difficult and a perilous matter for the soldier, while today the temptation is much stronger, the facility for doing so greater and the peril less. Modern combat, therefore, exacts increasingly more moral cohesion and greater unity than previously. Unfortunately, only a few officers are exceptional soldiers capable of devising or improvising battle methods which vary with the enemy and which are properly adapted to each individual case. "There is then need for prescribed tactics . . . which may serve to guide an ordinary officer . . . [much] as the perfectly clear and well-defined tactics of the Roman legion served the legion commander. The officer could not afford to neglect them without failing in his duty. Of course, they will not make him an outstanding leader. But, except in case of utter incapacity, they will keep him from entirely failing in his task, from making absurd mistakes."

Success in modern war depends upon the individual valor of the soldier and of small fighting units, and this in turn depends on mutual moral pressure and mutual supervision of men "who know each other well." Du Picq did not, of course, equate valor with physical strength. To him, valor is chiefly a moral category and largely identical with "will" and "resolution." Moreover, it is not the qualities of the individual soldier but the resolution and the psychological integration—that is, the collective will of the whole fighting unit—which is decisive. The fighting units and the individual soldiers must undergo a long and thorough military training not only with a view to learning the profession of arms, but also (and perhaps chiefly) to developing beyond a "mutual acquaintanceship which establishes pride," a kind of passionate unity, comradeship, and even amity among themselves which overshadows the individuality and animates every soldier with collectivistic or group "passion . . . religious fanaticism, national pride, a love for glory, a madness for possession." Only if an *esprit de corps* comes into existence can there be that firm and conscious, intimate confidence "which does not forget itself in the heat of action and which alone makes true combatants. Then we have an army; and it is no longer difficult to explain how men carried away by passion, even men who know how to die without flinching, without turning pale, really strong in the presence of death, but without discipline, without solid organization, are vanquished by others individually less valiant, but firmly, jointly, and severally knit into a fighting unit."

"It is time that we should understand the lack of power in mob armies." Ardant du Picq never tires of explaining the importance of drill, training, and military education and the necessity of having a psychologically integrated army. He points out rightly that an army is an artificial kind of society, a "collective man" and that therefore extraordinary means are required to maintain it. "Solidarity and confidence cannot be improvised." There is indeed little doubt that armies cannot be improvised and that, if necessity forces upon a country the speedy dispatch into battle of improvised forces, heroic fighting may be the result, but rarely victorious fighting. The French Revolutionary Wars entirely bear out this contention and Gambetta's experience, which Ardant du Picq did not live to witness, undoubtedly corroborates it.

Du Picq takes care to demonstrate that the traditional militaristic forms of society and command are by no means the right way of creating a true military spirit. He is strongly opposed to the frittering away of millions a year for "uniforms, tinsel, schakos, plumes, ribbons, colors" and other kinds of *"fla-fla."* Forty-five years before the First World War, he denounced the red trousers of the French soldiers, and although this warning was easy enough to understand, it went completely unheeded. Nor did he overlook the essential weakness of a military organization whose sole foundation is "military spirit," however good it may be. "When, in complete security, after dinner in full physical and moral contentment, men consider war and battle, they are animated by a noble ardor which has nothing in common with reality. How many of them, however, at that moment would be ready to risk their lives? But oblige them to march for days and weeks to arrive at the battle front, and on the day of battle oblige them to wait a few minutes or hours to deliver it. If they were honest, they would testify how much the physical fatigue and the mental anguish that preceded action have lowered their morale, how much less eager to fight they are than a month before, when they arose from the table in a generous mood." They will overcome this moral crisis only if they have a passionate belief in the righteousness of their cause; du Picq did not fail to point to the example of Cromwell's armies. Yet even so, the overcoming of such a crisis is not made possible by creed alone, however strong, but requires a perfect integration of the army which cannot but be the fruit of long and hard labor.

In his emphasis upon quality rather than quantity, du Picq foreshadowed the ideas of General von Seeckt and General de Gaulle. He foresaw a departure from the mass army of Napoleon and thought it possible that "in these days of perfected long-range arms of destruction a small force . . . by a happy combination of good sense or genius with morale or appliances may . . . secure heroic victories over a great force similarly armed." It was for this reason that he opposed Marshal Niel's projected reforms with their intention of calling up and training vast numbers of reservists and thereby democratizing the hitherto aristocratic and professional French army. From the start, du Picq argued that "a democratic society is antagonistic to the military spirit." "What good," he wrote, "is an army of two hundred thousand men of whom only one half really fight, while the other one hundred thousand disappear in a hundred ways?" The question implies at least that du Picq believed that reservists, on principle, are loath to fight, while professional soldiers lust after battle.

It is true, of course, that the French army of the period before 1870 needed other reforms as badly as the mere numerical increase upon which Niel was insisting. Twenty-five years previously, Marshal Bugeaud had declared to King Louis Philippe that too many incapable officers reached the top ranks of the army, and this deplorable situation had never been entirely abolished. At the same time, the cadres, which are the real force of any army, were in need of a much improved education, while the military spirit of the officers themselves left much to be desired. According to General Trochu, they were animated by a special kind of *esprit capitaliste* which made them parsimonious with their own lives and their good positions.

Still, du Picq's conclusions cannot be considered valid. It is by no means necessary to have an aristocratic society to kindle a true military spirit. Without denying that there has often been a very close relationship between both, it cannot be denied that democratic societies are able to wage war successfully. Nor is it true, as du Picq asserts, that only a military nobility likes to wage war. Examples abound showing that the warrior caste was frequently forced into war against its will by bellicose, democratic masses. Under modern conditions, the traditionalist spirit of the aristocracy, more often than not, has proved to be an obstacle to the development of military power. Ardant du

Picq recommends for officers the life of an aristocrat, that is, money, little work, and leisure: "We have no finer ideal of aristocratic life than one of leisure." It does not require many words to show that, at present, a leisurely officer will be a bad officer. History indicates that in the past, also, successful officers were more often than not hard working men, even if they did enjoy a good standard of living.

It must also be considered as highly doubtful that officers who are separated from their soldiers by a wide gulf of class will be able successfully to lead modern soldiers into battle. On the contrary, those officers are most successful who do not keep aloof from their men—a truth which has been recognized even by modern militaristic Germany. Obviously, Ardant du Picq still had in mind an army which was chiefly recruited from among the maladjusted members of society and the peasants. He completely failed to realize what an immense amount of military energy may be contributed by the industrial and intellectual strata of modern society. On the whole, he did not attempt to find out how the mind of soldiers could be influenced.

It is true that the value of an army largely depends upon the value of the officers and the cadres. Nevertheless, the privates are no less important than the officers and they require as much attention as the latter. If Ardant du Picq's precept were followed, if an attempt were made to create a military spirit by establishing a more or less closed, aggressive military nobility, it may be safely predicted that such an experiment would fail under modern conditions. The wars of the last decades clearly show that whenever military leadership has been made the privilege of certain classes, or whenever persons qualified for leadership turn voluntarily away from a military career to other professions, a decline of military power inevitably results. Restricted military castes no longer foster the true military spirit; at best they are militaristic (which is not at all identical with military minded) and care usually only for their own vested interests, as was shown time and again by the militaristic clique of du Picq's own country.

Under modern conditions merit must be the only basis of promotion and the only road to honors and, possibly, wealth. A military organization based exclusively upon merit is the very condition of victory. The enormous losses of the French army in 1914 could have been avoided if the French general staff had followed Ardant du Picq's counsel where it was wise and refused to follow it where it was unwise. The prejudice against reservists accounted for the fact that in 1914 Joffre threw against the Germans only half of the forces available and gave up prematurely important positions which he considered indefensible because they were held by reservists. A modern army must be built along rational lines and no nation can afford, either politically or militarily, not to use, in addition to its standing army, as many thoroughly trained reservists as its security demands or to overlook the possibility of molding the reservist into a high quality fighter.

The famous French school of the *offensive à outrance* took inspiration from Ardant du Picq; in particular from his *dictum* that "he will win who has the resolution to advance." They interpreted this phrase in the sense that the offensive, everywhere, at all times, and undertaken with whatever means, must necessarily lead to victory. It is hardly necessary to show that this mechanical and narrow interpretation of du Picq's doctrine is erroneous. What Ardant du Picq really had in mind was the superiority of maneuver, whether on the offensive or on the defensive. Instead of a rigid system such as preached by Colonel de Grandmaison, he asked for the utmost flexibility of military techniques. Besides, the will of the enemy should not be forgotten: it cannot be broken by actions which do not take it into account.

There is little doubt that the main ideas set forth by du Picq are sound and should not be overlooked. The human heart is indeed the foundation of war and, under the strain of battle and danger, this heart is ruled by fear. It is true also that, in war, quality does precede quantity. And it is true, finally, that the tremendously powerful weapons of today are not effective by virtue of the mere weight of steel they hurl against the enemy, for "the new arms are almost worthless in the hands of weak-hearted soldiers, no matter what their number may be." Ardant du Picq certainly throws light upon many problems of modern war which, up to now, have but rarely been discussed, let alone solved. If we are certain that we cannot accept the solution of the professional army, we are, nevertheless, still confronted with the difficult task of transforming *bourgeois* conscripts into an integrated and powerful army which will not lose its unity in the moment of crisis. We have also to find new methods of discipline which take into account the fact that today "one must swallow in five minutes that dose of fear which was taken in one hour in Turenne's day" and that, with the present dispersed fighting methods, effective control is more difficult than ever before.

II

If after the battle of Sadowa doubts had arisen in the minds of some of the French officer corps concerning the perfection of their military organization, the war of 1870 brought a rude awakening to all of them. There was now no doubt that France had lost that primacy in military science which had seemed to belong to her by right of inheritance.

Among the many causes of the disaster of 1870, none was more glaring than the incompetence of the high command. Confronted with Prussian generals who were carefully trained in theory and assisted by a well-organized staff system, the French leaders had displayed an ignorance, a confusion, and a recklessness for which the bravery of their troops could not atone. If the French wished to regain the position of a first-rate military power, it was clear that a reform had to take place and that this reform had to begin at the top. In 1874, the French general staff, *l'état major de l'armée*, was reorganized on the Prussian model. This, however, was not enough; it was necessary that the officers receive an education which would enable them to fulfill the functions of modern staff officers. Not the least important reason for the incompetence of the French staff was the fact that, before 1870, purely practical abilities like horsemanship had been considered the decisive criteria of a good officer, while learning and a knowledge of military theory had been ignored. But now the necessity of good theoretical grounding for the staff officer was realized. In 1878, the Ecole Militaire Supérieure was organized and after 1880, when its name was changed to the Ecole Supérieure de Guerre, it became the intellectual center of the army and the training place of the higher officers.[3]

What kind of military theory was taught at the Ecole Supérieure de Guerre? No army wants to give up its tradition and to abandon its past, but it was clear that the traditional French military theory could be revived only if it was revised and adapted to modern circumstances. Yet the renaissance of French military thought, which took place in the period between the Franco-Prussian war and the First World War, was not limited to a modernization of the traditional heritage. New influences also made themselves felt and were of extreme importance. Ardant du Picq's *Etudes sur le Combat* were published in book form in 1880 and seemed to open an entirely new perspective. Furthermore, in their search for an explanation of their defeat, the French turned to the study of German military thought and discovered for the first time the work of Clausewitz, which immediately began to have a powerful and revolutionary influence on French military thinking.

In the year 1885, when the first lectures on Clausewitz were given by Cardot at the Ecole de Guerre, a young officer, Ferdinand Foch, entered the school. Nine years later, in 1894, he was himself professor at the college. By blending the traditions of the past with the exciting new discoveries, which were made in the years of his apprenticeship, into a unique and original system, he became the re-creator of French military thought and the most important and influential figure in molding the intellectual outlook of the French officer before the First World War.

III

In the opening chapter of his first important book, *The Principles of War*,[4] Ferdinand Foch set out to refute the notion that war could be taught only on the battlefield. The old axiom that war can be taught only by war, he wrote, is a spurious one. For no study is possible on the battlefield where "one does simply what one *can* in order to apply what one *knows*."[5] Therefore, in order to *do* even a little, one already has to *know* a great deal and to know it well. This was the lesson of the successes of the Prussians who, after intensive academic training but with no experience of war after 1815, had beaten the Austrians in 1866, notwithstanding actual experience gained by the latter in 1859. The case of France in 1870 was a still better example.

Hence the necessity and possibility of teaching the theory of war, on the basis of definite historical cases. Foch never produced a systematic treatise on the art

---

[3] On this development, cf. Dallas D. Irvine, "The French Discovery of Napoleon and Clausewitz," *Journal of the American Military Institute*, IV (1940), 143-161 and "The Origin of Capital Staffs," *Journal of Modern History*, X (1938), 161-179.

[4] Ferdinand Foch, *The Principles of War*, trans. by Hilaire Belloc (New York, 1920).

[5] *Ibid.*, p. 5.

of war. "Shepherds' fires lit on a stormy coast to guide the uncertain seaman" were the words he used to describe his "Principles." But we can find in this work a discussion of "certain fundamental points in the leading of troops, and, above all, the direction which the mind must be given so that it may in every circumstance conceive a manoeuvre that is at least rational."[6]

It is clear, from the two books which Foch wrote before the war of 1914, that he had been under the influence of Clausewitz to a greater extent probably than of any other military theorist. As a consequence, most of his historical illustrations were taken either from the Napoleonic wars or from the campaign of 1870, of which he made a detailed study in his book *De la Conduite de la Guerre*. As Captain Liddell Hart has observed,[7] there is little evidence that he followed Napoleon's advice "to read and reread the campaigns of the great captains," from Alexander to Frederick. To this fragmentary historical knowledge, Captain Liddell Hart has attributed some of the shortcomings of Foch's strategy in 1914-1918; but it must be observed that even Clausewitz seldom employed historical illustrations which antedated the wars of the eighteenth century, and Foch, in his teachings, "acted as an amplifier of Clausewitz' more extreme notes."[8]

The originality of Foch, therefore, lies less in the expression of new principles of strategy than in the special emphasis attached to a few very simple notions which have remained the symbols of his tradition. They reflect the duality of his own character: the intellectual element and the philosophy of reason, the spiritual element and the exaltation of the will. It is true that they often appear as little more than platitudes; but the student of military thought must confess that the highest principles of strategy are made up of little else.[9]

Foch began his first book by asserting that principles of war of a permanent value did exist; but he hastened to add that these principles should be qualified by application to particular cases, for, "in war there are nothing but particular cases; everything has there an individual character; nothing ever repeats itself."[10] Here, at the very outset, we have the core of Foch's "doctrine," and at the same time the key to his future conduct, which enabled him to escape from the inadequacy of his teachings when faced with the realities of the battlefield: a reconciliation between fixed and permanent principles and the ever changing conditions of the art of war.

The formula was taken from the words of General Verdy du Vernois upon his arrival on the field of Nachod in 1866. "In the presence of the difficulties which faced him, he looked into his own memory for an instance of a doctrine that would supply him with a line of conduct. Nothing inspired him. Let history and principles, he said, go to the devil! After all, what is the problem?"[11] *De quoi s'agit-il?* This maxim has since been repeated *ad nauseam*; it is true, nevertheless, that it will stand as the expression of the Fochian paradox: the

---

[6] *Ibid.*, p. v.
[7] B. H. Liddell Hart, *Foch, The Man of Orleans* (London, 1931), pp. 21, 468.
[8] *Ibid.*, p. 23.
[9] "Monsieur de la Palisse est mon meilleur ami," Foch once declared to Major Bugnet.
[10] *Principles*, p. 11.   [11] *Principles*, p. 14.

combination of highly abstract, almost abstruse, metaphysical generalizations[12] with a common sense reduced to its barest rudiments and a freedom from ready-made solutions. This common sense is perhaps, after all, the ultimate secret of all the science of strategy. Yet it was the merit of Foch that he impressed upon his pupils, and no less upon himself, the constant necessity of freeing oneself from the shackles of preconceived theories.

The importance attached by Foch to this necessity of constant reflection and of perpetual improvisation and adaptation in the midst of action, found its expression in his criticism of the German campaign of 1870.[13] One of Napoleon's favorite maxims, and one which Foch quoted most frequently, was: "War is a simple art: its essence lies in its accomplishment." Foch did not minimize the value of careful preparation: the whole outcome of a war may depend on the manner in which the first battle had been engaged. But he believed that it was impossible to elaborate with any certainty a plan of operations beyond the first battle. Again quoting Napoleon, he showed that the emperor "never had a plan of operations; but this did not mean that he did not know where he was going; he had a plan of war, and a final aim. He marched on and selected, on his way, according to circumstances, the means of reaching this end."

Although Moltke had recognized this impossibility of abiding by a preconceived plan, the weakness of his campaign of 1870, observed Foch, lay precisely in the absence of action on the part of the high command, once the plan of operations had been left to the initiative of the generals. The Prussian plan was based upon "a constant and almost exclusive appeal to reason: the reactions of the enemy were conceived as inspired by a logical, rational consideration of his best interest. . . . Against this enemy, they elaborated an attack essentially preconceived. . . ."[14] If they did not behave according to plan, the plan would collapse, unless the commander in chief were always present, ready to adapt his decisions to the changing conditions. But the conduct of operations by the high command was "indirect, blind, and unreal . . . success came, not from a combination conceived by Moltke with precision and executed by his troops to the letter—the troops, rather, won victories where and when the commander was not expecting it."[15] The French Army, argued Foch, was not beaten by a faultless strategy (although he was otherwise lavish in his praise of Moltke), but because the French high command, in its incompetence, was unable to profit by the mistakes of its opponents, the chief of which was the rigidity in the plan of operations and the absence of continuous direction by the high command.

Retrospective criticism of military operations is open to all the shortcomings peculiar to hypothetical history based on the famous argument "if only."

---

[12] "This officer, during his professorship at the Ecole de Guerre taught metaphysics, and metaphysics so abstruse that it made idiots of a number of his pupils," said a police report to Clemenceau in 1908, when he was deciding whether to appoint Foch head of the Ecole de Guerre.

[13] *La Conduite de la Guerre* (Paris, 1905).

[14] *Ibid.*, p. 478.   [15] *Ibid.*, p. 481.

Foch's observation, however, is of interest, because today it is generally admitted that one of the causes of the German defeat at the Marne in 1914 was precisely this same aloofness on the part of the high command. "Victory," Foch remarked after the war, ". . . prevented the Germans from perceiving the faults they had committed; subsequently, it caused them to persist in their errors. . . . The plan conceived by Schlieffen was excellent, but it was badly executed. Imagine Napoleon at the head of the invading armies. He would not have stationed himself three or four hundred kilometers behind the lines . . . he would not have left to subordinates the initiative and the onus of decision; he would have controlled events instead of letting them happen. Moltke did not emulate his example; but Joffre did, and that is why he won the battle of the Marne."[16]

Foch's concept of the conduct of war therefore appears as a fine balance between rationalism and empiricism; the acquired habit of applying general principles and the faculty of adapting solutions to existing conditions are the secrets of successful strategy.

What were the general principles? And, first of all, to what type of war were they then to apply? Before reviewing Foch's principles of strategy, we must briefly examine his general concept of war.

Although Foch's concept of war followed almost exactly that of Clausewitz, it must be remembered that Clausewitz had merely summed up the effects of the transformation, brought on by the wars of the French Revolution, from limited to national warfare and that this trend had been predicted by Mirabeau even before the outbreak of the Revolution. The new character of "absolute war" was therefore no novelty, least of all to the French. Yet the neglect of this fact had been at the root of the defeat of 1870. "It is because we ignored that radical transformation among our neighbors and the consequences it was bound to bring about that we, who had created national war, became its victims. . . . It is because the whole of Europe has now come back to the national thesis and hence to nations in arms that we stand compelled today to take up again the *absolute* concept of war as it results from history."[17] It is for this very reason that Foch selected his historical references from a limited period, the modern period of the "people's wars."[18]

Although the adjective "absolute" was used by Foch, it is quite clear that he did not perceive all its implications, as they were to make themselves felt after 1914; that he had but a faint notion of the necessity of total economic mobilization; and that even the experience of four years of warfare did not impress upon him the true importance of naval operations. These shortcomings were

---

[16] R. Recouly, *Marshal Foch, His Own Words on Many Subjects* (London, 1929), p. 130.

[17] *Principles*, p. 25.

[18] Speaking of the War of 1870, Foch wrote: "Every German has a *share in the profits*, and is directly interested *in the form, in the constitution, and in the victory.* That is what is meant by a people's war." *Principles*, p. 36.

perhaps natural in a soldier trained for land warfare and they were common to almost all his professional contemporaries.

This neglect of the multiplicity of elements influencing the decision, with the emphasis on the strictly *military* aspect of war, is revealed in Foch's theory of the role of battle in warfare—of the relationship between strategy and tactics. The idea of national warfare and the necessity of armed combat, in contrast with the "chessboard" strategy of the eighteenth century, are probably the dominant themes in Clausewitzian military philosophy. Here Foch unreservedly adopted Clausewitz' point of view, namely, that battle is the only solution of war; that against the "tottering theories" and "degenerated forms" of the eighteenth century, the methods of Napoleon should rule. "Blood is the price of victory," Clausewitz said. "No victory without battle," Foch added; "no strategy can henceforward prevail over that which aims at ensuring tactical results, victory by fighting."[19] Here again, was a lesson from Napoleon, remembered by Prussia, and forgotten by the France of the Second Empire.[20] For the French generals were then impressed by the importance of good "positions," the holding of which it was thought would make up for the necessity of final engagement with the enemy or, at least, make the defensive so strong as to reduce very greatly the chances of the assailant. The new war, argued Foch, "more and more national in its origin and aims, more and more powerful in its means, more and more impassioned, . . . does away with all systems founded on positive quantities: ground, position, armament, supply; it is a war which relegates to the background the possession of territory, the capture of towns, the conquest and occupation of strong positions."[21] It was childish merely to rely on the advantages of terrain in order to escape ordeal by battle in the presence of superior forces. But it was no less childish (to use Foch's own words, after the World War) to jump to the other extreme and to rely, as the French high command had done in the first days of the war, merely on morale.

The consequence of this principle was a return to the "barbarism" of the Napoleonic battle, pushed to a maximum by modern technical developments. That this involved a "negation of the art of war" and a prevalence of hazard and improvisation, that it implied the "impossibility" of directing war and a "harking back to the confusion of the barbaric invasions"[22] was immediately perceived. Nevertheless, Foch believed that this absolute war could be formalized and he seemed, in a sense, to nullify his own prophetic insight by the

---

[19] *Principles*, p. 43.

[20] There is no contradiction, as has been suggested, between this doctrine and the general methods of Napoleon. Captain Liddell Hart quotes a saying of Napoleon at the opening of the campaign of 1805 to show that he was concerned with "gaining victory with the least shedding of blood," and then opposes to this the doctrine of the French general staff in 1913: "the result can only be obtained at the price of bloody sacrifices." This is to forget the antithesis drawn by Foch, between the eighteenth century method, illustrated by the Marshal de Saxe's maxim: "I do not favor pitched battles, especially at the beginning of a war, and I am convinced that a skillful general could make war all his life without being forced to one" and the Napoleonic method: "There is nothing I desire so much as a great battle."

[21] *Principles*, p. 41.   [22] *Principles*, pp. 49-50.

enunciation of his famous principles of war. These principles, already enumerated at the beginning of the book, were:

"The principle of economy of forces.
The principle of freedom of action.
The principle of free disposal of forces.
The principle of security, etc. . . ."

It has been observed, with good reason, that the "etc." exposed an indefiniteness not clarified in the rest of the book.[23] Foch, however, summed up elsewhere and with greater precision the essentials of Napoleon's art of war, thus somewhat illustrating the meaning of the ambiguous phrase. "I have given much thought to the matter," he declared to Recouly, "and it seems to me that his art consisted in a few principles of extraordinary simplicity and clarity. These he used with the touch of a master. To husband his troops; to use them judiciously so that the enemy might be attacked at his weakest point with superior forces; to keep control of his men, even when they were scattered, much as a coachman holds the reins, so that they could be concentrated at a moment's notice; to mark down that portion of the opposing army which he aimed at destroying; to discern the critical point where defeat might be turned into rout; to surprise the enemy by the rapidity of his conceptions and operations—those are a few of the essential elements of Napoleon's military genius."[24]

If we examine Foch's principles in detail, we find no clear distinction between the *principle of freedom of action* and that of *free disposal of forces*. Foch seems to have used them alternately to impress upon his students the supreme importance of initiative, of freedom from the will of the enemy. More important are the other principles—that of "economy of forces" which was the corollary of that freedom of action which he preached and that of "security" which was the condition of its application.

According to Foch it was *the principle of economy of forces* which enabled an "art of war" to persist, despite the danger of chaos and confusion resulting from the conditions of modern war. Foch himself never clearly defined this principle, but we may quote the most significant of his observations concerning it. "There is a proverb which says: 'You cannot hunt two hares at the same time.' You would catch neither of them. . . . Efforts must be concentrated. . . . Those who would say . . . economy means sparing one's own forces, being careful not to disperse one's own efforts, would only state part of the truth. Those would come closer to the truth who could assimilate it to the art of knowing how to expend, to expend usefully and profitably, to make the best possible use of all available resources."[25]

"Rational calculus" would perhaps be a satisfactory equivalent for this principle which, as is well known in economic theory, is of universal application in *ideal* human conduct—and not merely in the art of war. Foch ascribed its

[23] Liddell Hart, *op. cit.*, pp. 23, 460 ff.   [24] Recouly, *op. cit.*, pp. 126-127.
[25] *Principles*, pp. 50-51.

origin to the wars of the French Revolution, following in this respect Jomini's emphasis upon Napoleon's use of a maximum of forces against the most vulnerable point of the enemy's position.

It is clear that such a broad principle did not allow of rules of application to the innumerable variety of possible circumstances; it could not be said, for instance, that *all* forces should always be concentrated for the decisive attack.[26] But it was no less clear to what lesson the principle was pointing: the impracticability of being everywhere *sure* that the enemy might not be the stronger and hence the necessity of taking risks. "There are many fine generals," continues Foch, "but they try to keep an eye on too many things; they try to see, to keep, to defend everything: depots, lines of communication, the rear, such and such a strong position, etc. Using such methods, . . . in the end means dispersion, which prevents them from commanding, concentrating on one single affair, from striking hard; they end in impotence."[27]

Was there not, then, as a result of this concentration, a danger of being surprised by the enemy at the most unexpected place? The *principle of security*[28] was to obviate such risk. Neglect of this principle in the early days of 1870 had been a major source of disaster to the French armies. On the other hand, Foch had taken pains to establish, in *La Conduite de la Guerre*, that similar neglect had been frequent with Prussian generals; but, while the latter had known how to profit from the enemy's mistakes, no such advantage had been derived by the French. Nearly one half of the *Principles* is devoted to a study of this principle of security. Its significance can best be summed up in Foch's own words:

"That notion of security, which we express by means of a single word, divides itself . . . into:

"1. *Material security,* which makes it possible to avoid enemy blows when one does not desire to strike back or cannot do so; this is the means of *feeling secure* in the midst of danger, of halting and marching under shelter.

"2. *Tactical security,* which makes it possible to go on carrying out a programme, an order received, in spite of unfavourable circumstances produced by war; in spite of the unknown, of measures taken by the enemy of his own free will; *also to act securely and with certainty, whatever the enemy may do,* by safeguarding *one's own freedom of action.*"[29]

"The unknown," wrote Foch, "is the governing condition of war."[30] Piercing the unknown and securing the greatest amount of intelligence was, therefore, the first element of security. Since before any battle the main body of troops was inevitably scattered, the duty of securing this information devolved upon the advance guard and upon its success depended the army's freedom of action. The three-fold function of the advance guard was:

[26] *Principles*, p. 57.   [27] *Ibid.*
[28] In French *sûreté*. The word *security* is normally used in translation to express the idea, but no simple English word is adequate. For *sûreté* has the sense of sureness as well as of assurance of safety; it implies that the commander acts with secure knowledge as well as with physical protection. Liddell Hart, *op. cit.*, p. 482.
[29] *Principles*, pp. 138-139.   [30] *Ibid.*, p. 145.

"1. To *inform*, and therefore to reconnoitre up to the moment the main force goes into action;

"2. To *cover* the concentration of the main body and to prepare its entrance into the field;

"3. To *fix* the adversary one intends to attack."[31]

In stressing the importance of the advance guard, it is clear that Foch did not contemplate the situation which developed in the war of 1914-1918, when the two armies, on immobile fronts, had to keep constant watch, so that their whole activity, in the intervals between great engagements, was reduced to a service of security. But his emphasis on security was essentially sound in view of the errors of the campaign of 1870 and the turn which the next war was to take.

It is interesting to note that, while Foch warned his pupils against surprise by the enemy, he did not develop to an equal degree the principle of offensive surprise. He did, however, mention it as the major element of success in battle; and his whole theory of battle—set forth in the last three chapters of the *Principles*—was focused on the necessity of offensive action.

In turning to Foch's theory of battle, it should be emphasized at the outset that he did not preach offensive action in all cases. We shall see later how far his part in the French offensive doctrine can be reasonably ascertained. In his teachings, and subsequently in his practice, his concept of the offensive was qualified. Nevertheless, he wrote that the "offensive form alone, whether resorted to at once or only after the defensive, can lead to results, and must therefore *always* be adopted—at least in the end."[32] "In tactics, action becomes the governing rule of war."[33] The moral was that all efforts should be directed to the planning of the attack to be delivered at the most favorable opportunity. "The battle: decisive attack" was the title of the chapter in which Foch emphasized the value of offensive action as the ultimate goal of maneuver.

Under modern conditions, Foch thought that the battle might assume either of two specific forms: the battle maneuver, where one supreme effort, a decisive attack, achieves surprise and victory; or the parallel battle, or battle of lines, "in which one goes into action at all points, and in which the *commander-in-chief* expects a favorable circumstance, or a happy inspiration . . . to let him know the place and time when he must act;—unless he leaves all this to be decided by his lieutenants, while the latter, again, leave this to their own subordinates, so that in the end, the battle is won by the privates: an *anonymous battle*."[34]

This, surely, constitutes a remarkable forecast of the nature of the immobile warfare of 1914-1918; and what follows describes it even more accurately: "Troops go into action everywhere: once in action, they are supported everywhere. In proportion as forces are used up, they are renewed and replaced. Such a battle consists in putting up with a constant, a successive wear and tear, until the result ensues from one or more successful actions of particular com-

[31] *Ibid.*, p. 153.
[33] *Ibid.*, p. 284.
[32] *Ibid.*, p. 283.
[34] *Ibid.*, p. 296.

batants—subordinate commanders or troops." These actions were characteristic of the battles of Verdun, the Somme, and Passchendaele. But Foch at once
dismissed this form of battle as "inferior," since no appeal could be made to the
commander in chief's action and maneuvering ability. Little could he foresee
that this inferior form would force its acceptance on the unwilling commander.
In the last chapter of the *Principles*, where he forecast the shape of battles to
come, he retained his faith in the maneuver as a superior form of war.[35]

It has often been said that Foch had not the faintest idea of how modern
weapons were soon to affect the battlefield. That is not entirely correct. For
while in his preface to the second edition of the *Principles*, for instance, he
shows that he was not impressed, after the Manchurian campaign, by the effects
of machine-gun fire and barbed wire, it should nevertheless be recognized that
he showed relatively more foresight in technical matters than most of his
French contemporaries.

Thus, he recognized that new conditions would of necessity impose certain
modifications. "Arms," he wrote, "have a longer range; they are more deadly,
which compels forces to make their dispositions for attack at a greater distance,
and under better cover."[36] From the manner in which operations were conducted on the French side in 1914 we may conclude that he did not foresee the
peculiar form of trench warfare. Yet he insisted, in a later passage, that with
the ever increasing power of firearms "the necessity of cover increases every
day." Infantry must therefore "utilise all practicable defilades and follow them
for the longest time possible."[37]

Improvement in artillery technique also led him to perceive the important
role of artillery fire in the preparation of attack. In the course of such preparation, the advance elements must maintain the conditions of security for the
planning of the decisive attack by a double action of information and protection.
This would involve some fighting, and in this case artillery would be effective
because of its greater range and mobility, and its means of effecting surprise.
"Opening a way for the infantry on the whole front, so as to enable it to carry
out decisive acts; supporting it in these attacks, in these decisive acts," such are
the tactics of artillery in the course of preparation.[88]

In analyzing the role assigned by Foch to infantry, it is but fair to remark
that, in addition to his brief but emphatic reference to cover, he stressed the
importance of fire. "Fire has become the decisive argument."[39] He warned the
most ardent of troops that they would incur considerable losses "whenever
their partial offensive has not been prepared by offensive fire," and he asserted
that a superiority of fire would become the main factor upon which the effectiveness of a force would depend.

Yet, here again, Foch could not have perceived the real dimensions which
such superiority must achieve if the attacking infantry was to overcome the
withering fire of the defenders' automatic weapons. Infantry, throughout the

---

[35] Foch here acknowledged his indebtedness to Ardant du Picq, Cardot, Millet, and
Bonnal.
[36] *Principles*, p. 327.   [37] *Ibid.*, p. 335.
[88] *Ibid.*, p. 335.   [39] *Ibid.*, p. 337.

preparatory stage, must advance to within 600 to 800 yards of the enemy, utilizing all possible shelter and defiladed areas, and adopting such formations as will allow the best use of these covers. But when the moment for execution and decisive attack arrives, the formations must make the best possible use of the two means of action: firing power and striking power. "The consideration of what fire one may oneself receive now becomes a secondary matter." Hence, infantry in two ranks, with a particularly strong second line, must "march, and march quickly, preceded by a hail of bullets." The most powerful fire could not alone secure decision.

"They march straight on to the goal, each aiming at its own objective, speeding up their pace in proportion as they come nearer, preceded by violent fire, using also the bayonet, so as to close on the enemy, to be the first to assault the position, to throw themselves into the midst of enemy ranks and finish the contest by means of cold steel and superior courage and will. Artillery contributes to that result with all its power, following, supporting, and covering the attack."[40]

In his novel *Verdun*, Jules Romains puts this sentence in the mouth of an infantryman, as a derisive and bitter comment by the common soldier experienced in the realities of modern combat, on the lucubrations of their generals in their most fanciful moods.[41] On the whole, this passage, in spite of Foch's frequent references to the necessity of artillery preparation and accompaniment, is a characteristic proof of his underestimation of the role of modern firearms in particular, and of matériel in general. But throughout the First World War, the final infantry assault did persist as the decisive factor.

In his first lectures to the Ecole de Guerre, Foch had criticized the mechanistic theories which had prevailed in France before 1870 and according to which victory was thought to depend on the sheer accumulation of material factors—the moral factors being assumed to be equal on both sides. Hence his emphasis on the moral aspect of war and, perhaps as a consequence of this, his overlooking—at least in part—the need for material superiority. But it was in his treatment of morale that Foch was to leave the strong mark of his personality and lay the foundation of his achievements.

A very devout Catholic, Foch could not fail to be impressed by Joseph de Maistre's grandiose and terrible philosophy of war. This profound and prophetic genius—the arch enemy of the French Revolution, a Catholic equivalent, as it were, of Edmund Burke—has earned the title of "bellicist" for having

---

[40] *Ibid.*, p. 344.
[41] "The very words 'cold steel' conjured up a whole chapter of pre-war stupidity, with its love of big words, its entire lack of imagination and intellectual honesty, its refusal to face facts, its inherent vulgarity—an attitude, in fact, that had been common to officials of every degree of eminence and obscurity." (J. Romains, *Verdun*, New York, 1939, p. 67.) This is no doubt unfair to the efficient and positive side of the work of preparation achieved by the French staff before 1914; but it reflects a feeling that was widespread in the French army during the war, not merely in the rank and file, but among a large proportion of subaltern officers. In spite of this, the army never—with the exception of the Nivelle episode in 1917—lost confidence in its leaders.

proclaimed that war is divine. It followed that war was appointed by divine ordinance as the perpetual ordeal and expiation for man's sin. In the famous dialogue of the *Soirées de Saint-Petersbourg*, one of the interlocutors recalled how elusive, even to professional soldiers, are the elements that decide victory or defeat. To his question: "What is a lost battle?" a general, much perturbed, had first replied: "I do not know," and a little later: "It is a battle one *believes* one has lost; for a battle cannot be lost physically." "Therefore," argued Foch, "it can only be lost morally. But then, it is also morally that a battle is won, and we may extend the aphorism by saying: a battle won is a battle in which one will not confess oneself beaten."[42]

"Victory = Will" ("Victoire = Volonté"). This formula, almost as famous now as the "De quoi s'agit-il?" is the expression of the spiritual element in Foch. Having shown the rationalist side of his character, we may conclude this brief analysis of his doctrine with his portrait as a believer. This combination of reason and will, of intellect and faith, has been universally recognized as the peculiar mark of his genius as a leader of armies. No general, perhaps, was ever great who did not possess it in the highest degree. But what made Foch supreme in his time was his ability to communicate his burning faith and energy first to his pupils, and later to his armies. "War = the domain of moral force." "Victory = moral superiority of the victors, moral depression of the vanquished. Battle = a struggle between two wills."

"The will to conquer: such is victory's first condition and therefore every soldier's first duty; but it also amounts to a supreme resolve which the commander must, if need be, impart to the soldier's soul."[43]

In a materialistic age, when all eyes were dazzled by the achievements of technology, Foch, like du Picq before him, emphasized the importance of the moral factor in war and thus reminded his audience that, however stupendous the changes wrought in our lives by scientific progress, they cannot modify the laws of the human heart. In warfare, as in every other phase of the social process, it is man who remains the first and last actor. As we have already observed, this simple truth could lead to murderous absurdities if not placed in its proper perspective. "At the beginning of the last war," Foch later confessed, "we believed that morale alone counted, which is an infantile notion." These were brave and sensible words. Yet as long as there is any persistence in the notion, no less infantile, that war can be won by sheer force of "material superiority," the lesson of Foch will still possess a portentous significance.

It is perhaps possible to link his conviction that moral force will determine the event rather than submit to it with his ardent desire to assert the triumph of mind over matter, of reason and will over the chaos of battle. From his conception of the role of spiritual power, we can trace his idea of a leader of men.

There was considerable intellectual pride in Foch's bold assertion: "Battles are lost or won by generals, not by the rank and file."[44] But there was, at the same time, the frank acceptance of responsibility. "Great results in war are due

---

[42] *Principles*, p. 286.   [43] *Ibid.*, p. 287.   [44] *Ibid.*, p. 108.

to the commander. History is therefore right in making generals responsible for victories—in which case they are glorified; and for defeats—in which case they are disgraced. Without a commander, no battle and no victory is possible."[45] This was the foregone conclusion of a teaching inspired by the decision to redress the failings of the French high command in 1870. The inexorable weight of events was soon to reduce Foch to a position of greater humility.

When the war of 1914 came, did Foch apply the principles which he had preached or did he abandon them? Historians and military critics have long disputed whether his victories were won because, or in spite of, his principles.[46] The influence of Foch's teaching at the Ecole de Guerre was doubtless felt in the drawing up of the French plan of campaign in which the doctrine of the offensive asserted itself fully in 1913. Colonel de Grandmaison, leader of the "Young Turks" who succeeded in having the plan adopted, had been one of Foch's pupils. But Foch had no direct share in the war plans and it may be said that insufficient emphasis was placed upon his doctrine of security which, immediately before the outbreak of the war, would have seemed to imply "a want of faith in the irresistible *élan* of the French soldier."[47] In any event, the offensive plan as drawn up ended in the shambles of Morhange, Arlon, and Charleroi.

Foch's belief in the war of maneuver also gave way to the new conditions of trench warfare. After the Marne, when he was sent to the north to coordinate action between the French, British, and Belgian armies, he said to Tardieu: "They have sent me here late to maneuver, but things are not going very brightly. This eternal stretching out in a line is getting on my nerves."[48] The battle maneuver, which he had believed the superior form of war, had to give way increasingly to the battle of lines.

But, however much Foch's earlier principles were altered by the war, it was his moral fiber rather than his intellectual inventiveness which withstood the tide. Pierrefeu has left us a memorable picture of Foch in action "bursting like a whirlwind into every headquarter, his face contracted, his body all tense and contracted, gesticulating, fulminating in jerky ejaculations. To a general in agony, who tells him: 'My troops are yielding under superior numbers; if I do not get reinforcements, I cannot answer for anything,' he replies, with a sweeping and furious gesture: 'Attack!' 'But . . . ,' says the general. 'Attack!' The general will insist. . . . 'Attack, attack, attack!' bellows Foch, who dashes out, charged, like an electric battery, with fierce energy and unconscious ag-

---

[45] *Ibid.*, p. 288.

[46] After the stabilization of the front in 1914, Foch is reported to have said to his staff: "Gentlemen, it remains for you to forget what you have learned, and for me to do the opposite of what I have taught you." Aston, *The Biography of the Late Marshal Foch* (London, 1929), p. 83. Liddell Hart has written (*Foch: The Man of Orleans*, p. 482) that "Foch's handicap was that he had to forget so much before he could learn." Louis Madelin, however, insists that, nine times out of ten, Foch applied his doctrines, instead of throwing them overboard. See his *Foch* (Paris, 1929), p. 24.

[47] Liddell Hart, *op. cit.*, p. 68.

[48] Tardieu, *Avec Foch* (Paris, 1939), p. 107.

gressiveness, rushing elsewhere to harden other energies, to toughen other faltering wills."[49]

Foch summed up his own role in the battle of the Marne in these words: "The first day, I was beaten. The last day, it was a question of holding out. Yet I advanced six kilometers. Why? I don't know. Largely, because of my men; a little, because I had the will. And then,—God was there."[50]

Notwithstanding this confession of faith and humility, destiny was to confirm his former insistence on the supreme importance of command, by making him supreme Allied commander. At Doullens, on March 26, when the British Fifth Army was reeling under the blows of Ludendorff, when Haig and Pétain were contemplating their respective retreats, the heroic tenacity of Foch saved the situation. "Common sense indicates that when the enemy wishes to begin making a hole, you do not make it wider. You close it, or you try to close it. We have only got to try, and to have the will; the rest will be easy." "You are not fighting," he said abruptly when the Supreme Allied Council met, "I want to fight. I would fight in front of Amiens. I would fight behind Amiens. I would fight all the time."

Optimism may appear easy to onlookers, too easy, indeed, when it is carried to the point of recklessness and becomes an excuse for inaction and complacency; but in the agony of impending castastrophe, under the strain of immediate and infinite responsibility, it becomes as precious as it is rare; only those who have gone through such moments will realize its exceptional value. Foch appeared, then, as this unique embodiment of a faith that was near the vanishing point. He was given the role of coordinator of the Allied armies on the Western Front. By and by, this high office, first conceived almost as a diplomatic mission, was reinforced with more far-reaching powers. The story of its adventures has been told many times. It never became the exact equivalent of a rigid and disciplinary command and Foch, in later years, was to pride himself on the manner in which he used diplomatic persuasion rather than military authority.

From March 1918 onward, therefore, all operations on the Western Front came under his supervision and initiative. In the beginning the strategy was simple enough: hold out. The armies held their ground after one week of retreat and confusion. Then they rallied and reorganized. Once the Americans arrived, Foch knew that he would have only to hold out long enough and that the offensive power would again be his. But they had not arrived yet. Another blow fell, on May 27, at the Chemin des Dames. The French army was taken entirely by surprise; a wide breach was opened in its front and German armies marched on in the open, for the first time since 1914, making ten or more miles per day. On May 30 they reached the Marne. The French army had been surprised; it was not to be surprised again. When the third blow was delivered, on July 15, Foch was awaiting it with all his available reserves and, three days later, launched his first counteroffensive. Here, for the first time since the beginning of the war, tactical inventiveness contributed to success in both phases of battle.

[49] Pierrefeu, *Plutarque a menti*, p. 308.   [50] Tardieu, *op. cit.*, p. 32.

In the defensive stage the maneuver executed by General Gouraud's army left a void between the thin first line of defense and the main line of resistance, upon which the German troops were to rush without artillery support. Here, at last, was a real "strategic retreat" in the expectancy of an enemy offensive. But it had taken many months before Foch could be persuaded of its effectiveness, bent as he had been, until then, on the necessity of holding every inch of ground. This system of defense, simple as it was, became the permanent feature of the French army's defensive position until 1940.

The device used in the offensive was more striking in its audacity. Artillery preparation had been necessary, in the early stages of the war, to open the way for the attacking infantry across the deadly obstacles of the defenders' positions. But this considerably diminished the surprise effect. The Germans, by reducing to twenty minutes their artillery preparation, had been able on May 27 to restore completely the element of surprise. This time it was decided that not one shot should be fired before the tanks of Mangin and Degoutte's armies moved out of the forest of Villers-Cotterets upon the German right flank of the Marne bulge.

Foch had foreseen the offensive and prepared the counteroffensive. In his memorandum of July 24 he declared that the tide had now turned. "The moment has come to abandon the general defensive attitude forced upon us until now-by numerical inferiority, and to pass to the offensive."[51] Drawing on the lessons of recent operations, he reminded his generals that "First and above all surprise must be effected. Recent operations show that this is a condition indispensable for success."

In the next three months, Foch was to give the enemy no respite. Offensive followed upon offensive at every point. But now Foch had the *means*: a continuously increasing supply of men and material. Can we then trace his final victory to any particular method in his conduct of operations?

In his later comments, Foch explained how a superior strategy enabled him to beat Ludendorff. "In the tactical details of his operations," he said, commenting on the offensives of 1918, "Ludendorff planned his attacks admirably. The planning was perfect. It could not have been improved upon. But—there was no reserve plan . . . he had no notion of the ensemble, and no large-scale plan. . . ."[52] Why, after each of his massive strokes, could he not, in spite of extraordinarily brilliant results, achieve a decisive success? "There is but one answer: final victory cannot depend upon the success, however outstanding, of a single attack. That attack must be linked to a certain number of others. It must be part of a whole, not the whole itself. This is what Ludendorff forgot."[53]

So we hear Foch laying stress on the value of the planned ensemble of operations which led to the German surrender. What was necessary was a series of offensives "fitted in so as to embrace the whole front,"[54] "interlocked into one another (*s'emboitant les unes dans les autres*)," an expression used repeatedly

[51] *The Memoirs of Marshal Foch* (New York, 1931), p. 370.
[52] Recouly, *op. cit.*, p. 96.   [53] *Ibid.*, p. 98.
[54] *Ibid.*, p. 95.

in his talks with Recouly, and which emerges as the principal characterization
of Foch's grand offensive strategy in the year 1918. Elsewhere, even before
the war, Foch had described the true offensive strategy as a "parrot march"—
that is, a progress similar to that of a parrot as it climbs the bars of its cage,
using beak and claws alternatively and assuring the firm grip of two of the
three before hazarding the next step. "Gentlemen . . . the parrot . . . sublime
animal" was one of these elliptic, cryptic sentences, so very much in the
Fochian style, with which he once concluded a conference. This is no doubt in
harmony with the principle of economy of forces, "the art of making the
weight of all one's forces successively bear on the resistances which one may
meet, and therefore of organizing those forces by means of a system."[55] But it
is also a far cry from "The battle: decisive attack."

Another inconsistency between Foch's theory and his practice has been
pointed out in his handling of the armistice. Since victory could only be won by
decisive action, by the destruction of the enemy's armies, was it not a mistake
to conclude the armistice before the knockout blow and while the enemy armies
were still on Allied soil? As is well known today, Foch had answered these
criticisms in advance. "War," he said to the Supreme Council in the early
days of November, "means fighting for definite results. I am not waging war
for the sake of waging war. If I obtain through the armistice the conditions we
wish to impose upon Germany, I am satisfied. Once this object is attained,
nobody has the right to shed one drop of blood."[56]

Foch had every reason to believe his victory complete and it was indeed complete, so far as it was his own, in the field. Germany was henceforth reduced to
impotence. Not so, however, at the conference table. Even though Germany
was not present while the treaty was being prepared, the principles of the
Wilsonian program, accepted by the Allies themselves, were to restrain the
victors by limitations which Foch could never have even imagined while he was
preparing his armistice conditions. And now his role was over; he was no
longer in the game. When he tried to interfere, Clemenceau, full of wrath at
this intrusion of a soldier in politics, with characteristic bluntness put him back
in his proper place.

With his intense conviction of his patriotic duty Foch nevertheless insisted
on being heard. His policy, outlined in his several memoranda and speeches to
the Supreme Council between the armistice and the signing of the treaty, is
well known but should be indicated here as indispensable for a complete picture
of his concept of military security.[57]

The idea was simple. European security, he argued, could be guaranteed
neither by German disarmament, which could not be enforced in perpetuity, nor
by pledges of alliances, which were illusory. Only a material guarantee could
be satisfactory: the occupation of the bridgeheads of the Rhine. "The river

---

[55] *Principles*, p. 51.

[56] *Memoirs of Marshal Foch*, p. 463. See also Seymour, *The Intimate Papers of Colonel
House* (Boston and New York, 1928), IV, 91.

[57] See Recouly, *op. cit.*, pp. 165-249.

is the deciding factor. The master of the Rhine is the master of the surrounding country. Whichever side does not control the Rhine has lost."[58] This solution had many advantages; the occupation of a few selected points would, because of the small number of troops involved, be most economical; and since Foch proposed that the occupation should be carried out by international contingents, he could not see how France's allies could entertain suspicions or fears of her obtaining thereby a means to European hegemony. Foch seems to have intended this occupation to be perpetual and to be supplemented by the creation of an independent state on the left bank of the Rhine. After his plan had been rejected, and after the guarantees offered to France by her allies in its stead had lapsed, Foch could not fail to accuse the statesmen—Clemenceau first of all—of having compromised his victory. It was natural that in agreement with almost every one of his compatriots, he should have considered the security provisions of the treaty as insufficient. "That," he told Recouly, "is how you must visualize the problem of security. You must visualize it in all its magnitude and complexity. It is not limited to the Rhine barrier; far from that. It consists in the maintenance of peace in Europe at all costs—the peace established by the treaties concluded after our victory. Suppose that Germany exercised her sway over those [new] States. Even if it be only a moral influence, they will none the less be gravitating within her orbit. Can you imagine the tremendous power she would then have? It would be futile to fight against her. The cause would be lost before the battle started."[59]

[58] *Ibid.*, p. 213.   [59] *Ibid.*, p. 268.

CHAPTER 10. Bugeaud, Galliéni, Lyautey: The Development of French Colonial Warfare

BY JEAN GOTTMANN

THE colonial domain of France is playing an important role in the grand strategy of the present World War. This domain is vast, covering about 4,600,000 square miles with a population of nearly 65 million inhabitants, composed of many lands of various sizes and types scattered all over the world. The main block is in Africa, stretching from the Western Mediterranean, down to the mouth of the Congo River. With the exception of a few small outposts (French Guiana and some islands such as Martinique, Guadeloupe, Réunion) this vast empire, some parts of which have had a decisive place in the history of the present conflict, was conquered and consolidated into a single unit within the past century. The conquest began with the capture of Algiers in 1830 and ended with the submission of the last restive tribes in southern Morocco in 1934.

As a great French colonial expert put it, "While the British Empire was built by businessmen wanting to make money, the French Empire was built by bored officers looking for excitement." A detailed historical study of this century of colonial expansion proves the essential truth of this witty statement: until 1914 the French empire developed without plan or coordination, and largely according to the initiative and endeavor of local officers scattered in isolated places and bored with garrison life. But such a study reveals that these men were not solely men of action but men of ideas also. They left a literature on colonial warfare which, in the form of correspondence, military instructions, reports, speeches, magazine articles, and historical books, constitutes a most valuable contribution both to the science of warfare and to the art of colonization.

Colonial warfare is quite different from what is commonly known as continental warfare. It is generally fought in remote countries over large areas of unknown territory, against a foe superior in number and in his knowledge of the terrain but inferior in material organization and in means of supply from abroad. In colonial wars quality must therefore balance a probable inferiority in quantity, and a colonial war is, by its very nature, fought between adversaries of strikingly different levels of civilization.

Different in means, colonial warfare is also different in goal: it aims not at the destruction of the enemy but at the organization of the conquered peoples and territory under a particular control. As far as possible it must avoid destruction during the campaign; first, in order to preserve the productive potential of the theater of operations and thus economize the supplies coming from more distant initial bases; but more important, because the conquered country is to be integrated immediately after the conquest into the "imperial" whole, politically as well as economically. It is in all respects desirable, therefore, that the territory should be in the best possible condition when conquest has been effected. The problem is not so much "to defeat the enemy in the most decisive manner" as to subordinate him at the lowest cost and in a way to guarantee permanent pacification.

Given these aims, colonial warfare is intimately linked with the occupation and organization of the subjugated territories, since successful occupation is wholly dependent on successful organization. Colonial warfare is more akin, then, to the warfare by which a conqueror builds up an empire through absorption of the conquered peoples than it is to a clash between two competitive powers unconcerned with thoughts of postwar union and reconstruction.

During the first period of colonial expansion, French army chiefs were not fully aware of all the peculiarities of colonial warfare and it was more than half a century before the principles and methods were worked out and definitely accepted. The development of this chapter of French military thought can be divided into three main periods, each symbolized by an outstanding name. During the century 1830-1930 three great men, three marshals of France, in the process of building the new France overseas, developed a theory in strategy and tactics and founded a new school of thought. They were Marshals Bugeaud (1784-1849), Galliéni (1849-1916) and, perhaps the most brilliant of them all, Lyautey (1854-1934). Their ideas and teachings are important not only because of their conquests and of their writings but also because of the part their pupils, the heirs of their thinking, played in the wars of the twentieth century. Among Galliéni's pupils in Madagascar were Joffre, Roques, Lyautey. Among Lyautey's pupils in Morocco were Franchet d'Esperey, Gouraud, Mangin, Huré, Noguès, Catroux, Giraud, and many others perhaps too young to be well known as yet. Many of these junior officers of the French colonial armies of the 1920's must, however, be considered as Lyautey's pupils and must be looked to as a possible influence in the decades to come.

The deeds of all these marshals and generals were bound together by a continuous trend of thought. As an example, General Huré, commander in chief of the last operations in Morocco, writing in 1939, credits Marshal Bugeaud's instructions of the 1840's as the basis for his strategy in the 1930's.[1]

I

In June 1830, France started her new colonial expansion with the landing of an expeditionary force numbering 37,000 men on the African coast near Algiers—a force sent there to avenge an insult to the consul of France by the local ruler, the dey of Algiers. The town was quickly taken and on July 5, 1830 the dey surrendered. The military power of the Barbary pirates based on the port of Algiers, and linked merely by a theoretical allegiance to the Turkish sultan, had been greatly overestimated in Europe, for the expedition, by an

[1] General Huré, "Stratégie et Tactique marocaines," *Revue des Questions de Défense Nationale*, I, 3 (July 1939), pp. 397-412.

easy police operation, soon put an end to piracy in the Western Mediterranean. But after the occupation of Algiers, the French found themselves in direct and permanent contact with the tribes populating the interior. Warlike but unorganized peoples, living in a state of political anarchy, they refused obedience to the French as they previously had refused obedience to the sultan at Constantinople, or even to his theoretical representative the dey of Algiers. There were frequent raids on the French establishments on the coast, and at each step out of the city of Algiers itself, the French had to fight the aggressive opposition of either the nomadic tribes from the plateaus or the peasant Berber populations of the mountainous massifs.

A first reaction of the French generals commanding the African expeditionary force was to fight this enemy according to the consecrated "Napoleonic" principles, by mass maneuver, sending heavy columns with powerful artillery deep into the country. Overburdened with impedimenta, moving through a wild, strange, and empty country, the French columns progressed slowly in this enemy territory, and generally met disaster. The native forces were at home; their chief weapon was mobility. Gathering suddenly at unexpected points, they attacked columns, raided convoys, set French establishments afire; they attacked columns on the flanks and from the rear, inflicting heavy losses, destroying or stealing equipment. Then they disappeared, melting away into the landscape before the heavy European military machine had a chance to re-form and resume operations.

These tactics made the first ten years of the French experience in North Africa very costly and the results disappointing. In 1840, Marshal Bugeaud was appointed governor general and commander in chief in Algeria. In his six years in this post he achieved the conquest and definitive submission of most of the country. But he applied quite different methods.

Bugeaud was fully aware of the fact that the main advantage of the native tribes, then united under Emir Abd-el-Kader, was their mobility. He decided to make the French troops as mobile as the enemy, for in his opinion the occupation of isolated towns or spots in the hearts of a hostile country was not an important step toward final victory. To multiply the number of fortified places which remained constantly on the defensive did not give domination over the surrounding countryside. He wanted the natives to fear the action of his troops everywhere, thus giving his army a moral prestige which in itself would result in economy in the actual application of material force. In this and many other respects Bugeaud followed the lines of the ancient Roman strategy in Africa.

In the reorganization of the Algerian army Bugeaud endeavored to lighten and simplify the equipment so that troops of all arms could enjoy greater mobility and elasticity. Supplies were put on the backs of draft animals, horses, mules, or camels, instead of being transported in slower wagons. Troops were ordered, in so far as possible, to find their supplies on the spot, in the area of operations. Light and swift columns, consisting generally of 6000 men and 1200 horses, patrolled the country. The war became one of movement, the

French learning the methods of colonial warfare from the natives themselves. Feeling their main weapon lost, many tribes submitted. In every subjugated region, the first step of the French army was to build a network of blockhouses commanding the roads, and serving as warehouses and bases for further penetration by light columns.

This strategy showed definite results from the very first years of Bugeaud's command and ultimately led to the complete defeat of the restive Arab tribes of Emir Abd-el-Kader. But the most difficult task that faced Bugeaud was the penetration and pacification of the coastal mountains inhabited by Berber peasants, tough fighters who historically were always the last to be conquered in North Africa—by the Romans, by the Arabs, and later by the French. Bugeaud's report *"De la stratégie, de la tactique, des retraites et du passage des défilés dans les montagnes des Kabyles"* remains as a classic in French colonial warfare up to this day. Its main ideas are worth recapitulation.

The mountainous massifs, like the Kabylies, do not offer much facility for maneuvering. They are an almost continuous chain of natural strong points, easy to fortify and to defend. Tactics which are often decisive in battles on the plains here lose their efficiency. The topography favors individual fighting and not the establishment of continuous fronts by forces attacking from the outside. For this reason Bugeaud stresses as most important the direction toward which each column is sent: the advantage lost in the field of tactics must be regained through strategy. A first axiom of this strategy is to understand that to conquer the mountains one must be strong not so much because of the defenders as because of the terrain itself. Whenever possible one should operate with several columns which protect one another because they create anxiety for the enemy in several directions and because they can encircle the positions on which the enemy could establish a strong front or from which he could attack a single column on its flank and rear.

But planning the itinerary is not the total of strategy. To cross a range and defeat the mountaineers once or twice is not enough: the goal is to subdue them so that after a defeat they will not attempt to reorganize for battle at another time and place. The army's action must be directed against the natives' interests and must be felt heavily in the territory of each tribe, so that their morale will be broken and they will be discouraged from continuing the war. Strategy, therefore, must also be planned in the field of local economics and the endeavor made to crush the enemy's potential by disorganizing not only his armies but his economy as well.[2]

These principles were remembered and carefully developed by following generations of colonial officers. Bugeaud improved on tactics as well as strategy. He stressed the use of surprise to weaken the enemy's resistance from the very beginning of a battle. Owing to his reforms, the French columns were able to

[2] Bugeaud's report on mountain warfare is not usually to be found in American libraries. A good summary of its main principles is provided in General Huré's article (see footnote 1). See also: works by General Paul Azan, *Bugeaud et l'Algérie* (Paris, 1930) ; *L'Armée d'Afrique de 1830 à 1852* (Paris, 1936) ; *Conquête et pacification de l'Algérie* (Paris, 1931) ; *Les grands soldats de l'Algérie* (in *Cahiers du Centenaire de l'Algérie*, No. 4, Paris, 1931).

move rapidly and strike where the blow was not expected. When such initial blows were not sufficient to achieve victory, Bugeaud ordered his infantry to form squares on the model of the old "carré" of the Roman legions. In this formation they could repulse attacks from all sides, and from the rear as well as from the front. During offensive action, the infantry square was supported on both flanks by cavalry squadrons. This restoration of the tactics of ancient Rome in the nineteenth century proved wise and successful: since the epoch of Jugurtha, in defiance of time, neither the terrain nor the tactics of the natives had changed. The methods used by the Romans to conquer the province of Africa was used by the French with equal success. The thorough training in the classics given in French colleges thus proved an incalculable aid to French generals in Africa.

Bugeaud, utilizing the Roman battle formation of the square, did not forget the importance of political action in the ancient techniques of empire building. He endeavored to weaken the enemy by internal discord and division, playing on the antagonisms between varied interests, groups, and leaders. Political warfare remained for the French, and for all other expansionist powers, one of the main weapons. Thus Bugeaud laid the foundation of a new school of military thought which developed even more in the following half century. In the ranks of the French armies he was the first soldier of the nineteenth century to renounce Napoleon's teaching as unsuited to every particular environment. He revived old Roman methods which had yielded good results.

But these old-new methods had to be adapted to modern armies and to widely expanded theaters of warfare which shifted rapidly from Algeria and the Mediterranean area to distant and varied countries. This new school found its leader and master in Galliéni who represents a generation of officers with a purely colonial background. But when Galliéni's contemporaries commented favorably on his achievements in the colonial field, they could find no higher praise than to say: *"C'est du meilleur Bugeaud!"*

II

Marshal Galliéni is most generally known for the outstanding part he played as military governor of Paris in 1914, especially in the preparation of the battle of the Marne. But his real career was that of a great empire builder. Born in 1849, he started as a young lieutenant of marines during the war of 1870-1871 against the Germans in which he participated in some of the most heroic episodes. But after 1871, with French power seemingly crushed in Europe, only the colonial field remained open to young officers. In his first twenty years of colonial campaigns and administration, Galliéni acquired a rich experience in various areas.

His first assignment was on the island of Réunion. He was then sent to West Africa, fought in the Senegal and Niger provinces, then went to the French Caribbean islands, returning later to the Sudan. In 1892, now a colonel, Galliéni arrived in Tonkin, northern province of Indo-China, and was appointed commander of a difficult district on the Chinese border. Here was his first opportunity to display on a large scale his art and skill as field commander and colonizer under extremely difficult conditions. He left the Tonkin frontier in 1896 having achieved amazing results in pacification and development work. He was then appointed governor of Madagascar, at that moment a post of great importance.

The Tonkin period in Galliéni's career seems a decisive factor in the elaboration of his techniques of colonial warfare. He was aided by a staff of brilliant officers—if they outranked him, he inspired them; if they were subordinates, he taught them. Among them he found a man whom he later called to Madagascar as his associate. That man was Lyautey. Here on the Chinese border was started the association of these two soldiers, which had a profound effect on the present and future of so many countries. The two men were made of the same stuff and they understood one another. Their backgrounds seemed antithetical: Galliéni, the colonial and realistic man of action; Lyautey, a metropolitan officer, educated in aristocratic circles, following the regular army career, and now at the age of forty fed up with theory, books, courses, and regulations.

Some forty years later, Lyautey liked to tell how, at the very beginning, Galliéni took away from him all the technical literature, textbooks, manuals, and staff regulations that the Parisian had brought with him. He was told by his chief to stop reading about military matters and to look around him, to learn realities. He discovered that he shared with Galliéni a common hatred of red tape, and a common will to attain one goal—effective action. From that moment the opposed backgrounds became complementary and the two men worked out what will remain in the eyes of history the French theory of colonial warfare and administration of the nineteenth and twentieth centuries.

His appointment to Indo-China was a kind of exile for Lyautey, at that time a well known authority among the young officers of the French capital. It was a punishment for the writing of a harshly criticized article published in the *Revue des Deux Mondes*, the most influential of the great French literary and political periodicals. This article, *"Le rôle social de l'officier français,"* emphasized the role of the army officer in the shaping of French society. Bored with garrison life and worried about the total lack of understanding among the military of the profound political and social changes the French nation was undergoing at that time, Lyautey endeavored to call to the attention of the French elite the outstanding importance of the army officer in the education of the younger generation. This was a momentous matter in a country where almost all the young men spent a few years in military service. Furthermore the period of service coincided with the beginning of the soldier's adult life, when the influence of the officers on his intellectual development could be deepest. By teaching them more than drill and by observing more than red tape, by taking care of their intellectual as well as their physical activities, the officers could render both the young men and their country the invaluable service of preparing better developed, more consciously patriotic, more intelligent generations of citizens.[3]

These bold ideas frightened the conservative minds of the high command. The general staff considered it almost revolutionary to speak of a social role for army officers and as a consequence Lyautey was virtually ostracized. General de Boisdeffre, then chief of the general staff, who was personally partial to the brilliant young major, appointed him to Indo-China, explaining in friendly fashion that a temporary exile was necessary to let powerful people forget about his *"Rôle Social,"* and continued that Lyautey should not take too seriously this colonial episode in his career. The colonial army, he warned, was a caste closed to outsiders; furthermore, Lyautey at forty was too old to start a colonial career!

Lyautey reported to Galliéni's headquarters on the Chinese border in 1894. He found there the ideal environment for a man who refused to accept the idea that the military profession was restricted to drill, red tape, and some occasional intermissions of fighting. For him, as for Galliéni, the army's task was one of progress, of improving the existing conditions of life and civilization. It was a task of creating conditions of enduring peace, avoiding bloodshed, developing cultural and material standards, and thereby of increasing the grandeur and prestige of the mother country. This had been Lyautey's goal in writing the *"Rôle Social"* and this was precisely the task which lay ahead in the backward colonial countries. This mission of the army from then on became the foundation of the French concept of colonization as opposed to the Anglo-Saxon idea which stressed mainly the material profits—a clear interpretation of the anecdote quoted earlier about businessmen and bored officers.

III

By the end of 1899, General Galliéni, Governor of Madagascar, reported in person to the government in Paris on what had been done or remained to be done in the newly organized colony, the Great Island of the Indian Ocean. Lyautey accompanied him, as his chief of staff and, at the end of their stay in Paris, he formulated the main ideas and doctrine of the colonial military school in a brilliant article, once more in the *Revue des Deux Mondes*, entitled *"Du rôle colonial de l'Armée."*[4] This text remains basic to the whole theory of action of the builders of the French empire.

Lyautey starts by stating that he does not intend to make a plea in favor of military administration in the colonies. He considers nonsense the old dispute between the civilian and military authorities. It is not the *label* but the *man* which matters. In colonial countries, where "the unforeseen is the rule and decision is an everyday necessity, one formula tops all others: and it is the right

[3] See Carl V. Confer, "The Social Influence of the Officer in the Third French Republic," *Journal of the American Military Institute*, III (Fall 1939), 157-165.

[4] Lieutenant Colonel Lyautey, "Du rôle colonial de l'Armée," *Revue des Deux Mondes*, CLVII (February 15, 1900), 308-328. This article was republished later in booklet form by Librairie Armand Colin, Paris.

man in the right place." The principal qualities which make the good officer and those which make the good colonial administrator are very much the same. When a man has these abilities it does not matter at all to what service he belongs—army officer or civil servant, he will do a good job. An introduction of this sort was obviously directed to the solution of an old and bitter struggle which had waged in every occupied area. Nothing was stranger to these men of action than these theoretical questions of authority and jurisdiction. In an address to the Réunion des Voyageurs Français,[5] a very select group of Frenchmen interested in foreign problems that met monthly in Paris, Lyautey emphasized once more the same point of the necessity of subordinating all such considerations to the "right man in the right place" policy.

The colonies, said Lyautey, are the best school of practical life, efficient command, everyday responsibility. At such a school diplomas, rank, elaborate hierarchies mean nothing. Only the solution of problems matters. If it helped, he would on occasion give civilian authority to an army officer or command of an army unit to a civilian. He quoted examples of civil servants compelled to act in Madagascar in critical circumstances as military commanders in the field. He mentioned particularly E. F. Gautier, then director of education in Madagascar, who later, as a professor at the University of Algiers, was one of the main advisers of the French administration in Africa. To the eternal question —military regime or civilian regime—Lyautey answered that he did not care for formulae; "the régime is called Doumer or Feillet, Faidherbe or Galliéni, and it is good because they are good." In fact, Lyautey thinks that each pioneer in the colonies is, in a sense, a soldier: he organizes his own security and the security of the land around him. He becomes the leader of the friendly natives —uniforms or the rote of field manuals can add nothing of worth to such an experience. The primary factor in colonial administration is the man.

But the abilities of Frenchmen abroad still require techniques for their effective utilization and, after the first principle ("the right man in the right place"), Lyautey carefully explains the main elements of such techniques in his *"Rôle colonial de l'Armée."*

Many soldiers at this time were shocked by his new principle of strategy— *in the use of the armed forces, avoid as far as possible the column and replace it by progressive occupation.* This might appear merely a development of Bugeaud's idea that it was not enough to defeat the enemy, that he could be beaten and later, at the first favorable opportunity, reconstitute his forces for a new battle. For this reason Lyautey wanted his front to be formed not by the spearheads of columns but by the regularly progressing tide of a well organized occupation. There was no intention, of course, of suppressing completely the column of attacking troops: such an operation is generally indispensable at the outset to impress the enemy with his inferiority to the military force of the colonizing power. But no definite and lasting achievement results from the *"coup de force"* alone, occupation must follow and here we have Lyautey's

---

[5] This address, delivered on February 19, 1900, opens the volume of Lyautey's speeches published under the title *Paroles d'Action* (Paris, 1927).

famous statement: *"Military occupation consists less in military operations than in an organization on the march."*

What is meant by this "marching organization?" It is an organization of the conquered territory set up, not behind the active front, but marching step by step with the armies as they advance. This organization must not be simply a new hierarchy imposed on the area but a network covering it, worked out in advance in the most minute detail and with the greatest care. Before each step in the progressive occupation the men who will have to do the job are trained for the task: each officer and private studies the area and knows exactly what his post and mission will be immediately after the occupation. "The occupation," says Lyautey, "deposits the units on the soil like sedimentary strata." Thus the real work of occupation does not start with the actual occupation. On the contrary, the operations start when the organization is ready; only the scope of the operations remains to be provided.

The method is one of thorough preparation, without stirring action or heroic deeds, which economizes men and minimizes hatred because, as a general rule, it avoids the spectacular clash of arms. It creates, on the contrary, a new terrain more favorable for the consolidation of the results of military operations. To illustrate this doctrine, Lyautey quotes at length texts written by his chiefs in Indo-China and, in particular, a report sent in 1895 to the governor general of Indo-China by General Duchemin, commander in chief of the occupation forces. Strongly advocating Galliéni's methods, General Duchemin protests against the criticism that progressive occupation only pushes the rebels farther outside the frontiers instead of annihilating them, so that they can always return to disturb the peace of the occupied territory. He emphasized that there is no possibility of completely annihilating by force a group of pirates—as rebels were called in Tonkin.

"The pirate is a plant which grows only on certain grounds. . . . The most efficient method is to render the ground unsuitable to him. . . . There are no pirates in completely organized countries. To pluck wild plants is not sufficient: one must plough the conquered soil, enclose it, and then sow it with the good grain, which is the only means to make it unsuitable to the tares. The same happens on the land desolated by piracy: armed occupation, with or without armed combat, ploughs it; the establishment of a military belt encloses and isolates it; finally the reconstitution and equipment of the population, the installation of markets and cultures, the construction of roads, sow the good grain and make the conquered region unsuitable to the pirate, if it is not the latter himself who, transformed, cooperates in this evolutionary process."

General Duchemin's report found a very favorable audience. The governor general of Indo-China, M. Rousseau, sent a memorandum to the French government in Paris expressing his own conviction that the methods of Galliéni's school were by far those best fitted to the conditions of the Chinese border. He described the task of the forces of occupation as consisting primarily in the protection of the frontier and the social and economic reconstruction of the occupied area; military operations and the use of force were consigned to the background. Aside from this method, M. Rousseau added, "piracy can only be met by either dubious compromises or costly and futile expeditions."

Lyautey stresses the contrast between such a progressive occupation and the classical strategy of strong columns in uninterrupted march across a country, striking forward at an almost constantly fleeing target. The support of the column exhausts the country, especially since the conqueror has no direct interest in its preservation. Progressive occupation or occupation by "organization on the march" is based on a system of conservation and rehabilitation. This fact gives rise to the following essential principle: *"A colonial expedition should always be under the command of the chief appointed to be the first administrator of the country after its conquest."* Citing examples from his own experience, Lyautey adds a sentence which remains a classic in French colonial circles: "If in taking a native den one thinks chiefly of the market that he will establish there on the morrow, one does not take it in the ordinary way." He could not have stressed better the sharp differences between ordinary warfare and colonial warfare.

The best illustration of the general techniques of their warfare is provided by Lyautey through extensive quotations from Galliéni's instructions of May 22, 1898, at Madagascar:

"The best means for achieving pacification in our new colony is provided by combined application of force and politics. It must be remembered that, in the course of colonial struggles, we should turn to destruction only as a last resort and only as a preliminary to better reconstruction. We must always treat the country and its inhabitants with consideration, since the former is destined to receive our future colonial enterprises and the latter will be our main agents and collaborators in the development of our enterprises. Every time that the necessities of war force one of our colonial officers to take action against a village or an inhabited center, his first concern, once submission of the inhabitants has been achieved, should be reconstruction of the village, creation of a market, and establishment of a school. It is by combined use of politics and force that pacification of a country and its future organization will be achieved. *Political action is by far the more important.* It derives its greater power from the organization of the country and its inhabitants.

"As pacification gains ground, the country becomes more civilized, markets are reopened, trade is reestablished. The role of the soldier becomes of secondary importance. The activity of the administrator begins. It is necessary, on the one hand, to study and satisfy the social requirements of the subject people and, on the other hand, to promote the development of colonization, which will utilize the natural resources of the soil and open the outlets for European trade. . . . The executives of the territory should conceive their administrative functions with the least possible conventionalism. Regulations provide only general rules applicable to a sum total of factors but are often inapplicable in individual cases. Our administrators and officers must protect, with common sense, the interests entrusted to them and not fight against these interests for the sake of the regulations."

The task of the army during the conquest itself is thus fixed. Galliéni then defines the part it has to play in the already occupied and pacified countries:

"The soldier is primarily a soldier as long as may be necessary for the submission of populations which have not yet been subdued. But, once peace is achieved, he puts down his weapons. He becomes an administrator. .... At first, these administrative functions may seem incompatible with the concepts certain people have formed about military men. However, this is the real role of the colonial officer, and of his devoted and intelligent collaborators, the noncommissioned officers and privates he commands. It is also a more delicate function, requiring greater diligence, effort and high personal qualities, since reconstruction is far more difficult than destruction.

"Moreover, circumstances inevitably impose these obligations. A country is not conquered and pacified when a military operation has decimated and terrorized its people. Once the initial shock passes, a spirit of revolt will arise among the masses, fanned by a feeling of resentment which has been created by the application of brutal force.

"During the period following the conquest, the part of the troops is reduced to policing, a function which is soon taken over by special troops, the military and civilian police. But it is wise to make use of the boundless qualities of devotion and ingenuity of the French soldier. As work superintendent, teacher, craftsman, chief of a small post—wherever an appeal is made to his initiative, his self respect, his intelligence—he proves equal to his task. And it would be misleading to believe that temporary neglect of military drill might harm his spirit of discipline and of military duty. The soldier of colonial troops is usually old enough to have gone through the sequence of exercises several times. He has little left to learn from the theoretical and physical training to which recruits in France are subjected.

"The services he is required to perform involve moral and physical activity which is stimulated through interest in the task entrusted to him. Moreover, by arousing the interest of the soldier in our task in the country, he becomes interested in the country itself. He observes, he calculates, and often, at the end of his term of service, he may decide to till some plot of ground, to utilize his skill and knowledge, and to make the colony profit from his devotion and good will. He becomes one of the most valuable elements of the small settlement which is an indispensable complement to colonization in the larger sense of the word."[6]

These instructions go beyond the scope of warfare in its strict sense. But they deal with the problem of settlement which provides the only possible guarantee that the conquest over colonial people and country can be lastingly and peacefully maintained. Such colonization, obviously, cannot be achieved

---

[6] These passages of Galliéni's instructions of May 22, 1898, are translated from the extensive quotations in Lyautey's article, *Revue des Deux Mondes*, CLVII (1900), 316-317.

by mere force of arms. The "marching organization" requires thorough preparation before action; there is no room for improvisation.

Lyautey therefore stresses the fundamental principle that the colonial army cannot be just any army sent to a colonial territory. It must be a highly specialized corps. "It is essential," he writes, "that the colonial army have autonomy and that it have its own chiefs, actually distinct, to whom the colonial idea, and the adaptation of the [military] tool to its use, surpasses any other consideration." Colonial warfare requires a greater continuity of work than European war; services cannot be gauged in the same manner. "Do you think," asks Lyautey, "that it does not need more authority, more *sang-froid*, more judgment, more firmness of character, to maintain in submission, without firing a single shot, a hostile and excitable population than to subdue it through gunfire once it has arisen?"

This was obviously a plea for a profound reorganization of the entire system of the French colonial armies. At least, Galliéni and Lyautey could claim that in Madagascar, through reorganization with only the means at their own disposal, they had indeed obtained remarkable results. Lyautey conquered most of the restive south of Madagascar by means of what could be called "pacific occupation." It may seem paradoxical to speak of the "pacific action" of an army but in fact the goal was achieved practically without fighting. And the ideal utilization of armed forces has always been to prevent fighting rather than to provoke it.

The broad meaning thus given to colonial warfare led Lyautey into the discussion of military administration in the colonies. He advocated Galliéni's system of uniting administrative and military authority in the same hands not only at the summit of the hierarchy but also on lower levels. Bugeaud created in Algeria the system known under the name of "Bureaux Arabes" in which specialized officers, belonging to a distinct corps, specializing in administration, were given government functions. But the officers of the Bureau Arabe did not command any military unit in their territory. Lyautey insists that such a division of powers is only a weakness in the colonial field: "To dispose constantly and directly of the armed force is a necessity in these immense colonial countries where security must be insured with a handful of men, against entire peoples." Galliéni's system of unified territorial command, known in colonial history under the name of "Cercles Militaires," asked, of course, for highly trained men but with a less narrow specialization. They had to be competent both in field command and local government.[7]

Lyautey and Galliéni also opposed Bugeaud's tradition, begun in Algeria, of settling former soldiers in the conquered area in large military villages where everyday life and agricultural work were carried on to the rhythm of orders and music like drill in the barracks. They wanted the soldier, if he decided to become a settler, to go into this work as a free individual and to merge in the

---

[7] With reference to these problems see: Albert Ringel, *Les bureaux arabes de Bugeaud et les cercles militaires de Galliéni* (Paris, 1903) and Arthur Girault, *Principes de colonisation et de legislation coloniale* (Paris, 1921-1923, 4th ed.).

local environment, physical and human. Thus they tried to eradicate most of the barriers that had grown up between settlers and natives under the Bugeaud system of military colonization.

At the dawn of the twentieth century, Lyautey's article on the "Colonial Role of the Army" summed up in a few pages the long experience of the French in military warfare, practice, and thought. The theory of the school was well formulated. In 1940 it was still alive for all colonial-minded Frenchmen.

Thus Bugeaud's teaching was largely left behind. Still the influence of the conqueror of Algeria remained an important one and, through the Latin classics, that of Roman imperial warfare as well. But during more than fifty years of experience, the French vastly improved and enriched their knowledge of colonial affairs and colonial warfare. They had profited also from certain foreign experiences, chiefly the Russian conquest of the Caucasus and of Turkestan. Skobelef certainly was not unknown to the French colonial officers. Lyautey frequently refers to various episodes of Russian Asiatic warfare as models for colonial officers in general. Still the doctrine and technique of "progressive occupation" remains an invention of this small group of Frenchmen. Viewing warfare as a whole from a broader perspective, Lyautey was able to define the essential difference between colonial warfare as such and all other kinds of military action: instead of bringing death to the theater of operations, the aim is to create life within it.

IV

In 1900, when Lyautey's basic article was published, most of the territories that constitute France Overseas were already flying the Tricolor. Already the army's main function in these countries was that of occupation and the study of its administration is beyond the scope of the present volume. One further conquest was made in the twentieth century at a cost of considerable warfare and with remarkable results. It was the conquest of Morocco, indissolubly associated with the name of Lyautey. Here, indeed, the theory worked out in Indo-China and Madagascar by Lyautey under Galliéni's orders and direction was best applied on a large scale.

In March 1902, Lyautey left his post as governor of southern Madagascar to fill a new assignment in France as colonel in command of a cavalry regiment, the Fourteenth Hussars. But he stayed only one year in this capacity. Grave troubles were developing in southwestern Algeria, on the border of Morocco, and at the edge of the great Sahara Desert. Algerian tribesmen, supported by Moroccan nomads, were inflicting serious casualties on the local French garrisons and threatening the region with invasion. This was a threat to the whole of Algeria and was, moreover, a source of serious international complications. Everything that concerned Morocco had serious repercussions in the capitals of various interested powers, particularly in Berlin. The French government decided to send to the area the best available specialist in the new doctrine of "pacific occupation," so that order might be restored with the least possible disturbance. On October 1, 1903 Lyautey, promoted to the rank of brigadier general, took command of the subdivision and territory of Aïn-Sefra, where he remained for three years (until December 1906). This was his introduction to Morocco. Aïn-Sefra, at the gates of the Sherifian empire, served as a laboratory where men, principles, and techniques were finally elaborated. The young officers who served there under his orders (such as Mangin, Henrys, and Laperrine) later became some of his most brilliant associates.

Lyautey's chief struggle was not with any native chief but with his superiors of the North African Army, who, like the general staff in Paris, were far from approving the methods of Galliéni's school. Lyautey's correspondence from Aïn-Sefra shows the progress of his battle with the high command as well as his pacification of the area. Particularly interesting is his long letter to his former chief and master, Galliéni, dated November 14, 1903, in which Lyautey set forth what he wanted to do.[8]

Lyautey explains that he found the army in Algeria (which was considered as part of the metropolitan French territory) organized exactly on the European pattern—red tape and all. He asked first for a special autonomy of his command, the right to report directly to the governor general in Algiers without passing through the army hierarchy. He wanted the right to effect a reorganization of his troops for the special purpose of making the army "really mobile." Since Bugeaud, mobility had been the great aspiration of all colonial chiefs for, as Lyautey later remarked, "in Africa, one defends himself by moving."[9]

Lyautey wanted also a real organization of the country, which, in a desert, meant essentially the control of the water points and the protection of markets and railroads so that trade could be carried on in safety. He pointed out, in particular, that in order to protect a given line (*e.g.*, the railway), it was useless to keep fortified posts on this line itself. Such a system of posts would not prevent groups of raiders, aiming for this line, from reaching their goal. The defense must be put at some distance ahead, to check an attack before it could impede traffic.

The French establishments of the disputed area (described as the region west of the Jebel Bechar) should be a *center of attraction* and not of repulsion. It should not be feared. The natives must find profit in the neighboring French protection. This protection should have, therefore, a definite economic orientation, encouraging trade, attracting caravans, bringing not only a promise of security but also material prosperity. In fact, the problem was not to occupy one point but a zone, to create not a military post but a center of action and influence. Political warfare was not neglected: the main hostile chief in the Moroccan mountains had to be attacked not by columns but by native auxiliaries and undermined by the systematic disintegration of his own authority, in his own territory, among his own supporters.

[8] Several volumes of Lyautey's correspondence have been published (see bibliography). The Aïn-Sefra period takes a whole volume: *Vers le Maroc. Lettres du Sud-Oranais* (Paris, 1937). The quoted letter to Galliéni is on pp. 12-28.
[9] In French: *En Afrique, on se garde par le mouvement.*

Two points in particular deserve special comment for they were to remain the bases of Lyautey's Moroccan strategy and policy. 1. In the field of diplomacy he advocated a loyal alliance with the sultan's government and representatives. No action was to be taken in Moroccan territory except in agreement with the official Moroccan authorities and with their help. This *"entente cordiale"* was the basis of the protectorate. 2. In the field of strategy one paragraph of the letter is fundamental: "In fact, the final establishment of the system of protection that I project will be accomplished very gradually; it would be impossible for me to assign even an approximate date for its realization, although I incline to believe that the result can be achieved more rapidly than most people think. It will advance not by columns, nor by mighty blows, but as a patch of oil spreads, through a step by step progression, playing alternately on all the local elements, utilizing the divisions and rivalries between tribes and between their chiefs."[10] The strategy of the "oil patch," the famous *"tâche d'huile,"* will take its place in history as the phrase which best characterizes the French penetration and pacification of Morocco.

The program was put into effect. Pacific occupation, bringing peace and prosperity to the occupied regions, bettered all of Lyautey's expectations in its progress southward and westward. From the headquarters at Aïn-Sefra the conquest of the Sahara was considerably extended. As a symbol of the economic development of the region stands Colomb-Bechar, which, founded by Lyautey, is now an active center of caravan trade and the head of the main transdesert road and of the Transsaharan Railroad. But the creation of Colomb-Bechar threatened to provoke a Franco-German incident. As it advanced westward, French colonization was coming into closer and closer contact with the territory and hence with the problems of Morocco. At the same time the Moroccan crisis was assuming increasing and grave importance on the European diplomatic scene, especially as a result of incidents which occurred on the Atlantic shores of Morocco.

From December 1906 to 1910, Lyautey, leaving Aïn-Sefra, commanded the Division of Oran (Western Algeria). There he was in still closer touch with the Moroccan frontier, even though he was further away. He contributed to the progressive reshaping of the French North African Army according to the ideas he had always advocated. During these years the pacific occupation, progressing from the Algerian base southward, conquered the immense open spaces of the Sahara Desert, achieved territorial continuity of the French African block from the Mediterranean to the Niger and the Gulf of Guinea. Analyzing the results of that conquest, Professor Gautier, who worked with Galliéni and Lyautey at Madagascar and who later became one of the main explorers of the Sahara, wrote that the pacification had a psychological basis: "the profession of policeman has become there [in the Sahara] more profitable than the profession of bandit."[11]

At Christmas, 1910 Lieutenant General Lyautey left Algeria to report to

---

[10] Lyautey, *Vers le Maroc. Lettres du Sud-Oranais*, p. 26.
[11] E. F. Gautier, *La Conquête du Sahara* (Paris, 1910), p. 121.

his new post as commander of the Tenth Army Corps, at Rennes, in western France. Once more he spent a year of relative inactivity in garrison life. Meanwhile the Moroccan crisis was developing and threatening international complications as well as military defeat in Africa. Gautier wrote in 1910: "All our recent conquests, in Tunisia, in Tonkin, in Madagascar developed through two phases: that of the initial war, with the winning of apparently decisive victories, followed by that of insurrection, inevitable, painful, and of which the issue was more administrative than military, the organization of the country. This would seem to point to the uselessness of battles in wild countries. It is plain, in any case, that they play a secondary part. Needless to say, we are not naïve enough to believe force useless, but the question is to know how to use it. In a European war victory is the goal, because you have to crush an organization, a military machine. It has no sense in an anarchic country where, as nothing exists, there is nothing to destroy and where the difficulty is, on the contrary, to create. That it is possible, despite appearances, to suppress the initial period of great war, in the tutelary relations of a civilized State with its barbaric neighbors, seems to be proved by the example of the Sahara."[12] Morocco was at the stage of entering active war. Ten years of French endeavor there, and even more, were at stake. To avoid the repetition of the previous bloody experiences, the government in Paris appointed Lyautey as resident general of Morocco.

V

From May 1912 to October 1925, Lyautey remained the resident general and commander in chief in Morocco. He created there the system of the French protectorate. His unification and development of the country is now reputed to be the masterpiece of French colonization. This was a more civilized and more warlike country than Tonkin or Madagascar. Bloodshed could not be avoided entirely, but Lyautey certainly reduced it to a minimum and won not only the territory but also support of the inhabitants to the new regime he was building. It is almost certain that no sultan has ever organized in Morocco a regime so popular with all the people as that of the French protectorate, which Lyautey conceived, organized, and administered.

Upon Lyautey's arrival, the country was in complete revolt, not only against the French but against its legal sultan and government as well. By two expeditions, one in the region of Fez, led by Gouraud, and the other to Marrakech, led by Mangin, Lyautey reestablished control of the main cities of the Sherifian empire. These were swift and daring blows, frequently studied since and described by colonial and military historians as models.[13] The speed of the initial success was largely due to Lyautey's policy with respect to the natives which was put into effect from the first day. Its ultimate success depended, of course, on the period that followed. Once more, instead of resting on his laurels,

---

[12] *Ibid.*, p. 122.
[13] See especially General Paul Azan, *L'expédition de Fez* (Paris, 1924), with a foreword by Lyautey.

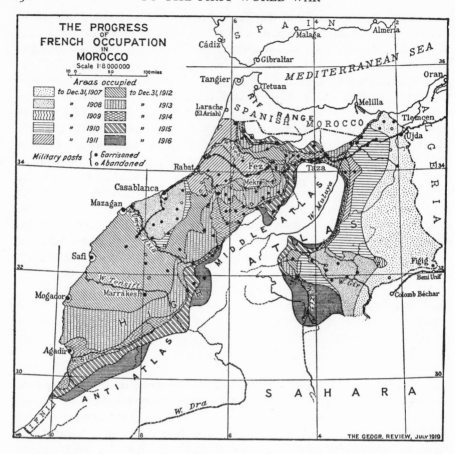

THE PROGRESS
OF
FRENCH OCCUPATION
IN
MOROCCO

Scale 1:8 000 000

Areas occupied

	to Dec.31,1907		to Dec.31,1912
	" 1908		" 1913
	" 1909		" 1914
	" 1910		" 1915
	" 1911		" 1916

Military posts { • Garrisoned
{ ○ Abandoned

THE GEOGR. REVIEW, JULY 1919

Lyautey decided that the narrow corridor which he now held from the coast to Fez and to Marrakech could be safe only if the defense line were pushed forward. As in the case of the railroad at Aïn-Sefra, the occupied area was secure only if the real front line was actually situated at some distance ahead. The French forces, until then driven into corners, proceeded in different directions. In the western part of High Atlas Range, the alliance with the Great Caïds guaranteed peace. That region has remained constantly loyal to the French since 1912, collaborating with them and covering the right wing of the forces penetrating inland.

To support the advancing front, a large scale and costly policy of economic development was immediately started in the rear: the hostile tribes had to be convinced of the advantages of French rule. In two years appreciable results were obtained. Then, in the summer of 1914, war broke out in Europe. The French government, needing all the available forces to defend frontiers of metropolitan France, ordered Lyautey to send most of his troops to Europe and evacuate all the territory he could not hold with forces reduced to only twenty-three infantry battalions and a few cavalry squadrons. At his discretion he was authorized to retreat to the coast, however, and preserve merely a bridge-head there for the duration. Lyautey executed the first part of the orders and immediately sent two divisions (about two-thirds of his troops) to France but refused to retreat.

"It appeared to one on the spot—and the government accepted this conception—that the true formula to provide France with the greatest and most effective help, was [not to retreat but] on the contrary, to preserve entirely the apparent *contour* of our occupation, and empty the rear, behind the screen of our maintained fronts, 'like an egg behind its shell.' It was the best way, the only way, to safeguard, on the one hand, our situation in Morocco, to preserve for us the benefit of its resources of all kinds, supplies and workers, and, on the other hand, to prevent the still restive populations, observing the retreat of our advanced troops, from rushing on them, so that they could reach the coast only at the price of ceaseless fighting and decimation; this would excite the whole country, disturb all our lines of communications, make mobilization impossible, and end by blockading us in the ports in front of a completely insurgent country."[14]

With incredibly limited forces, Lyautey held his fronts. He mobilized all possible resources of man power and, through four and a half years of world war, carried on his determination both to send as many reinforcements as he could to France and to push the conquest inland. Such a plan succeeded as a result of his tremendous prestige, supported by the same clever and daring policy. He himself called it the "policy of the smile." Through the darkest moments of the war, he never ceased to show his confidence in the final victory. He pushed the economic development of the country with speed. In the first year of the war he inaugurated at Casablanca large department stores, agricultural exhibitions, a large fair. The latter, opened in September 1915, made a deep impression on the government in Paris as well as on the native populations. Lyautey called it a "war gesture" and, indeed, one rebel tribal chief asked for a truce and special passports in order to visit the Casablanca fair. Both were granted. Returning to his tribesmen, the chief was so impressed by the power and prosperity displayed at the fair that, instead of resuming the fight, he decided to submit.[15] The war was thus, as Lyautey intended that it should be, generating a new life, new towns and harvests, while the fighting, which was, of course, conducted with great skill, was still in process.

The strategy that produced such results even with small forces from 1914 to 1918, was the same which, later, with greater means and higher speed, achieved the conquest of all the vast country. The best principles of Bugeaud's and Galliéni's teaching were put into force and improved through adaptation to the local conditions. Certain special factors determined the course of the conquest: the Protectorate extended over the plains; the remaining task was the occupation of high mountain ranges which made up most of Morocco. The mountains were densely populated by proud and warlike tribes (the Moroccan

[14] Lyautey, *Paroles d'Action* (Paris, 1927), p. 128.
[15] *Ibid.*, p. 143.

warrior had always been highly esteemed in Islam). These tribes accepted no rule, not even that of the sultan, and they were determined to fight to death against the foreigners. During the World War help from Germany was regularly sent to them.

Lyautey's tactics in conquering the mountain ranges somewhat resembles those of siege warfare. Thorough preparation, in which political warfare played a large part, preceded the launching of each attack. As a first step one section or massif was encircled and isolated. To do this a long continuous front was necessary. Lyautey renounced definitely the strategy of columns in favor of the wide front as the only means of avoiding enemy infiltration on the flanks and in the rear of advancing troops. But mountainous topography, as Bugeaud rightly observed, does not favor the establishment of continuous attacking front lines. Lyautey established his front in the lowlands, along the foot of the whole restive range and pushed forward into the gaps, penetrating into the mouths of valleys and other depressions, occupying first the lower regions and rising slowly in altitude like a tide on a cliffy shore. When contact with the mountain was closely established all along its foot, preferably on three and at least on two sides, the assault was launched by two strong columns converging in a pincer movement from both sides on the main passes that separated this massif from the next one. When the two spearheads met, the mountaineers found themselves actually besieged, isolated from the outside, cut off from supplies.

Then the belt encircling the massif was tightened. Infiltration continued along valleys and depressions, like a new rise of the tide. On a given day the final assault was ordered, several strong columns converging and striking rapidly forward against the isolated, tired, demoralized tribesmen. In most cases this final attack yielded immediate surrender, practically without casualties. Along with the armies, the organization of the territory, carefully prepared in advance, marched forward, occupying ridge by ridge, massif by massif, expanding in the "oil patch" manner.

The operations, given the peculiarities of the terrain, were governed by geography, planned on the map according to the relief as well as to the tribal geography which determined the political preparations. This strategy, expanded on a larger scale after 1919, was applied without any appreciable change to all the stages of the Moroccan pacification, including the violent episode of Abd-el-Krim's uprising, the so-called Riff war (1925-1926). Lyautey, then seventy-one years old and seriously ill, retired in the midst of the war with the Riffs, after having checked Abd-el-Krim's offensive and having established a front from which to base a victorious counterattack. His pupils completed his work in the years 1926-1934.

The principal improvements added to Lyautey's strategy and tactics after 1925 were largely due to the extensive use by his pupils of the newest weapons which advancing military technology put at their disposal: the motor car and the airplane. Both fitted admirably into the Moroccan picture, for the dominant trend of colonial warfare was toward increased mobility. Henceforth the tools

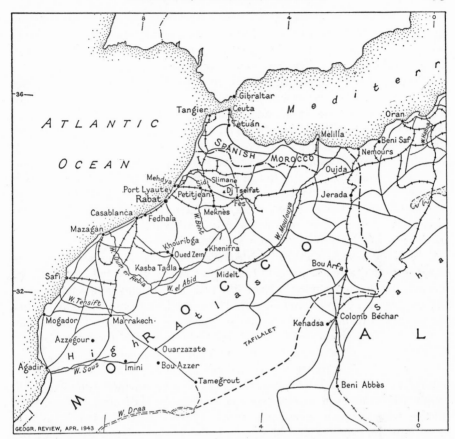

were at hand. Motorization of the columns and of the services of supply greatly increased the speed and effectiveness of encircling movements and surprise blows. Bombing from the air robbed the natives of their chief trump card: fire from dominating positions in the mountains. These modern methods were especially employed in the last steps of the Moroccan pacification in 1931-1934.

These operations developed under the superior command of General Huré who explained the principles of his warfare in his already quoted article on "Moroccan Strategy and Tactics." His *Instruction Générale* of February 19, 1932 sums up the directives controlling the movements of the mobile groups (as the motorized columns were then called). It shows the application of both Bugeaud's and Lyautey's lessons: attack is made on large fronts ensuring the safety of the rear; in the mountains, action is through parallel or convergent valley; attack is by surprise from bases carefully prepared in the rear and progressing with rapidity. The terrain is conquered by auxiliary units, artillery and air force, then occupied by the regular troops (native troops have a better knowledge of the terrain and a greater mobility but, as they are unable to hold the area taken, this is done by the regular troops which thus will have to fight only in defensive positions). The terrain must be organized as soon as conquered—shovels and pick axes are as necessary as rifles and guns; every conquered position must be linked to the rear by a road as soon as possible; it is by means of roads that the country is controlled.[16]

The preparation of these last campaigns and the operations themselves have been well described by General Catroux, one of the most brilliant officers of Lyautey's school, who played an outstanding part in the last decade of the Moroccan conquest.[17] In 1930 almost all Morocco was pacified and organized. It remained to bring under the Protectorate's rule the south of the country, mainly a mountainous arc of almost one thousand miles in length, extending from the Central High Atlas Range to the Atlantic shore through the massifs of the Tafilalet, Jebel Sagho, and the Anti-Atlas Range. South of this high barrier a large section of the Sahara Desert stretching down to the Wadi Draa was Moroccan territory and still the refuge of unmanageable nomads whose raids rendered the whole northwestern Sahara insecure.

The operations began in the High Atlas and progressed southward. The general plan was drawn according to Lyautey's strategy of the "oil patch." Under General Huré, commander in chief, the two main prongs of the pincer were led by Generals Catroux and Giraud. The campaigns themselves lasted only a few weeks at a time, while long months each year were spent in making the preparations: political warfare had certainly not been forgotten. General Catroux stressed once more in his review of the campaign that the goal of pacification was "to transform the dissidents into associates." Decisive action started in 1933: the Jebel Sagho was attacked and conquered in March, the High Atlas massifs in July and August. A new period of preparations ensued during which was organized a Catroux-Giraud pincer directed toward the southwest. In February and March 1934, the Western Anti-Atlas Range was surrounded and while Catroux subdued the mountain, Giraud, rushing southward with rapid motorized columns, surrounded and forced the last restive tribes in the desert to surrender.

The operations of 1933-1934 were probably the first in history when relatively large numbers of highly mechanized troops succeeded in crushing a strong defense in depth with a few powerful blows. From the point of view of warfare, mountains may be compared to an area strongly fortified in depth; the fortifications are built by nature and the mountaineers have an ancient technique for the organization of defense. The success of these blitzlike campaigns was attributed by General Catroux "to the cleverness of the command's combinations. They had the skill to make the maximum of forces concur in the minimum of time for manoeuvres objectively planned; and, without swerving from the general plan, they varied, with flexibility and according to the needs of the moment, the missions of the different generals in command of operational groups."

[16] See footnote 1.
[17] General Catroux, "L'achèvement de la pacification marocaine," *Revue politique et parlementaire* (Paris, 1934), CLXI, No. 479, pp. 24-46.

In March 1934, the Moroccan conquest was finished and the second French colonial empire erected in its final and complete form. Henceforth the role of the colonial army was reduced to the policing of vast empty lands and to the defense of the frontiers. The main threat of the twentieth century was not to colonial territories but to the European boundaries of France. Colonial troops and generals of the Galliéni-Lyautey school played a considerable part in World War I; they are playing an even greater part in the present World War and in the liberation of France.

VI

After 1934, the task of defending France Overseas from outside enemies seemed a simple one. Almost all the neighboring countries were friendly to France and a serious threat could come only from farther abroad. Thus imperial defense seemed to be chiefly a matter for the French navy, especially in the Far East. In the late 1930's the only questionable point with respect to actual frontiers was the frontier between southern Tunisia and Italian Libya. Franco-Italian relations were deteriorating rapidly and Fascist Italy was taking a more and more aggressive stand against France, particularly with regard to Italian designs on Tunis. Military preparations on a large scale were being conducted in Libya. Frequent incidents occurred, especially after 1937. The French then built a line of fortifications along the Libyan frontier, the now well known Mareth line, and strongly organized the territory to the rear.

Early in 1939 the general situation in Europe became acute. If war broke out, what could and would Italy do from her Libyan bases in northern Africa against the French and the British? Here lay a major problem of Mediterranean strategy. In a short article, General Catroux analyzed Italy's strategical position in North Africa and outlined the possible operations around Libya.[18] He had just retired from the command of the North African forces in Algiers. Events since 1940 have confirmed the accuracy of his views.

Catroux stressed that Italy's Libyan colonization was a capital investment for war purposes, the establishment of an operational base in the midst of African territory controlled by Britain and France. An offensive bastion had been erected there from which to strike out toward the fulfillment of the *Mare Nostrum* project. But from what direction would the first blow fall? The main lines of the preparations had to be considered:

"The Italian forces in this country [Libya], by their organization, by their numbers, and by their equipment in all kinds of matériel, have all the requirements, in fact, for an offensive action in depth and,—an important feature which should not be overlooked—one conducted through desert areas. . . . The two motorized army corps are fitted, by their nature, for shock operations and for rapid movements. . . ." Catroux emphasized the increased use of motorized transport for supplying water, fuel, and ammunition, the training of the command for rapid and massive offensive action through wide areas of desert, the

---

[18] General Catroux, "La position stratégique de l'Italie en Afrique du Nord," *Politique Etrangère* (Paris, June 1939), pp. 271-281.

interest displayed in air-borne and paratroops to be sown in the enemy's rear. Such tactical features revealed, in Catroux's opinion, the strategical plan into which they fitted. There are no large desert spaces between Tripoli and Tunisia. Neither Italian infantry nor its artillery equipment were fitted for an attack on the fortifications of the Mareth line, behind which were strong reserves in depth. Although the Italian aerial superiority was considerable, the French navy was strong enough to prevent any landing in the rear of the Mareth system from the sea. Marshal Balbo would definitely not be able to break through the French defenses, nor did this seem to worry him. On the other hand, his army was especially fitted to cross the Libyan Desert eastward and attack Egypt probably with Suez as a target.

In this connection, General Catroux stressed the youth and inexperience of the Egyptian army, the very small numbers of the British forces, the weakness of the newly begun fortifications in the coastal region between the Great Libyan Erg and the sea. Though narrow, the corridor between the sea and the ocean of sand dunes that leads to the Nile Delta offered opportunity for outflanking movements: "One can always find a way through in the Sahara, when the space factor is in your favor" said Catroux. He suspected then that Balbo's plan was for an offensive against Egypt. He weighed what the loss of Suez would mean for the Allies. But the operation was a risky one: an offensive from Tunisia might advance in Libya and take the Italian bases, thus isolating the armies in the desert. Anyhow the bases could not function without supplies from the Italian mainland. Sea power, therefore, would be decisive and the Italian submarine fleet and air force might well prove unable to prevent the Allied navies from controlling the Mediterranean.

With forces trained and equipped for remote desert fighting, where else could Balbo be planning to strike? The main French centers in the Sahara, such as Ouargla or Fort-Flatters might seem to be within his reach. "A course of 1,000 kilometers is certainly not prohibitive for a strong motorized column adapted to the Sahara and protected by aviation." But the oases were carefully protected by French mobile units. The most interesting target at which to aim a blow against the French from Libya would be the Lake Chad region, focal tie of the French system of communications in Africa and a base for any further action against the Niger and Guinea, or the Anglo-Egyptian Sudan. Lake Chad is only 800 kilometers from the Libyan southern frontier and the region between offers no great obstacle to motor transportation. Still a French counteroffensive could be organized with comparative ease and the enemy pushed back from the Chad to the Fezzan.

From its new Ethiopian colony, Italy could attempt an offensive in the direction of Kassala-Khartoum in the Anglo-Egyptian Sudan. But Catroux doubted that the Ethiopian base was strong enough for such an offensive to make much progress. Thus his conclusion was plainly that the Libyan preparations were for an assault on Egypt, and his article was largely a plea for the reinforcement of the British and Allied forces in the Middle East to protect the Nile and Suez. Still he did not believe that Italian successes in Africa could be lasting: "Italy's offensive system is based on the supremacy of the motor employed both in the air and on the ground, a system capable of considerable strategic results on the condition that the renewal of matériel and the supply of fuel is not to be interrupted. Libya cannot meet these imperative needs. Italy is enslaved by the sea, both in the homeland and in her possessions, yet she does not control it. This servitude, which she cannot break, necessarily prepares an unlucky end to the enterprises she may attempt in wartime in Africa."

General Catroux foresaw in the spring of 1939 all the general lines of development of the war in Africa except the collapse of France and its African consequences. In the fall of 1939, he had been appointed governor general of Indo-China, but he left this post in 1940 when the agreement between Vichy and Japan deprived him of any chance for resistance. He led the Fighting French forces which, with the help of the British forces, reoccupied Syria and the Lebanon in 1941. He was in command of the Fighting French in the Middle East when, with the British Eighth Army, they started a victorious march that ended in Tunisia after a breach of the Mareth line, which was this time held by Axis forces. One may find in General Catroux's article even a prediction of the part to be played by the base organized on Lake Chad by General de Gaulle's forces. It was from this base that columns set out which, under General Jacques Leclerc's orders, first raided and ultimately conquered the Libyan oases of Fezzan, Koufra, and Ghadames, before joining the Eighth Army. Few Frenchman realized the full meaning of the broadcast in which, in the fall of 1940, General de Gaulle announced that the Chad Colony had rallied to his standard, prophesying that this adherence would prove to be a great event in the history of French participation in this war.

VII

The general theory of French colonial warfare was elaborated at the end of the nineteenth century. After 1900, few innovations of any importance were added to the ideas expressed in Lyautey's "Colonial Role of the Army." The experience in Morocco led Lyautey to develop his strategy of the "oil patch" and his tactics for conquering the mountainous regions. Some modifications followed, dictated by the technical changes in the means of transportation. Colonial officers were always eager to try any new method: they were the first group among French officials to be air and machine minded.

But according to the French school, warfare in the colonies appeared inseparable from administration, at least during the early periods of occupation and pacification. The work was carried out with thoroughness, to judge from results: within one century, with extremely limited forces and almost in the face of opposition by the central government, one of the largest empires in history was conquered and organized; and its organization was such that peace and greater prosperity were brought to all its parts. The best proof of the success of the conquest and of this type of colonization is certainly the empire's attitude in the present war. As in 1914-1918, colonial troops came to France to help defend its European frontiers. When, in June 1940, France experienced the greatest and most humiliating military defeat of its history, a defeat that hundreds of thousands of natives from the colonies saw and felt deeply, not a single one of the parts that constituted France Overseas attempted to overthrow the French regime. The native populations showed their determination to share the French fate. In each country further policy was decided by a small group of Frenchmen at the top. They chose to join the Vichy rule or the Fighting French movement of General de Gaulle. Through her colonies, France continued the struggle after the collapse in Europe. From the Chad her troops invaded enemy territory. And, in Algiers, after the Allied victories in North Africa, a new form of the French might arose, ready to participate in the liberation of European France. Whatever the balance of payments, the colonial conquest has rendered to France the greatest possible service in her darkest hour.

Such a success is not to be explained merely by a theory of warfare and administration. It would not have been possible had the men not proved worthy of the theory. A large part of the theory itself was devoted to the training of able men. In his address upon the occasion of his admission to the *Académie Française*, July 8, 1920, Marshal Lyautey rated high the importance to a military commander of his staff and particularly of his chief of staff. "The first condition," he said, "of supreme command is the full freedom of mind of the chief, the certainty, guaranteed to him, that his thought, thrown in the air, will recover its shape immediately and be transmitted without any loss of time, without any distortion, to the farthest extremities." Such is the indispensable, critical task of the commander's staff. To this need, Lyautey added two main axioms: "A general must never be bothered with political anxieties," and "Faith in victory determines victory." The last principle puts Lyautey in the same category of men as a Foch or a Clemenceau: they believed more in faith than in the plan.

The principle of "the right man in the right place" remained a dominating one in Lyautey's thought throughout his life. He gave a great deal of his time to the education of his associates and he wanted them to have the widest possible experience, culture, and outlook. He constantly fought against specialization and left a tradition in the colonial army that academic degrees for officers were to be valued no less than military deeds. One of his best pupils, General Freydenberg, among other things, was a noted geologist. Many young officers of the Moroccan army endeavored to add diplomas to their medals.

After the great success of his Casablanca Fair in 1915, Lyautey was invited as a guest of honor to the Fair of Lyon, which had been organized by Edouard Herriot, then Mayor of Lyon and in normal times professor of French literature at the university of that city. Lyautey took advantage of this circumstance to give in the conclusion of his address delivered on February 29, 1916 his definition of the right type of man:

"He who is only a soldier is a bad soldier, he who is only a professor is a bad professor, he who is only an industrialist is a bad industrialist. The complete man, he who wants to fulfil his entire destiny and to be worthy of leading men—in short, to be a chief—this man must have an open mind on everything that honors mankind."[19]

But the highest quality for a man in the opinion of the colonial school was undoubtedly to be a driving man, a man of action. Galliéni said: "It is in action that I always found the greatest satisfactions and the true *raison d'être* of life."[20] On the ring that he used as his personal seal in Morocco, Lyautey had engraved Shelley's verse: "The soul's joy lies in doing." And they wanted driving men that believed in their own action: "Faith in victory determines victory." Such they were themselves, and they still may be classified among the most interesting thinkers of their time.

[19] Lyautey, *Paroles d'Action*, p. 176.
[20] Quoted in René Musset, "Galliéni et Madagascar," *Mélanges de Géographie et d'Orientalisme offerts à E. F. Gautier* (Tours, 1937), pp. 388-390.

CHAPTER 11. Delbrück: The Military Historian

BY GORDON A. CRAIG

HANS DELBRÜCK, whose active life coincided almost exactly with that of the Second German Empire, was at once military historian, interpreter of military affairs to the German people, and civilian critic of the general staff. In each of these roles his contribution to modern military thought was noteworthy. His *History of the Art of War* was not only a monument to German scholarship but also a mine of valuable information for the military theorists of his day. His commentaries on military affairs, written in the pages of the *Preussische Jahrbücher*, contributed to the military education of the German public and, during the first World War especially, helped them comprehend the underlying strategic problems which confronted the general staff. His criticisms of the high command, written during the war and in the period following it, did much to stimulate a reappraisal of the type of strategical thinking which had ruled the German army since the days of Moltke.

The military leaders of Germany have always placed great emphasis upon the lessons which can be drawn from military history. This was especially true in the nineteenth century. It had been Clausewitz' ideal to teach war from purely historical examples; and both Moltke and Schlieffen had made the study of military history one of the responsibilities of the general staff.[1] But if history was to serve the soldier, it was necessary that the military record be an accurate one and that past military events be divested of the misconceptions and myths which had grown up around them. Throughout the nineteenth century, thanks to the influence of Leopold von Ranke, German scholars were engaged in the task of clearing away the underbrush of legend which obscured historical truth. But it was not until Delbrück had written his *History of the Art of War* that the new scientific method was applied to the military records of the past, and it is this which constitutes Delbrück's major contribution to military thought.

It was not, however, his sole contribution. In the course of the nineteenth century the basis of government was broadened and in the western world generally the voice of the people was felt increasingly in every branch of governmental administration. The control of military affairs could no longer remain the prerogative of a small ruling class. In Prussia, the embittered struggle over the military budget in 1862 was an indication that the wishes of the people and their representatives with regard to matters of military administration would at least have to be given serious consideration in the future. It seemed important therefore for the safety of the state and the maintenance of its military institutions that the general public should be educated to a proper appreciation of military problems. The military publications of the general staff were designed not only for use in the army but also for more general consumption. But the writings of professional soldiers, devoted as they were to accounts of

[1] See Chapter 8.

single wars and campaigns, were in general too technical in style and content to fulfill the latter function. There was a genuine need for instruction in the elements of military affairs on a popular level.[2] Delbrück felt this need and tried to supply it. In all of his writings he considered himself as interpreter of military affairs to the German people. This phase of his work was most marked during the first World War, when in the pages of the *Preussische Jahrbücher,* Delbrück wrote monthly commentaries on the course of the war, explaining on the basis of available materials the strategy of the high command and of Germany's opponents.

Finally, especially in his later years, Delbrück became a valuable critic of the military institutions and the strategical thinking of his time. His study of the military institutions of the past had shown him, in every age, the intimate relationship of war and politics, and had taught him that military and political strategy must go hand in hand. Clausewitz had already asserted that truth in his statement that "war admittedly has its own grammar, but not its own logic" and in his insistence that war is "the continuation of state policy by other means." But the Clausewitz dictum was too often forgotten by men who remembered that Clausewitz had also argued for the freedom of military leadership from political restrictions.[3] Delbrück returned to the Clausewitz doctrine and argued that the conduct of war and the planning of strategy must be conditioned by the aims of state policy and that once strategical thinking becomes inflexible and self-sufficient even the most brilliant tactical successes may lead to political disaster. In Delbrück's writings in the war years, the critic outgrew the historian. When he became convinced that the strategical thinking of the high command had become antithetical to the political needs of the state, he became one of the foremost advocates of a negotiated peace. After the war, when the Reichstag undertook to investigate the causes of the German collapse in 1918, Delbrück was the most cogent critic of Ludendorff's strategy and his criticism grew naturally from the precepts which he had drawn from history.

I

The details of Delbrück's life may be passed over quickly.[4] He himself summed them up tersely in 1920 with the words: "I derived from official and scholarly circles, on my mother's side from a Berlin family; I had war service and was a reserve officer; for five years I lived at the court of Emperor Frederick, when he was Crown Prince. I was a parliamentarian; as editor of the *Preussische Jahrbücher,* I belonged to the press; I became an academic teacher."

---

[2] See Hans Delbrück, "Etwas Kriegsgeschichtliches," *Preussische Jahrbücher,* LX (1887), 607.

[3] See Chapters 5 and 8.

[4] Delbrück himself has written brief autobiographical sketches in *Über die Glaubwürdigkeit Lamberts von Hersfeld* (Bonn, 1873), 78, *Die Geschichte der Kriegskunst im Rahmen der politischen Geschichte* (Berlin, 1900-1920), I, vii f., and *Krieg und Politik* (Berlin, 1918-1920), III, 225 ff. See also J. Ziekursch in *Deutsches biographisches Jahrbuch* (1929). An excellent account of Delbrück's life is given in Richard H. Bauer's article on Delbrück in Bernadotte Schmitt, ed., *Some Historians of Modern Europe* (Chicago, 1942), 100-127.

Delbrück was born in November 1848 in Bergen. His father was a district judge; his mother, the daughter of a professor of philosophy at the University of Berlin. Among his ancestors were theologians, jurists, and academicians. He received his education at a preparatory school in Greifswald and later at the universities of Heidelberg, Greifswald, and Bonn. He showed an early interest in history and attended the lectures of Noorden, Schäfer, and Sybel, all men deeply inspired by the new scientific tendency which was Ranke's contribution to scholarship. The influence of Ranke was clearly evident in Delbrück's doctoral dissertation, which was a critical and highly devastating appraisal of the writings of a German chronicler of the eleventh century.[5] Delbrück showed that these writings, long accepted as genuine by historians, were in large part unreliable and, in doing so, revealed for the first time the critical acumen which was to distinguish his later work.

As a student Delbrück was keenly interested in political problems and was a stanch advocate of German unity. It was not, however, until after 1870 that he was convinced that Bismarck's policy would attain that unity.[6] Nevertheless, feeling that war with France was inevitable, he enlisted in the army in 1867, saw active duty in the war of 1870, and remained a reserve officer until 1885.

From 1874 to 1879 Delbrück was tutor of Prince Waldemar, the son of the crown prince. His position not only placed him on terms of intimacy with the members of Frederick's court but gave him an excellent insight into the political problems of his time. Meanwhile, he had remained true to his early determination to become a historian and in 1881 was successful in obtaining a post at the University of Berlin, beginning a distinguished academic career which was to last until 1920. Although his research and his lectures occupied most of his time, Delbrück found an opportunity to take an active part in politics as well. From 1882 to 1885 he was a member of the Prussian Landtag and from 1884 to 1890 of the German Reichstag. As a parliamentarian, he was always, however, more of an observer than an active participant, and he regarded himself as "the scholar in politics."[7]

In addition to these activities, he was a publicist of weight and reputation. He served as editor of the *Staatsarchiv*, an annual collection of official and diplomatic documents, and of Schulthess' *Europäischer Geschichtskalender*, a publication which reviewed annually the events of the preceding year. In 1883 he was appointed to the editorial board of the *Preussische Jahrbücher* and, after 1890, he became the sole editor of that vigorous publication. It was in the pages of this journal that Delbrück wrote his military commentaries during the war and, in the post war years, his bitter attacks on the war-guilt clause of the Treaty of Versailles.

Even before he had begun his career at the University of Berlin, Delbrück had turned his attention to the study of military history. As a soldier, during spring maneuvers at Wittenberg in 1874, he had read Rüstow's *History of*

---

[5] *Über die Glaubwürdigkeit Lamberts von Hersfeld* (Bonn, 1873). See Richard H. Bauer in Schmitt, *op. cit.*, 101 f.

[6] *Krieg und Politik*, III, 226.　　　　　[7] *Ibid.*

Infantry and he later spoke of that event as having determined his choice of career. It was not, however, until 1877 that he turned seriously to the study of warfare. In that year he was given the opportunity of completing the editing of the memoirs and papers of Gneisenau which had been begun by Georg Heinrich Pertz. As he immersed himself in the history of the War of Liberation he was struck by what seemed to be a fundamental difference in the strategical thinking of Napoleon and Gneisenau on the one hand and Archduke Charles, Wellington, and Schwarzenberg on the other. As he carried his investigations further in the biography of Gneisenau with which he followed his editorial task,[8] the difference seemed more marked, and he sensed that nineteenth century strategy in general was markedly different from that of the previous century. He read Clausewitz for the first time and held long conversations with the officers attached to Frederick's court. While he did so, his interest was heightened and he determined to seek the basic and determining elements of strategy and of military operations.

His first lectures at the University of Berlin were on the campaign of 1866. But thereafter, he turned his mind to the past, lecturing first on the history of the art of war from the beginning of the feudal system, and then pushing his researches even further back into the period between the Persian Wars and the decline of Rome. He began a systematic study of the sources in the ancient and medieval periods and published short studies of the Persian Wars, the strategy of Pericles and Cleon, the tactics of the Roman maniple, the military institutions of the early Germans, the wars between the Swiss and the Burgundians, and the strategy of Frederick the Great and Napoleon. Meanwhile, he encouraged his students to make equally detailed studies of special periods. Out of these lectures and monographs grew Delbrück's *History of the Art of War in the Framework of Political History*, the first volume of which appeared in 1900.[9]

II

From the date of the publication of the first volume, the *History of the Art of War* was the butt of angry critics. Classical scholars resented the way in which Delbrück manhandled Herodotus; medievalists attacked Delbrück's section on the origin of the feudal system; patriotic English scholars were furious at his slighting of the Wars of the Roses. Many of the resultant controversies have been written into the footnotes of the later editions of the work, where the fires of academic wrath still smolder. But in its main outlines the

[8] H. Delbrück, *Das Leben des Feldmarschalls Grafen Neidhardt von Gneisenau* (Berlin, 1882).
[9] *Geschichte der Kriegskunst im Rahmen der politischen Geschichte* (Berlin, 1900). The work is in seven volumes but only the first four can be considered Delbrück's own. The fifth volume (1928) and the sixth (1932) were written by Emil Daniels; a seventh volume (1936) was written by Daniels and Otto Haintz. The first four volumes will be treated here. All citations will be made from the first edition. A second edition of the first two volumes appeared in 1908 and a third edition of the first volume in 1920. None of the corrections or additions in these later editions made essential differences in the original work.

book stands unaffected by the attacks of the specialists and it has received its meed of praise from such widely separated readers as General Groener, war minister under the Weimar Republic, and Franz Mehring, the great socialist publicist. The former referred to it as "simply unique";[10] the latter as "the most significant work produced by the historical writing of bourgeois Germany in the new century."[11]

Of the four volumes written by Delbrück, the first discusses the art of war from the period of the Persian Wars to the high point of Roman warfare under Julius Caesar. The second volume, which is largely concerned with the early Germans, treats also the decline of Roman military institutions, the military organization of the Byzantine Empire, and the origins of the feudal system. The third volume is devoted to the decline and near disappearance of tactics and strategy in the Middle Ages and concludes with an account of the revival of tactical bodies in the Swiss-Burgundian Wars. The fourth volume carries the story of the development of tactical methods and strategical thinking to the age of Napoleon.

In Proust's novel *The Guermantes Way*, a young officer remarks that "in the narrative of a military historian, the smallest facts, the most trivial happenings, are only the outward signs of an idea which has to be analyzed and which often brings to light other ideas, like a palimpsest." These words are a reasonably accurate description of Delbrück's conception of military history. He was interested in general ideas and tendencies rather than in the minutiae which had crowded the pages of earlier military histories. In his introduction to the first volume of his work, he specifically disclaimed any intention of writing a completely comprehensive history of the art of war. Such a work, he pointed out, would necessarily include such things as "details of drill with its commands, the technique of weapons and of the care of horses, and finally the whole subject of naval affairs—matters on which I have either nothing new to say or which I don't for a moment comprehend." The purpose of the history was stated in its title; it was to be a history of the art of war in the framework of political history.[12]

In the introduction to his fourth volume, Delbrück explained this in greater detail. The basic purpose of the work was to establish the connection between the constitution of the state, and tactics and strategy. "The recognition of the interrelationship between tactics, strategy, the constitution of the state and policy reflects upon the relationship [between military history and] world history and has brought to light much which until now has been hidden in darkness or left without recognition. This work has been written not for the sake of the art of war, but for the sake of world history. If military men read it and are

[10] Wilhelm Groener, "Delbrück und die Kriegswissenschaften" in Emil Daniels and Paul Rühlmann, ed., *Am Webstuhl der Zeit, eine Erinnerungsgabe Hans Delbrück dem Achtzigjährigen ... dargebracht* (Berlin, 1928), p. 35.

[11] Franz Mehring, "Eine Geschichte der Kriegskunst," *Die Neue Zeit* (Erganzungsheft, No. 4, 16 October 1908), p. 2.

[12] *Geschichte der Kriegskunst,* I, p. xi.

stimulated by it, I am pleased and regard that as an honor; but it was written for friends of history by a historian."[13]

At the same time, however, Delbrück realized that, before any general conclusions could be drawn from the wars of the past, the historian must determine as accurately as possible how those wars had been fought. It was precisely because he was intent on finding general ideas which would be of interest to other historians that Delbrück was forced to grapple with the "trivial happenings," "the smallest facts" of past campaigns; and, despite his own disclaimer, his reappraisal of those facts was of great value not to historians alone but to soldiers as well.

The "facts" were to be found in the great volume of source material which had been handed down by the past. But many of the sources of military history were obviously unreliable and were no better than "wash-room prattle and adjutants' gossip."[14] How was the modern historian to check these ancient records?

Delbrück believed that this could be done in several ways. Provided the historian knew the terrain in which past battles were fought, he could use all the resources of modern geographical science to check the reports which were handed down. Provided he knew the type of weapons and equipment used, he could reconstruct the tactics of the battle in a logical manner, since the laws of tactics for every kind of weapon could be ascertained. A study of modern warfare would supply the historian with further tools, for in modern campaigns he could judge the marching powers of the average soldier, the weight-carrying capacity of the average horse, the maneuverability of large masses of men. Finally, it was often possible to discover campaigns or battles, for which reliable reports existed, in which the conditions of earlier battles were reproduced almost exactly. Both the battles of the Swiss-Burgundian Wars, for which accurate records exist, and the battle of Marathon, for which Herodotus was the only source, were fought between mounted knights and bowmen on the one side and foot soldiers armed with weapons for hand-to-hand fighting on the other; in both cases, the foot soldiers were victorious. It should be possible, therefore, to draw conclusions from the battles of Granson, Murten, and Nancy which could be applied to the battle of Marathon.[15] The combination of all of these methods, Delbrück called *Sachkritik*.[16]

Only a few applications of the *Sachkritik* need be mentioned. Delbrück's most startling results were attained by his investigations of the numbers of troops employed in the great wars of the past. According to Herodotus, for instance, the Persian army which fought against Athens in the fifth century B.C. numbered over four million men. Delbrück pointed out that this figure could not be considered reliable.

"According to the German order of march, an army corps, that is 30,000

---

[13] *Ibid.*, IV, Preface.   [14] *Ibid.*, I, 377.

[15] Delbrück used this last method in his first account of the Persian Wars, *Die Perserkriege und die Burgunderkriege: zwei combinierte kriegsgeschichtliche Studien* (Berlin, 1887).

[16] *Geschichte der Kriegskunst*, I, Introduction.

men, occupies about three miles, without the baggage trains. The marching column of the Persians would therefore have been 420 miles long and as the first troops were arriving before Thermopylae the last would have just marched out of Susa on the other side of the Tigris."[17]

Even if this awkward fact could be explained away, none of the fields on which battles were fought were big enough to hold armies as large as those in Herodotus' accounts. The plain of Marathon, for instance, "is so small that some fifty years ago a Prussian staff officer who visited it wrote with some astonishment that a Prussian brigade would scarcely have room enough there for its exercises."[18]

On the basis of modern studies of the population of ancient Greece, Delbrück estimated the size of the Greek army which faced Xerxes as about 12,000 men. It was a citizen army trained to fight in a rude phalanx but incapable of tactical maneuvering. The Persian army was a professional army, and the bravery of its soldiers was admitted even in the Greek account. "If both things were true, the size (of the Persian army) as well as its military bravery, then the ever-repeated victory of the Greeks would remain inexplicable. Only one of the two things can be true; hence, it is clear that the advantage of the Persians is to be sought not in numbers but in quality."[19] Delbrück concludes that, far from having the mass army described by Herodotus, the Persians were actually inferior in numbers to the Greeks throughout the Persian Wars.

The account of Herodotus had long been suspect, and Delbrück's criticism was by no means wholly original. But his real contribution lay in the fact that he applied the same systematic methods to the numerical records of every war from the Persian Wars to those of Napoleon. Thus, in his discussion of Caesar's campaigns in Gaul, he clearly demonstrated that Caesar's estimates of the forces pitted against him were, for political reasons, grossly exaggerated. According to Caesar, the Helvetians, in their great trek, numbered 368,000 persons and carried three months' provisions with them. To Delbrück the numerical estimate smacked of the fabulous; but it was Caesar's remarks on the Helvetian food supply which enabled him to prove it so. He pointed out that some 8,500 wagons would be required to carry such provisions and, in the condition of roads in Caesar's time, it would be quite impossible for such a column to move.[20] Again, in his discussion of the invasion of Europe by the Huns, Delbrück effectively disposed of the belief that Attila had an army of 700,000 men, by describing the difficulties which Moltke experienced in maneuvering an army of 500,000 men in the campaign of 1870. "To direct such a mass unitedly is, even with railroads, roads, telegraphs and a general staff an exceedingly difficult task. . . . How could Attila have led 700,000 men from Germany over the Rhine into France to the Plain of Chalons, if Moltke moved

[17] *Ibid.*, I, 10.
[18] H. Delbrück, *Numbers in History: Two Lectures Delivered before the University of London* (London, 1913), p. 24.
[19] *Geschichte der Kriegskunst*, I, 39.   [20] *Ibid.*, I, 427.

500,000 with such difficulty over the same road? The one number acts as a check on the other."[21]

Delbrück's investigations of numbers have more than a mere antiquarian interest. At a time when the German army was being taught to seek lessons in history, the destroyer of myths helped it avoid the drawing of false conclusions. In war and the study of war, numbers were of the highest importance.[22] Delbrück himself pointed out that "a movement which a troop of 1,000 men executes without difficulty is a hard task for 10,000 men, a work of art for 50,000, an impossibility for 100,000."[23] No lessons can be drawn from past campaigns unless an accurate statement of the numbers involved is available.

Sachkritik had other uses. By means of it, Delbrück was able to reconstruct the details of single battles in a logical manner, and his success in doing so made a profound impression upon the historical section of the German general staff. General Groener has attested to the value of Delbrück's investigation of the origins of that oblique battle order which made flanking possible;[24] while it is well known that his scientific description of the encircling movement at Cannae strongly influenced the theories of Count Schlieffen.[25] But it is his account of the battle of Marathon which is perhaps the best example of the skill with which Delbrück reconstructed the details of past battles, the more so because it most clearly illustrates his belief that "if one knows the armament and the manner of fighting of the contending armies, then the terrain is such an important and eloquent authority for the character of a battle, that one may dare, provided there is no doubt as to the outcome, to reconstruct its course in general outline."[26]

The Greek army at Marathon was composed of heavily armed foot soldiers, formed in the primitive phalanx, the maneuverability of which was restricted to slow forward movement. It was opposed by an army inferior in numbers but made up of highly trained bowmen and cavalry. Herodotus had written that the Greeks had won the battle by charging across the plain of Marathon some 4,800 feet and crushing the center of the Persian line. Delbrück pointed out that this was a physical impossibility. According to the modern German drill book, soldiers with full pack could be expected to run for only two minutes, some 1,080 to 1,150 feet. The Athenians were no more lightly armed than the modern German soldier and they suffered from two additional disadvantages. They were not professional soldiers, but civilians, and many of them exceeded the age limit required in modern armies. Moreover, the phalanx was a closely massed body of men which made quick movement of any kind impossible. An

[21] *Numbers in History*, p. 18.
[22] General Groener made explicit acknowledgment of Delbrück's contribution. See *Am Webstuhl der Zeit*, p. 38.
[23] *Geschichte der Kriegskunst*, I, 7.
[24] *Am Webstuhl der Zeit*, p. 38. The oblique battle order, first used by the Theban Epaminondas, bears a striking resemblance to that used by Frederick the Great at Leuthen in 1757. On Epaminondas, see *Geschichte der Kriegskunst*, I, 130-135.
[25] *Geschichte der Kriegskunst*, I, 281-302. Graf Schlieffen, *Cannae* (Berlin, 1925), p. 3. See also Chapter 8.
[26] *Geschichte der Kriegskunst*, II, 80. Delbrück used the method not only for the battle of Marathon but also in his reconstruction of the battle of the Teutoburger Wald.

attempted charge over such a distance would have reduced the phalanx to a disorganized mob which would have been cut down by the Persian professionals without difficulty.[27]

The tactics described by Herodotus were obviously impossible, the more so because the Greek phalanx was weak on the flanks and, in any encounter on an open field, could have been surrounded by Persian cavalry. It seemed obvious to Delbrück that the battle was not fought on the plain of Marathon proper but in a small valley to the southeast where the Greeks were protected by mountains and forest from any flanking movement. The fact that Herodotus speaks of the opposing armies delaying the engagement for days shows that Miltiades, the Athenian commander, had chosen a strong position and, given the tactical form of the Greek army, the position in the Brana Valley was the only one possible. Moreover, that position dominated the only road to Athens. To reach the city, the Persians were forced to dispose of Miltiades' army, or give up the whole campaign, and they chose the former alternative. The only logical explanation of the battle, then, is that the Persians, despite their numerical inferiority and inability to use flanking tactics, made the initial attack; and Miltiades, shifting at the crucial moment from the defensive to the offensive, crushed the Persian center and swept the field.[28]

To the casual reader, the *History of the Art of War*, like many a work before it, is a mere collection of such battle pieces. But the care with which Delbrück reconstructed battles was necessary to his main purpose. He felt that by the study of key battles the student could acquire a picture of the tactics of an age and from that he could proceed to the investigation of broader problems.[29] For the key battles are important not only as typical manifestations of their age but as mileposts in the progressive development of military science. In a sense, Delbrück, like Proust's young officer, believed that past battles were "the literature, the learning, the etymology, the aristocracy of the battles of today." By reconstructing single battles he sought continuity in military history, and thus his *Sachkritik* enabled him to develop the three major themes which give his work a meaning and a unity found in no previous book on the subject: namely, the evolution of tactical forms from the Persians to Napoleon, the interrelationship of war and politics throughout history, and the division of all strategy into two basic forms.

Delbrück's description of the evolution of tactical bodies has been called one of his most significant contributions to military thought.[30] He was interested in

---

[27] Ulrich von Wilamowitz upheld the Herodotus story by arguing that the goddess Artemis had given the Greeks sufficient strength to make the charge and he issued a stern reproof to scholars who minimized the military contributions of the deities. Delbrück's refusal to accept this explanation led to a controversy, not with Wilamowitz, but with J. Kromayer who came to Wilamowitz' aid. The battle was fought out in the pages of the *Historische Zeitschrift* (XCV, 1 ff., 514 f.) and the *Preussische Jahrbücher* (CXXI, 158 f.) and, like most of Delbrück's controversies, left the parties involved unreconciled.

[28] *Geschichte der Kriegskunst*, I, 41-59.

[29] *Ibid.*, I, 417.

[30] F. J. Schmidt, Konrad Molinski, and Siegfried Mette, *Hans Delbrueck: der Historiker und Politiker* (Berlin, 1928), p. 96. Eugen von Frauenholz, *Entwicklungsgeschichte des deutschen Heerwesens*, II, p. vii.

discovering the reasons for the military supremacy of the Romans in the ancient world. Searching for a key to this problem, he came to the conclusion that their success rested on the excellence of their tactical forms. It was the gradual evolution of the primitive Greek phalanx to the highly coordinated tactical body used by the Romans which comprised "the essential meaning of the ancient art of war."[31] Turning then to the modern period, Delbrück argued that it was the revival of tactical bodies, not unlike the Roman, in the Swiss-Burgundian Wars and their improvement and perfection to the age of Napoleon which gave unity to modern military history.

The turning point in the history of ancient warfare was the battle of Cannae,[32] where the Carthaginians under Hannibal overwhelmed the Romans in the most perfect tactical battle ever fought. How were the Romans able to recover from that disaster, to defeat the Carthaginians and eventually to exercise military supremacy over the whole of the ancient world? The answer is to be found in the evolution of the phalanx. At Cannae the Roman infantry was formed in a body which was essentially the same as that which had won the battle of Marathon. The basic weaknesses of the phalanx had delivered the Roman army into Hannibal's arms. The exposed flanks and the inability of the Roman rear to maneuver independently of the mass of the army made it impossible for the Romans to prevent the encircling tactics employed by the Carthaginian cavalry. But in the years following Cannae, striking changes were introduced into the Roman battle form. "The Romans first articulated the phalanx, then divided it into columns (*Treffen*) and finally split it up into a great number of small tactical bodies which were capable, now of closing together in a compact impenetrable union, now of changing the pattern with consummate flexibility, of separating one from the other and of turning in this or that direction."[33] To modern students of warfare this development seems so natural as to be hardly worthy of notice. To accomplish it, however, was extremely difficult and only the Romans, of all the ancient peoples, succeeded. In their case it was made possible only by a hundred years of experimentation —in the course of which the army changed from a civilian to a professional army—and by the emphasis upon military discipline which characterized the Roman system.[34]

The Romans conquered the world, then, not because their troops "were braver than all their opponents, but because, thanks to their discipline, they had stronger tactical bodies."[35] The only people who successfully avoided conquest by the Romans were the Germans and their resistance was made possible by a natural discipline inherent in their political institutions, and by the fact that the German fighting column, the *Gevierthaufe*, was a tactical group of great effectiveness.[36] Indeed, in the course of their wars with the Romans, the Germans

---

[31] *Geschichte der Kriegskunst*, II, 43.   [32] *Ibid.*, I, 330 ff.
[33] *Ibid.*, I, 380.
[34] *Ibid.*, I, 381. See also I, 253. "The meaning and power of discipline was first fully recognized and realized by the Romans."
[35] *Ibid.*, II, 43.   [36] *Ibid.*, II, 45 ff.

learned to imitate the articulation of the Roman legion, maneuvering their *Gevierthaufe* independently or in union as the occasion required.[37]

With the decline of the Roman state and the barbarization of the empire, the tactical progress which had been made since the days of Miltiades came to an end. The political disorders of the age following the reign of the Severi weakened the discipline of the Roman army, and gradually undermined the excellence of its tactical forms.[38] At the same time, as large numbers of barbarians were admitted into the ranks, it was impossible to cling to the highly integrated battle order which had been devised over the course of centuries. History had shown that infantry was superior to cavalry only if the foot soldiers were organized in strong tactical bodies. Now, with the decline of the state and the consequent degeneration of tactics, there was a growing tendency, in the new barbarian empires of the west and in Justinian's army as well, to replace infantry with heavily armed mounted soldiers.[39] As that tendency gained the upper hand, the days when battles were decided by infantry tactics died away and Europe entered a long period in which military history was dominated by the figure of the armed knight.[40]

Delbrück has been accused of maintaining that the development of military science stops with the decline of Rome and starts again with the Renaissance,[41] and the accusation is justified. The essential element in all warfare from the days of Charlemagne to the emergence of the Swiss infantry in the Burgundian Wars was the feudal army. This, in Delbrück's opinion, was no tactical body. It depended upon the fighting quality of the single warrior; there was no discipline, no unity of command, no effective differentiation of arms. In this whole period, no tactical progress was made, and Delbrück seems inclined to agree with Mark Twain's Connecticut Yankee, that "when you come to figure up results, you can't tell one fight from another, nor who whipped." It is true that at Crecy, the English knights dismounted and fought a defensive battle on foot and that, at Agincourt, dismounted knights actually took the offensive; but these were mere episodes and cannot be considered as forecasts of the development of modern infantry.[42]

It was among the Swiss in the fifteenth century that the independent infantry was reborn. "With the battles of Laupen and Sempach, Granson, Murten and Nancy we have again a foot soldiery comparable to the phalanx and the legions."[43] The Swiss pikemen formed themselves in bodies similar to the German *Gevierthaufe*;[44] and, in the course of their wars against the Burgundians, they perfected the articulated tactics used by the Roman legions. At Sempach, for instance, the Swiss infantry was divided into two bodies, one

---

[37] *Ibid.*, II, 52 f.

[38] *Ibid.*, II, 205 ff. This chapter, entitled *Niedergang und Auflösung des römischen Kriegswesens* is the key chapter of the second volume.

[39] *Ibid.*, II, 424 ff.   [40] *Ibid.*, II, 433.

[41] T. F. Tout in *English Historical Review*, XXII (1907), 344-348.

[42] *Geschichte der Kriegskunst*, III, 483. For a penetrating criticism of Delbrück's discussion of medieval warfare, see Tout, *loc. cit.*

[43] *Geschichte der Kriegskunst*, III, 661. See Chapter I.

[44] *Ibid.*, III, 609 ff.

holding a defensive position against the mounted enemy, the other delivering a decisive blow on the enemy's flank.[45]

The revival of tactical bodies was a military revolution comparable to that which followed Cannae. It was this revival, rather than the introduction of firearms, which brought feudal warfare to an end. At Murten, Granson, and Nancy the new weapons were employed by the knights, but had no effect upon the outcome of the battle.[46] With the restoration of the tactical body of infantry as the decisive one in warfare, the mounted soldiers became a mere cavalry, a highly useful but supplementary part of the army. In his fourth volume, Delbrück discussed this development and the evolution of the modern infantry to the age of the standing army and concluded with an account of the revolution in tactics made possible by the French Revolution.[47]

The attention which Delbrück pays to the emergence of tactical bodies serves not only to give a sense of continuity to his military history but also to illustrate the theme which he considered basic to his book, namely, the interrelationship of politics and war. In every period of history, he pointed out, the development of politics and the evolution of tactics were closely related. "The Hopliten-Phalanx developed in quite a different manner under the Macedonian Kings than it did in the aristocratic Roman *Beamten-Republik*, and the tactics of the cohort were developed only in relationship with constitutional change. Again, according to their nature, the German *hundreds* fought quite differently from the Roman cohorts."[48]

The Roman army at Cannae, for example, was defeated because of the weakness of its tactics. But contributory to that weakness was the fact that the army was composed of untrained civilians rather than professional soldiers and the fact that the constitution of the state required that the high command alternate between the two consuls.[49] In the years following Cannae the necessity of a unified command was generally recognized. After various political experiments were tried, P. C. Scipio was in the year 211 B.C. made general in chief of the Roman armies in Africa and assured of continued tenure for the duration of the war. The appointment was in direct violation of the state constitution and it marked the beginning of the decline of republican institutions. The interrelationship of politics and warfare is in this case apparent. "The importance of the Second Punic War in world history," Delbrück writes, "is that Rome effected an internal transformation which increased her military potentiality enormously,"[50] but which at the same time changed the whole character of the state.

Just as the political element was predominant in the perfection of Roman tactics, so also the breakdown of tactical forms can be explained only by a careful study of the political institutions of the later empire. The political and economic disorders of the third century had a direct effect upon Roman military institutions. "Permanent civil war destroyed the cement which till now

[45] *Ibid.*, III, 594.
[47] See Chapter 3.
[49] *Ibid.*, I, 305.
[46] *Ibid.*, IV, 55.
[48] *Geschichte der Kriegskunst*, II, 424.
[50] *Ibid.*, I, 333.

had held the strong walls of the Roman army together, the discipline which constituted the military worth of the legions."[51]

In no part of the *History of the Art of War* does Delbrück include a general discussion of the relationship of politics and war. But, as he moves from one historical epoch to another, he fits the purely military into its general background, illustrating the close connection of political and military institutions and showing how changes in one sphere led of necessity to corresponding reactions in the other. He showed that the German *Gevierthaufe* was the military expression of the village organization of the German tribes and demonstrated the way in which the dissolution of German communal life led to the disappearance of the *Gevierthaufe* as a tactical body.[52] He showed how the victories of the Swiss in the fifteenth century were made possible by the fusion of the democratic and aristocratic elements in the various cantons, and the union of the urban nobility with the peasant masses.[53] And in the period of the French Revolution he showed how the political factor, in this case "the new idea of defending the fatherland, inspired the mass [of the soldiers] with such an improved will, that new tactics could be developed. . . ."[54]

That politics and war were closely related had been accepted as a truism even before Delbrück's time. But it was a truism which had to be studied from every angle and illustrated by actual events. Delbrück's service to military theorists lay in the systematic manner in which he illustrated the interplay of political and military factors in every age.

The most striking of all of Delbrück's military theories was that which held that all military strategy can be divided into two basic forms. This theory, formulated long before the publication of the *History of the Art of War*, is conveniently summarized in the first and fourth volumes of that work.[55]

Under the influence of Clausewitz' book *On War*, the great majority of military thinkers in Delbrück's day believed that the aim of war is the complete destruction of the enemy's forces and that, consequently, the battle which accomplishes this is the end of all strategy. Delbrück's first researches in military history convinced him that this type of strategical thinking had not always been generally accepted; and that there were long periods in history in which a completely different strategy ruled the field. He discovered, moreover, that Clausewitz himself had admitted the possibility of the existence of more than one strategical system. In a note written in 1827, Clausewitz had suggested that there were two sharply distinct methods of conducting war: one which was bent solely on the annihilation of the enemy; the other, a limited warfare, in which such annihilation was impossible, either because the political aims or political tensions involved in the war were small or because the military means were inadequate to accomplish annihilation.[56]

Clausewitz did not live long enough to do more than suggest the existence of the two forms; Delbrück determined to accept the distinction and expound

[51] *Ibid.*, II, 209.
[53] *Ibid.*, III, 614 f.
[55] *Ibid.*, I, 100 ff.; IV, 333-363, 426-444.

[52] *Ibid.*, II, 25-38, 424 ff.
[54] *Ibid.*, IV, 474.
[56] See Chapter 5, section vi.

the principles inherent in each. The first form of warfare, to which Clausewitz had devoted the book *On War*, he named *Niederwerfungsstrategie* (the strategy of annihilation). Its sole aim was the decisive battle, and the commanding general was called upon only to estimate the possibility of fighting such a battle in a given situation.

The second type of strategy Delbrück called variously *Ermattungsstrategie* (the strategy of exhaustion) and two-pole strategy. It was distinguished from the strategy of annihilation by the fact "that the *Niederwerfungsstrategie* has only one pole, the battle, whereas the *Ermattungsstrategie* has two poles, battle and maneuver, between which the decisions of the general move." In *Ermattungsstrategie*, the battle is no longer the sole aim of strategy; it is merely one of several equally effective means of attaining the political ends of the war and is essentially no more important than the occupation of territory, the destruction of crops or commerce, and the blockade. This second form of strategy is neither a mere variation of the first nor an inferior form. In certain periods of history, because of political factors or the smallness of armies, it has been the only form of strategy which could be employed. The task it imposes on the commander is quite as difficult as that required of the exponent of the strategy of annihilation. With limited resources at his disposal, the *Ermattungsstratege* must decide which of several means of conducting war will best suit his purpose, when to fight and when to maneuver, when to obey the law of "daring" and when to obey that of "economy of forces." "The decision is therefore a subjective one, the more so because at no time are all circumstances and conditions, especially what is going on in the enemy camp, known completely and authoritatively. After a careful consideration of all circumstances—the aim of the war, the combat forces, the political repercussions, the individuality of the enemy commander, and of the government and people of the enemy, as well as his own—the general must decide whether a battle is advisable or not. He can reach the conclusion that any greater actions must be avoided at all cost; he can also determine to seek [battle] on every occasion so that there is no essential difference between his conduct and that of one-pole strategy."[57]

Among the great commanders of the past who had been strategists of annihilation were Alexander, Caesar, and Napoleon. But equally great generals had been exponents of *Ermattungsstrategie*. Among them, Delbrück listed Pericles, Belisarius, Wallenstein, Gustavus Adolphus, and Frederick the Great. The inclusion of the last name brought down upon the historian a flood of angry criticism. The most vocal of his critics were the historians of the general staff who, convinced that the strategy of annihilation was the only correct strategy, insisted that Frederick was a precursor of Napoleon. Delbrück answered that to hold this view was to do Frederick a grave disservice. If Frederick was a strategist of annihilation, how was one to explain away the fact

---

[57] H. Delbrück, *Die Strategie des Perikles erläutert durch die Strategie Friedrichs des Grossen* (Berlin, 1890), 27-28. This work is Delbrück's most systematic exposition of the two forms of strategy.

that in 1741, with 60,000 men under his command, he refused to attack an already beaten army of only 25,000, or that, in 1745, after his great victory at Hohenfriedberg, he preferred to resort again to a war of maneuver?[58] If the principles of *Niederwerfungsstrategie* were to be considered the sole criteria in judging the qualities of a general, Frederick would cut a very poor figure.[59] Yet Frederick's greatness lay in the fact that although he realized that his resources were not great enough to enable him to seek battle on every occasion he was nevertheless able to make effective use of other strategical principles in order to win his wars.

Delbrück's arguments did not convince his critics. Both Colmar von der Goltz and Friedrich von Bernhardi entered the lists against him, and a paper warfare ensued which lasted for over twenty years.[60] Delbrück, who loved controversy, was indefatigable in answering refutations of his theory. But his concept of *Ermattungsstrategie* was rejected by an officer corps trained in the tradition of Napoleon and Moltke and convinced of the feasibility of the short, decisive war.

Yet the military critics completely missed the deeper significance of Delbrück's strategical theory. History showed that there could be no single theory of strategy, correct for every age. Like all phases of warfare, strategy was intimately connected with politics, with the life and the strength of the state. In the Peloponnesian War, the political weakness of Athens in comparison with that of the League which faced her, determined the kind of strategy which Pericles followed. Had he attempted to follow the principles of *Niederwerfungsstrategie*, as Cleon did later, disaster would have followed automatically.[61] The strategy of Belisarius' wars in Italy was determined by the uneasy political relations between the Byzantine Empire and the Persians. "Here as always it was politics which determined the administration of the war and which prescribed to strategy its course."[62] Again, "the strategy of the Thirty Years War was determined by the extremely complicated, repeatedly changing political relationships," and generals like Gustavus Adolphus, whose personal bravery and inclination toward battle were unquestioned, were nevertheless compelled to make limited war.[63] It was not the battles won by Frederick the Great which

---

[58] *Preussische Jahrbücher*, CXV (1904), 348 f.

[59] In the *Strategie des Perikles*, Delbrück wrote a parody which showed that the application of such criteria to Frederick's campaigns would prove him a third-rate general. For this he was accused in the Prussian Landtag of maligning a national hero.

[60] A full account of the controversy, with bibliography, appears in *Geschichte der Kriegskunst*, IV, 439-444. See also Friedrich von Bernhardi, *Denkwürdigkeiten aus meinem Leben* (Berlin, 1927), pp. 126, 133, 143.

The most thorough and judicious criticism of Delbrück's strategical theory is that of Otto Hintze, "Delbrück, Clausewitz und die Strategie Friedrichs des Grossen,"*Forsuchungen zur Brandenburgischen und Preussischen Geschichte*, XXXIII (1920), 131-177. Hintze objects to the sharp distinction which Delbrück draws between the strategy of Frederick's age and that of Napoleon and insists that Frederick was at once a *Niederwerfung* and an *Ermattung*-strategist. He also questions Delbrück's interpretation of Clausewitz' intentions, as does H. Rosinski in *Historische Zeitschrift*, CLI (1938). See Delbrück's answer to Hintze, *Forschungen zur Brandenburgischen und Preussischen Geschichte*, XXXIII (1920), 412-417.

[61] *Geschichte der Kriegskunst*, I, 101 f.   [62] *Ibid.*, II, 394.

[63] *Ibid.*, IV, 341.

made him a great general, but rather his political acumen and the conformity of his strategy with political reality. No strategical system can become self-sufficient; once an attempt is made to make it so, to divorce it from its political context, the strategist becomes a menace to the state.

The transition from dynastic to national war, the victories of 1864, 1866, and 1870, the immense increase in the war potential of the nation seemed to prove that *Niederwerfungsstrategie* was the natural form of war for the modern age. As late as 1890, Delbrück himself, despite his insistence on the relativity of strategy, seems to have believed that this was true.[64] Yet in the last years of the nineteenth century, the mass army of the 'sixties was being transformed to the *Millionenheer* which fought in the World War. Might not that transformation make impossible the application of the strategy of annihilation and herald a return to the principles of Pericles and Frederick? Was not the state in grave danger as long as the general staff refused to admit the existence of alternate systems of strategy? These questions, implicit in all of Delbrück's military writings, were constantly on his lips as Germany entered the World War.

III

Since Delbrück was Germany's leading civilian expert on military affairs, his writings in the war years, 1914-1918, are of considerable interest. As a military commentator, his sources of information were in no way superior to those of other members of the newspaper and periodical press. Like them he was forced to rely on the communiqués issued by the general staff, the stories which appeared in the daily press and reports from neutral countries. If his accounts of the war were distinguished by a breadth of vision and understanding not usually found in the lucubrations of civilian commentators, it was due to his technical knowledge of modern war and the sense of perspective which he had gained from his study of history. In his monthly commentaries in the *Preussische Jahrbücher* one can find a further exposition of the principles delineated in his historical works and especially of his theory of strategy and his emphasis upon the interrelationship of war and politics.[65]

In accordance with the Schlieffen strategy, the German army swept into Belgium in 1914 with the purpose of crushing French resistance in short order and then bringing the full weight of its power against Russia. This was *Niederwerfungsstrategie* in its ultimate form, and Delbrück himself, in the first month of the war, felt that it was justified. Like most of his fellows, he had little fear of effective French opposition. The instability of French politics could not but have a deleterious effect upon France's military institutions. "It is impossible

---

[64] *Strategie des Perikles*, chapter 1.

[65] The articles which Delbrück wrote in the *Preussische Jahrbücher* are collected in the three volume work called *Krieg und Politik* (Berlin, 1918-1919). To the articles as they originally appeared Delbrück has added occasional explanatory notes and a highly interesting summary statement. The best article on Delbrück's war writings is that by General Ernst Buchfinck, "Delbrücks Lehre, das Heer und der Weltkrieg," in *Am Webstuhl der Zeit*, pp. 41-49. See also M. von Hobohm, "Delbrück, Clausewitz und die Kritik des Weltkrieges," *Preussische Jahrbücher*, CLXXXI (1920), 203-232.

that an army which has had forty-two war ministers in forty-three years will be capable of an effectively functioning organization."[66] Nor did he feel that England was capable of continued resistance. Her past political development, he believed, would make it impossible for her to raise more than a token force. England had always relied on small professional armies; the institution of universal conscription would be psychologically and politically impossible. "Every people is the child of its history, its past, and can no more break away from it than a man can separate himself from his youth."[67]

When the first great German drive fell short of its goal, however, and the long period of trench warfare set in, Delbrück sensed a strategical revolution of the first importance. As the stalemate in the west continued, and especially after the failure of the Verdun offensive, he became increasingly convinced that the strategical thinking of the high command would have to be modified. In the west at least, defensive warfare was the order of the day, a fact "the more significant since, before the war, the preeminence of the offensive was always proclaimed and expounded with quite exceptional partiality in the theory of strategy fostered in Germany."[68] Now, it was apparent that conditions on the Western Front approximated those of the age of *Ermattungs-strategie*. "Although this war has already brought us much that is new, nevertheless it is possible to find in it certain historical analogies: for example, the Frederician strategy with its impregnable positions, its increasingly strengthened artillery, its field fortifications, its infrequent tactical decisions and its consequent long withdrawals presents unmistakable similarities with today's war of position and exhaustion (*Stellungs- und Ermattungskrieg*)."[69] In the west, reliance upon the decisive battle was no longer possible. Germany would have to find other means of imposing her will upon the enemy.

By December 1916 Delbrück was pointing out that "however favorable our military position is, the continuation of the war will scarcely bring us so far that we can simply dictate the peace."[70] A complete and crushing victory of German arms was unlikely, if not impossible. That did not mean, however, that Germany could not "win the war." Her inner position not only separated her opponents but enabled her to retain the initiative. Her strength was so formidable that it should not be difficult to convince her opponents that Germany could not be defeated. While a firm defensive in the west was sapping the will of Allied troops, the high command would be well advised to throw its strongest forces again the weakest links in the Allied coalition—against Russia and Italy. A concentrated offensive against Russia would complete the demoralization of the armies of the Czar and might very well precipitate a revolution in St. Petersburg. A successful Austro-German offensive against Italy would not only have a tremendous moral effect in England and France but would threaten France's communications with North Africa.[71]

---

[66] *Krieg und Politik*, I, 35.
[67] Delbrück's views on England's weakness as a military power were most clearly developed in an article in April 1916. See *Krieg und Politik*, I, 243 ff.
[68] *Ibid.*, II, 242.   [69] *Ibid.*, II, 164. See also II, 17.   [70] *Ibid.*, II, 97.
[71] E. Buchfinck, "Delbrücks Lehre, das Heer und der Weltkrieg," *loc. cit.*, p. 48.

In Delbrück's opinion, then, Germany's strategy must be directed toward the destruction of the enemy coalition and the consequent isolation of England and France. In this connection, it was equally important that no measures be adopted which might bring new allies to the western powers. Delbrück was always firmly opposed to the submarine campaign, which he rightly feared would bring the United States into the war.[72]

But in the last analysis, if the war was to be won by Germany, the government would have to show a clear comprehension of the political realities implicit in the conflict. Since the war in the west had become an *Ermattungskrieg*, the political aspect of the conflict had increased in importance. "Politics is the ruling and limiting factor; military operations is only one of its means."[73] A political strategy must be devised to weaken the will of the people of France and England.

In the political field, Delbrück had felt from the beginning of the war that Germany suffered from a very real strategical weakness. "Because of our narrow policy of Germanization in the Polish and Danish districts of Prussia, we have given ourselves the reputation in the world of being not the protectors but the oppressors of small nationalities."[74] If this reputation were confirmed in the course of the war, it would give moral encouragement to Germany's enemies and would jeopardize the hope of ultimate victory. Turning to history, Delbrück argued that the example of Napoleon should serve as a warning to Germany's political leaders. The emperor's most overwhelming victories had served only to strengthen the will of his opponents and to pave the way for his ultimate defeat. "May God forbid that Germany enter upon the path of Napoleonic policy. . . . Europe stands united in this one conviction: it will never submit to a hegemony enforced upon it by a single state."[75]

Delbrück believed that the invasion of Belgium had been a strategical necessity[76]; but it was nonetheless an unfortunate move, for it seemed to confirm the suspicion that Germany was bent upon the subjugation and annexation of small states. From September 1914 until the end of the war, Delbrück continued to insist that the German government must issue a categorical disclaimer of any intention of annexing Belgium at the conclusion of hostilities. England, he argued, would never make peace as long as there was danger of German retention of the Flanders coast. The first step in weakening the resistance of western powers was to state clearly that Germany had no territorial desires in the west and that her war aims would "prejudice in no way the freedom and honor of other peoples."[77]

Perhaps the best way to convince the western powers that Germany was not seeking world domination was to make it apparent that Germany had no objection to a negotiated peace. Delbrück had favored such a peace ever since the successful Allied counteroffensive on the Marne in September 1914. He firmly

---

[72] *Krieg und Politik*, I, 90, 227 ff., 261.   [73] *Ibid.*, II, 95.
[74] *Ibid.*, I, 3 f.
[75] See *Krieg und Politik*, I, 59, and the article entitled "Das Beispiel Napoleons," *ibid.*, II, 122 ff.
[76] *Ibid.*, I, 33.   [77] *Ibid.*, II, 97.

believed that the war had been caused by Russian aggression and saw no reason why England and France should continue to fight the one power which was "guarding Europe and Asia from the domination of *Moskowitertum*."[78] As the war was prolonged, he was strengthened in his conviction that a sincere willingness to negotiate would win for Germany a victory which arms alone would be powerless to effect; and after the entrance of the United States into the war he openly predicted defeat unless Germany's leaders used that weapon. He was, therefore, enthusiastic about the passage by the Reichstag of the Peace Resolution of July 1917,[79] for he felt that it would do more to weaken the resistance of the western powers than any possible new offensive upon the western front.

Delbrück never for a moment wavered in his belief that the German army was the best in the world but he saw that that best was not good enough. Throughout 1917 he hammered away at one constant theme: "We must look the facts in the face—that we have in a sense the whole world leagued against us—and we must not conceal from ourselves the fact that, if we try to penetrate to the basic reasons for this world coalition, we will ever and again stumble over the motive of fear of German world hegemony. . . . Fear of German despotism is one of the weightiest facts with which we have to reckon, one of the strongest factors in the enemy's power."[80] Until that fear was overcome, the war would continue. It could be overcome only by a political strategy based upon a disclaimer of territorial ambitions in the west and a willingness to negotiate.

Just as the conditions of the present war were, to Delbrück, comparable in some ways to those of the eighteenth century, so was this heightened emphasis upon the political aspects of the war in full accordance with the principles of *Ermattungsstrategie* as practiced by Frederick the Great. When the German army had taken the field in 1914 it had staked all on the decisive battle and had failed. Delbrück would now relegate military operations to a subordinate position. The battle was no longer an end in itself but a means. If Germany's political professions failed at first to convince the western powers that peace was desirable, a new military offensive could be undertaken and would serve to break down that hesitation. But only such a coordination of the military effort with the political program would bring the war to a successful issue.

[78] *Ibid*, I, 18.

[79] The Peace Resolution, passed by the Reichstag by 212 votes to 126, stated in part: "The Reichstag strives for a peace of understanding and a lasting reconciliation among peoples. Violations of territory and political, economic and financial persecutions are incompatible with such a peace. The Reichstag rejects every scheme which has for its purpose the imposition of economic barriers or the perpetuation of national hatreds after the war. The freedom of the seas must be secured. Economic peace alone will prepare the ground for the friendly association of the peoples. The Reichstag will actively promote the creation of international organizations of justice. But so long as the enemy governments dissociate themselves from such a peace, so long as they threaten Germany and her allies with conquest and domination, then so long will the German people stand united and unshaken, and fight till their right and the right of their allies to live and grow is made secure. United thus, the German people is unconquerable."

[80] *Krieg und Politik*, II, 187.

In his desire for a political strategy which would be effective in weakening the resistance of the enemy, Delbrück was bitterly disappointed. It became apparent as early as 1915 that strong sections of German public opinion regarded the war as a means of acquiring new territory not only in the east but in the west of Europe. When Delbrück called for a declaration of willingness to evacuate Belgium, he was greeted by a howl of abuse and was accused by the *Deutsche Tageszeitung* of being "subservient to our enemies in foreign countries."[81] The changing fortunes of war did not diminish the desire for booty and the powerful *Vaterlandspartei*, the most important of the annexationist groups, exercised a strong influence on the governments which ruled Germany. Not only did the German government not make any declaration concerning Belgium but it never made its position clear on the question of a negotiated peace. When the Peace Resolution was being debated in 1917, Hindenburg and Ludendorff threatened to resign if the Reichstag adopted the measure. After the passage of the resolution, the influence of the high command was exerted so effectively that the government did not dare to make the resolution the keystone of its policy. As a result of the so-called crisis of July 1917, the western powers were encouraged to believe that the Reichstag's professions were insincere and that Germany's leaders were still bent on world domination.

To Delbrück the crisis of July had a deeper significance. It showed within the government a dearth of political leadership and a growing tendency on the part of the military to dominate the formulation of policy. Germany's military leaders had never been known for their political acumen, but in the past they had followed the advice of the political head of the state. Gneisenau had willingly subordinated his views to those of Hardenberg; Moltke—although at times reluctantly—had bowed to Bismarck's political judgment. Now, in the time of Germany's greatest crisis, the military were taking over completely and there was among them no man with a proper appreciation of the political necessities of the day. For all their military gifts, Hindenburg and Ludendorff still thought solely in terms of a decisive military victory over the western powers, a *Niederwerfung* that would deliver western Europe into their hands. It was with a growing sense of despair that Delbrück wrote; "Athens went to her doom in the Peloponnesian War because Pericles had no successor. We have fiery Cleons enough in Germany. Whoever believes in the German people will be confident that it has not only great strategists among its sons but also that gifted statesman in whose hands the necessity of the time will place the reins for the direction of foreign policy."[82] But that gifted statesman never appeared; and the fiery Cleons prevailed.

It was, consequently, with little confidence that Delbrück watched the opening of the German offensive of 1918. "It is obvious," he wrote, "that no change can be made in the principles which I have expounded here since the beginning

[81] See R. H. Lutz, ed., *Fall of the German Empire* (Hoover War Library Publications, No. 1), p. 307.
[82] *Krieg und Politik*, III, 123.

of the war, and the dissension with regard to our western war aims remains."[83] Strategy, he insisted, is not something in the abstract; it cannot be divorced from political considerations. "The great strategical offensive should have been accompanied and reinforced by a similar political offensive, which would have worked upon the home front of our enemies in the same way as Hindenburg and the men in field gray worked upon the front lines." If only the German government had announced, fourteen days before the opening of the offensive, that they firmly desired a negotiated peace and that, after such peace, Belgium would be evacuated, what would the result have been? Lloyd George and Clemenceau might have regarded these claims as signs of German weakness. But now, as the offensive rolled forward, "would Lloyd George and Clemenceau still be at the helm? I doubt it very much. We might even now be sitting at the conference table."[84]

Because of the failure to coordinate the military and political aspects of the war, Delbrück felt that the offensive, at most, would lead to mere tactical successes and would have no great strategical importance. But even he did not suspect that this was the last gamble of the strategists of annihilation, and the suddenness and completeness of the German collapse surprised him completely. In the November 1918 issue of the *Preussische Jahrbücher* he made a curious and revealing apology to his readers. "How greatly I have erred," he wrote. "However bad things looked four weeks ago, I still would not give up the hope that the front, however wavering, would hold and would force the enemy to an armistice which would protect our boundaries." In a sentence which illustrates the responsibility which he felt as a military commentator to the German people, he added, "I admit that I often expressed myself more confidently than I felt at heart. On more than one occasion, I allowed myself to be deceived by the confident tone of the announcements and reports of the army and the navy." But despite these mistakes in judgment, he could, he said, be proud of the fact that he had always insisted that the German people had a right to hear the truth even when it was bad and, in his constant preaching of political moderation, he had tried to show them the road to victory.[85]

It was in this spirit also that Delbrück made his most complete review and most searching criticism of the military operations of the last phase of the war. This was in the two reports which he made in 1922 before the Fourth Subcommittee of the commission set up by the Reichstag after the war to investigate the causes of the German collapse in 1918. In his testimony before the subcommittee, Delbrück repeated the arguments which he had made in the pages of the *Preussische Jahrbücher*, but the removal of censorship restrictions enabled him to give a much more detailed criticism of the military aspect of the 1918 offensive than had been possible during the war.[86]

[83] *Ibid.*, III, 63.   [84] *Ibid.*, III, 73.
[85] *Ibid.*, III, 203-206.
[86] The Delbrück testimony is reproduced completely in *Das Werk des Untersuchungsausschusses der Deutschen Verfassunggebenden Nationalversammlung und des Deutschen Reichstages 1919-1926. Die Ursachen des Deutschen Zusammenbruchs im Jahre 1918.* (Vierte Reihe im Werk des Untersuchungsausschusses), III, 239-273. Selections from the

The main weight of Delbrück's criticism was directed against Ludendorff, who conceived and directed the 1918 offensive. In only one respect, he felt, had Ludendorff shown even military proficiency. He had "prepared the attack, as regards both the previous training of the troops and the moment for taking the enemy by surprise, in a masterly manner with the greatest energy and circumspection."[87] But the advantages of this preliminary preparation were outweighed by several fundamental weaknesses and by gross mistakes in strategical thinking. In the first place, the German army on the eve of the offensive was in no position to strike a knockout blow against the enemy. Its numerical superiority was slight and, in reserves, it was vastly inferior to the enemy. Its equipment was in many respects equally inferior, and it was greatly handicapped by a faulty supply system and by insufficient stocks of fuel for its motorized units. These disadvantages were apparent before the opening of the offensive but were disregarded by the high command.[88]

Ludendorff was sufficiently aware of these weaknesses, however, to admit the impossibility of striking the enemy at that point where the greatest strategical success could have been won. In his own words, "tactics were to be valued more than pure strategy." That meant, in effect, that he attacked at those points where it was easiest to break through and not at those points where the announced aim of the offensive could best be served. The strategical goal of the campaign was the annihilation of the enemy. "In order to attain the strategical goal—the separation of the English army from the French and the consequent rolling-up of the former—the attack would have best been arranged so that it followed the course of the Somme. Ludendorff, however, had stretched the offensive front some four miles further to the south because the enemy seemed especially weak there."[89] The defensive wing of the army under Hutier broke through at this point, but its very success handicapped the development of the offensive, for its advance outpaced the real offensive wing under Below which was operating against Arras. When Below's forces were checked "we were forced with a certain amount of compulsion to follow the line of [Hutier's] success . . . thereby the idea of the offensive was altered and the danger of dispersing our forces evoked."[90]

In short, by following the tactical line of least resistance Ludendorff began a disastrous policy of improvisation, violating the first principle of that *Niederwerfungsstrategie* which he professed to be following. "A strategy which is not predicated upon an absolute decision, upon the annihilation of the enemy, but is satisfied with single blows, may execute these now in this place, now in that. But a strategy which intends to force the decision, must do it where the first successful blow was struck." Far from obeying this precept, Ludendorff and Hindenburg operated on the principle that, when difficulties developed in one

Commission's report, but only a very small portion of the Delbrück testimony, found in R. H. Lutz, ed., *The Causes of the German Collapse in 1918* (Hoover War Library Publications, No. 4).

[87] *Die Ursachen des Deutschen Zusammenbruches*, III, 345. Lutz, *op. cit.*, p. 90.
[88] *Die Ursachen des Deutschen Zusammenbruches*, III, 246.
[89] *Ibid.*, III, 247.   [90] *Ibid.*, III, 346.

sector, new blows could be struck in another.[91] As a result, the grand offensive degenerated into a series of separate thrusts, uncoordinated and unproductive.

The cardinal fault of the offensive was the failure of the high command to see clearly what could be accomplished by the German army in 1918 and the failure to adapt its strategy to its potentialities. Here Delbrück returned to the major theme of all his work as historian and publicist. The relative strength of the opposing forces was such that the high command should have realized that the annihilation of the enemy was no longer possible. The aim of the 1918 offensive, therefore, should have been to make the enemy so tired that he would be willing to negotiate a peace. This in itself would have been possible only if the German government had expressed its own willingness to make such a peace. But once this declaration had been clearly made, the German army in opening its offensive would have won a great strategical advantage. Its offensive could now be geared to the strength at its disposal. It could safely attack at the points of tactical advantage—that is, where success was easiest—since even minor victories would now have a redoubled moral effect in the enemy capitals.[92] The high command had failed in 1918 and had lost the war because it had disregarded the most important lesson of history, the interrelationship of politics and war. "To come back once more to that fundamental sentence of Clausewitz, no strategical idea can be considered completely without considering the political goal."[93]

IV

The military historian has generally been a kind of misfit, regarded with suspicion by both his professional colleagues and by the military men whose activities he seeks to portray. The suspicion of the military is not difficult to explain. It springs in large part from the natural scorn of the professional for the amateur. But the distrust with which academicians have looked on the military historians in their midst has deeper roots. In democratic countries especially, it arises from the belief that war is an aberration in the historical process and that, consequently, the study of war is neither fruitful nor seemly. It is significant that in his general work *On the Writing of History*, the dean of living military historians, Sir Charles Oman, should entitle the chapter dealing with his own field "A Plea for Military History." Sir Charles remarks that the civilian historian dabbling in military affairs has been an exceptional phenomenon, and he explains this by writing:

"Both the medieval monastic chroniclers and the modern liberal historiographers had often no closer notion of the meaning of war than that it involves various horrors and is attended by a lamentable loss of life. Both classes strove to disguise their personal ignorance or dislike of military matters by deprecating their importance and significance in history."[94]

The prejudice which Oman resented was felt equally keenly, throughout his life, by Hans Delbrück. When, as a relatively young man, he turned his talents

---

[91] *Ibid.*, III, 250-251.   [92] *Ibid.*, III, 253 f.   [93] *Ibid.*, III, 253.
[94] Sir Charles Oman, *On the Writing of History* (New York, n.d.), pp. 159 f.

to the study of military history, he found that the members of his craft too often regarded his specialty as one not worthy of the energy he expended upon it. While the Prussian academicians were not so ready as the English liberal historians to regard war as an unnatural occurrence, they were not convinced that excessive absorption in the study of military affairs entitled a man to academic recognition and the promotions and emoluments that went with it. Delbrück's advancement to a full professorship was certainly delayed by his insistence that the history of war was quite as important as the deciphering of Roman inscriptions[95] and, throughout his life, he was constantly arguing the legitimacy of his historical field. At the beginning of his career he pleaded that there was a crying need for historians "to turn not only an incidental but a professional interest to the history of war."[96] In his last years, long after he had won a secure position in academic circles, he lashed out once again in the pages of his *World History* at those who persisted in believing "that battles and wars can be regarded as unimportant by-products of world history."[97]

Delbrück's importance in the history of military thought has been contested almost as bitterly as were his academic privileges. Many of the discoveries of his *Sachkritik* have been questioned or dismissed as unoriginal, while his theory of strategy has never been generally accepted either by historians or by military men. But there is no doubt that the *History of the Art of War* will remain one of the finest examples of the application of modern science to the heritage of the past and, however modified in detail, the bulk of the work stands unchallenged. Moreover, in an age in which war has become the concern of every man, the major theme of Delbrück's work as historian and publicist is at once a reminder and a warning. The coordination of politics and war is as important today as it was in the age of Pericles, and strategical thinking which becomes self-sufficient or which neglects the political aspect of war can lead only to disaster.

[95] J. Ziekursch, "Delbrück," *Deutsches Biographisches Jahrbuch* (1929).
[96] H. Delbrück, "Etwas Kriegsgeschichtliches," *Preussische Jahrbücher*, LX (1887), 610.
[97] H. Delbrück, *Weltgeschichte* (Berlin, 1924-1928), I, 321.

SECTION IV

From the First to the Second World War

CHAPTER 12. Churchill, Lloyd George, Clemenceau: The Emergence of the Civilian

BY HARVEY A. DE WEERD

THE democratic form of government and philosophy of life pose problems of military responsibility, organization, and control which were largely nonexistent under monarchies. For one thing, the best minds and most enterprising men in democratic states seldom are attracted to the profession of arms. Writing about the United States at an early stage in our history de Tocqueville observed that "the best part of the nation shuns the military profession because that profession is not honored and the profession is not honored because the best part of the nation has ceased to follow it."[1] Then, too, the separation of powers and the system of checks and balances, so essential to safeguard against the possible rise of a tyrant, involved handicaps to the successful administration and conduct of war which increased as military operations grew more complex. Thus, while the civilian elements of democratic states for the most part avoided military life and concerned themselves only reluctantly with military problems, each major war in which they were engaged brought with it such formidable difficulties and revealed such glaring defects in their war administration that sweeping reforms generally accompanied or followed these military trials.[2]

The ultimate responsibility of the civilian element in a democracy for supervision of military matters rested, as Madison and other early American thinkers pointed out, on the primary duty of the government to protect the state.[3] Civilian control of military affairs in a democracy could not be abdicated without ultimately endangering the sovereignty of the people. The gradual evolution of congressional control of military policy in the United States from the ratification of the Constitution to the passage of the National Defense Act of 1916, however, shows how reluctant the national government was to act upon the authority and responsibility implied in the constitutional power granted Congress "to raise and support armies."[4]

If civilian elements were to control military policy, this entailed an inescapable responsibility to go beyond the stage of mere financial support and legislation to a more or less direct participation in or control of military opera-

---

[1] Alexis de Tocqueville, *Democracy in America*. Translated by Henry Reeve (London, 1889), II, 243.
[2] In the case of the United States these tendencies are clearly revealed in E. Upton's *The Military Policy of the United States* (Washington, 1904), F. V. Greene's *The Revolutionary War and the Military Policy of the United States* (New York, 1911), and J. M. Palmer's *America in Arms: The Experience of the United States With Military Organization* (New Haven, 1941).
[3] E. M. Earle, "American Military Policy and National Security" in *Political Science Quarterly*, LIII (1938), 4-5.
[4] H. White, *Executive Influence in Determining Military Policy in the United States* (Urbana, Illinois, 1925), pp. 26-51.

tions themselves. Here the civilian was to encroach upon the traditional sphere of the professional soldier and open a vast field for civilian-military controversy and friction.

If political considerations alone were not sufficiently powerful in themselves to force a growing civilian participation in military matters, other factors such as technological trends were at work to achieve the same ends. Oswald Spengler observed that the technique of war hesitatingly followed the advance of craftsmanship until the beginning of modern times, then it suddenly pressed all mechanical possibilities into its service. He saw a close relation between the fact that gunpowder and the printing press came into use at about the same time, that the reformation saw the first flysheets and field guns, that the first rain of political pamphlets came at about the same time as the mass fire of artillery at Valmy.[5]

An even closer relationship between war and technical advances followed the industrial revolution which made it possible to clothe, arm, and supply the mass armies inspired by the French Revolution. Machine manufacture made the existence of these armies possible; the railway made them mobile to a degree hitherto undreamed of. In the words of one observer, "Their development effected a fairly complete revolution in the physical basis of the management of armies—no less far reaching than that brought about at sea by the shift from sail to steam."[6] This revolution made itself felt in the Italian war of 1859, but was unmistakably revealed in 1862 by Haupt's feat of transferring two corps from the army of the Potomac by rail to Nashville in a week. Railways enabled the elder Moltke to deploy 400,000 men on the French frontier in sixteen days in 1870, and the younger Moltke to concentrate four times that number within the same period of time in 1914.[7]

While it was not immediately apparent, the increasingly dependent relationship between military and industrial functions in the nineteenth and twentieth century involved growing participation of the civilian elements of society in the preparation for and conduct of war. It was a prevision of this development which prompted Engels to ask: "Who introduced the victories of the French Revolution into the army? Not the generals—but the civilian power."[8]

Professor Wright has divided the history of warfare into convenient periods on the following pattern: (a) The adaptation of firearms (1450-1648); (b) The period of professional armies and dynastic wars (1648-1789); (c) The capitalization of war (1789-1914); (d) The totalitarianization of war (1914-1942).[9]

During the last two periods there was a definite trend toward the mechanization of war, toward the increased size of armies, toward the militarization of

---

[5] Oswald Spengler, *The Decline of the West* (New York, 1926), II, 460.

[6] T. H. Thomas, "Armies and the Railway Revolution," in *War as a Social Institution* (New York, 1941), p. 88.

[7] *Ibid.*, p. 93.

[8] Letter to Marx, September 26, 1851. *Der Briefwechsel zwischen Friedrich Engels und Karl Marx, 1844-1883* (Stuttgart, 1919), p. 252.

[9] Quincy Wright, *A Study of War* (Chicago, 1942), I, 294-300.

population, toward the nationalization of war effort, and toward the intensification of military operations.[10]

Whereas the conduct of war in earlier periods was primarily the concern of admirals and generals, the problem of directing the whole resources of a nation toward a specific objective was too vast to be handled effectively by one class of leaders. It became the responsibility of the whole people and the government.[11] Yet the lack of interest of civilian leaders in military matters, together with the widespread assumption that peace was the permanent and normal state of society, caused the civilian elements to be ill-prepared to assume the increasingly active role in war which technical and industrial progress made inevitable.

Since the capacity to wage war is bound up inextricably with the nature of government itself, it was inescapable that democratic countries with loose methods of control and less centralized organization would suffer disadvantages when pitted against more tightly organized, less democratic states. The philosophy of peace which motivated some of the democracies militated against steps which would enable them to mobilize and direct their military efforts along the lines indicated by advancing technology and modern conditions. Thus, while many continental European states modeled their central agency for the direction of military activity around the model of the Prussian-German general staff, Britain and America delayed in the establishment of such a system of military planning and control until the twentieth century. The fear of setting up an agency deemed likely to involve them in specific preparations for war—and hence, they believed, in war—was instrumental in preventing an earlier adoption of the general staff system despite its obvious advantages.[12] Thus in Britain as well as in America, there was no body for planning and coordinating military action which enjoyed anything like the historical background or authority of the great German general staff.

In many ways the American Civil War foreshadowed the Great War of 1914-1918. It saw the employment of mass armies, of railroads, telegraphs, armored ships, railway artillery, balloons, Gatling guns, repeating rifles, trenches, and wire entanglements. In one sense, the Civil War was the first modern war of materiel. The industries of the north more than offset the valor and military qualities of the south. Machines like the reaper which freed agricultural laborers for military service while, at the same time, allowing the north to capture the grain markets of Europe, were factors of considerable importance in the final outcome of the war. The rapidly developing fire power of modern weapons promised to drive all soldiers into trenches.[13] Unless decisions were reached in the initial stages of war, a struggle of exhaustion would

---

[10] *Ibid.*, I, 302-310.

[11] Major General Sir F. K. Maurice, *Government and War* (London, 1926), p. 123. As Clemenceau put it in the First World War, "La guerre est une chose trop sérieuse pour qu'on la laisse diriger par les militaires."

[12] *Annual Report of the Secretary of War, 1902* (Washington, 1903), pp. 44-49; *Richard Burdon Haldane: An Autobiography* (London, 1929), pp. 212-213; and E. Root, *Military and Colonial Policy of the United States* (Cambridge, Mass., 1916), pp. 108-110.

[13] I. J. de Bloch, *The Future of War* (Boston, 1914).

follow. These lessons were reinforced by the Boer War and the Russo-Japanese conflict.

The military plans of the European general staffs of 1914, however, contemplated a traditional war of movement, in which maneuver would bring about a decision. With the sole exception of Lord Kitchener, leading professional men were unanimous in their belief that a European war would be ended in a matter of months. This view persisted until after the battle of the Marne (September 6 to 9, 1914) which was followed by resort to trench warfare in the west. By early 1915, the line of entrenchments stretched from the sea to Switzerland. This presented the belligerents with a situation in which there were no flanks to turn in the conventional method and in which a war of movement was impossible. The entrenchments, guarded by machine guns and artillery, were soon swathed in lanes of barbed wire which had to be destroyed by prolonged artillery bombardment and which precluded surprise. With both movement and surprise gone on the Western Front, a war of position followed in which attrition and the costly expedient of the frontal attack were the sole alternatives.[14]

II

By 1915 materiel became the decisive factor in the war. The demands of trench warfare for all kinds of military equipment and supplies exceeded all previous expectations. It soon became clear that the professional military leaders could not organize and direct the national resources efficiently in such a struggle.[15] It required cooperation between both civilian and military leaders to a degree hitherto unattained in war. Because it was a war of coalitions waged over vast theaters, it also demanded civilian thought and energy in its direction. At the risk of indulging in oversimplification, one may say that the war of 1914-1918 presented two difficult problems for solution. One was how to prepare effectively for a war of materiel. The other was how to coordinate military effort in a modern coalition war. Both of these tasks could in the nature of things be carried out more effectively and expeditiously by the civilian element of the state rather than the military.

Even a monarchy such as the German empire found that it could not carry on the war of 1914-1918 without the aid of important civilian elements. Among the first in Germany to see the full implications of trench warfare and the struggle of materiel which followed it was Dr. Walther Rathenau. Director of the *Electrochemische Werke*, the *Allgemeine Elektrizitäts-Gesellschaft* and a hundred other firms, he had supervised the construction of vast enterprises in many parts of the world. His knowledge of manufacturing processes and his

---

[14] P. J. L. Azan, *The War of Positions* (Cambridge, 1917) ; for examples of typical trench warfare operations in 1916, see J. E. Edmunds, *Military Operations, France and Belgium 1916* (An Official History of the Great War) (London, 1932), vol. I.

[15] As late as December 31, 1914, Sir John French, commander of the British army in France, set the ammunition requirements of his force at the then high rate of 50 rounds per day for the 18-pounder field guns, 40 rounds for the 4.5 inch howitzers, and 25 for the 4.7 inch guns. By late September 1918 the B.E.F. was expending artillery ammunition at the rate of 21,000 tons daily. *Sir Douglas Haig's Despatches* (London, 1920), p. 333.

wide acquaintance with lands and peoples outside Germany gave him an infinitely broader outlook than that of the average staff officer. He saw that in its essence war was really a gigantic project of moving, supplying, feeding, and caring for a host of men engaged in a destructive rather than a constructive task. Modern war used up material much more rapidly than previous wars. When the battle of the Marne ended Germany's hopes for a quick victory in the west, Rathenau saw the need of conserving and organizing her raw materials for the longer struggle impending. As Professor Shotwell put it: "Instead of troops on the march and the clash of armies in the field he had a vision of tall chimneys pouring out smoke and of flaring furnaces lighting the sky all the way from Berlin to the Rhine. This, as he saw it, was the vital element in modern war."[16] He was not the only industrialist or economist who saw the handwriting on the wall. Arthur Dix is reported to have sent a memorandum to Moltke early in 1914 setting forth the need of an economic general staff. Moltke is said to have replied: "Don't bother me with economics, I am busy conducting the war."[17]

Rathenau was more successful with General Falkenhayn, who was minister of war in 1914 and became Moltke's successor after the Marne defeat. He convinced Falkenhayn that a thoroughgoing mobilization and a systematic use of Germany's raw materials were prerequisites to a successful continuation of the war. Setting up an organization called the *Kriegsrohstoff-Abteilung*, with an initial staff of three late in 1914, the work of his branch became so important that in 1918 it was the largest unit in the war ministry.[18] His work, together with that of Dr. Fritz Haber, whose chemical and engineering skill enabled Germany to meet its high explosive requirements after the imports of Chilean nitrates were cut off by the British blockade, were extremely important factors in Germany's four-year resistance against a hostile coalition vastly superior in resources and man power.

Yet the achievements of German industry from 1914-1918 were disappointing to Rathenau and failed to prevent her collapse. The obvious lesson to be drawn from Germany's defeat was that a much more complete control of all industry and resources, including man power, would be required in any future war of revenge. The economics of war was studied zealously in post-war Germany, and even before the end of the war Rathenau saw that great changes in the socioeconomic structure of Germany were impending.[19] These pointed to the direction of state socialism in peace and military autarchy in war. The lessons of the First World War were embodied in the economic general staff set up and headed by General Georg Thomas and were the principal motives behind Goering's famous Four Year Plan.[20]

[16] J. T. Shotwell, *What Germany Forgot* (New York, 1940), p. 20.
[17] Quoted in *Axis Grand Strategy*, edited by L. Farago et al. (New York, 1941), p. 499.
[18] For details of Rathenau's career and war work see H. Kessler, *Walther Rathenau, sein Leben und sein Werk* (Berlin, 1928).
[19] For a discussion of Rathenau's views see A. T. Lauterbach, "Roots and Implications of the German Idea of Military Society," in *Military Affairs*, V (Spring 1941), 5-8.
[20] Discussed in G. Thomas, *Wehrwirtschaft* (Berlin, 1939).

Certain men in England such as David Lloyd George and Winston Churchill struggled to solve the military, industrial, and political problems involved in the coalition war of materiel from 1914-1918. They were operating in a parliamentary democracy under all the restraints and handicaps inherent in that system of war administration. Accordingly, their contribution to the thought and conduct of the war is worth setting down in somewhat greater detail than that of Rathenau.

Lloyd George had no education or experience in war; his life had been devoted entirely to dealing with social, legal, and political problems. Mr. Churchill had enjoyed a brief period of training at Sandhurst and service in India, but this led to a career as war correspondent, historian, and politician. During the Boer War Lloyd George had shown himself to be a critic of imperialism, and after this war his passion for social legislation led him to oppose swollen military and naval budgets.[21] Churchill, on the contrary, was directly concerned with naval affairs after 1911 and felt competent to write a memorandum for the Committee of Imperial Defense on the probable course of a German invasion of France at the time of the Agadir crisis.[22] Both men were involved in the events of 1914-1918 through their official positions. Lloyd George was Chancellor of the Exchequer and Churchill First Lord of the Admiralty in 1914.

When war came in 1914 Churchill made the broadest possible interpretation of his functions as First Lord of the Admiralty. He wrote: "I accepted full responsibility for bringing about successful results, and in that spirit I exercised a close general supervision over everything that was done or proposed. Further, I claimed and exercised an unlimited power of suggestion and initiative over the whole field, subject only to the approval and agreement of the First Sea Lord on all operative orders."[23] Churchill's readiness to accept responsibility, his quickness of perception, and his brilliant language caused him to move among the slower members of the cabinet "like a panther among seals." His position brought him at once into close contact with all phases of the war. He was not content with controlling the seas but ventured into the domain of military strategy as well. On September 5, 1914, he wrote a memorandum to Lord Kitchener suggesting the transport of two Russian army corps from Archangel to Ostend to strike at German communications.[24] He took a personal part in the defense of Antwerp and urged a program to gain allies in the Balkans.[25] Thus an enterprising civilian head of a department of government became, in a sense, a volunteer general staff.

Both Churchill and Lloyd George recognized the significance of the 1915 trench stalemate in France but drew slightly different conclusions about it. To Churchill's mind the problem resolved itself into a restoration of maneuver.

[21] *The War Memoirs of David Lloyd George* (London, 1933-1937), I, 10. Hereafter cited as Lloyd George.
[22] W. S. Churchill, *The World Crisis 1911-1914* (London, 1923), pp. 58-62. The four volume series: *The World Crisis 1911-1914, The World Crisis 1915,* and *The World Crisis 1916-1918,* will hereafter be cited as Churchill, vols. I-IV.
[23] *Ibid.,* p. 259.   [24] *Ibid.,* I, 293.   [25] *Ibid.,* I, 390; II, 2.

The entrenched machine gun presented a mechanical problem in the face of which the existing military doctrines were obsolete. He wrote:

"Battles are won by slaughter and manoeuvre. The greater the general, the more he contributes in manoeuvre, the less he demands in slaughter. The theory which has exalted the 'battaille d'usure' or 'battle of wearing down' into a foremost position, is contradicted by history and would be repulsed by the greatest captains of the past. Nearly all the battles which are regarded as masterpieces of the military art, from which have been derived the foundation of states and the fame of commanders, have been battles of manoeuvre in which very often the enemy has found himself defeated by some novel expedient or device, some queer, swift, unexpected thrust or stratagem. In many such battles the losses of the victors have been small. There is required for the composition of a great commander not only massive common sense and reasoning power, not only imagination, but also an element of legerdemain, an original and sinister touch, which leaves the enemy puzzled as well as beaten. It is because military leaders are credited with gifts of this order which enable them to ensure victory and save slaughter that their profession is held in such high honour. For if their art were nothing more than a dreary process of exchanging lives, and counting heads at the end, they would rank much lower in the scale of human esteem. . . .

"The mechanical danger (the torpedo and the machine gun) must be overcome by a mechanical remedy. Once this is done, both the stronger fleet and the stronger armies will regain their normal offensive rights. Until this is done, both will be baffled and all will suffer."[26]

Churchill came to the conclusion that the solution to the problem lay in interposing a thin steel shield between the ship and the torpedo and between the breast of the soldier and the machine gun bullet. He sought to escape the deadlock in France by opening new theaters of war in the Middle East and Balkans.

Lloyd George did not concern himself with the military aspects of the war until he had completed the financial steps made necessary by the transition from peace to war. Then his active mind began to explore the military organization and program which his department was financing. What he discovered led him to take a more and more direct part in the preparation for and finally in the direction of the war itself.

The war appeared to him as a battle of materiel, and to his way of thinking the procurement policy of the war office was hampered by traditional reactionary methods. He soon exhibited a skepticism of British military leadership which increased throughout the war. At the outset of the war he found that the whole business of supplying the British army, even down to the tailoring contracts, was "jealously retained by the War Office."[27] Lord French, the first commander of the B.E.F., and Haig, his successor, were cavalrymen; Kitchener was an engineer officer. None of these generals, he felt, or few in

Europe for that matter, had a prevision of the magnitude of the supply requirements of trench warfare.[28] They were slow to recognize the changes in weapons and methods it made inevitable. At first they rejected the demands for high explosive shell in favor of shrapnel. Lloyd George did not blame the war office for failing to have an adequate supply of ammunition and equipment for the large forces which were found necessary to carry on the war, but he charged them with "mental obtuseness in failing to keep abreast of the modern developments in pattern of munitions and machinery for munitions production."[29] He held them responsible for failing to see that twentieth century warfare was to a considerable extent a conflict between chemists and manufacturers. When he urged in October 1914 that the capacity of existing armament firms should be extended, the master general of ordnance declined to make use of the funds made available on the grounds that firms had not asked for assistance.[30] The war office retained their prewar belief that only the arsenals and a few experienced firms were capable of producing dependable military equipment.

In a memorandum of February 22, 1915, Lloyd George pointed out that the great hope of the Allied cause lay in their superiority of industrial resources.[31] If these resources were quickly and fully utilized and the military effort of the Allies coordinated, victory over the central powers was assured. He repeatedly emphasized the necessity of mobilizing all industrial resources for war production. To all his suggestions for improving the munitions situation there were countless military objections and obstacles. These continued until the Dardanelles controversy in 1915 brought about a political shake-up which resulted in the creation of a munitions ministry with Lloyd George at its head.

Mr. Churchill's ideas about the trench stalemate and lack of coordination among the Allies were embodied in a memorandum to the prime minister on December 24, 1914. He wrote:

". . . I think it quite possible that neither side will have the strength to penetrate the other's lines in the Western theatre. . . . Without attempting to take a final view, my impression is that the position of both armies is not likely to undergo any decisive change—although no doubt several hundred thousand men will be spent to satisfy the military mind on the point.

"On the assumption that these views are correct, the question arises, how ought we to apply our growing military power? Are there not other alternatives than sending our armies to chew barbed wire in Flanders? Further, cannot the power of the Navy be brought more directly to bear upon the enemy? If it is impossible or unduly costly to pierce the German lines on existing fronts, ought

[28] *Ibid.*, I, 126. A notable exception, but one whose low rank prevented proper attention being given to his views, was Colonel E. Mayer of the French army. Cf. *La France Militaire*, December 15, 1938.

[29] Lloyd George, I, 129. Even the French, who were much better prepared from the standpoint of supplies in 1914 than the British, found that their program laid down before the war was pitifully inadequate. *Ibid.*, I, 152-153. Whereas all the German armies in the war of 1870-1871 expended a total of 817,000 rounds of artillery ammunition, the French armies in the year 1918 alone expended 81,000,000 rounds. U.S. General Staff, Statistical Branch, Report D-2-153.

[30] Lloyd George, I, 133.

[31] *Ibid.*, I, 168.

we not, as new forces come to hand, to engage him on new frontiers, and enable the Russians to do so too?

". . . The action of the Allies proceeds almost independently. Plans could be made now for April and May which would offer good prospects of bringing the war to its decisive stage by land and sea. We ought not to drift. We ought now to consider while time remains the scope and character we wish to impart to the war in the early summer. We ought to concert our action with our allies, and particularly with Russia. We ought to form a scheme for a continuous and progressive offensive, and be ready with this new alternative when and if the direct frontal attacks in France on the German lines and Belgium have failed, as fail I fear they will. . . ."[32]

When no other immediate alternative seemed at hand, Churchill committed the navy to an attack on the Dardanelles, which led to a costly naval reverse on March 18, 1915, and ultimately to a bloody and unsuccessful attempt by the army to take the Gallipoli peninsula. This failure gave rise to a heated controversy over the relationship between the First Lord of the Admiralty and his professional adviser. The charge was widely made that Churchill had disregarded the advice of Admiral Lord John Fisher, the First Sea Lord, who resigned in protest against the losses at the Dardanelles. The whole problem of how to implement civilian thought in the military sphere was bound up with the relationship between civilian heads of the government and their professional military advisers. The system of civilian responsibility for military results, with full military control of operations, was workable only when the military advisers were technically competent and abreast of the times. When they refused on narrow professional grounds to realize the full potentialities of science and industry, and held to a strategical program which they could not defend in debate with the more articulate politicians, the system broke down and bitter controversies followed.[33]

III

The Dardanelles disaster cost Churchill his position as First Lord of the Admiralty, and for a time he was relegated to a sinecure post as Chancellor of the Duchy of Lancaster. Before leaving the admiralty, however, he had put in train a project which was to produce the major tactical innovation of the war: the development of the tank. Using admiralty funds, he had ordered the construction of a machine-gun-bearing, armored, caterpillar-tread tractor, designed to cross the broken terrain of the battlefield. He discussed the employ-

[32] Churchill, II, 5-6.
[33] As an important factor in the story of civilian-military relationships and responsibility, it should be pointed out that Lord Fisher's position was seriously weakened in this controversy by the fact that he remained silent in council when a proposal was accepted against which he felt a strong opposition. In a later controversy with Lloyd George, Sir William Robertson spoke out openly in council when he was expected to remain silent, and was roundly abused for his action. The question of whether the silence of a professional military adviser gave consent to the plans of his civilian chief was never clearly determined. Cf. Field Marshal Sir William Robertson, *From Private to Field Marshal* (London, 1921), pp. 255, 317-318.

ment of such a vehicle in a memorandum addressed to Sir John French on December 3, 1915.[34] The first tank had all the weakness of a new invention, but in time it produced a revolution in tactics. Thus, despite the importance of Churchill's service as minister of munitions from 1916-1918, the tank and all that stemmed from it must be considered his greatest contribution to the military history of this period. Many men played a part in the development of the tank; but it is clear that the main impulse to utilize the machine, in this case the internal combustion engine, to surmount the barrier of battlefield conditions in 1914-1918, came from a civilian mind.

During the war years Mr. Churchill wrote to the military and political leaders of Britain numberless memoranda which demonstrated an unusual grasp of strategic problems and a notable vision for future developments. In 1917, for example, he wrote an appreciation of the limitations and possibilities of military aircraft, too long to be quoted at this point, which showed prophetic insight into the future uses of that weapon.[35]

Looking back, many years later, from the vantage point of 10 Downing Street, in the midst of a war more replete with military surprises and problems than that of 1914-1918, Mr. Churchill summed up his experience in the field of military administration in the following words:

"Modern war is total, and it is necessary for its conduct that technical and professional authorities should be sustained and if necessary directed by heads of government, who have the knowledge which enables them to comprehend not only the military but the political and economic forces at work and who have the power to focus them all upon the goal."[36]

Once the decision was made to attack the Dardanelles in 1915, Lloyd George supported it with all his energy, not because he thought it was the most promising area for Allied military action, but because it avoided the error of attacking the enemy at his strongest point. He opposed Western Front offensives on the ground that the Allies did not have strength enough to succeed in such actions. The Allied failure to relieve Serbia in 1915 increased his doubts about Lord Kitchener's competence in the strategical sphere. As minister for munitions in 1915, he felt that the war office tables of organization and weapons for the new divisions were wholly inadequate. He set out to provide guns on a scale ranging up to 25 per cent above the war office estimates and projected this table of equipment for 100 divisions instead of the war office program of 70.[37] When Lord Kitchener set the number of machine guns required per battalion at two, Lloyd George told Geddes to "square that number, multiply the result by two, and when in sight of that, double it for good luck."[38] He further interfered with the normal routine of adopting a weapon for the army by ordering 1000 Stokes mortars in spite of war office opposition. Thus we find civilian leaders, both Churchill and Lloyd George, forcing a reluctant

---

[34] *Ibid.*, II, 78-81.   [35] *Ibid.*, IV, 309-313.
[36] Address before the Joint Session of the Congress of the United States, May 19, 1943, as reported in the *New York Times*, May 20, 1943.
[37] Lloyd George, II, 557.   [38] *Ibid.*, II, 605.

military institution to modify its body of doctrine by dumping new weapons in its lap.

Lloyd George repeatedly stressed the necessity of coordinated inter-Allied military action. As he wrote after the war: "The real weakness of Allied strategy was that it never existed. Instead of one great war with a united front, there were at least six separate and distinct wars with a separate, distinct and independent strategy for each. There was some pretence at timing the desperate blows with a rough approach to simultaneity. The calendar was the sole foundation of inter-Allied strategy. . . . There was no real unity of conception, co-ordination of effort or pooling of resources in such a way as to deal the enemy the hardest knocks at his weakest point. There were so many national armies, each with its own strategy and its own resources to carry it through. Neither in men, guns or ammunition was there any notion of distributing them in such a way as to produce the greatest results with the available resources of the Alliance as a whole. There had been no genuine endeavor to pool brains with a view to surveying the whole vast battlefield and to deciding where and how the most effective blows could be struck at the enemy. Before 1917 no General that mattered in the East had ever met a military leader who counted in the West. The two-day conferences of great generals which were held late each autumn to determine the campaign for the ensuing year, were an elaborate handshaking reunion. They had all of them come to the meeting with their plans in their pockets. There was nothing to discuss. It was essential that a body should be set up for common thinking. . . ."[39]

Efforts to set up a unified command or to place British commanders under French control for specific operations, however, met with the stonewall opposition of the chief of the imperial general staff and the British commander in chief in France. This controversy, which went on concurrently with the Lloyd George-Haig-Robertson struggle over the strategic control of the imperial war effort, further embittered the relations between the prime minister and the high command. The generals viewed any attempt at establishing a unified command as a concealed effort to lessen their own authority. By appealing to national sentiments and even to constitutional grounds they were able to avoid it. Finally the disaster of March 1918 forced the appointment of a unified command in France.[40]

Professional military leaders opposed many of Lloyd George's suggestions with flat statements that what he proposed was technically impossible. The use of convoys for antisubmarine protection was an example. He wrote:

"The difficulties experienced by the War Cabinet in handling this problem are inherent in all war operations when civilian opinion clashes with that of the experts. Naval science and strategy are matters very remote from the lay comprehension, and the aura of authority glistened round the heads of the Naval

---

[39] *Ibid.*, IV, 2347-2348.
[40] The military side of the controversy appears in Sir William Robertson's *From Private to Field Marshal* (London, 1921) and *Soldiers and Statesmen*, 2 vols. (London, 1926). A critical evaluation of Sir William Robertson's methods appears in Captain Peter Wright's work. Cf., *infra*, note 42.

High Command. Whenever I urged the adoption of the convoy system, I was met, as I have related, with the blank wall of assertion that the experts of the Admiralty knew on technical grounds that it was impossible. That is a very difficult argument to counter.

"A persistence of a few more weeks in their refusal to listen to advice from outside would have meant irretrievable ruin for the Allies. . . . It was not the first time in this War that the lesson was driven home . . . that no great national enterprise can be carried through successfully in peace or in war except by a trustful co-operation between expert and layman—tendered freely by both, welcomed cordially by both."[41]

As the war went on there was a tendency, observed even by soldiers, for the professional military directorates to consider themselves not only free from interference by civilian agencies with whom they had to cooperate in order to achieve success, but in a sense, as being above the nation itself. Captain Peter Wright wrote:

"This great deceit—at last emancipates all General Staffs from all control. They no longer live for the nation; the nation lives, or rather dies, for them. . . . What matters to these semi-sovereign corporations is whether dear old Willie or poor old Harry is going to be at their head, or the Chantilly party prevail over the Boulevard des Invalides party. . . . Two branches of a staff can get more hostile to each other than to the enemy. . . ."[42]

As Lloyd George summed up the Allied situation in 1917, he concluded that: "The fundamental error of the Allied strategy up to the present has been the refusal of their war direction to recognize the fact that the European battle-field is one and indivisible. A corollary to this error has been the concentration of the strongest armies on the attacking of the strongest fronts, whilst the weakest fronts have been left to the less well-equipped armies."[43]

Sir William Robertson, who was chief of the imperial general staff from 1916-1918, represented the Western Front school of military thought. He and Haig, whose views were similar, were reinforced in all their clashes with Lloyd George by the French general staff, which was happy to have France regarded as the principal theater of war. Members of the French staff had little interest in or realization of the importance of sea power. Their stock reply to all suggestions for opening a front elsewhere was: "C'est toujours une question de tonnage."

Two persons of more divergent mind could hardly be imagined than Robertson and Lloyd George. A tough, burly, orderly man, "Wully" Robertson had risen from the ranks on sheer merit and tenacity. He had met every problem in his long and honorable career by applying his full energy in the most direct fashion on the objective to be won or the obstacle to be overcome. Faced with the German entrenchments in France his reaction was characteristic. He would

[41] Lloyd George, III, 1169.
[42] Captain Peter Wright, *At the Supreme War Council* (New York, 1921), p. 104. Hereafter cited as Wright.
[43] Lloyd George, IV, 2169.

concentrate the British military effort in France. He would do this on the conviction that if the German army in France was destroyed the war was won. Having decided upon his theater of operations, he wisely and stubbornly resisted attempts to divert British resources. On the final point, at least, he was on sound ground. All of Lloyd George's efforts to convince him that victory could be achieved elsewhere at less cost, he rejected as violating the fundamental principle of concentration on the decisive point. Slow in speech, his mind was keen and inflexible. Given time, he could turn out massive and logical memoranda. He was capable of tremendous rages, at which time "his face went the color of mahogany, his eyes became perfectly round, his eyebrows slanted outward like a forest of bayonets held at the charge." He could terrify his subordinates but could not convince the prime minister that the only way to win the war was to arm every man and boy and see that they were set to "killing Germans" on the Western Front. His counterpart in France, Sir Douglas Haig was "well versed in his profession, with a clear vision over a limited field, but he suffered from the handicap inherent in men who know what they want and see their goal, but are conscious they lack the gift of persuasion. He expressed himself clearly and forcibly on paper. A hesitant speech, ending in silences reinforced by a forward movement of the jaw giving the impression that obstinacy rather than reason dictated his decisions, stood him in poor stead when dealing with Ministers."[44]

Both of these men placed a great deal of importance on the military concept of loyalty, although their interpretation of it has been repeatedly subjected to criticism.[45] They both found Lloyd George's personality and mentality annoying and felt that his methods of conducting the war endangered the nation. They could not deny his energy and determination. In fact, no one could. As Captain Wright wrote: "In spite of his [Lloyd George's] oblique and subterranean methods; his inveterate taste for low and unscrupulous men; of the distrust felt for him by his favorites, even at the height of their favor; of his superficial, slipshod, and hasty mind; this determination of character made him, without any assumption on his part, the leader of the Alliance."[46]

Such, then, were the men who personified the struggle between the civilian and military elements of the state in England from 1916-1918. Both Lloyd George and the generals worked for a victory with all their might and mind, but each party was convinced that the other was wrong, and inadequate language prevented the soldiers from placing their case to the best advantage.

The vital implications of the Lloyd George-Robertson-Haig controversy lay in the fact that the civilian head of the British democracy found himself unable to impose his will on the generals. In some measure this was due to political causes, but since they were paralleled in other countries, particularly in France at the same period, the fault seems inherent in the system. Lloyd George strenuously opposed the sterile British offensive at Passchendaele in 1917, but he was not able to stop it, nor was he strong enough to risk the dismissal of Haig

[44] Brigadier General E. L. Spears, *Prelude to Victory* (London, 1939), p. 266.
[45] *Ibid.*, p. 270.   [46] Wright, p. 27.

and Robertson. Because the official communiqués and the newspapers tend to build up soldiers in the minds of the average citizen, Haig and Robertson were too strongly entrenched in popular support to be relieved. Thus Lloyd George was forced to be party to a program which he knew in advance would fail. Not until disaster forced the Allies to improve their machinery for waging coordinated coalition warfare, were statesmen in Britain and France able to achieve full military implementation of their policies.

The whole complex and unbelievably involved system of pressures operating in a democratic state war is illustrated by the ill-fated Nivelle offensive in France in 1917. All its attendant clashes of civilian and military authority are revealed in General Spears' brilliant book *Prelude to Victory*. Speaking of the Compiègne conference which was only one of the many necessary to put Nivelle's ill-fated plan into operation, Spears wrote: "The Compiègne conference stands as a monument to the inefficiency of democracy at war, to the helplessness of Ministers facing technicians, and their total inability to decide between professional opinions. . . . The Prime Minister and Painlevé controlled the War Cabinet. Painlevé ruled the army. They had power to over-ride the Commander-in-chief in whose plan they had no faith; yet they were incapable of pointing out the failings of that plan or suggesting alternatives, impotent even to call a halt. The cabinet was supreme in name only. It was hobbled by its lack of technical knowledge and fettered by public opinion, which, aware of its ignorance in military matters, would have been intolerant of civilian intrusion in the military sphere. April 6, 1917, epitomizes the terrible disability from which democracies, even when fighting for their existence, are unable to free themselves."[47]

The disasters of 1918 finally forced the Allies to accept a kind of unity of command under Foch, and Robertson was replaced by General Wilson, who was more amenable to Lloyd George's persuasion. A start toward a working machinery for the direction of war was set up at Versailles, but the war ended before it could make its weight felt in the scales of military decision.

The Allied victory of 1918 caused the failure of Allied war machinery to be forgotten. The basic problems which embittered all of Lloyd George's relations with the British high command in the years of 1916-1918 remained unsolved. When, and under what circumstances should the civilian head of a state overrule the professional military leaders? What course should be taken when the civilian and military leaders are at complete variance as to proper procedure? What course should the civilian heads of the state take when professional military men are not in agreement among themselves?

That these questions continued to perplex Lloyd George after the war appears in the final volume of his *War Memoirs*:

"Ought we to have interfered in the realm of strategy? This is one of the most perplexing anxieties of the Government of a nation at war. Civilians have had no instruction, training or experience in the principles of war, and to that extent are complete amateurs in the methods of waging war. It is idle, however,

[47] Spears, p. 376.

to pretend that intelligent men whose minds are concentrated for years on one task learn nothing about it by daily contact with its difficulties and the way to overcome them. . . . But strategy is not entirely a military problem. There is in it a considerable element of high politics. . . .

"Generally speaking, the argument of the high Commands in the War for their claim to be the sole judges of military policy was put far too high by them and their partisans. War is not an exact science like chemistry or mathematics where it would be presumption on the part of anyone ignorant of its first rudiments to express an opinion contrary to those who had thoroughly mastered its principles. War is an art, proficiency in which depends more on experience than on study, and more on natural aptitude and judgment than on either. . . .

"Looking back on this devastating war and surveying the part played in it by statesmen and soldiers respectively in its direction, I have come definitely to the conclusion that the former showed too much caution in exerting their authority over the military leaders."[48]

IV

The French legislature had far greater control over the army in war than its British counterpart. Both the Chamber of Deputies and the Senate had army commissions, and by 1916 some of these had become parliamentary inspectors or delegates to the army. This in a sense was a revival of the revolutionary practice of sending deputies on mission to see that the wishes of the government were carried out. The relationship of the civilian and military elements of the state was established by the Decree of October 28, 1913, which read as follows: "The Government, which is responsible for the vital interests of the country, is the only authority competent to fix the political aims of the war. If operations extend to more than one front, it designates the principal adversary, against which the greater part of the national power is to be directed. It distributes accordingly the means of action and resources of all kinds, and places them under the full control of the generals commanding in chief in the various theaters of operations." This decree made the minister of war a *de jure* generalissimo of all French forces, but this formula broke down in practice.[49]

In addition to the commissions of the Senate and Deputies, a Supreme Council of National Defense was established in 1906 composed of the ministers of war, marine, colonies, foreign affairs, and finance. The original powers of this body were extended by the Decree of July 28, 1911, which organized what amounted to investigating committees of the Council of National Defense to supplement the parliamentary commissions.[50]

From the very outset of war the zones of authority of the army and the civil

[48] Lloyd George, VI, 3409, 3416, 3421.

[49] Captain Commandant J. Fraeys, "Relations Between the Government and the Command: French Experiences, 1914-1918," translated for the Army War College by Colonel O. L. Spaulding from the *Bulletin Belge des Sciences Militaires*, December 1937. Hereafter cited as Fraeys. See also J. M. Bourget, *Gouvernement et commandement: les leçons de la guerre mondial* (Paris, 1930), pp. 7-24.

[50] S. C. Davis, *The French War Machine* (London, 1937), p. 101 and P. Renouvin, *The Forms of War Government in France* (New Haven, 1927), pp. 80-91.

government presented grounds for conflict. Questions arose over the authority of the minister of war to act on matters pertaining to mobilization, to the ten-kilometer withdrawal from the frontier, to the defense of Dijon and Paris.[51] Messimy, minister of war, resigned on August 27, 1914. He was followed by Millerand who let Joffre have his own way, sometimes giving advice but never an order.[52]

The relationship between the war minister and the commander in chief underwent a profound change when General Galliéni became war minister on October 29, 1915. Then a military man as minister of war faced a military man as commander in chief. The strained relationship was heightened by the faint praise given in Joffre's Marne dispatch to the part played by Galliéni as military governor of Paris in the victory of the Marne. Although Briand, the premier, found Joffre's practical dictatorship irksome, he wisely pointed out that either Galliéni or Joffre had to rule, and that since Galliéni had to answer questions in the Chamber of Deputies on past and future operations, he could not also run the army.[53] Joffre had his powers augmented by the Decree of December 2, 1915, which gave him "control" of all the armies in France. When Galliéni protested against Joffre's handling of the Verdun battle, he was not sustained by the Chamber and resigned.[54] His successor, General Roques (March 17-December 9, 1915), a "conciliatory, friendly, affable" man, was unable to dominate anyone.[55] Accordingly Joffre was unchallenged until his removal from power on December 26, 1916.

Under the strong hand of General Lyautey, who served as minister of war until March 15, 1917, the war ministry and parliament set up what Bugnet calls a "parliamentary dictatorship." This period coincided with the preparations for General Nivelle's spring offensive. Lyautey's resignation on March 15, 1917, after a clash with the Chamber, left the military situation in a considerable muddle, and Painlevé, his successor, had the unenviable task of supporting a general whose plans he did not approve.

The failure of Nivelle's offensive in the spring of 1917 led to mutiny in many French divisions. Morale and discipline were restored by General Pétain, and the defeatist movement in political circles was crushed by Clemenceau. Assuming the position of minister of war in addition to the premiership, Clemenceau not only strengthened the home front in the darkest hours of the war, but was able to achieve a more effective working arrangement with the army command than his predecessors had attained.

The stormy petrel of French politics, Clemenceau's acquaintance with French government dated back to the defeat of 1871. A politician, orator, newspaper editor, and philosopher, he brought to the task of saving France the support of all factions of the Chamber except the socialists.[56] In the person

[51] See C. Bugnet, *Rue St. Dominique et G.Q.G.* (Paris, 1937), and *The Personal Memoirs of Joffre*, translated by Colonel T. B. Mott (New York, 1932), vol. I.
[52] Bugnet, p. 50.   [53] *Ibid.*, p. 107.
[54] Fraeys, p. 11.   [55] Bugnet, p. 135.
[56] For details of Clemenceau's career see H. M. Hyndman, *Clemenceau and His Times* (London, 1919).

of his military aide, General Mordacq, Clemenceau had an adviser who had studied the relationship of civilian and military elements in other wars, particularly in the American Civil War.[57] Clemenceau replaced the dictatorship of parliament by a personal supervision of military affairs which included not only broad questions of policy but even the details of defense arrangements.[58] Not all French critics are agreed that Clemenceau's direct intervention in military operations was always helpful. A French officer, writing under the cover of anonymity, asserts that many of Clemenceau's proposals were harmful and that Pétain merely made a pretense of carrying out the directives which he felt were unwise.[59] In particular he held that Clemenceau's insistence that the major part of the French army spend its time digging trenches upset Pétain's schedule of reliefs and furloughs with a resultant decline of morale among the troops, and that his emphasis on defense left the French Army ill-prepared for open warfare in 1918.[60]

The great German offensives of 1918 revealed the weaknesses of inter-Allied machinery for coordinating the armies in France. When disaster fell upon the British Fifth Army, Pétain was prepared to break off connection with the British in the north and fall back for a defense of Paris. In the long and rather complicated story of the formation of the unified command under Foch, Clemenceau played a leading role. Long convinced that war was too important an affair to leave solely to the generals, he now saw that if Pétain and Haig had their way, the defeat of the Entente was sealed. The high regard which the British had at that time for Clemenceau's capacity to direct military operations as well as civilian affairs, was shown in Lord Milner's proposal on March 25, 1918, that Clemenceau be nominated as generalissimo.[61] This arrangement was impossible since in that position Clemenceau would have been torn between the demands and views of Pétain and Foch, both French army officers. Having chosen Foch for the supreme command, Clemenceau supported him loyally during the period of the Chemin des Dames disaster.

When the tide of battle turned against the Germans, Clemenceau along with Haig saw the desirability of coordinating the Allied offensive along a strategic pattern. He wanted to aim at the German lines of communication rather than merely to push the enemy back simultaneously all along the front.[62]

It was in this period that the good relations existing between Clemenceau and Foch broke down. Many irritating incidents combined to set up a hostility between the two men which later flamed into bitter controversy.[63] These disa-

[57] J. H. Mordacq, *Politique et stratégie dans une démocratie* (Paris, 1912).
[58] Fraeys, p. 17.
[59] General X, *La crise du commandement unique: Clemenceau, Foch, Haig, et Pétain* (Paris, 1931), pp. 50-55.
[60] A defense of Clemenceau's assumption of military control can be found in J. H. Mordacq, *Le ministère Clemenceau: Journal du témoin* (Paris, 1930), I, 89-111.
[61] T. H. Bliss, "The Unified Command" in *Foreign Affairs*, December 15, 1922 (I, 27).
[62] Mordacq, *Le ministère Clemenceau*, II, 309-340.
[63] See R. Recouly, *Mémorial de Foch* (Paris, 1930) and G. Clemenceau, *Grandeurs et misères d'une victoire* (Paris, 1930). That Clemenceau insisted on the supremacy of the

greements hinged on the age-old question of military or civilian supremacy, though ironically enough Clemenceau, in at least one of the disagreements, was indirectly attempting to increase the authority of Foch. An additional confusing factor was that Poincaré, president of the Republic, who might normally be expected to fight for the supremacy of the civilian element over the military, carried his political feud with Clemenceau to the point of supporting Foch against the premier.[64]

Clemenceau, eager to spare France further blood losses, found Pershing's program of an independent American army distasteful. He was extremely critical of American staff work in the St. Mihiel and Argonne offensives.[65] When the progress of American divisions into the trenches seemed excessively slow, he directed Foch by letter on October 21, 1918 to order Pershing to send his troops into action. This Foch wisely refused to do and was supported in his position by Poincaré who advanced the curious constitutional argument that Foch's post as generalissimo exempted him from the kind of control that Wilson and Lloyd George exercised over their national troops.[66]

After many minor controversies, the relations between Foch and Clemenceau reached a new point of bitterness on April 17, 1919 when Foch refused to send a telegram to the German peace delegation informing them officially that they would be received at Versailles. He based his refusal on the grounds that he had been deprived of a promised opportunity to make his views on the peace known to the council of ministers.[67] According to Clemenceau, Foch had to retreat from this rather absurd position when preparations were made to dismiss him and when Wilson declared that he would no longer entrust the control of the American army to a general who did not obey his own government.[68]

Clemenceau's contribution to the victory of 1918 was so distinctive that the phrase used in France to describe him was "l'animateur de la victoire." According to his close associate, General Mordacq, Clemenceau's achievements in the military sphere were: a reorganization of the war ministry, an abolition of many military sinecures and useless commissions, the choice of new and energetic leaders, a reorganization of the general staff on a logical basis, a great expansion of the tank and armored car program for the French army, a reorganization and revitalization of the French high commands in Italy and Salonika, and personal contributions toward the strategic offensive of July 18-November 11, 1918.[69]

Like Lloyd George in 1916-1918, and Churchill in 1940-1943, Clemenceau represented the civilian administrator in the role which modern warfare

civilian element is illustrated in his rebuke to Foch at the meeting of the Supreme War Council at London on March 14, 1918, when he said: "Taisez-vous! C'est moi qui représente la France ici!" Bugnet, p. 273.

[64] Clemenceau, p. 83.
[65] Mordacq, *Le ministère Clemenceau*, II, 244-249.
[66] Clemenceau, p. 124.
[67] *Ibid.*, p. 132.
[68] *Ibid.*, p. 133.
[69] Mordacq, *Le ministère Clemenceau*, II, 363-367.

imposes on the heads of states. In the present era of complex and all-embracing war, they alone have the information, breadth of view, detachment, and power necessary for the successful conduct of war. If they succeed in establishing a sound basis for cooperation with the professional military leaders, their work is greatly facilitated. If they fail, then bitter friction, loss of efficiency, and even disaster may follow.

CHAPTER 13. Ludendorff: The German Concept of Total War

BY HANS SPEIER

ERICH LUDENDORFF'S contribution to the development of military thought is that of a general who lost a war. He began to write almost immediately after the defeat of the German armies in 1918. Although his books are born of a rich strategical and organizational experience, they are full of conceit and resentment, and are apologetic in character. They attempt to prove that, in a military sense, Germany did not lose the first World War. Considering the vital importance of this opinion, known under the slogan of "the stab in the back" in German domestic politics in the time of the Weimar republic, one is justified in regarding Ludendorff's literary activity as political pamphleteering. Certainly, his writings are not distinguished by detachment or subtlety. It was Ludendorff's fame as a great general rather than the intrinsic merit of his works which accounted for his amazing literary success in republican Germany.

Ludendorff wrote on three main subjects, of which only one is of immediate interest to this study. He specialized in reminiscing, debunking, and forecasting. In reminiscing about the military events of the first World War, he tried to enhance his historical stature as a general and he polemicized against those who belittled his generalship. Amplifying his prejudices against Freemasons, Jews, Jesuits, and Christianity at large, and advocating under the influence of his wife a martial religion of his own, he tried to expose those sinister forces which he held responsible for Germany's defeat in the first World War. He was convinced that both his political enemies and his National Socialist competitors after the war were also under the influence of those forces. Finally, in putting forth his ideas on total war, Ludendorff outlined the conditions which in his opinion would have enabled him to operate more effectively as a general in the first World War.

Ludendorff's theory of total war is not based on a study of military developments between the two world wars. Nor is it derived from a careful consideration of the interrelations between politics, warfare, technology, economics, and popular morale. In fact, there are few military writers to whose historical works Friedrich Schlegel's statement that "history is retrospective prophecy" can be applied with more justice. And in appraising Ludendorff's writings on total war one is sometimes tempted to modify Schlegel's aphorism by saying that the general's prophecies were history projected into the future.

For the historian of military thought Ludendorff's criticism of Clausewitz' ideas on war is particularly arresting. The carelessness of this criticism makes it easy to note Ludendorff's intellectual inferiority to the master of German military thought, but the point of interest is the political motive of the criticism rather than its content. Not as a military scientist and historian, but as a politician, did Ludendorff, the advocate of total war, renounce Clausewitz, the theoretician of "absolute war." After stating his unqualified demand of complete authority of the supreme military leader in all *political* matters as well, Ludendorff adds, "I can hear how politicians will get excited about such an opinion, as they will about the idea in general that politics is to serve the conduct of war, as though Clausewitz had not taught that war is but the continuation of politics with different means. Let the politicians get excited and let them regard my opinions as those of a hopeless 'militarist.' This does not change any of the demands of reality, which require precisely what I demand for the conduct of war and thus for the preservation of the life of the people."[1] In such phrases, which abound in his work, Ludendorff discarded the principles of Prussian statecraft and militarism as they existed in the latter half of the nineteenth century, and suggested a return to the days of his hero, Frederick the Great.[2]

Like Hitler, Ludendorff was not only opposed to the German republic, but also to the political structure of the Reich under Wilhelm II. The resemblance between the two men, however, does not go very far. Ludendorff was a reactionary who differed from other reactionaries at the time of the republic only in two respects. He was violently anti-Christian, and he was not concerned with years or decades but with more than a century in his desire to turn back the wheel of history. At the same time, he had a keen understanding of the advantages of centralized power in the direction and administration of modern politics, but measured these advantages only with the yardstick of bureaucratic efficiency. His understanding of the masses in modern society was limited and so was his experience as a demagogue. By contrast, Hitler is the political leader of a modern, subversive, mass movement. He comes from nowhere, that is, from one of the many interstitial groups which modern industrial society produces. He conquered power as a political upstart and transformed the society which allowed him to ascend. His rise to power and his conduct of the second World War have been predicated in part upon his ability to sway the German masses as a demagogue. Hitler's regime is a political dictatorship over all social institutions and groups, including the military ones. Ludendorff, on the other hand, drew up the blueprint of a military dictatorship to eliminate politics for purposes of total war; mass movements would have been crushed by this dictatorship. In National Socialist Germany, the corps of officers is altogether relegated to positions where they exert violence in an expert fashion but without political momentum of their own. Certainly Hitler is not the general who Ludendorff thought would control politics in the war of the future, but rather the plebiscitarian mass leader who, after a prolonged struggle, has succeeded in forcing generals to obey his orders.

Thus, while Ludendorff was a reactionary, military bureaucrat, who advo-

[1] General Ludendorff, *Der totale Krieg* (Munich, 1935), p. 115, note.
[2] Ludendorff was of the opinion that Frederick the Great was "on his side," even in his attitude toward religion. Ludendorff put out a pamphlet in his own publishing house entitled *Friedrich der Grosse auf Seiten Ludendorffs. Friedrichs des Grossen Gedanken über Religion. Aus seinen Werken.* (Munich, 1935.)

cated what may be called a technical dictatorship for purposes of the conduct of mass warfare, Hitler is a political dictator thriving on, rather than disregarding, the social tensions of modern mass society.

I

To understand Ludendorff's ideas on total war one must keep in mind the nature of German militarism before the outbreak of the first World War. It was a widely accepted class militarism on a half feudal basis within the bounds of an otherwise capitalistically stratified society. The conflict between the military traditions of the monarchy and the aspirations of the industrial middle classes was not resolved but institutionalized. Military power and social prestige were so distributed as to favor the aristocracy and the owners of large and medium sized landed estates while economic power was concentrated in the hands of the politically inexperienced leaders of industry, trade, and finance.

The tradition of German militarism was rooted in the social structure of eighteenth century Prussia in which the armed forces had been exempt from industry and the productive classes, in turn, exempt from military service. In the second half of the nineteenth century with its widespread literacy and industrialization such conditions no longer prevailed. The monarchy and the preindustrial political institutions were modified by a liberal constitution.

In 1934 Carl Schmitt put it thus: "The liberal movement of the year 1848 had forced the Prussian state to come to terms with a 'constitution' and to expose itself to the danger of losing no less than its character, as its government became parliamentarian and its army a parliamentary army."[3] According to Schmitt, this represented "a victory of the citizen over the soldier." In point of fact, this was hardly true, since the political leaders did not succeed in assuming political leadership, in molding the state according to their interests, and in determining foreign and domestic policies.

Germany had no class that produced political leaders with foresight and with experience in international affairs. Germany's most prominent statesmen succeeded in corrupting the bourgeoisie politically by forcing it to endorse their military successes after they had been accomplished and by exploiting its fear of the working class movement.

Wilhelm II, while admitting representatives of the economic elite to his court, tried to maintain political independence and military control. Until 1914, the military cabinet, through its influence on the selection of the highest military personnel, and the general staff, through its jealously guarded competence in planning future campaigns, were actually in control of military policy. The naval armament, on the other hand, was carried out with the help of tremendous propaganda campaigns conducted by the German Navy League. The younger branch of Germany's armed forces, under the influence of von Tirpitz with his keen sense for patriotic publicity, was socially more aggressive than the conservative army. The Navy League, founded in 1898, constituted the first

[3] Carl Schmitt, *Staatsgefüge und Zusammenbruch des zweiten Reiches* (Hamburg, 1934), p. 9.

social organization in Germany whose propaganda activities regarding armaments can be compared with those of the National Socialist Party after 1933. The League was sponsored by heavy industry.[4]

A fair indication of the political power wielded by the various agencies among which military authority was distributed is the access each of them had to the monarch. Wilhelm II preferred to listen to advisers who were not responsible to the Reichstag, i.e. to the chiefs of the civil, military, and naval cabinets, and to the chiefs of the general staff and of the admiralty. The heads of the various ministries who were responsible to the Reichstag did not report regularly to the kaiser. They could get the ear of the monarch only through the civil cabinet. Even the chancellor of the Reich did not see the kaiser regularly. An exception to the emperor's preference for irresponsible advisers rather than constitutionally responsible ministers existed only with regard to the minister of war. The war minister's influence, however, was contested by the military cabinet and the general staff. During the reign of Wilhelm II the power of the war department was more and more restricted, paradoxically enough with the help of the secretaries of war. While the opposition in the Reichstag tried to strengthen the position of the war department and to curb that of the military cabinet, the secretaries of war refused to accept this help and regarded themselves as generals who took their orders from the emperor.[5]

The time schedule of Wilhelm II clearly showed the predominance of the military over the civil authorities, and among the military authorities the predominance of the parliamentarily irresponsible agencies. The chief of the military cabinet saw the kaiser three times a week while the secretary of war reported only once. In addition, the representative of the military cabinet was present when the kaiser received the secretary of war, while the latter had no right to attend the audience for the military cabinet. When the kaiser was not in Berlin the heads of the cabinets took over all the reports.

The power of the military cabinet, a small body of men growing from four persons in 1872 to seventeen after 1900, rested not only on close contact with

---

[4] In 1910, the League had 300,000 individual members and 740,000 additional members who had joined collectively through no less than 1700 other associations. See Konteradmiral A. D. Weber, "Der deutsche Flottenverein," in *Deutsche Revue*, XXXV (1910), 177. It had about 100 branches in foreign countries, and spent about £50,000 per year as compared with £3000 spent by the British Navy League, according to an anonymous article, "The German Navy League," in *National Review*, XLVI (1905), 639. *Die Flotte*, the official organ of the League, distributed 320,000 copies in 1905, at a time when the four principal German newspapers sold 152,000 copies together. On July 1, 1933 Hitler sent a telegram to the general meeting of the reconstituted League declaring "that since his youth the League had been known and familiar to him, that he had always read its literature with the greatest interest and that he welcomed and desired the continued work of the association in its well proved manner." See Admiral A. D. Bauer, "Der deutsche Flottenverein 1898 bis 1934," in *Marine Rundschau*, XL (1935), 64. The article in the *National Review* quoted above says of the League in 1905: "Its political power and influence in Germany is exceedingly great and probably greater than that wielded by any of the German political parties."

[5] In 1889 General von Verdy in an attempt to recommend himself as a desirable candidate for the position of secretary of war pointed out that according to his opinion the secretary of war should act "as a kind of suicide in relation to his department." R. Schmidt-Bückeburg, *Das Militär-Kabinett der preussischen Könige und deutschen Kaiser* (Berlin, 1933), pp. 174 ff.

the kaiser but also on its watertight control of the personnel policy in the army. The Reichstag had no influence whatever on promotions, resignations, and transfers in the corps of officers. By the same token, the inferior authority and restricted competence of the ministry of war was reflected in the composition of its personnel. With very few exceptions the secretaries of war were generals of no military distinction. The most energetic and most intelligent officers had the ambition to get into the general staff. Only second rate staff officers were transferred to the war department and the rest of the personnel came from the Military-Technical Academy (known in the snobbish circles of the officers corps as the "plumbers' academy"), from the Artillery-Inspection Commission, and from officers of the line.

The military elite was primarily interested in retaining its high social status and in defending its political independence against civilian control. Both their status and power were jeopardized by the industrialization of German society, which created an economically powerful counterelite and urban masses which would not fit in with the preindustrial pattern of politics. To illustrate, from 1870 to 1910 the production of coal had increased from 34,000,000 to 190,000,000 tons, that of iron ore from 5,300,000 to 2,900,000,000 tons. In 1870 Germany had 18,560 km. of railways and about 4000 telegraph offices as compared with 59,031 km. of railways and 45,000 telegraph offices in 1910. In 1871 almost two-thirds of the German population were rural, living in places with less than 2000 inhabitants; in 1910 about three-fifths of the population were urban, and one-fifth lived in large cities with more than 100,000 inhabitants. From 1882 to 1907 the number of independent persons among the gainfully employed had increased only slightly, from 4,995,000 to 5,332,000, while their position in relation to other classes had decreased sharply from 45 per cent to 20 per cent of all gainfully employed. The number of workers, on the other hand, had increased from 10,705,000 to 17,836,000 in the same period, a gain of 60 per cent.

In this context it should be noted that the military elite in the imperial era was not imperialistic. They expended their energy in the struggle to maintain their prestige and power in a process of economic change which endangered their privileged position. The military elite was not responsible for developing schemes of national aggrandizement and conquest. Social organizations like the German Navy League, the Pan-German League and the various societies interested in the acquisition and development of colonies were controlled and sponsored by the economic counterelite and by intellectuals coming from the middle classes.[6]

[6] For example, in 1904 no less than 128 of the 276 members of the officials of the Pan-German League were academicians, according to Lothar Werner, *Der Alldeutsche Verband 1890-1918* (Berlin, 1935), p. 64. The proclamation founding the German Colonial Society in 1882 was signed by representatives of the National Liberal Party, university professors, presidents of the chambers of commerce, and industrialists from the Rhineland and from southern Germany. Cf. *Die Deutsche Kolonialgesellschaft 1882-1907* (Berlin, 1908). The German Navy League was founded in 1898 by persons "most of whom belonged to the Central Association of German Industrialists." Cf. Eugen Richter, *Zur Flottenfrage* (Berlin, 1900), p. 31.

The peculiar class structure of German militarism in the imperial era stood in the way of wholly efficient preparedness and even of that measure of efficiency in the conduct of war which the progress of technology had rendered possible. These limitations manifested themselves chiefly in four sectors of military life: in the social composition of the army, in war economics, in the attitudes of the military toward technology and in the lack of comprehension of the importance of propaganda in the case of armed conflict.

German militarism extended its social basis by half measures in order not to isolate itself from the economically and ideologically powerful groups. Apart from the frequent occurrence of intermarriages between the aristocracy and the bourgeoisie and from a tight control of positions in the higher bureaucracy through the alumni of socially exclusive student fraternities, the prevailing system of education was utilized in the interest of the military elite. The shorter one-year service in the army was contingent upon a certain amount of high school education and thus upon economic status. Those privileged in the economic hierarchy were given moderate privileges in the army. This was particularly important for the social composition of the corps of reserve officers, an institution which forced upon the sons of the bourgeoisie the standards of the aristocratic military elite. In the years before the war there were about 15,000 officer candidates—the so-called *Einjährige*—in Germany every year, while the total number of reserve officers was only 29,000 in 1914.[7] The actual control of the selection of officers was exercised in the mess of the officers of the line, where the aspirants were inspected as to their social background and political opinions. No liberal was allowed to pass.

The increase in the ratio of commoners to aristocrats in the corps of officers which proceeded apace under Wilhelm II created a special problem. The fear of destroying the conservative character of the officers corps and the apprehension that the political stability of the Reich would be endangered by enlarging the army from the urban areas functioned as brakes to the full utilization of the available man power. The demands of efficiency were overruled by considerations of status and power in the state. For example, in 1904, von Einem, secretary of war, wrote to Count von Schlieffen, chief of the general staff, that the scarcity of officers could be remedied if requirements were lowered. He added, "This cannot be recommended because we could then not help accepting in a large measure democratic and other elements unfit for this profession."[8] Similar opinions were expressed with regard to both commissioned and non-commissioned officers by the secretary of war, von Heeringen, in a memorandum to the chief of the general staff, von Moltke, in January 1913.

The limitations of the prewar German military caste is also apparent in the fact that there was little understanding of the economic aspects of war in Germany before 1914. Most military and financial circles held the opinion that

[7] Herbert Rosinski, *The German Army* (New York, 1940), p. 112.
[8] Reichsarchiv, *Der Weltkrieg 1914 bis 1918. Kriegsrüstung und Kriegswirtschaft*, I (Berlin, 1930), 91.

a future war would not and could not last long.[9] The inquiries into the food supply of the army in case of war, which were conducted by the ministry of war in 1884, 1906, and 1911, were based on the assumption that a war would not last longer than nine months. This notion was based partly on the experiences of the preceding wars in the nineteenth century which had lasted only weeks or months and partly on considerations of the destructiveness of modern armament. Modern industrialized nations were not expected to be able to conduct a long war.

It is noteworthy that the military elite had a marked professional interest in a short war. The general staff was afraid that preparation for a *long* war would lead to a curtailment of the military monopoly in the control of war, or would at least increase the importance of the economic and social factors in war. Economic mobilization for a long war would have required a new budgetary and a new social policy : the money to be invested in an adequate storage of food and raw materials would not have been available for military purposes in the narrower sense of the term, and the assignment of armies of skilled workers to war industries might have meant the reduction of military man power. Thus, it appeared that the preparation of a long war would reduce the military strength with which the general staff hoped quickly to force a military decision. The plans of the general staff were focused on the necessity and feasibility of shock strategy. Schlieffen rejected the idea of a strategy of attrition because it would endanger the existence of the nation by the disruption of all commercial and industrial activities. Similar opinions were expressed by General von Blume, a pupil of von Moltke, in his book on strategy in 1912.

The technological limitations of German prewar militarism are perhaps best reflected in the early development of air power. The following figures speak for themselves. From 1909 to 1912 France spent 30,610,000 francs for military aircraft, while the German expenditures amounted to only 6,486,000 marks. In 1912 the French army had 390 airplanes and 234 pilots, the German army 100 planes and 90 pilots.[10] Thus the German-French ratio for expenditure for aircraft in the period 1909-1912 was approximately 1 :4; with regard to the number of planes, the ratio in 1912 was 1 :4; and, in number of pilots, the ratio was 1 :2.5. Shortly before the outbreak of the war, the German general staff became convinced that airplanes were superior to lighter-than-air craft. The ministry of war, however, and especially the treasury, were opposed to spending more money for airplanes and pointed to the difficulties of proper training.[11] In addition, the lighter-than-air craft command was of the opinion that airplanes could never be usable in war.[12]

---

[9] There were only few exceptions: von Moltke in 1890 and Max Warburg in 1907 expressed the opinion that a future war would last many years.

[10] *Kriegsrüstung, op. cit.*, I (Documents), p. 441, note 1. W. O'D. Pierce, *Air-War* (New York, 1939), p. 92, gives the following figures for certified pilots in 1911 : France, 353; England, 57; Germany, 46; Italy, 32; Belgium, 27; U.S.A., 26; Austria, 19.

[11] In France, these difficulties were overcome by making army orders of planes conditional upon the training by the airplane factory of one pilot per plane.

[12] A. Hildebrandt, "Die Luftfahrertruppe," *Handbuch der Politik*, III (2nd ed.; Berlin, 1914), p. 305.

There was a similar reluctance to strengthen and modernize the technical branches of the army, particularly in the fields of communication and engineering.[13] In 1900 Lieutenant General von der Goltz made a number of important suggestions regarding the sappers in the German army. Their number, he held, should be increased to three companies per army corps; the technical training of the officers should be intensified; and a closer tactical connection between infantry and technical troops should be effected by exchanging officers and by the formation of a special engineering staff composed of technically trained officers from all arms. The ministry of war and the general staff declined to accept these recommendations. Financial difficulties served as a most convenient excuse for lack of foresight. The general staff was apprehensive of having its competence restricted by the formation of such a technical staff. The Russo-Japanese war taught a number of lessons on the importance of technical troops in modern warfare which could not be lightheartedly disregarded. But again the general staff declared simply that an increase of sappers was superfluous.

Limited by its social composition, by its ignorance of the economic aspects of modern war and by its prejudice against technological innovation, the military caste was also insufficiently aware of the importance of the propaganda factor. At the outbreak of the war, the military authorities were completely unprepared in this field. Despite their fear that war might cause unemployment and social unrest they did not conceive the possibility of upholding morale at home by a concerted effort of the government. Since they had little contact with, or understanding of, modern business methods, they could not learn anything from commercial advertising. Besides, propaganda, like technology, was stained with the spirit of middle class civilization toward which the military elite had a condescending attitude. The sword, not the pen, was to decide future wars.

Shortly after the outbreak of the first World War the limitations of German class militarism became all too apparent. Theoretically the kaiser was the German commander in chief, with the chief of the general staff as his strategic adviser and the chancellor as his political adviser. In practice, there existed in Germany during the war a military dictatorship from the moment in which von Falkenhayn was replaced by Hindenburg and Ludendorff. In the conflict between military and political authorities Ludendorff was put into a position where he became the advocate of the supremacy of generals over statesmen. Wherever he found constitutional limitations to his power, he disregarded them. That he did feel such limitations, however, is shown by the fact that all of his later reflections on the relation of military and political authorities were inspired by these wartime experiences.

The influence of the military cabinet waned under the impact of war. The

---

[13] "Frequently before the war very great artillery ranges were rejected on the ground that no observation was possible over long distances. One forgot that the efficiency of telescopes was constantly increasing and that the telephone permitted an extension of observation." Max Ludwig, "Heerestechnik," *Die deutsche Wehrmacht 1914-1939* (Berlin, n.d.), p. 87.

restricting influence which the Reichstag through its right of budgetary control might have exerted was effectively counterbalanced by threats of resignation on the part of Hindenburg and Ludendorff. In all decisive conflicts between the politicians and the generals, Ludendorff determined the outcome; not so much because he wanted power for himself as because, when efficiency was the demand of the hour, there was nobody else who wanted it.

Ludendorff did not belong to the socially leading aristocracy. He came from an obscure family and had advanced to a position which was usually reserved for men of noble blood. Considering his background and the newly created position of *Erster Generalquartiermeister*, which he held from August 1916 on, he was socially an upstart. Perhaps for that very reason he felt the political limitations—however ineffective—to his authority and the lack of central responsibility for the far-flung war effort more keenly than an aristocrat with more firmly rooted traditions might have done. His will power was greater than that of the chancellor or the kaiser, and Ludendorff managed to balance his incompetence as a statesman and economist by his will power.

Conflicts between statesmen and generals arise easily in war. Even mediocre statesmen may be able to hold their own against generals if the political institutions and their entrenched traditions operate against the inevitable wartime trend toward redistribution of power in favor of the military elite. Strong personalities are capable of asserting themselves even against heavy institutional odds. Bismarck succeeded in doing so in 1866 and 1871. In a passage of his memoirs which, incidentally, may be taken as the verdict on Ludendorff's political ideas, he declared that "to set up and to delimit the aims to be reached by means of war, and to advise the sovereign upon this matter, is a political function while the war is in progress no less than in times of peace."

The only political statesman of note in Germany during the first World War was Bethmann-Hollweg, and he had neither the personality nor the will to assert himself against the military. Curiously enough, he did not hesitate to check the independence of the admirals. In the most important naval decision of the first World War Bethmann-Hollweg shifted responsibility to general headquarters; and Ludendorff's judgment, which was influenced by his desire for a peace through military victory, was decisive in the introduction of unrestricted submarine warfare.[14] On all other questions, Bethmann surrendered statesmanship to the generals and made no effective attempt to curb their political power.

As a result, Ludendorff emerged as the military dictator of Germany. His role as such is too well known to need more than a bare listing of the events which he determined or influenced by wielding his unconstitutional power over kaiser, chancellor and Reichstag: the dismissal of Bethmann-Hollweg and the appointment of Michaelis as Reich chancellor; the dismissal of von Kuehl-

---

[14] For Ludendorff's role as a military dictator during the war cf. Arthur Rosenberg, *Die Entstehung der deutschen Republik* (Berlin, 1928), especially Chapters IV-VI, and the literature cited there; Karl Tschuppik, *Ludendorff: the Tragedy of a Military Mind*, transl. by W. H. Johnston (Boston, New York, 1932); K. von Oertzen, "Politik und Wehrmacht," *Wissen und Wehr*, V (1928).

mann as state secretary of the foreign office; the decisive influence upon the peace treaties with Russia and Rumania, and the dismissal of von Valentini as chief of the civil cabinet. In addition, Ludendorff exerted a decisive influence in every phase of German economic and social policy during the war.

II

Ludendorff's idea of total war can be stated in the form of five basic propositions. War is total; first, because the theater of war extends over the whole territory of the belligerent nations. In addition to this diffusion of risks, total war also involves the active participation of the whole population in the war effort. Not armies but nations wage total war. Thus, the effective persecution of total war necessitates the adaptation of the economic system to the purposes of war. Thirdly, the participation of large masses in war makes it imperative to devote special efforts, by means of propaganda, to the strengthening of morale at home and to the weakening of the political cohesion of the enemy nation. Fourth, the preparation of total war must begin before the outbreak of overt fighting. Military, economic, and psychological warfare influence the so-called peacetime pursuits in modern societies. Finally, in order to achieve an integrated and efficient war effort, total war must be directed by one supreme authority, that of the commander in chief.

Ludendorff's relatively simple idea of total war is not devoid of a few interesting details. The geographical extension of the theaters of total war is a consequence of the technical progress of the means of destruction and of the increasing functional interdependence of modern nations. Not only are the fighting zones widened by the technical improvement of long range weapons of all kinds, but the regions behind the actual fighting zones are also affected "by hunger blockade and propaganda." The nation at war can thus be compared to the people in a besieged fortress. As the besiegers try to force a fortress to surrender not only by directing strictly military means against its military defenders, but also by starving its civil inhabitants, so total warfare implements the military assault upon the armed forces of a nation by the use of nonmilitary weapons directed against the noncombatant parts of the enemy population. The distinction between combatants and noncombatants loses its former significance.

In order to secure the necessary military supplies and foodstuffs for the sustenance of the besieged nation, Ludendorff advocated economic self-sufficiency. His ideas on war economics, however, are little else than textbook generalities. He dealt with the organizational aspect of war economics rather than with the strategic possibilities of improving the raw material, food, and labor supply of the nation at war through conquest.[15]

Surprisingly enough, the most original contribution General Ludendorff made to the theory of total war does not lie in the field of military warfare, but in the realm of what is often inadequately called "psychological warfare."

[15] These aspects of war economics were discussed in Germany by the adherents of geopolitics. See Chapter 16.

Ludendorff is almost excessively concerned with the problem of the "cohesion" of the people. It is in this regard that he differs most strikingly from National Socialist writers on total war. He despised, and regarded as ineffective, any attempt to achieve social unity by force and drill. Such methods he called "mechanical" or "external." "An external unity of the people, achieved by compulsion—a unity in which the soul of the people has no share by common and conscious racial and religious experience—is not a unity which people and army need in war, but a mechanical phantom dangerous to the government and the state."[16] Similarly, he spoke with unconcealed derision of such measures as the Fascists and National Socialists had taken in the field of the premilitary training of youth. He compared this training with that of dogs and doubted that mass drill, which "deprives youth of personalities," prepared young men satisfactorily for military service.[17]

His model of a closely knit social unity is therefore not the old Prussia, nor is it the new Germany of Hitler. Ludendorff thought of Japan when he spoke of unity and cohesion. "Entirely different [from mechanical unity] is the unity of the Japanese people; it is a spiritual one, essentially resting upon Shinto religion, which compels the Japanese to serve their Emperor in order thus to preserve the road to the life of their ancestors. For the Japanese, service for the Emperor and thus for the state is prescribed by his experience of God. Shintoism, stemming from the racial heritage of the Japanese, corresponds to the needs of the people and the state. . . . In the unity of racial heritage and faith and in the philosophy of life erected upon them resides the strength of the Japanese people."[18]

Ludendorff's own racial religion was to provide the Germans with a faith corresponding to Shintoism in Japan. However fantastic this may sound, Ludendorff must be credited with understanding the fact that something more profound than merely a clever propaganda policy is needed in order to produce a state of popular morale which enables people in modern industrial society to endure the hardships of total war. Like Ernst Juenger, Ludendorff realized dimly that "a mobilization may organize the technical abilities of a man without penetrating to the core of his faith,"[19] and that the spirit of sacrifice cannot be injected into the body politic by a clever doctor. Ludendorff realized that the source of concord in society lies in deep rooted traditions rather than in an efficient organization of the police. In fact, he did not advocate violence against dissenters in his book on *Total War*, and when he spoke in his memoirs of the fact that the government may use force against those who jeopardize the common war effort he did so in an almost apologetic fashion.

On the whole, Ludendorff's ideas on the role of propaganda were sounder than Hitler's. In the realm of propaganda techniques as well, Ludendorff's opinions revealed surprising expertness. He deplored the German govern-

---

[16] *Der totale Krieg*, p. 17.   [17] *Ibid.*, p. 58.
[18] *Ibid.*, p. 17.
[19] In this way Ernst Juenger characterized Walter Rathenau's great contribution to the organization of German war economy in the first World War. Cf. his *Die totale Mobilmachung* (Berlin, 1931), p. 16.

ment's concealment of the defeat in the battle of the Marne from the German people and advocated a policy of frankness in order not to give "free reign to the 'discontented' and the rumor-mongers."[20] Similarly, Ludendorff wrote in his memoirs that every German, whether man or woman, should have been told every day what a lost war meant for the fatherland. Pictures and films should have broadcast the same story. The presentation of the dangers would have had a different effect than the thinking about profits, or talking and writing about a peace by negotiation.[21] Goebbels, in the propaganda strategy of gloom upon which he embarked a few months after the invasion of Russia, seems to have taken a few leaves from Ludendorff's book. Whenever possible, Goebbels also follows that advice of Ludendorff which several National Socialist authors on propaganda have repeated: "A good propaganda must anticipate the development of the real events."[22] Finally, Ludendorff regarded the circulation of rumors, in which the National Socialists were to become past masters, "the best means of propaganda" against the enemy.[23]

Ludendorff's theory of total war culminates in the role assigned to the supreme military commander. In addition to conducting the military operations he is to direct the foreign and economic policies of the nation and also its propaganda policy. "The military staff must be adequately composed: it must contain the best brains in the fields of land, air, and sea warfare, propaganda, war technology, economics, politics and also those who know the people's life. They have to inform the Chief-of-Staff, and if required, the Commander-in-Chief, about their respective fields. They have no policy-making function."[24] Thus, in Ludendorff's total war there is no place for the civilian statesman. The general rules supreme. And Ludendorff concludes: "All theories [sic] of Clausewitz have to be thrown overboard. War and politics serve the survival of the people, but war is the highest expression of the racial will to life."[25]

Clausewitz had been of the opinion that the French Revolution had removed many of the limitations which in the *Ancien Régime*, when cabinets rather than nations waged war, had prevented war from assuming its "abstract" or "absolute" character. Clausewitz had rejected the erroneous idea that war had "emancipated" itself from politics in consequence of the revolution. Instead, he insisted—in a passage which Ludendorff later quoted—that the political forces of the French Revolution had unleashed energies which subsequently changed the type of war. By thus attributing the change in the type of war to politics, Clausewitz defined the prevailing type of war in terms of the structure of the political community that wages it.

According to Ludendorff, on the other hand, total war is a product of demographic and technological developments. The increased size of populations and the improved efficiency of the means of destruction have inevitably created the totality of war. Total war, which has no political cause, absorbs politics.

There is not the slightest suggestion in Ludendorff's writings that he prefers

[20] *Der totale Krieg*, p. 26.   [21] *Meine Kriegserinnerungen* (Berlin, 1919), p. 296.
[22] *Ibid.*, p. 300.   [23] *Ibid.*, p. 302.
[24] *Der totale Krieg*, p. 111 f.   [25] *Ibid.*, p. 10.

total to limited war on moral or metaphysical grounds. Nor is there any explicit justification of total war in terms of an imperialistic doctrine,[26] or of a value system in which pugnacity, heroism, and the love of sacrifice are so supreme as to demand war for their realization and glorification. Instead, Ludendorff goes so far as to contend that total war is essentially defensive. The people will not cooperate in waging it unless they know that war is waged to preserve their existence. To be sure, it would be expedient to make such contentions for the mere sake of appearance in a culture which, in E. M. Forster's words, "preaches idealism and practices brutality." Yet Ludendorff did not shrink from shocking the public by unorthodox opinions in the field of religion, and it would do injustice to his character to doubt that he meant what he said when he talked about the defensive character of total war. He insists that "the nature of total war requires that it be waged only if the whole people is really threatened in its existence, and determined to wage it."[27] If this insistence upon the defensive nature of war were a mere attempt on the part of Ludendorff to conceal his true opinion as to its nature, he would have to be credited with a Machiavellian attitude toward the masses. There is no trace of such an attitude in his writings. In fact he rejects explicitly the opinion, characteristically to be found in the circles of National Socialist intellectuals, that the masses can and should be psychologically manipulated in the interest of the power holders. As has been pointed out, Ludendorff regarded attempts to manage the masses in this way as futile.

III

The National Socialists have not only organized German society for total war but have also written profusely about it. Their contributions to the development of the theory of total war are built upon Clausewitz' and Ludendorff's basic contention that in modern war all material and moral resources of the nation must be mobilized. The main difference between National Socialist literature and Ludendorff's writings on total war lies in the fact that the National Socialists have attempted to produce ideological justifications of modern war. Racial superiority and the law of nature, Darwinistically and geopolitically understood, are supposed to provide German war and the new militarism with a moral halo. Moreover, some National Socialist writers have pushed Ludendorff's theory to its logical end by denying the *existence* of peace altogether. They no longer regard war as a *phase* in the interrelations of states, preceded and followed by phases of peace, but as "the expression of a new political and social development in the life of peoples."[28] Similarly, geopoliticians have written books about the forms which warfare assumes in times which according to common usage are designated as times of peace. Instead of

---

[26] It is useless to look in Ludendorff's writings for statements like this: "The genuine International is that of imperialism, of rule over the Faustlike civilization, hence over the whole earth, to be exerted by one singly forming principle, and not through compromise and concessions, but through conquest and destruction." Oswald Spengler, *Preussenthum und Sozialismus* (Berlin, 1919), p. 84.

[27] *Der totale Krieg*, p. 6.   [28] Guido Fischer, *Wehrwirtschaft* (Leipzig, 1936), p. 23.

speaking about peace between wars, they have found the formula of "the war between wars."[29]

One of the most important changes of warfare, without which blitzkrieg methods and the coordination of different arms would have been impossible, was predicated upon a complete removal of resistance to technological change on the part of the new corps of officers. Karl Justrow, who criticized sharply the scarcity of engineers in the German armies of the first World War and contended that engineers had "not the slightest influence" on the conduct of the war, wrote shortly before the outbreak of the second World War, that "technology—once a stepchild in all organizations—is today treated with more and more understanding in the conduct of war."[30]

This breakdown of resistance to technology has inevitably involved a greater equalization of society. The higher the demands for technological skill and physical fitness on the part of the expert operators of the means of destruction, the less important must be the respect paid to the status qualifications of the soldiers. Status consideration must be sacrificed in favor of high skill requirements, especially when there is a shortage of available personnel. Thus, the technology of total war favors the "egalitarian" militarism of Hitler's Germany. When Hitler reintroduced universal military service, he abolished the privilege of shorter service in the army which boys with a high school education had enjoyed under the kaiser. In this respect and also with respect to promotion from the ranks the military system of this modern despot is more egalitarian than was that of Imperial Germany. Hitler has proceeded in the spirit of what Oswald Spengler has called "Prussian Socialism,"[31] adapting this kind of socialism to the postdemocratic structure of his plebiscitarian dictatorship. With this resolution to liberate destructive techniques from humanitarian fetters and to sacrifice any tradition to military efficiency he has brought to a head the martial equalization of society that began in eighteenth century Prussia but required the rise of modern political mass movements to become truly effective.

When, in 1733, the so-called canton system of recruitment was introduced in Prussia, the enlargement of the social basis of the army was officially justified by the argument that the inequality among those enrolled should be abolished. Later the introduction of universal military service in Prussia was announced by a cabinet order which stated that "all privileges based on social status cease to exist with the army." This was in 1808, when the liberal Barthold Georg Niebuhr wrote that universal conscription, "this equality which disgusts the true friend of liberty," will lead to "... the demoralization and degeneration of the whole nation, universal brutality, the destruction of civilization and of the educated classes."

The two developments which have most incisively changed the social structure of German economy in this war were not discussed at great length before

[29] Cf. Rupert von Schumacher and Hans Hummel, *Vom Kriege zwischen den Kriegen* (Stuttgart, 1937).
[30] Karl Justrow, *Der technische Krieg*, II (Berlin, 1939), 13.
[31] Oswald Spengler, *Preussenthum und Sozialismus* (Berlin, 1919).

war broke out.[32] The ruthless principle of turning the conquered territories into a reservoir of labor for the conqueror, from which, according to German claims, no less than twelve million foreign workers had been forced to migrate to Germany by 1943, was not fully anticipated in the *Wehrwirtschaft* discussions before 1939. Nor did any German economist dare to reveal the possibility of a virtual destruction of the German middle classes in the war of the future.[33]

The contribution of German intellectuals to the literature of "psychological warfare" has probably been grossly overrated. The contribution of German psychologists to the theories of political mass behavior is smaller than the many volumes written on that subject would have us believe. Nazi literature on political propaganda contains little that has not been known for centuries to less pedantic students of rhetoric and to modern specialists in advertising. The recent overestimation of the German contribution is to a large extent a consequence of the fact that talk about German "fifth columns" and propaganda gave intellectuals in the democratic countries a thrilling opportunity to offer an excuse for the military inferiority of the democracies at the beginning of this war. The prestige of propagand waxes and wanes with military success and failure. The Nazis as leaders of a modern mass movement have undoubtedly realized more clearly than did the leaders of Imperial Germany in the last war the importance of propaganda. They have invested more money and talent in it and have organized it efficiently. They have not produced any new "theory."

As to their propaganda practice, it is often overlooked that in some respects the National Socialists are in a less favorable position than was Germany under the kaiser. For example, any attempt to create discord in the camp of the enemy coalition must necessarily appear as a move of the German propaganda machine, because the world knows that all German propaganda is centrally planned. The German anti-Bolshevist campaign can at once be recognized as Dr. Goebbels' campaign. It is interesting to compare this predicament with the greater freedom of operation which Ludendorff had in the first World War. In June 1918, Colonel Haeften, the representative of the supreme command at the foreign office, presented the plan of an anti-Bolshevist campaign to Ludendorff for approval. In order to strengthen Lansdowne's peace party in England, influential Germans, acting to all appearances in complete independence from the government, were to advocate in public speeches a united European front against Bolshevism.[34] The less centralized setup of the German propaganda machine at that time permitted maneuvers in political warfare upon which National Socialist propaganda cannot embark without giving itself away.

---

[32] Henry William Spiegel in a discussion of *Wehrwirtschaft*, makes the interesting observation that "there is no academic economist of repute among the godfathers of the new discipline." See his "Wehrwirtschaft. Economics of the Military State" in *The American Economic Review*, XXX (1940), 715.

[33] On the middle classes in Germany's total war economy cf. *The Fate of Small Business in Nazi Germany* (Washington, 1943), prepared by A. R. L. Gurland, Otto Kirchheimer and Franz Neumann for the Special Senate Committee to Study Problems of American Small Business.

[34] Cf. Arthur Rosenberg, *op. cit.*, pp. 210 ff.

Ludendorff's ideas on the supreme position of the general in total war, however, were buried with him. German generals are under the domination of the National Socialist Party led by Hitler, a charismatic corporal. So complete was that domination that on January 30, 1943 Goering could dare to refer in public to the weak German military leaders who "were whining" before Hitler when he "held" the Eastern front against the onslaught of the Russians.

CHAPTER 14. Lenin, Trotsky, Stalin: Soviet Concepts of War

BY EDWARD MEAD EARLE

IN ANY social upheaval as violent as the November Revolution of 1917 in Russia, military organizations and concepts of strategy are swept away as ruthlessly as political institutions and social classes, for the army is intimately associated with the former structure of society, and national strategy is related to the political ambitions of the old order. As the French Revolution had suppressed the aristocratic army of the Bourbons, substituted the *levée en masse,* and adopted war as an instrumentality for the promotion of Liberty, Equality, Fraternity,[1] so the Bolsheviks scrapped the demoralized and derelict army of the czars and established an army of the peasantry and the proletariat —the Red Army, born February 23, 1918.

Originally this new army was to be a weapon of the "toiling and exploited masses," for the defense of the revolution against domestic and foreign enemies. But the passage of time and the exigencies of domestic and international politics compelled a reappraisal of its character and purposes, with the result that it has gradually been transformed into a truly national institution. It is still the child of the Revolution, but it is also the idol of the fatherland. After a quarter century of experience as the rulers of a great power, the leaders of the Soviet Union are now less concerned with the military dogmas and formulas of earlier days than with the bitter realities of war which govern modern armies, whatever their ideological background. There can be little doubt that the magnificent morale of the Red Army in the war with Nazi Germany is attributable in some measure to the Communist ideology and the "classless society" of the Soviet fatherland, as well as to a renascent Russian patriotism. But success in battle cannot be explained by morale alone, however heavily it may weigh in the scales of war. Victories are won also by materiel, by discipline and tactical excellence of the troops, by high standards of professional competence among field and staff officers, and by adherence to sound strategical principles. The wellsprings of Soviet resistance to the German invasion are to be sought deep in Russian soil and Russian historical tradition, as well as in the dynamics of the Bolshevik Revolution.

I

When Lenin seized power in Petrograd (now Leningrad) in 1917, he was far from being illiterate in military and strategical matters. A keen student of socialism—as well as revolutionary anarchism, syndicalism, and terrorism— he was unusually appreciative of the role in human affairs of violence and

NOTE. In the preparation of this chapter I have had generous editorial and research assistance from Dr. Felix Gilbert and Mrs. V. Tschebotareff Bill. I am happy to acknowledge my debt to each of them.—E.M.E.
[1] See Chapters 3 and 4.

armed force. Like Engels he had read, annotated, and pondered Clausewitz. Speaking of the latter's "famous dictum" that "war is politics continued by other [i.e. forcible] means," Lenin said: "The Marxists have always considered this axiom as the theoretical foundation for the meaning of every war."[2] He believed, furthermore, that there was an intimate connection between the structure of the state and the system of government, on the one hand, and military organization and the conduct of war, on the other.

From Marx and Engels, among others, Lenin acquired a perception of the realities of power politics. He was aware that warfare is not only military, but also diplomatic, psychological, and economic in character.[3] He believed that war and revolution were in continuous and fundamental relationship with one another—in particular, to paraphrase Marx, that war could be midwife to revolution. Just as the government of Nicholas II had been discredited by incompetence and defeat in the War with Japan, so, thought Lenin, the Russian empire would be brought down by defeat and revolution in the war of 1914 with Germany. The critical question for him was to find means of transforming the national, imperialist war into a civil war—not only a civil war within Russia but one which would cut across national lines and precipitate widespread social revolution.[4] In similar manner every war crisis in Europe since 1848 had been regarded expectantly by the followers of Marx, in the hope that it would usher in the rule of the proletariat.

The revolutionary movement in Russia before 1917 had been militant and activist, not—as in France, Britain, Germany, and the United States—pacifist and parliamentary. Lenin had little sympathy with peaceful methods as such, for he believed in the forceful seizure of power by workers and peasants and the establishment of a dictatorship of the proletariat. He was contemptuous of those Marxists who turned "patriotic" after 1914 and supported their respective governments in prosecution of the war. After having helped bring about the March Revolution of 1917 in Russia, he immediately set about undermining the liberal, bourgeois regime which it brought into being—the fundamental goal of the revolution, he said, was the overthrow of capitalism, as a preliminary step towards peace. "His was the policy and his was the strategy that made the Bolsheviks the one party in the chaos of Russian politics [from April to October, 1917] which both knew its own mind and could convey its purposes to [the] people."[5] Believing in force as opposed to parliamentary maneuver, Lenin came to power by that favorite device of dictators, the *coup d'état*.

Having won the first round of the civil war in Russia, Lenin was now faced

---

[2] V. I. Lenin, *Works* (English translation, New York, 1929), XVIII, 224. For Lenin's interest in military theory and his views on Clausewitz see S. Rabinowitch, "War Questions in the Lenin-Magazine," in *The Bolshevist*, 1930, pp. 70-79. The Soviet Government published Lenin's annotations to Clausewitz (Moscow, 1933).

[3] See Chapter 7 on Engels and Marx.

[4] See a letter which Lenin, an exile in Switzerland from 1914 to 1917, wrote a Russian friend a few months after the outbreak of the First World War, *op. cit.*, XVIII, 74.

[5] Harold J. Laski, "Vladimir Ilich Ulyanov (Nicolay Lenin)," in the *Encyclopedia of the Social Sciences* (1937).

with German invasion. There was never any question of resuming military operations, for the Russian army was thoroughly demoralized—partly, indeed, as a result of revolutionary propaganda which had been carried on within the ranks for a period of two years or more, and which could not be readily reversed. As Lord Balfour said somewhat later, "An army cannot be made by fine words, though they can easily destroy it. The Bolsheviks have with complete success endeavoured to shatter the fighting spirit of Russia, and they can hardly revive it in the same way."[6] No one knew this better than Lenin, for from first to last he was uncompromisingly opposed to continuing the war with Germany *by military means*, whatever the circumstances and whatever the provocation. On the other hand, Lenin was convinced that diplomatic and psychological warfare could be waged against both the central powers and the Allies. If so, he could attain the twin objectives of defending revolutionary Russia and of turning the international war into a European civil war. Lenin and his associates hoped and believed that, just as propaganda had subverted the discipline and the will to fight of the Russian army, so could morale in the armies of the "capitalist" and "imperialist" powers be undermined. The enemies of Russia were the governments of Europe and America; these governments could be defeated only by proletarian revolution within their own countries, not by any conceivable military force which the Soviets might muster. Russia would wage an offensive, but it would be an offensive of words and nerves. The ammunition to be used would be those tried and true methods of revolutionary propaganda of which the Bolsheviks were past masters.

The peace offensive was waged against Russia's allies as well as Russia's enemies. The Great War was described by the Bolsheviks as an imperialist war, a war in defense of the capitalist order. Little hope, they said, was to be placed in formal diplomatic negotiations; rather, appeals must be made over the heads of governments direct to their peoples. The secret treaties with the Allies were published as a proof of their imperialistic designs, and the slogan "No annexations, no indemnities" was offered as the basis of peace. Peace on such terms "will not have the good fortune to please the capitalists," wrote Lenin in mid-September, 1917, "but it will receive such a warm welcome from the people, will evoke such an explosion of enthusiasm in the whole world, such indignation against the interminable war of plunder waged by the bourgeoisie, that very probably we shall obtain at one slide both an armistice and the opportunity to broach peace negotiations." To this program Lenin adhered after assuming power. It was based on his conviction that "revolution will break out in all the belligerent countries."[7]

As against the Allies, Lenin made little progress. This was partly because in Woodrow Wilson he was meeting another master of psychological warfare. Wilson's speech to Congress of January 8, 1918, which first announced the Fourteen Points as a basis for peace, stole a good deal of the Soviet thunder—

[6] *Foreign Relations of the United States*, 1919, Russia, I (Washington, 1931), 393.
[7] *Collected Works*, XXI, 259, "The Aims of Revolution"; see also C. K. Cumming and W. W. Pettit, *Russian-American Relations* (New York, 1920), pp. 44-45, and John Reed, *Ten Days That Shook the World*, pp. 125-130.

as it was intended to do—and was much more comprehensible to both the Allies and the Germans. Lenin was an astute observer of Russian psychology, but he could not understand the western European and American mind. "Its lack of Slav mysticism, its failure to appreciate the grandiloquencies of Marxist phraseology, seem to have created a non-receptive mentality which failed to respond to Lenin's undoubted revolutionary genius."[8] Hence the thunder of Wilson's Fourteen Points had more practical effect than the lightning of Lenin's revolutionary slogans.

There was retributive justice in Lenin's diplomatic and psychological offensive against the central powers. The German government had allowed Lenin to cross Germany from Switzerland into Russia in 1917 in the hope that he would successfully sap the remaining foundations of Russian army morale. The German commander in Russia, one of the most uncompromising opponents of Bolshevism, said in this connection, "We naturally tried, by means of propaganda, to increase the disintegration that the Russian Revolution had introduced into the army. . . . In the same way as I send shells into the enemy trenches, or I discharge poison gas at him, I, as an enemy, have the right to use propaganda against him. . . . I personally knew nothing of the transport of Lenin through Germany. However, if I had been asked, I would scarcely have made any objections to it. . . ."[9] What revolutionary agitation had done to the Russian army Lenin now determined to do to the German army and, in addition, to the German government. And he had a degree of success which was not altogether understood until after the collapse of Germany a year later.

The Bolsheviks used the Brest Litovsk armistice and peace conferences as a public forum for the propagation of their ideas, thus converting diplomatic relations with the enemy into a virtual Trojan horse. Throughout the negotiations the foreign office under Trotsky, the press bureau under Karl Radek, and the Bureau of International Revolutionary Propaganda under Boris Reinstein turned the full blast of their power against the German army. German language newspapers, *Die Fackel* (The Torch) and *Der Völkerfriede* (The Peoples' Peace), were distributed to the soldiers of the central powers by the hundreds of thousands. Although themselves contemptuous of Wilson's Fourteen Points, the Bolsheviks nevertheless distributed among German troops over a million copies of the speech in translation. German prisoners of war in Russia were harangued and indoctrinated so effectively that upon their return to the fatherland they were confined for thirty days in "political quarantine camps" and mentally "deloused" with patriotic literature. Nevertheless, throughout the spring and summer of 1918 prisoners brought back to Germany "the infection bred of the revolutionary propaganda to which they had been subjected in

[8] John Wheeler-Bennett, *Brest-Litovsk, the Forgotten Peace* (London, 1938), p. 138. This book by Mr. Wheeler-Bennett is indispensable to an understanding of Lenin's psychological warfare against the rest of Europe and, more especially, against Germany. Wilson's speech of January 8 clearly was intended to be an answer to Bolshevik propaganda. *Ibid.*, pp. 144-145.
[9] Major General Max Hoffmann, *War Diaries and Other Papers* (English translation, 2 vols., London, 1929), II, 176-177.

Russian prison camps since the November Revolution. They had drunk deeply of the heavy wine of freedom and sedition, they had seen [the Russian] army melt away before their eyes, and now they returned to their depots speaking a new language of peace and bread—and bringing with them a spirit of general insubordination." General Hoffmann said that the German army in Russia was so "rotten with Bolshevism" that he did not dare transfer some of his divisions to the Western Front."[10]

Lenin and Trotsky knew that, in conventional diplomatic and political terms they were doomed to defeat at Brest Litovsk. They knew that, despite all the talk about no annexations and no indemnities, they would have to make territorial concessions to Germany. But they were playing for bigger stakes than frontiers. "He is no Socialist," Lenin wrote to American workers, "who does not understand that the victory over the bourgeoisie may require losses of territory and defeats. He is no Socialist who will not sacrifice his fatherland for the triumph of the social revolution." However, this triumph required time—time for the infant Bolshevist regime to stabilize its position in Russia, time for revolutionary propaganda to do its work on German soldiers and workers. Hence Trotsky—who became head of the German delegation at Brest Litovsk in January, after the conference had already been under way for a month—resorted to a strategy of procrastination and succeeded in dragging out the negotiations for six more precious weeks. Even the most experienced United States senator would pay tribute to his brilliant "filibuster." Caustic, arrogant, adroit, he maneuvered the German delegates into discussing all sorts of extraneous subjects. "Their debates travelled from Dan to Beersheba, and from China to Peru, embracing such apparent irrelevancies as the degree of dependence of the Nizam of Hyderabad upon the British Crown, and the scope and powers of the Supreme Court of the United States."[11]

Dialectics cannot indefinitely be a substitute for military force, however. Trotsky was having considerable success, but it was of necessity a limited success. Faced with a decisive offensive on the Western Front, the German general staff needed security in the east—a security to be attained either by peace or by destruction of the feeble remnants of Russian military strength. Furthermore, General Hoffmann was determined to put a stop to the demoralization of his troops by Bolshevik propaganda. By the first of February Ludendorff, Hindenburg, and Hoffmann made it perfectly clear to the German civilian authorities both at Berlin and Brest Litovsk that there could be no further temporizing with Russia. Trotsky must be presented with a virtual ultimatum. If he did not accept German terms, military operations must be resumed. If even then the existing government in Petrograd refused a dictated peace, the German army would again invade Russia, drive the Bolsheviks out, and compel another Russian government to sign a treaty.

Meanwhile, Lenin was beginning to reexamine his strategy. The *Blitzrevolution* in Germany and Austria-Hungary, upon which all Bolshevik policy had

[10] Wheeler-Bennett, *op. cit.*, pp. 90-95, 147, 351-352.
[11] *Ibid.*, pp. 116, 158.

been predicated, had failed to come off. Lenin was reluctantly forced to the conclusion that the war party in Germany had the upper hand and that revolution in central Europe, if it came at all, would come too late to prevent further German invasion of Russian soil. With the Russian army in no condition to resist effectively, the whole success of the revolution would be imperiled by German arms. "It would be very bad policy to risk the fate of the Socialist Revolution on the chance that a revolution might break out in Germany by a certain date. Such a policy would be adventurous. We have no right to take chances," Lenin told Bolshevik leaders on January 20, 1918. The Soviets needed time to effect economic recovery and to build an army. "The formation of a Socialist army, with the Red Guard as its nucleus, has only just begun. To attempt now, with the present democratization of the army, to force a war against the wishes of a majority of the soldiers would be hazardous. It will take months and months to create an army imbued with Socialist principles." Russia clearly was not in a position to wage a revolutionary war. In order to gain security for the revolution in Russia, a treaty with Germany, however humiliating, must be signed. In short, Russia must execute a strategic retreat. To those of his associates who urged continuance of the war rather than acceptance of German "robber" terms, Lenin replied contemptuously that he would not be interested in a "patriotic" peace even if it could be assured.[12]

Lenin was a convinced believer in *Realpolitik*. Peace to him was not an end in itself; on the contrary, peace, like war, was an instrument of policy. Now Russia needed peace to consolidate the revolution, and the revolution in Russia could not be jeopardized on vague hopes of achieving revolution elsewhere. With the greatest reluctance, therefore, he consented to try Trotsky's plan of "No War, No Peace" vis-à-vis Germany. To the very end he considered it a dangerous and futile "international political demonstration." And he was right. For as soon as they recovered from what General Hoffmann apoplectically called the "unheard of" procedure of refusing either to fight or to make peace, the Germans marched so far into Russia that they menaced both Russian independence and the Russian revolution. Hence the Bolsheviks were forced, on March 3, 1918, to sign the treaty of Brest Litovsk, by which Russia lost 34 per cent of her population, 32 per cent of her agricultural land, 54 per cent of her industry, and 89 per cent of her coal mines.[13] This outrageous settlement, be it noted, was set aside not by Soviet arms but by the Allies under the Armistice of November 11, 1918, and the ensuing treaty of Versailles.[14]

Trotsky's *coup de théâtre* at Brest Litovsk had all the dire consequences which Lenin feared. But it taught the Bolsheviks a lesson, for in their subsequent strategy they piloted a course "between war and peace, without losing

---

[12] "Lenin's Twenty-one Theses for Peace," in *Pravda*, February 24, 1918. English translation in James Bunyan and H. H. Fisher, *The Bolshevik Revolution, 1917-1918: Documents and Materials* (Stanford, California, 1934), pp. 500-505. The term *Blitzrevolution* has been taken from T. A. Taracouzio, *War and Peace in Soviet Diplomacy* (Cambridge, Mass., 1940), Chap. III.

[13] *Foreign Relations of the United States, op. cit.*, p. 490.

[14] The Soviets annulled the treaty on November 13, 1918, thus ratifying a *fait accompli*.

sight of the dangers and benefits potentially inherent in both."[15] But Lenin had faced realities even earlier. Two weeks before the signature of the treaty with Germany, Lenin had warned his colleagues, "We cannot joke with war." If Russia really had meant to wage war, she should not have demobilized. Five days later (February 23) he said, "It is time to put an end to revolutionary phrases and get down to real work. If this is not done I resign from the Government. To carry on a revolutionary war, an army, which we do not now have, is necessary."[16] The Red Army was born as a result of these words. Soviet Russia has never given up the waging of war by all available means—political, economic, psychological, and military—but after Brest Litovsk she never forgot the bitter experiences of February and March, 1918, which showed that without sufficient armed force other methods of warfare can hope for only ephemeral success.

Nevertheless, Lenin did not believe that the Brest Litovsk campaign of psychological and political warfare had been fought in vain. "This war will be settled in the rear, not in the trenches," he told an unofficial British representative on the eve of the signature of the treaty. "As a result of this robber peace Germany will have to maintain larger, not fewer, forces in the East. As to her being able to obtain supplies in larger quantities from Russia, you may set your fears at rest. Passive resistance . . . is a more potent weapon than an army that cannot fight."[17]

In eloquent speeches advocating ratification of the Brest Litovsk treaty Lenin nevertheless warned that experiments of the sort must not be repeated. "Let us beware of becoming slaves of our own phrases. In our day wars are won not by mere enthusiasm, but by technical superiority." The settlement with the Germans must be regarded as a "Tilsit Peace"—the signal for a national renaissance similar to the German revival after Prussia's humiliation at the hands of Napoleon. Brest Litovsk merely offered a breathing space during which the revolution could be entrenched in Russia, an armistice during which war could still be carried on against imperial Germany by revolutionary methods.[18]

The breathing space never materialized. The ink was hardly dry on the treaty of Brest Litovsk before the Soviet regime was confronted with foreign intervention and civil war, on fronts scattered from the Baltic to the Black Sea and from Murmansk and Archangel to Vladivostok. The ensuing wars lasted for more than two years and threatened to cause not merely the defeat of the revolution, but also the complete disintegration of the Russian empire. That the revolution was saved is a miracle of sorts, which must be attributed in part, at least, to the determination of the Bolsheviks to wage war with the sword as well as with words. The policy of arming the revolution went steadily

---

[15] Taracouzio, op. cit., p. 28.
[16] Bunyan and Fisher, op. cit., pp. 512-513, 519.
[17] Bruce Lockhart, Memoirs of a British Agent (London, 1932), pp. 239-240.
[18] Wheeler-Bennett, op. cit., pp. 260 and 409-426. The treaty provided for complete cessation of Russian propaganda in Germany but the Russians never had any intention of observing this provision in other than "perfect bad faith."

forward and bore its full fruit when the Nazi panzer divisions thrust deep into Soviet territory in 1941. Soviet leaders since Lenin have constantly warned that the experience of foreign intervention might be repeated, that war between the Socialist fatherland and the capitalist or fascist states was to be expected and prepared for. "The Bolshevist Revolution has proved to every honest communist the absolute necessity of the armament of the proletariat," read a statement of the Sixth World Congress of the Communist Party in 1928. And the duty of every citizen of the Soviet Union, as "a warrior for socialism consists of making all the necessary preparations for such a war—political, economic, and military; of strengthening the Red Army, the power weapon of the proletariat, and of training the working masses at large in military science."[19]

II

The period of foreign intervention and civil war was the great school of military affairs for the Soviet Union. Out of it came the Red Army, hastily improvised but nevertheless the strong nucleus of a future military establishment. Out of it, too, came a group of civilians—among them Trotsky, Frunze, Timoshenko, Voroshilov, and Stalin—who were to have a large part not only in organizing the army and fighting the war but in determining Russia's strategy in world politics. There also emerged from the civil wars some former members of the Imperial Army—especially Tukhachevsky and Shaposhnikov—who were to win names for themselves under the red banner of the Soviet Union. Finally, these wars were to demonstrate anew that revolutionary propaganda, however useful, cannot in itself win military victories.

The war between the Bolshevists and the various White armies reached its climax in the autumn of 1919 and winter of 1920. Attacks converged on Moscow from all sides and for a time it seemed impossible that the Soviet regime could survive. But Kolchak, sweeping down from his main base in Siberia, was defeated by Frunze and driven back across the Urals. Then Yudenitch, advancing from the Baltic states to the suburbs of Petrograd, was halted by a last-ditch defense by the city's workers. Denikin's southern army, which penetrated the Bolshevik defenses to Voronezh and beyond Orel before it was halted, presented perhaps the greatest threat of all. The course of its advance was delayed by the famous siege of Tsaritsyn (now Stalingrad)—which became known as the Verdun of the Civil War as it was in fact the Verdun of the war with Nazi Germany. Here, according to subsequent official histories, Stalin, Voroshilov, and Budenny inspired the Soviet defenders and so sapped Denikin's strength as to end all hope of his overthrowing the Bolshevik regime. The war with Poland was terminated about a year later, after Maxim Weygand—loaned to Poland by France—had halted Tukhachevsky's spectacular drive at the gates of Warsaw. By the spring of 1921 the Bolsheviks were once more in undisputed control of Russia, Allied armed intervention had ceased, and peaceful relations with the Soviets' neighbors had been established.

[19] Cited by Taracouzio, *op. cit.*, p. 271.

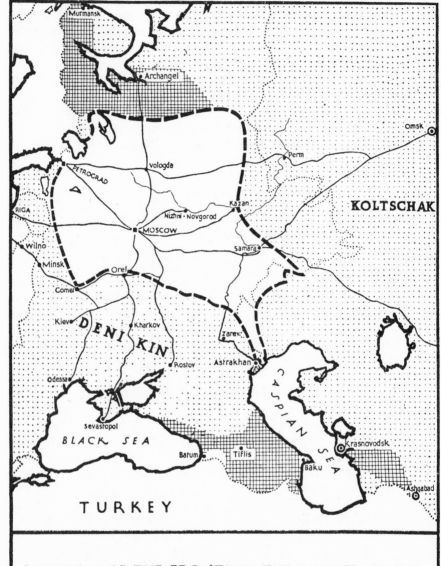

POSITION OF THE FRONTS IN THE SUMMER OF 1919

▬ ▬ ▬ ▬ ▬ . *the counter-Revolutionary Army*

⬚ : *counter-Revolutionary governments assisted by the Entente*

▦ : *territories occupied by English troops*   ◉ : *English garrisons*

Because it was a class war growing out of generations of grievances on the part of the peasants and workers, and because the middle class and the aristocracy were fighting for their very existence, the civil war was fought with a bitterness and savagery which left their mark on Russia for a generation.[20] Despite the bitterness and horrors of the Russian civil war, it cannot be compared, however, with the First World War as regards the size of the forces involved, the intensity of the fighting, the magnitude of military operations, or the broad sweep of strategy. Furthermore, as Winston Churchill has said, it was "a strange war." It was ". . . a war in areas so vast that considerable armies, armies indeed of hundreds of thousands of men, were lost—dispersed, melted, evaporated; a war in which there were no real battles, only raids and affrays and massacres, as the result of which countries as large as England or France changed hands to and fro; a war of flags on the map, of picket lines, of cavalry screens advancing or receding by hundreds of miles without solid cause or durable consequence; a war with little valour and no mercy. Whoever could advance found it easy to continue; whoever was forced to retire found it difficult to stop. On paper it looked like the Great War on the Western and Eastern fronts. In fact it was only its ghost: a thin, cold, insubstantial conflict in the Realms of Dis. Kolchak first and then Denikin advanced in what were called offensives over enormous territories. As they advanced they spread their lines ever wider and ever thinner. It seemed that they would go on till they had scarcely one man to the mile. When the moment came the Bolsheviks lying in the centre, equally feeble but at any rate tending willy-nilly constantly towards compression, gave a prick or a punch at this point or that. Thereupon the balloon burst and all the flags moved back and the cities changed hands and found it convenient to change opinions, and horrible vengeances were wrecked on helpless people, vengeances perseveringly paid over months of fine-spun inquisition. Mighty natural or strategic barriers, like the line of the Volga River or the line of the Ural Mountains, were found to be no resting places; no strategic consequences followed from their loss or gain. A war of few casualties and unnumbered executions!"[21]

Leon Trotsky, the directing genius of the Soviet armies in the struggle, agreed that the civil war was a miniature war, judged by the size of the armies engaged. But, "the small war differed from a big one only in scale. It was like a living model of war. . . . The small war was a big school."[22]

Although the revolutionary ardor of the peasants and workmen played a large part in the ultimate triumph of the Bolsheviks, so did the military competence of former officers, noncommissioned officers, privates, and technicians taken over from the former Imperial Army. Propaganda behind the White

[20] For vivid descriptions see W. H. Chamberlin, *The Russian Revolution, 1917-1921* (2 vols., New York, 1935), II, 335.

[21] W. S. Churchill, *The World Crisis, 1918-1928: The Aftermath* (New York, 1929), pp. 240-241. Regarding Russia's appalling loss of life and other casualties of the First World War see Stanislas Kohn and A. E. Meyendorff, *The Cost of the War to Russia* (New Haven, 1932).

[22] L. Trotsky, *My Life* (English translation, New York, 1930), p. 407.

lines undoubtedly helped undermine the morale of the counterrevolutionists, but it was of no avail so long as the Allied powers continued to supply Denikin and the others with adequate quantities of modern equipment and supplies. Weygand's skillful handling of the Polish forces showed that an experienced and able professional officer, even if he were a bourgeois and representative of a capitalist and imperialist state, could successfully parry the most astute political warfare if such warfare were not supplemented by enough men, guns, and discipline. In short, the civil war demonstrated anew that revolutionary strategy must be related to military efficiency and even to orthodox military methods.

Foremost among the concessions to military orthodoxy was the restoration of discipline, without which an army is but a disorganized mob, a "rabble in arms." By its very nature a revolution is a disruptive force, possessed in its initial phases of elements of complete anarchy. Such anarchy cannot long continue, for, as Lenin said in 1917, it is only one step from anarchy to counter-revolution. Despite all their denials to the contrary, the Bolsheviks had been a principal influence in the disintegration of the old army. They had spread defeatism, advocated insubordination, undermined the authority of officers, and otherwise helped bring about a state of virtual chaos. Now the Soviet leaders faced the imperative necessity of a complete change of front: "from disorganizers of the old army, they had to become organizers of the new." Lenin and Trotsky undertook with determination the task of establishing a disciplined force. Democratization of the army, including election of officers, was severely restricted; heavy penalties were inflicted for desertion and cowardice; army regulations were enforced and made stricter. Real authority was vested in "commanders," who succeeded the officers of all ranks. And the establishment of the Order of the Red Banner restored the formerly detested custom of military decorations.

Trotsky showed that even the typical army drill sergeant might have his place in the new order: "Every army unit must receive its rations regularly, foodstuffs must not be allowed to rot, and meals must be cooked properly. We must teach our soldiers personal cleanliness and see that they exterminate vermin. They must learn their drill properly and perform it in the open air as much as possible. They must be taught to make their political speeches short and sensible, to clean their rifles and grease their boots. They must learn to shoot, and must help their officers to ensure strict observance of the regulations for keeping in touch with other units in the field, reconnaissance work, reports and sentry duty. They must learn and teach the art of adaptation to local conditions, they must learn to wind their puttees properly so as to prevent sores on their legs, and once again they must learn to grease their boots. That is our programme for next year [1919] in general and next spring in particular, and if anyone wants to take advantage of any solemn occasion to describe this practical programme as 'military doctrine,' he's welcome to do so."[23]

[23] Quoted by Erich Wollenberg, *The Red Army: A Study of the Growth of Soviet Imperialism* (London, 1940), pp. 157-158.

Thus the Soviets had the experience—so typical of revolutionary groups—of learning that institutions of the past have more vitality and *raison d'être* than had been believed. Power is a sobering influence in most instances, and adjustment to some of the institutions and traditions of the past is therefore inherent in the establishment of a new regime. Revolutions will sweep away many of the forms of army organization, but there are certain fundamentals of military discipline and order which cannot be ignored by any system of statecraft, however radical. That Trotsky was aware of this is evident from his statement that "most of the questions of principle and the difficulties in connection with the constructive work of the Soviets during the years that followed [1917] were encountered first of all in the military sphere, and in most concentrated form at that."[24]

Nevertheless, the Bolsheviks did not altogether abandon revolutionary strategy in fighting the counterrevolution. The army which they raised was a peasant and proletarian army, the war which they fought was a class war. The workers and the peasants, on whom the Bolsheviks relied, were the groups who had the most to lose by a return of the old regime and they rallied to the support of the Soviets. The partisan war which revolutionaries conducted behind the White front was perhaps as effective in defeating Denikin and Kolchak as the organized military resistance of the Red Army. The *levée-en-masse* of the proletariat saved Petrograd after the resistance of the army had collapsed. There was a good deal of truth in what Lenin said on March 23, 1919, to the Eighth Congress of the Communist Party: ". . . if this war is waged with much greater energy and with exalted gallantry, it is only because for the first time in history an army has been created which knows what it is fighting for, [because it is] for the first time in the world that the workers and peasants, amidst unprecedented sacrifices, clearly understand that they are defending a Soviet Socialist Republic, the supremacy of the toiling masses over the capitalist, and the cause of the world socialist revolution of the proletariat."[25]

Behind the army stood the disciplined phalanxes of the Communist Party—a dynamic force crusading for a new idea. Some of them were political commissars with the troops in the field. But whoever they were they provided "the electric current that charged the vast machine of the Red Army, and supplied the driving power of the leaders' will, translating it into action. . . . Commissars, commanders, rank-and-file soldiers, their will to victory provided the unifying cement, the iron frame, that held the Red Army together."[26]

The Bolsheviks emerged from the bitter experiences of the civil war with an almost unique grasp of the role of war in society. Lenin, in particular, evolved a complicated philosophical system dealing with the subject. The attitude of the working classes toward war, he said, cannot be categorical—

---

[24] *My Life*, p. 437. For the experiences of the French Revolution as regards military affairs cf. R. R. Palmer, *Twelve Who Ruled: The Committee of Public Safety during the Terror* (Princeton, 1941), especially pp. 23-24, 78 ff., 81 ff., 96 ff., 182-185. Some of the parallels are suggestive.

[25] Cited by Taracouzio, *op. cit.*, p. 95.

[26] D. F. White, *The Growth of the Red Army* (Princeton, 1943), p. 126.

pacifism, nonresistance, opposition to military service, the general strike against mobilization, and other devices of western European socialism are meaningless demonstrations in and of themselves. The reaction of peoples to war must be determined by the kind of war in question and the purposes for which it is being waged. And all such matters are intimately related to the idea of Clausewitz that war is simply an instrument of policy.[27] Long before he came to power in Russia, Lenin was firmly convinced that war—both defensive and offensive—must be used by the proletariat to bring about the social revolution. As a result of what he learned in the "big school" of intervention and civil war, he was even more convinced.

The Sixth World Congress of the Communist International (1928) made it clear that Lenin's views were part of his bequest to his followers. One of the theses accepted by the congress stated that: "The overthrow of capitalism is impossible without violence, i.e., without armed uprisings and wars against the bourgeoisie. In our era of imperialistic wars and world revolution, revolutionary civil wars of the proletarian dictatorship against the bourgeoisie, wars of the proletariat against the bourgeois states and world capitalism, as well as national revolutionary wars of oppressed peoples against imperialism are unavoidable, as has been shown by Lenin."[28]

Ever since the period of the civil war and foreign intervention the Soviet Union and its people have visualized the world as divided into two armed camps—the camp of socialism and the camp of capitalism. The Declaration of Union in 1922 stated solemnly that "the instability of the international situation and the danger of new attacks render inevitable the creation of a common front by the Soviet Republics against capitalist encirclement." This was a natural reaction to armed intervention by Britain, France, Japan, and the United States during the critical years 1918-1919—an intervention which Russians have never been able to understand and which they still resent with a consuming bitterness. It was also an effect of the *cordon sanitaire* and the persistent refusal for many years of many nations to grant diplomatic recognition to the Soviet government. The resulting feeling of isolation and encirclement has been responsible for the existence of a war mentality within the Soviet Union almost from the beginning. Even the day-to-day language of Russians reflects a fundamental militancy. They speak of the "battle of production," "advances on the agricultural sector," fights on the "coal front," spies and saboteurs in industry.[29] This war-mindedness has been an important factor in the successful Soviet resistance to Nazi aggression, but it is also a factor to be reckoned with if we are to understand the policies of the Soviet Union in world affairs. It is the consequence of an unsuccessful policy of twenty-five years ago.[30]

[27] The Soviet ideas of war are developed at length and with considerable skill by Taracouzio, *op. cit.* See also D. F. White, "Soviet Philosophy of War," in *Political Science Quarterly*, LI (1936), 321-353.

[28] Taracouzio, *op. cit.*, p. 26.

[29] On this point see A. R. Williams, *The Russians: The Land, the People, Why They Fight* (New York, 1943), p. 43 ff.

[30] Winston Churchill, an advocate of intervention in 1918 and 1919, subsequently criticized our policy as vacillating, uncertain of its purposes, insincere—in fact, he doubted

In our kind of world it is shortsighted and dangerous for a nation to be unprepared for the eventuality of war—as democratic peoples found out to their sorrow during the harrowing years which followed the Munich "settlement." For a country like Russia, which considered itself *ipso facto* in danger because of the nature of its social system, it would have been folly. Hence the determination of the Soviet Union to maintain a Red Army, a Red Navy, and a Red Air Force adequate to protect itself against the attack which it believed would certainly come. Hence the industrialization of Russia, with its Five Year Plans, which gave precedence to capital goods over the needs of consumers in order that back of the armed forces might stand a fully mobilized economy capable of withstanding the severe strains which modern war of necessity imposes. As a result, the failures of the czars were not repeated on the steppes of Russia in the second war with the German empire. The foundations for the victory of Stalingrad were laid decades ago, when the Red Army was founded and then carefully nurtured to maturity.

III

Leon Trotsky was a living refutation of Karl Kautsky's statement that warfare is not the strong point of the proletariat. Trotsky was the father of the Red Army, the organizer of victory during the civil war, and the author of much of the doctrine upon which Soviet military policy is founded. If he had had any previous military experience, he might with justice be called the Carnot of the Russian Revolution—a title which he generously conferred upon his principal assistant, Skliansky. Only thirty-seven years old when he assumed the post of War Commissar in March, 1918, Trotsky nevertheless had back of him a record of twenty years of militant revolutionary activity. Unlike Lenin, he was not essentially a scholar, but a journalist, agitator, and organizer. Arrogant, egocentric, histrionic, he nevertheless was tireless, courageous, clear-headed, and devoted to duty. If his judgment on military matters was not always faultless, it was certainly above average and was based, in general, upon correct appraisals of the situations which confronted him. His task of bringing military order out of the chaos of revolution, civil war, and intervention would have broken the back and deranged the sanity of a less sturdy soul. "In that vast panorama of confusion and disorder," writes an American journalist, "the comet-like figure of Trotsky, storming up and down the Red lines, distributing new revolutionary military honors and orders for execution with equal prodigality, exhorting and denouncing, always organizing for victory, was certainly one of the decisive factors in finally bringing the whole Russian land under the

whether we had a policy at all. In 1929, he wrote: "Were they [the Allies] at war with Soviet Russia? Certainly not; but they shot Russians at sight. They stood as invaders on Russian soil. They armed the enemies of the Soviet Government. They blockaded its ports, and sunk its battleships. They earnestly desired and schemed its downfall. But war— shocking! Interference—shame! It was, they repeated, a matter of indifference to them how Russians settled their own internal affairs." *Op. cit.*, pp. 243-244. Another severely critical account is by General W. S. Graves, commander of the American expeditionary force in the Soviet Far East: *America's Siberian Adventure* (New York, 1931).

Red flag of the Soviets."[81] Trotsky's long feud with Stalin has led to a ukase in orthodox Soviet circles that Trotsky shall be virtually ignored in telling the story of the Revolution and the Red Army. In addition to Trotsky's very gifted polemics in his own defense, there exists plenty of other evidence to the effect that Trotsky is one of the outstanding figures of modern military history. If justice has not been done him, it is because in totalitarian countries history is what the dictator says it is.

Trotsky laid no claim to being a military expert or a strategic genius. He pointed out, correctly enough, that in parliamentary countries "war and navy ministries are often given over to lawyers and journalists who, like myself, see the army chiefly from the window of their editorial offices." His sole acquaintance with military affairs was residence in Serbia, Bulgaria, and Rumania during the Balkan Wars. "But my approach to these questions was by nature still political rather than military. The World War brought everyone—myself included—close to the questions of militarism," which Trotsky analyzed with care. But even there "the important thing was war as the continuation of politics, and the army as the instrument of the latter. The problems of military organization and technic were still in the background, as far as I was concerned. On the other hand, the psychology of an army, in its barracks, trenches, battles, hospitals, and the like deeply stirred my interest." Thus far Trotsky seems to be a compound of Clausewitz, Engels, and Lenin.

Trotsky did not think of himself "as in any sense a strategist, and had little patience with the sort of strategist-dilettantism that flooded the party as a result of the revolution." In so far as his duties as War Commissar required him to make strategic decisions—as they did in at least three important instances—his "strategic position was determined by political and economic considerations, rather than by those related to pure strategy. It may be pointed out, however, that questions of high strategy cannot be solved in any other way." He realized fully that he needed the assistance of those to whom military affairs were a life work. Hence, "in the technical sphere and in that of operations, I saw my task chiefly as a matter of putting the right man in the right place, and then letting him exercise his abilities." But, as the army was an instrumentality of the Revolution, he understood that an indispensable condition to success in his official duties was that he should completely merge his work of organizing the army with his work as a member of the party, thus demonstrating political astuteness of a high order.[32]

For the Bolsheviks there could be no question of reforming the Imperial Army. The army of the czars was, in fact, dead beyond hope of resurrection. During the years 1914-1916 it suffered grievous casualties, fought without adequate food or munitions or supplies, and was scandalously lacking in leadership. Later, it became afflicted with "the deadly psychology of retreat," riddled by desertion, indiscipline, and war-weariness. To all of these ailments, serious enough in themselves, was added left-wing revolutionary propaganda of pacifism, defeatism, and class hatred, with the result that the army reached a state

[81] Chamberlin, op. cit., II, 40.   [32] Trotsky, op. cit., pp. 349-350, 358.

of complete disintegration and finally melted away. When the attempt was made to revive it by an offensive during the summer of 1917, the army, as Lenin cynically remarked, "voted for peace with its legs."

Corrosion of the fighting spirit of the army was under way by the end of 1916, as General Brussilov testified. There are therefore some critics who assert that Bolshevik propaganda could have had little effect upon the ultimate result, that it was merely the straw which broke the camel's back. But these critics overlook the fact that the *speed* and *thoroughness* of the demoralization of the army were definitely related to the pacifist and revolutionary propaganda which was waged so zealously by the Bolsheviks. And it overlooks the further fact that, in the absence of the revolution, or in the presence of a strong government in its early phases, the army might have recovered a semblance of discipline, just as the French army recovered from the great mutiny of 1917.[33]

Trotsky met the issue directly by agreeing that the Revolution, and the Bolshevik propaganda which preceded it, had destroyed the army. Professor Milyukov, minister of foreign affairs in the government of Prince Lvov, had asserted that the army was ruined by "the conflict between 'revolutionary' ideas and normal military discipline, between 'democratization of the army' and the 'preservation of its fighting power.' " To this Trotsky replied: "A historian ought to know, it would seem, that every great revolution brings ruin to the old army, a result of the clash, not of abstract disciplinary principles, but of living classes. . . . 'Surely the fact is evident,' wrote one wise German [Engels] to another [Marx] on September 26, 1851, 'that a disorganized army and a complete breakdown of discipline has been the condition as well as the result of every victorious revolution.' The whole history of humanity proves this simple and indubitable law."[34]

Furthermore, said Trotsky, it was in the interest of the nation that the old army be completely liquidated. "The instrument of war is the army." But under the czars "this army was a serious force only against semi-barbaric peoples, small neighbors, and disintegrating states; in the European arena it could act only as part of a coalition; in the matter of defense it could fulfill its task only by the help of the vastness of spaces, the sparsity of population, and the impassability of the roads. . . . The semi-annulment of serfdom and the introduction of universal military service had modernized the army only as far as it had the country—that is, it introduced into the army all the contradictions proper to a nation which still has its bourgeois revolution to accomplish. It is true that the Tsar's army was constructed and armed upon Western models; but this was more form than essence. There was no correspondence between the cultural level of the peasant-soldier and modern military technique."[35]

[33] M. T. Florinsky, *The End of the Russian Empire* (New Haven, 1931), Chap. IX, makes a persuasive case for the point of view that Bolshevik propaganda was simply the final step in the collapse of the army. *Contra* see White, *op. cit.*, Chap. I. Neither Mr. Florinsky nor Mr. White is a Communist or a Communist apologist. They simply draw different conclusions from substantially the same evidence. For the view of a czarist general see A. A. Brussilov, *A Soldier's Notebook, 1914-1918* (London, 1930), Chaps. X-XII.
[34] *History of the Russian Revolution*, I, 374-375.   [35] *Ibid.*, p. 17.

This statement shows a keen perception of the realities of Russia's military position in Europe at the time. Of particular interest is the statement that "there was no correspondence between the cultural level of the peasant-soldier and modern military technique." The controlling—but by no means the only—factor in this respect was the appalling degree of illiteracy in the Imperial Army. General Sir Alfred Knox, British military representative in Russia during the First World War, and one of the most sympathetic students of Russian military affairs, has commented on the fact that about fifty per cent of the soldiers of 1914 could neither read nor write and that the smattering of education which others possessed in no way "expanded [the soldier's] mind or made him a civilized, thinking being." It was impossible, therefore, to give him any but the most elementary training or responsibility, to inform him concerning maps, weapons, and tactics, or to educate him concerning the issues of the war.[36] Tolstoy in *War and Peace* has idealized this type of soldier, submitting without complaint to nature, tyranny, defeat, and death. But, whatever his qualities of courage and stoicism, the advance of military technology and the increasingly complicated character of military operations made him a liability to an army rather than an asset. It was with full realization of this fact that the Red Army later waged so effective a war against illiteracy and ignorance.

There were other reasons, too, why Trotsky believed that the dead past should bury its dead. "The physiognomy of the army was determined by the old Russia, and this physiognomy was completely feudal. The officers still considered the best soldier to be a humble and unthinking lad, in whom no consciousness of human personality had yet awakened. Such was the 'national' tradition of the Russian army. . . . In the eighteenth century [Field Marshal] Suvorov was still creating miracles out of this material," but throughout the nineteenth and twentieth centuries the czars' army was continually defeated "because it was a feudal army. Having been formed on that 'national' basis, the commanding staff was distinguished by a scorn for the personality of the soldier, a spirit of passive Mandarinism, an ignorance of its own trade, a complete absence of heroic principles, and an exceptional disposition toward petty larceny. The authority of the officers rested upon the exterior signs of superiority, the ritual of caste, the system of suppression, and even a special caste language. . . ."[37]

No such state of affairs could continue under a socialist state, which was designed to be the antithesis of the czarist regime. Neither could the new army be built upon bourgeois, parliamentary, democratic foundations. "The Red Army," said Trotsky, "was an institution built from the top on the principle of the dictatorship of the working class, with officers selected by the Soviet Government and the Communist Party."[38] Hence Trotsky's task, as he saw it, was almost unique: "In capitalist countries the problem is that of maintaining the

[36] Major General Sir Alfred Knox, *With the Russian Army, 1914-1917* (2 vols., New York, 1921), I, pp. xxxi-xxiv. Beginning with 1935 the Red Army rejected all illiterate recruits—something which would have been impossible in czarist Russia.
[37] Trotsky, *History of the Russian Revolution*, I, 253.
[38] Bunyan and Fisher, *op. cit.*, pp. 569-570.

existing army—strictly speaking, of maintaining a cover for a self-sustaining system of militarism. With us, the problem was to make a clean sweep of the remains of the old army, and in its place to build, under fire, a new army, whose plan was not to be discovered in any book. This explains sufficiently why I felt uncertain about military work, and consented to take it over only because there was no one else to do it."[39]

However diffident he may have been about undertaking the task of War Commissar, Trotsky showed no vacillation in the exacting work which lay before him. In many ways the most difficult problem confronting him was to undo the pacifist, defeatist, and disruptive propaganda which the Soviets had waged against the army and against war before they came to power. He was determined, whatever the opposition or the obstacles, to create a proletarian and peasant army, to convert the nation into "a Spartan war camp," and to create within the framework of the Revolution "a regime of military communism." Everything which had gone before must be put in reverse. "Revolutionary violence is the means of attaining the freedom of the toilers. From the moment of assuming power revolutionary violence assumes the form of an organized army." There must be no more nonsense about peace and the brotherhood of man, no Tolstoyan pacifism. "We are trying to create a communist society . . . in which there will be no conflicts, because there will be no classes, because nations will not be separated by state fences but will live together and work for a common cause." But that utopia is still far from realization, because the ruling classes "will not yield an inch without a struggle." The revolution believes in the use of force, because there can be no change in social relations "without a bloody war." The rehabilitation of "the armed forces of the Russian Republic" is a question of life and death for the country and the working class. Therefore, "every young worker, instead of drinking home brew on Sundays, should learn to shoot," for in learning to shoot the worker is "accomplishing the most important task of his own class." The strictest discipline must be restored and must be enforced, whenever necessary, by the death penalty for cowardice, desertion, and other military offenses, for "a revolution not only permits strict discipline in an army, but creates it." An army cannot be run by committees, elected officers, and dialecticians, but only by the firm hand of military authority.[40]

This was stern medicine, which had to be rammed down unwilling throats. But Trotsky seems to have had Lenin's complete confidence which, together with his own considerable abilities as an orator and administrator, helped bring order out of chaos and victory out of impending defeat. Lenin understood that there are times when, as he said, "history causes the military problem to become the essence of the political problem."[41]

[39] *My Life*, p. 349.
[40] See statement by Trotsky in Bunyan and Fisher, *op. cit.*, pp. 569-572; his speech of April 21, 1918, *ibid.*, p. 572; his speech of June 29, 1918, in James Bunyan, *Intervention, Civil War, and Communism in Russia, April-December, 1918: Documents and Materials* (Baltimore, 1936), pp. 268-270; Trotsky, *My Life*, p. 411, and *History of the Russian Revolution*, I, 375, 461.
[41] Wollenberg, *op. cit.*, pp. 1, 3-4.

It took courage of a high order for Trotsky to take into the service of the Red Army "technicians" and "specialists"—euphemisms for officers and noncommissioned officers—of the old army. They were largely the hated "wearers of the gold epaulets," the nobility-intelligentsia, the sons of the well-to-do bourgeoisie and kulaks. "We had to consider these tsarist officers for reasons of their professional training," said Trotsky. "Without them we should have had to begin from the beginning, and it was not likely that our enemies would have given us the time needed to carry our self-education to the necessary level." To the charge that there was danger in employing retainers of the old regime, Trotsky replied: "There is danger in everything. We must have teachers who know something about the science of war. We talk to these generals with complete frankness. We tell them: 'There is a new master in the land—the working class. He needs instructors to teach the toilers . . . how to fight the bourgeoisie.' If these generals serve us honestly we shall give them our full support; if they attempt counter-revolution, we shall find a way to deal with them."[42] This determination of Trotsky to utilize the services of officers of the former army was to the very end a bone of bitter contention with his colleagues. But by adhering to his program despite opposition and criticism, he not only built an army within a relatively short time, but he enlisted the services of some distinguished soldiers—among them Tukhachevsky and Shaposhnikov—who gave to the Red Army over the coming years a degree of professional competence which it could hardly have attained otherwise.

Between April, 1918, and August, 1920, the number of former officers who served in the Red Army, including indispensable medical and veterinary personnel, exceeded 48,000. To these must be added about 250,000 noncommissioned officers, whose role in maintaining discipline and morale can hardly be overestimated. Many of the latter rose to commissioned status, and some of them rose to high command—the most notable being the picturesque but far from brilliant Budenny, whose reputation of civil war days did not survive the tests of the Second World War. A considerable number of czarist officers remained in the army after the civil war and won the respect even of those who had originally objected to their being used at all. Thus Voroshilov, speaking at the eighth anniversary of the Red Army, said that as regards fidelity they were indistinguishable from party members. "They are comrades alike and would die in the same manner at the first order of the Workers' and Peasants' Government at their combat posts, as their other, party, comrades would."[43]

To keep a constant vigil over officers of the old army, to carry on propaganda and party work among the rank and file, to educate the peasantry concerning the objects of the revolution and the civil war, and to do other non-

---

[42] Bunyan and Fisher, *op. cit.*, pp. 570-571; Bunyan, *op. cit.*, p. 267; *My Life*, p. 438. The weapon with which Trotsky threatened the officers was to hold their families as hostages. Those who consider betraying the revolution, he said, should "know that they betray their own families: fathers, mothers, sisters, brothers, wives and children." To give effect to the threat the necessary measures were taken to "detain" the families of suspected traitors. If Trotsky was cruel in this respect, says an American journalist, it was because the revolution itself was cruel. Chamberlin, *op. cit.*, II, 30-32.

[43] White, *op. cit.*, p. 207.

military work, Trotsky appointed political commissars to each Red Army unit. These commissars were not supposed to interfere with military operations, but they were responsible for the morale and loyalty of their commands. If a unit failed, the commissar was held responsible. Trotsky believed in the good old revolutionary principle of shooting unsuccessful commissars *pour encourager les autres*.[44] The position of commissar was, therefore, no sinecure for a rising young communist. Aside from Trotsky's wrath, the commissars faced ceaseless hard work and constant exposure to death in combat. Trotsky paid them a well deserved tribute when he said that they constituted "a new Communist Order of Samurai who—without caste privileges—are able to die and to teach others to die for the cause of the working class." To supplement the leavening influence of the commissars, young communist officers were trained at Soviet military schools. An energetic New York trade-union organizer, Goldfarb-Petrovsky, was at the head of the network of these training schools, which supplied not only commanders but also picked shock troops for the Red Army. After short courses of two to six months, frequently interrupted by service at the front, these young Red cadets exerted influence far beyond the adequacy of their military training.[45]

Seen in retrospect, Trotsky's work of organizing, supplying, officering, and even personally commanding the Red Army is one of the outstanding achievements of modern military history. So many toes had to be stepped on that it is small wonder that he aroused bitter opposition within the party ranks. Trotsky was more than equal to his critics in the give and take of debate—he could demolish them with epigrams and by the sheer force of his eloquence. But they were strong men, the members of this opposition, not to be put off by logic or dialectics. For involved in the controversy over military policies was a fundamental and ruthless struggle for power—a struggle in which Trotsky was destined to lose to the less facile but equally determined. Among the many able men among the anti-Trotsky faction was one who was to play an even larger role in the history of the Red Army. His name was Joseph Stalin, and he now bears the proud title of Marshal of the Soviet Union.

IV

The opposition to Trotsky fought him on a great variety of issues, especially the employment of former czarist officers, the degree of autonomy which should be possessed by military units in the field, guerrilla warfare, methods of military discipline, and the relative merits of offensive and defensive warfare. Trotsky summarized the position of his opponents objectively when he wrote: "The opposition tried to find some general theoretical formula for their stand. They insisted that a centralized army was characteristic of a capitalist state; revolution had to blot out not only positional war, but a centralized army as well. The

---

[44] His enthusiasm for such shootings had to be curbed by higher authority. The institution of political commissar was originated by the earlier bourgeois provisional government but had no success in the thoroughly demoralized army of 1917.

[45] Chamberlin, *op. cit.*, II, 34, 36.

very essence of revolution was its ability to move about, to deliver swift attacks, and to carry out manoeuvres; its fighting force was embodied in a small, independent detachment made up of various arms; it was not bound to a base; in its operations it relied wholly on the support of a sympathetic populace; it could emerge freely in the enemy's rear, etc. In short, the tactics of a *small war* were proclaimed the tactics of revolution."

Trotsky believed that the doctrine of his opponents was "really nothing but an idealization of our weakness." He thought it doubtful that the lessons of the civil war could be made applicable to future wars of indeterminate character.[46] Stalin and Voroshilov were determined, for their part, not to submit to control from Moscow on questions of discipline involving their own forces. They knew that in attacking the employment of officers of the old army they would have widespread support in the party and among younger elements of the new army. Also, they knew that guerrilla warfare was so strongly rooted in popular tradition that they could gain further adherents by urging it in preference to orthodox European methods. But the controversy went far beyond the relatively narrow range of such problems.

A debate now ensued over military theories, concepts of strategy, and forms of army organization. In the forefront was the question whether there was or was not a distinctive Marxist military doctrine. Some of the brilliant, self-made commanders of the Red Army—notably Frunze, Voroshilov, and Gusev, along with the former officer Tukhachevsky—thought they had detected in the civil war the makings of a new military theory of the revolutionary proletariat, which relegated to limbo all former war doctrine. If such a new military theory did, in fact, exist or could be formulated, then it must be formally adopted for the guidance of the Red Army. "One of the basic conditions for securing the maximum strength of the Red Army," wrote Frunze and Gusev in 1921, "is to transform it into a monolithic organization, welded from top to bottom not only by a common political ideology, but also by unity of views on the character of the military problems facing the Republic, on the method of solving these problems, as well as on the system of combat training of troops."[47] Those concerned with the actual administration of the army saw in any such thesis an attempt to impose party censorship on the staff, to interfere with the development of military thought, and to impose penalties upon those who did not accept the official dogma. Fears on this account were well justified, for the usual repressions of a totalitarian state were later extended to military affairs.

Trotsky denied that there was any Marxian military theory. "The Marxian method," he said, "is a method of historical and social science. There is no 'science' of war, and there never will be any. There are many sciences war is concerned with. But war itself is not a science; war is practical art and skill. How could it be possible to shape principles of military art with the help of Marxian method? It is as impossible as it is impossible to create a theory of

---

[46] *My Life*, p. 438.
[47] M. Frunze, *Selected Works* (Moscow, 1934), p. 7, quoted by White, *op. cit.*, p. 161. I have made slight changes in the translation.

architecture or to write a veterinary textbook with the help of Marxism."[48] Furthermore, even if one should agree "that 'military science' is a science, it is impossible to assume that this science could be built according to the method of Marxism. Historical materialism is by no means a universal method for all sciences. . . . To attempt to apply it in the special domain of military affairs would be the greatest fallacy." Trotsky went on to warn that discussions on such matters must not be encumbered with "trite Marxist terminology, high-sounding words, grandiloquent problems, which as often as not would turn out to be shells without kernel or content."[49]

Trotsky readily admitted that Marxian doctrines might determine the broad strategy of the Bolshevik Revolution in world politics. But he was opposed to amateur and ideological approaches to military theory, without reference to the experience of other countries and of the old regime in Russia. "When strategy is developed from the viewpoint of young revolutionaries, the result is chaos." There are certain constant factors in war—especially man and geography—which operate even when technology, social structure, and political organization change. These relatively permanent phenomena must be studied as carefully as transitory conditions at home and abroad. The fundamental rule of strategy for the Soviet republic, he concluded, "is to be on the alert and to keep our eyes open."[50]

However sensible Trotsky's view may seem to the objective student of military affairs, it enraged ardent Communists. Their principal spokesman was Mikhail V. Frunze, behind whom stood the rugged figure of Stalin. Frunze had had an active revolutionary career since 1904 and had been the directing genius of an illegal soldiers' organization in the old army. But he first achieved widespread recognition as the commander of the armies operating in the east against Kolchak and later in the Crimea against Wrangel. He emerged from the civil war with the reputation of being a first-rate tactician and strategist. Although he was no match for Trotsky in debate, he had a dogged determination which offset, in part, the brilliance of his opponent. And, above all, he was on the side of the angels—that is to say, he was supported by the man who was to assume Lenin's power. Early in 1924 he was made vice president of the Revolutionary Military Council and the *de facto* head of the Red Army. In January, 1925, he was made People's Commissar for Military and Naval Affairs, and shortly thereafter Trotsky ceased to have any determining influence on Soviet policy. Frunze died before the end of the year, at the early age of forty, but not before he had introduced substantial reforms in army organization and procedure.

"According to Frunze, the principal condition for the formulation of an adequate military doctrine was its strict co-ordination with the general aims

---

[48] Trotsky, *How the Revolution Developed Its Military Power* (Moscow, 1924), quoted by S. Gusev, "Our Disagreements in Questions of War," in *The Bolshevist*, 1924, No. 15, pp. 34-49 (translated by V. T. Bill) ; also cited by White, "Soviet Philosophy of War," *loc. cit.*, p. 339.

[49] Trotsky, *ibid.*, cited by White, *Growth of the Red Army*, p. 167.

[50] White, *op. cit.*, pp. 164, 165.

of the state and the material and spiritual resources at its disposal. He admitted that it was impossible to invent such a doctrine"; its elements already existed, and the work of military theorists was to appraise these elements and to bring them together into a coherent system in accordance with "the basic teachings of military science and the requirements of military art." In this respect he differed little from Trotsky, who, in fact, found some difficulty for that very reason in refuting Frunze and Voroshilov in the debate which took place before the Eleventh Party Congress in 1922. But Frunze went further. Taking everything into account, he said—the size of the Soviet Republic, the relatively few troops at its disposal, the inadequate transportation, and the backward industrial state of the country—the basic tactical and strategical concepts of the Red Army should be based on the offensive, on a war of maneuver, as opposed to the war of trenches and position which had characterized the operations of 1914-1917. The experiences of the civil war had shown that revolutionary troops had a special adaptability as regards a war of maneuver. Besides, the very spirit of the Revolution demanded sustained activity, a sense of daring, guerrilla operations, and other means of carrying the war to the enemy. As the defensive was not to be recommended on either political or military grounds, fortifications should play little or no part in the Soviet system.[51]

As opposed to this Trotsky and his professional entourage (with the notable exception of Tukhachevsky) contended that no system, offensive or defensive, is *ipso facto* good or bad. Victory, to be sure, can be achieved only by means of the offensive. But the revolution is concerned primarily with the political offensive, which is quite consistent with the strategical defensive. The exhaustion of Russia after the World and civil wars, the reluctance of the peasantry to undertake further military adventures, the shattered and undeveloped economy of the nation, the imperative necessity for the early creation of a large and well trained armed force—all these recommended caution as regards the military offensive. A defensive war of attrition was the only kind of war upon which all classes in the republic could agree; it was also the only kind of war which would exploit to the full the size, geographic position, and economic backwardness of the country. Until Russia could develop its resources and build up its industrial strength this was the only strategy which held out hope of success. Furthermore, no military doctrine will fit every unforeseen emergency or every set of conditions. Svechin, a former czarist officer serving as professor in the Staff College, put it very shrewdly in these terms: "For each war it is necessary to develop a special line of strategic behavior; each war represents a particular case, which calls for the establishment of its own peculiar logic, instead of applying the same pattern, albeit a red one. . . . In the broad framework of the general theory of contemporary warfare, dialectics permit a clearer characterization of the line of strategic conduct which should

[51] Frunze, cited and summarized by White, *op. cit.*, Chap. VI, the best account in English of the great debate of 1921-1922. Based upon the writings of the participants and upon the stenographic proceedings of the Eleventh Congress of the Soviets of Workers, Peasants, Cossacks, and Red Army Representatives, Moscow, 1922.

be chosen in a given instance than could be achieved even in a theory specially formed to cover that specific instance."[52] In short, a wise strategy is an elastic strategy.

Another subject for heated debate was the character of the permanent military establishment of the Soviet republic. Should the Red Army be a professional army or a militia? In answering this question—which was closely related to the controversy over the existence of a Marxist military doctrine— the Communists had the guidance of Marx and Engels, who had taught that the ideal form of military organization for a communist state was the militia. They were also influenced, in all probability, by Jean Jaurès, whose book L'armée nouvelle, published in 1914, had had a marked impact on socialist and radical thought throughout Europe.[53] On the question of the militia Trotsky's stand was less categorical than it had been on the question of military theory as a whole. He took no clear-cut position, indicating that with some reservations territorial armies might be acceptable to him. Judged by the character of the military reforms of 1924-1925, Frunze also was unwilling to adopt an extremist position. The professional officers, however, took an uncompromising attitude toward anything but a regular army, largely on the ground of military efficiency. Their beliefs were stated with eloquence and effectiveness by Tukhachevsky, who succeeded in amalgamating the case for a regular army with a passionate plea for an aggressive Communist imperialism.

Mikhail N. Tukhachevsky (1893-1937) was a youthful lieutenant of the czarist army who went over to the Bolsheviks in 1917 and became the hero of the war with Poland. He was ambitious, vain, clever, intelligent, and courageous, and was possessed of a passionate, primitive Russian patriotism, which he somehow or other intensified by ardent espousal of the Revolution. A descendant of the Count of Flanders, he may well have aspired to be the Red Napoleon. The army was his consuming interest, and he won deserved success in a great variety of posts from director of the Military Academy to chief of the general staff. He met his death by execution in 1937, during the great purge, on charges of treason, the specific details of which are as yet among the unrevealed secrets of the Kremlin. In studying Tukhachevsky it is not always easy to determine whether he is more the passionate zealot for Communism, the ardent Russian nationalist, or the clever military careerist. Perhaps the answer would help determine why he ultimately faced a firing squad.[54]

Tukhachevsky believed in taking the political and military offensive. Communism could then be carried to the rest of the world by what one might call military imperialism. He was opposed to the traditional policy of retreat into the vast spaces of Russia, since loss of territory necessarily affects the nation's

[52] A. Svechin, Strategy (second edition, Moscow, 1927), pp. 8-9, cited in translation by White, p. 163.
[53] Concerning Marx and Engels see Chapter 7; concerning Jaurès, Chapter 15. The Communists frequently disavowed all interest in the opinions of those connected with the Second International, but despite such denials they could not altogether renounce their intellectual heritage.
[54] For a brief biographical sketch of Tukhachevsky see M. Berchin and E. Ben-Horin, The Red Army (New York, 1942), pp. 110-114.

war potential. Furthermore, to a revolutionary state, conquests of territory are essential since they enable the advancing army to spread revolutionary ideas and gain recruits from among the exploited classes of the enemy country. Inasmuch as the class struggle is an important phase of every war, Tukhachevsky proposed that the Communist International have a military general staff charged with preparing plans for using the armed proletariat in the cause of the world revolution. But quite aside from political and ideological considerations, the objective of all war is victory through the destruction of the enemy's armed forces which, said Tukhachevsky, could be achieved only by a relentless offensive.[55]

If a revolutionary state is to follow an offensive strategy, it can hope to succeed only if it has an army of the highest standard of professional excellence. This a militia cannot be. In a brilliant pamphlet, written when he was only twenty-seven, Tukhachevsky made a frontal attack on the militia system as being unsuitable to the Soviet government and the Communist Party, which sought both to preserve a revolution at home and to spread the class struggle abroad. On the contrary, it is "the most logical, the most formidable military system for a bourgeois state which has reached the maximum of its capitalist development." In a bourgeois society the advanced state of the railways is the determining factor as regards the size of the army and the development of military techniques. The training of the army then becomes unimportant, the training of the staff paramount. "If we presuppose the most perfect military technique, the quality of the army can be entirely replaced by its quantity." In capitalist armies, as in bourgeois society in general, a small ruling class can control the masses by a monopoly of material resources. Hence the militia fits into the general political concepts of capitalism. For a revolutionary state, however, the conditions are different. The socialist revolution will face continuous war until it is destroyed or until the revolution has spread throughout the world. In either event no further debate over the militia system would be required. In the meantime, however, the Soviet Republic must have a workers' army, based upon obligatory service. Furthermore, recruitment should be international, since the class struggle cuts across national boundaries. When the flag of the revolution is carried outside Soviet territory, it would then rally to it the trained worker-soldiers of the invaded countries, thus assuring success to a daring policy of the offensive.[56]

This was a bold program on all counts, and it was too bold for the Soviet government. Orthodox Communists continued to defend the militia system, decry the use of bourgeois military specialists, and warn against "men on horseback." The military reforms of 1924 failed to resolve the conflict, for they provided for both a cadre army of professionals numbering about 560,000 and a territorial militia of 43 divisions. Thus Frunze, who had originally sponsored the militia as against a regular army, had fathered a compromise by which the

[55] White, op. cit., Chap. VI, passim. See also Tukhachevsky's article "War as a Problem of Military Struggle," in the Great Soviet Encyclopedia, Vol. XII.
[56] M. Tukhachevsky, Die Rote Armee und die Miliz (German translation, Leipzig, 1921), especially pp. 9-10, 18, 19, 25, 31.

Red Army was neither the one nor the other. But it was a fruitful compromise because it permitted the youthful Soviet Union during its critical years to make full use of tradition and technical competence in its military establishment, while it continued to retain those new elements in the army which formed a nucleus of advanced political and strategical ideas.[57]

In emphasizing the differences between the participants in the great debate of 1921-1924 on Soviet military policy, it would be easy to overlook the points on which they were in substantial agreement. The Red Army was to be a class, and class-conscious, army of workers and peasants; in so far as they were permitted to serve at all, the bourgeoisie and other disfranchised groups were reduced to what Trotsky called "dirty work" in labor battalions and otherwise. The oath which every soldier took was an oath of loyalty not only to the Soviet republic but to the working class. "In the presence of the laboring classes of Russia and of the whole world," it read, "I bind myself to uphold honorably this title [of warrior in the Worker-Peasant Army]."[58] Discipline must be preserved and unity of command achieved by theoretical, if not actual, curtailment of the authority of political commissars in purely military matters. It was agreed, also, that the Revolution must be defended at all costs against "an international, armed, capitalist conspiracy." The military policy of the Soviet government, whatever form it might take, was to build upon communist foundations and the revolutionary outlook on life. The resources of the nation must be developed, and its industry expanded, so that purely material limitations should not be imposed upon the growth and efficiency of the Red Army. Upon these things all were agreed, whatever their other differences.

V

The uncertainties and controversies which characterized Soviet military affairs during the years 1918-1924 were attributable to three fundamental factors. First, they were almost inherent in the chaos of revolution itself, which disorganized all agencies of administration and catapulted inexperienced civilians into positions of grave responsibility in the conduct of war. That Lenin, Trotsky, and the others accomplished as much as they did is remarkable, for the qualities which make a revolutionary agitator and those which make a capable military strategist are not necessarily the same.

Second, the debate concerning military policies was conducted in abstract terms and not within the essential framework of Soviet foreign policy. It was never clearly understood by the participants, for example, whether the first concern of the government was to promote international revolution or to guarantee the success of communism in one country. During the early phases of the Brest Litovsk negotiations, it apparently was Lenin's intention to bring about a European civil war, but thereafter policy varied from time to time and from

---

[57] An informative volume on Russian military theory and history is N. Basseches, *The Unknown Army* (New York, 1943). For the reforms of 1924-1925 see White, *op. cit.*, Chapter VIII, "Frunze's Army."

[58] For text of the oath see Bunyan and Fisher, *op. cit.*, p. 574.

individual to individual. But until this fundamental question was resolved, it was futile to discuss such questions as the merits of the defensive *vs.* the offensive, of maneuver and mobility *vs.* position and fortification. The importance of the interrelation of foreign and military policies is something which needs to be reiterated in every nation in the world, and it was therefore not to be expected that it should have been understood from the outset by the Bolsheviks.

Third, as has already been said, much of the argument concerning military affairs was only a rationalization of individual ambitions, a cloak thinly concealing a fundamental struggle for control of the party and the government. For almost two years before his death in January, 1924, Lenin was ill and growing progressively worse. Throughout these years the choice of his successor was clearly involved in all political discussion and affected individual opinion as well as public policy.

The assumption of power by Stalin and his ruthless suppression of dissidence within the communist ranks went a long way toward clarifying the issues. Although, excepting recent dissolution of the Comintern, Stalin never completely disavowed the international character of the communist movement, the general direction of his policy has been increasingly nationalistic. Certainly in foreign affairs his principal concern seems to have been the security of the Soviet Union rather than a quixotic zeal in carrying the Revolution beyond its boundaries. In part Stalin's aims are to be explained by his own life and character; in part they have been determined by developments outside Russia, particularly the rise of Nazi Germany and Hitler's frankly avowed intention of seizing Soviet territory. There have been deviations from the main trend of Stalin's policy but its most powerful forces have been centripetal.

Unlike Lenin, Trotsky, and many other Bolsheviks, Joseph Stalin never served a period of exile abroad. He waged a relentless struggle against the czarist regime from 1896 to 1917 and has an impressive list of arrests and imprisonments to his credit on the revolutionary ledger. But his warfare was underground, on home soil, and it is on home soil that he has continued to fight for communism and the fatherland. He is not an intellectual or predominantly a theorist, although he has shown a rugged determination to carry out Marxist principles as he understands them. Through his control of the Communist Party machine, he has gained wide experience in politics, and he is a "realist" if there ever was one. It is this man, still something of a mystery and an enigma to the rest of the world, who for a generation has determined Soviet military policy and prepared the Red Army, technically and morally, for the great struggle against Germany which began in June, 1941.

For fifteen years after Frunze's death in 1925 Stalin's principal deputy in military matters was Marshal Klementy Voroshilov. He is a typical revolutionary of the old school. Born in Lugansk (renamed Voroshilovgrad in his honor) the son of a coal miner, he was raised in poverty and drifted into subversive activities in his early youth. He joined the Bolsheviks in 1917 and during the civil war was associated with Stalin on the southern front, especially in the defense of Tsaritsyn. As he was one of Trotsky's bitterest opponents, as well as an official apologist for the anti-Trotsky faction, it was natural enough that Stalin should make him Commissar of War. Under him the Red Army became one of the strongest in the world. "Its development must be credited to efforts of the entire machinery of the state, but there is no doubt that Voroshilov's share in these efforts was considerable."[59] He was a believer in discipline, initiative, and professional competence, rather than in theories of warfare. He helped direct the military aspects of the Five Year Plans and must therefore be credited with some of the success of the Soviet war economy. Among other things, he realized the importance of decentralization of industry, the construction of shadow factories, and the eastward migration of war plants. Presumably because of failures in the Finnish war, he was replaced as Commissar of War in 1940. He was charged with the defense of the Leningrad area in 1941 but more recently has not had a field command.

Always more the politician than the soldier, Voroshilov devoted himself largely to administrative work. As an old enemy of Tukhachevsky, he was at first not inclined to give the latter any considerable role in Red Army affairs. However, Tukhachevsky's conspicuous ability led, after a banishment of about five years in relatively minor posts, to his appointment as Assistant Commissar of War and chief of the general staff. It was Tukhachevsky rather than Voroshilov who encouraged the theoretical study of military affairs, and under his general direction the Red Army developed a really first-rate military literature.[60] Tukhachevsky somewhat altered his earlier views of Soviet strategy in the light of newer developments abroad and of the growing industrialization of the U.S.S.R. In his introduction to the Field Service Regulations of 1936, for example, he rejected the theory of the special maneuverability of the Soviet soldier and hence his special capacity to take the offensive. "How the fire power of the modern opponent is to be overcome is not to be learned from this theory," he wrote. "There were some comrades, for example, who asserted that in training a Red Army man for attack, fewer shells have to be expended than in the training of a soldier of a capitalist army—a difference which they explained by the superiority of the spirit of the Red Army man. In actual warfare this self-deception can lead only to unnecessary losses and serious setbacks. The theory is incompatible with the demand made by Comrade Voroshilov with regard to military training, namely, to learn to win with 'small loss of blood.' "[61]

Although he admitted that the power of the defense was growing, Tukhachevsky nevertheless continued to believe in the offense as the only strategy which gives victory. He was well acquainted with foreign military criticism and

---

[59] Berchin and Ben-Horin, *op. cit.*, p. 104; see *ibid.*, pp. 100-104 for further biographical data.

[60] Tukhachevsky and Shaposhnikov wrote admirable studies of the Polish war. In 1927 Shaposhnikov published his book on the general staff, *The Brain Trust of the Army*. Triandafilov's work on *The Character and Operations of Modern Armies* appeared in 1929. The periodical *Voyennaia Mysl* (Military Thought) was initiated in 1937 and has sponsored excellent articles on strategy and tactics.

[61] Berchin and Ben-Horin, *op. cit.*, p. 132.

referred to De Gaulle as that "brilliant French military writer" and spoke unfavorably of the defensive doctrines of Liddell Hart. Writing in 1927, he expressed the opinion that the coming European war would be on an even greater scale and of even greater intensity than the war of 1914-1918 and admitted that it might involve trench warfare and campaigns of attrition. Nevertheless, the goal of all warfare is the destruction of the enemy's forces and the capture of his sources of economic strength; these objectives could be attained only by the offensive. In a speech in January, 1935, he said that perhaps the most important qualities for an army are flexibility and resourcefulness. "We need men who, facing an entirely new technique [such as mechanization], are capable of changing their concepts with lightning rapidity. . . . It is difficult to discard [cavalry and other concepts of the civil war] and to utilize correctly the mobility of airplanes and mechanized troops."[62]

Stalin likewise appears to have been convinced that the offensive alone can give victory. He is quoted as having told H. G. Wells: "Who wants a military leader incapable of understanding that the opponent is not going to surrender —that he must be crushed?"[63] But along with all other persons connected with Soviet military affairs, he knew that no strategy could be successful unless supported by the full mobilization of the nation's economic and moral resources. Hence the significance of Stalin's role in Soviet war doctrine and in modern military history is to be found not in statements of tactical and strategical theories but in his achievements in industrializing the Soviet Union, in training its men, women, and children for industry and modern mechanized war, and in fostering in the population as a whole that psychological preparedness which has been so invaluable in the resistance to the Nazis. Stalin's regime prepared for total war on a scale which few persons in the outside world even remotely suspected or comprehended. Stalin did not earn the self-conferred title of marshal on the field of combat but in his study, devising ways and means by which the economic resources and the man power of his country could be so utilized as to realize fully its vast military potential.

That the principal motivating force of the Five Year Plans was the fear of war and defeat is highly probable, if not certain. As far back as 1924 Frunze had pointed out that because of the backwardness of Russian industry as a whole, and the primitive character of its automotive industries in particular, the Red Army could not be increased in size or improved in quality and could hardly hope therefore to compete with others on anything like equal terms. Furthermore, the Russian soldier was almost entirely without mechanical training or mechanical sense, which only the large-scale industrialization of the nation could provide. Voroshilov was even more aware of the technical backwardness of the Red Army; he described the state of war industry before 1928 as "chaos and disorganization" and as "the sore spot of our economy," a poten-

---

[62] Concerning the evolution of Tukhachevsky's views see White, "Soviet Philosophy of War," *loc. cit.*, pp. 342-346. For the quotation see Michael Freund (ed.), *Weltrüstung. Geschichte der Umwälzungen des Wehrwesens der Nationen im Jahre 1934-1935 in Dokumenten* (Essen, 1934-1935), p. 76.

[63] Berchin and Ben-Horin, *op. cit.*, p. 130.

tial cause of military defeat. He warned, also, that the railways and other internal communications in Russia were altogether inadequate to the needs of modern war. He wanted the Red Army raised to the efficiency of other armies without any increase in numbers, because quantity in war is no adequate substitute for quality. Hence, in his judgment, the first and principal aim of the First Five Year Plan should be to build those basic industries which were related to the production of war materials and to lay the foundation for the technical education of Soviet manhood. Throughout the period of industrialization, the equipment of the Red Army was given priority over all other demands for manufactured and semimanufactured goods, raw material priorities, and the allotment of skilled labor. A great many observers of the Russian scene during the years 1928-1938 thought that the scarcity of consumers' goods in the U.S.S.R. was due to inefficiency in the administration of industry. As events proved, however, the primary cause was a war economy which sacrificed everything to the interests of the army and military preparedness.[64]

Almost innumerable figures could be used to show the intensity and extent of industrialization during the years 1928-1941. Perhaps the most graphic single fact is the movement of people from country to town, from agriculture to industry, between 1926 and 1939, the greater part of which came after the initiation of the First Five Year Plan in 1928. In about a decade, as is shown in the accompanying chart, the industrial population of the Soviet Union increased from about sixteen per cent of the total to about forty-six per cent—almost threefold. This was made possible by a decline in farming groups from almost seventy-seven per cent of the total to less than forty-seven per cent. No such drastic shift in the economic center of gravity of a nation, in so brief a time, is recorded in the whole history of mankind, certainly not in modern times. It was made possible by collectivization of the farms, with an ensuing technological unemployment in agriculture which compelled migration to the cities. It was accomplished by the most ruthless sort of compulsion on the part of the government in Moscow. That it resulted in widespread suffering and social dislocations need hardly be said, and it is doubtful if any other country in the world could have persisted in a state policy of similar consequences to its people. It is also improbable that any leader in the U.S.S.R. other than Stalin would have possessed the iron will required to give effect to a planned economy which so thoroughly uprooted humanity. The Russian people made a terrifying investment in their future during those awful years, but, seen in the light of later developments, they undoubtedly saved the revolution and their national independence. For the war potential of the Soviet Union is built upon its geographical position, its resources, and the quantity and quality of its man power.

---

[64] Max Werner, *The Military Strength of the Powers* (New York, 1939) was one of the very first books to emphasize the economic basis of Soviet military strength; see Chap. II. More recent and probably the best general summary in English of the military implications of the Five Year Plans (1928-1932, 1933-1937, and 1938-1942) is White, *Red Army*, Chapter IX, "The Impact of Industrialization." In Russian, there is an article by A. Baikov, "The Economic Basis of the Defense System of the U.S.S.R.," in *Voyennaia Mysl* (Military Thought), November, 1939, pp. 21-36, abstracted for me in English by Mrs. V. T. Bill.

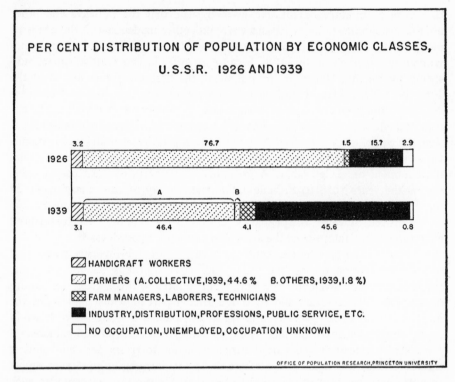

PER CENT DISTRIBUTION OF POPULATION BY ECONOMIC CLASSES,
U.S.S.R. 1926 AND 1939

HANDICRAFT WORKERS
FARMERS (A. COLLECTIVE, 1939, 44.6 %   B. OTHERS, 1939, 1.8 %)
FARM MANAGERS, LABORERS, TECHNICIANS
INDUSTRY, DISTRIBUTION, PROFESSIONS, PUBLIC SERVICE, ETC.
NO OCCUPATION, UNEMPLOYED, OCCUPATION UNKNOWN

OFFICE OF POPULATION RESEARCH, PRINCETON UNIVERSITY

But its resources would be of no military value unless converted by modern industry into the instruments of war, and its man power would be ineffective without the mechanical aptitudes which only an industrialized country can transmit to its youth, the raw material of the armed forces.[65]

From the military point of view, industrialization in Russia was not only quantitative but also qualitative. During the Second Five Year Plan defense industries were expanded about two and a half times as rapidly as industry as a whole, and under the Third Plan the proportion was to be more than three to one. The equipment of war plants and the quality of their output was definitely superior to others. In some particularly critical industries—such as iron and steel, metallurgy, and automotive—the percentages of increase and the improvement of manufactured products were spectacular. Impressive progress was made in improving and even extending railway transportation. This was possible because, as Marshal Voroshilov said, "Our country is a coordinated economic organism. Therefore it is easier for us to relate the parts, to coordinate and direct this organism according to our will in the necessary direction."[66]

[65] For population data on the U.S.S.R. I am deeply indebted to Professor Frank Notestein and his colleagues of the Office of Population Research of Princeton University. On the social consequences of the Five Year Plans see W. H. Chamberlin, *Russia's Iron Age* (Boston, 1934).

[66] Baikov, *loc. cit.* Also *The Land of Socialism Today and Tomorrow*, Reports of the Eighteenth Congress of The Communist Party of the U.S.S.R., March, 1939 (Moscow, 1939).

But even on the quantitative side, the showing was phenomenal; by 1938 the Soviet Union had lifted itself by its bootstraps from the status of a predominantly agricultural country to that of a first-rank industrial power.

The Five Year Plans involved other objectives of importance to the Soviet military effort. A vast reservoir of skilled and semiskilled labor was created, partly by industrial conscription; war industries were dispersed and thereby rendered less vulnerable to an invading army; ghost factories were brought into being and whole new cities sprang up east of the capital and even beyond the Urals; plans were laid for the eastward migration of industrial plants in wartime; the largest possible measure of self-sufficiency was sought. Out of the expanded national income, an increasingly large proportion went to expenditures for the armed forces. All of this and more was achieved at enormous sacrifice to the population as a whole, for not even in Nazi Germany was the butter of civilians so completely converted into guns for the army.[67] The test of any such policy is, of course, its ultimate results. Only the Russians themselves can say whether their survival as a nation was worth the price they paid over the years 1928-1941.

It should be noted, however, that the Soviet Union did not lift its production levels, even of war materials, to those of Nazi Germany or raise its war potential to that of the United States. "This meant that for a long war against Germany, or a combination of Germany and Japan, the U.S.S.R. was severely handicapped. . . . While it had accumulated a tremendous reservoir of military engines and weapons, its power to replenish this reservoir was still insufficient to enable it to wage a long war without importation of war materials in large quantities from abroad."[68] Hence Stalin is wise enough to avoid a two-front war by preventing, at all costs, a break with Japan, since war with another major power would enormously increase the demand for materiel and would menace Russia's continental and overseas communications. Hence, also, he knows that, as Trotsky said, Russia can fight best as a member of a coalition and that the ultimate triumph of the Soviet Union, by means of a Red Army offensive, can be achieved only if lend-lease materiel continues to arrive in large amounts from the United States and Great Britain. The insistence upon a second front is an understandable demand for even further implementation of Russo-Anglo-American coalition warfare.

Stalin seems to have realized from the very beginning that back of the new industry of the Soviet Union must stand a united nation keenly conscious of the imminence of war and prepared to take an active part in it to the last man and last woman. As already has been said, the Russian people have been war-minded ever since the foreign interventions of 1919-1920. This war-mindedness has been sedulously cultivated by the government through the press, the radio, the party organization, the Red Army, and the great annual celebrations

---

[67] Ambassador Joseph E. Davies was in some ways a naive observer of the Soviet scene. But he was quick to see that total preparation for total war was the key which opened many otherwise insoluble riddles. His *Mission to Moscow* (New York, 1941) is for that reason an important book.

[68] White, *op. cit.*, p. 352. I have omitted italics from the original.

of November 7 in Red Square. The theme of the official propaganda has been that the U.S.S.R., a nation of workers and peasants, is encircled by capitalist states. The "capitalists and imperialists"—that is, the whole non-Soviet world —are, by the nature of things, hostile to Soviet society and the Soviet state. Naturally enough, the existing fear of war was enormously increased after the advent of the Nazis, with their revival of German imperialism, their anti-Bolshevik propaganda, their anti-Comintern Pact, and their ascendant military power. By 1938 Hitler had at his command the most formidable army in Europe, which he stated repeatedly was for use in securing *Lebensraum* at the expense of the Soviet Union. Every reader of *Mein Kampf* knew that Hitler had denounced the leaders of Russia as "common bloodstained criminals . . . the scum of humanity," belonging to "a nation which combines a rare mixture of bestial horror with an inconceivable gift of lying." Strident voices shouted these and similar sentiments over the German radio from 1933 to 1943, with the exception of the brief interval of the Hitler-Stalin agreement of 1939-1941.

But even before the advent of Hitler the psychological preparation of the Soviet peoples for war was well under way. Building on fear and its accompanying communist zeal and Russian nationalism, Stalin was able to make "military preparedness, the art of warfare, and the science of war the everyday occupation of Russia's workers, peasants, students, and civil servants."[69] A quasi official organization—*Osoaviakhim*—formed by the merger in 1927 of *Oso* (for defense) with *Aviakhim* (aviation and chemistry)—was the principal agency through which the mobilization of the civil population was effected. *Osoaviakhim* was founded on the principle that, as the entire population must take part in the coming war, the entire population must be actively and adequately prepared for it. It helped make the nation mechanically minded for the era of mechanized warfare and defense minded for the tasks of active resistance to an enemy. It had a membership of about eleven millions in 1931, and the goal for the following year was almost twice that number. It taught courses in technical warfare, in marksmanship (including sniping), various phases of military aviation, gas warfare, air raid defense, meteorology, gliding and parachuting, military communication and administration, and almost every other subject which could conceivably be related to the war effort—all of this for civilians. To cite specific achievements, literally hundreds of thousands of Russians were instructed to handle firearms, and hundreds of thousands more were taught to drive motor cars and trucks. In short, "a general knowledge of warfare was provided to the whole population, and specialized knowledge was made available to substantial numbers of Soviet citizens through organized instruction and training." Nothing quite like *Osoaviakhim* exists outside of the U.S.S.R. To it must be assigned a large share of the credit for the heroic resistance of the entire population of the Soviet Union to the German invasion. Without it, it is difficult to believe that the bright pages of Leningrad, Stalin-

---

[69] Berchin and Ben-Horin, *op. cit.*, p. 149. Chap. X, "Total Preparedness" is an excellent discussion of the subject.

grad, and Sevastopol could have been written or that mobilization and defense against air raids could have proceeded so efficiently and smoothly.[70]

It would be impossible to pay adequate tribute to the magnificent contribution which women have made to the total defense of the U.S.S.R. By the statute of August 8, 1928, they were accepted as volunteers in the armed forces and were subject to conscription for specialized duties. Although their greatest service has been in noncombatant work, they have served in the ranks of the army on the same basis with men in several branches of the service. And as about sixty per cent of all the physicians and surgeons in Russia are women, their contribution to the medical corps has been indispensable. In no country of the world have women done so much, so soon, and so efficiently for the cause of national security.[71] From the very beginning the Soviet concept of total war has recognized no barriers of sex. The enlistment of women in war activities was in accordance with the Marxist ideal of the nation-in-arms and the Marxist belief that the army must be inseparable from the whole people.

To mobilize for total war it was imperative that the whole nation, not merely its workers and peasants, be recognized as having a stake in the Soviet Union. To be sure, Molotov told the Party Congress of March, 1939—within a few days of the Nazi rape of Czechoslovakia—that there no longer were any classes, that the policies of Stalin had for all time done away with the division of society into the exploiters and exploited. This may not have been true. But whatever the facts, the Soviet government proceeded to remove certain of the legal remnants of class discrimination and class consciousness. For example, the Constitution of 1936 had made military service an obligation of all citizens, in contradistinction to the earlier statutes of 1925 and 1928 which had restricted the army to workers and peasants. The universal military service act of 1939 now gave effect to the constitutional mandate: all male citizens, ages 18 to 40, were made liable to service in the armed forces, regardless of race, nationality, creed, education, or social origin.[72] Unless one reads the legislation, decrees, proclamations, reports, and speeches which had gone before, one cannot fully appreciate the fundamental changes involved in this new legislation. In part, the new order of things was necessitated by the growth of the Red Army and Red Air Force and their demands for increasing numbers of noncommissioned officers, officers, and enlisted men with broader educational background. But the factor of growing national patriotism must not be overlooked, particularly in view of the general European war which was in the process of materializing.

There were definitive indications even before 1939 that Stalin had ordered increasing emphasis upon Russian nationalism as the emotional driving power

---

[70] On mobilization and air raid defense see H. C. Cassidy, *Moscow Dateline, 1941-1943* (Boston, 1943). For *Osoaviakhim* see Berchin and Ben-Horin, *ibid.*, and White, *op. cit.*, Chap. IX.

[71] For a brief but effective statement see Berchin and Ben-Horin, *op. cit.*, pp. 152-153.

[72] For the new Soviet constitution of 1936 see Anna Louise Strong, *The New Soviet Constitution* (New York, 1937). For the statute of 1939: White, *op. cit.*, pp. 363-366; K. Voroshilov in *Voyennaia Mysl*, September, 1939.

of total mobilization. The exigencies of the coming war transcended the gospel of international revolution as preached in earlier days. As far back as 1934 the patriotic note had been sounded by *Pravda*: "For the fatherland! That cry kindles the flames of heroism, the flame of the creative initiative in all fields in all the realms of our rich, our many-sided country. . . . The defense of the fatherland is the supreme law of life. . . . For the fatherland, for its honor, glory, might, and prosperity."[73] In 1937 there was a nationalistic field day in celebration of the 125th anniversary of the battle of Borodino; from then on there was almost a "cult of 1812," fostered by the army paper *Red Star* and the magazine *Military Thought*, as well as the more popular press. The goal of all this, as was revealed frankly later, was the stimulation of national unity, the revival of interest in the glorious traditions of Russia, and the exaltation of a new patriotism.[74] The new oath required of soldiers of the Red Army eliminated internationalism and soft-pedaled the class struggle. It bound the soldier to "learn the art of war conscientiously," to be "an honest, brave, disciplined and zealous soldier," and to respond at the first call "to defend my homeland, the Union of Socialist Soviet Republics" with "manliness and judgment, with dignity and honor, sparing neither my blood nor my life for the cause of complete victory over the enemy." This was a far cry from the earlier oath of allegiance, which implied a large measure of loyalty to the international proletariat.[75]

Since the outbreak of war with Germany the patriotic motif appears in all official pronouncements and in all of Stalin's speeches. The daily communiques appear in *Pravda* and *Izvestia* under the heading "Lastest news from the front of the Patriotic War." In his broadcast of July 3, 1941, Stalin spoke of the war with Germany as a "war of liberation," a "war for the freedom of our country." On November 6, 1942, he spoke again of the "patriotic war against the German invaders" and "the independence and liberty of our glorious Soviet motherland." In an order of the day of August 5, 1943, he hailed the troops who died during the recapture of Orel and Belgorod as "the heroes who fell in the struggle for the freedom of the motherland." The field marshals of the eighteenth century, Russia's military golden age, are constantly held up as inspirations for the present generation: Suvorov and Kutuzov now receive more notice in official pronouncements than Marx and Engels. Stalin the nationalist, leader in the war of liberation, marshal of the Soviet Union, has traveled a long way since the days of Lenin, who ridiculed patriotism. Stalin, who announces that not an inch of Soviet soil shall be yielded to Nazi aggressors, is not the same brand of leader as the Lenin who said that "he is no socialist who will not sacrifice his fatherland for the triumph of the social revolution." This is no mere ideological somersault on the part of Stalin. It is rather his recognition of the patent fact that the total resources of the U.S.S.R. could be mobilized only if an appeal were made to *all* the people and to the

[73] English translation from Max Eastman, *Stalin's Russia* (New York, 1934), p. 26.
[74] Footnotes 283-287 of White, *op. cit.*, Chap. X.
[75] Text of English translation in Wollenberg, *op. cit.*, p. 264.

same Russian patriotism as animated the men of 1812.[76] Stalin is not the type to be a modern Bonaparte, carrying the revolution to all Europe by force of arms.

Meanwhile, radical changes were being effected in the Red Army. In March, 1934, as soon as the first material results of the Five Year Plan permitted, the number of troops in the standing army was increased from 560,000 to 940,000. The following year there was a further increase to 1,300,000. In 1935, also, the Far Eastern Army was made an autonomous and self-sufficient force. By January, 1936, seventy-seven per cent of the Red Army were in the regular forces and only twenty-three per cent in the militia—reversing the ratio of 1924. In March, 1939, the Red Army was put entirely upon a regular basis. "The territorial system, as the foundation of our armed forces," Voroshilov told the Party Congress, "came into contradiction with the requirements of the defense of the state, as soon as the principal imperialist countries began to increase their armies in size and to place them on a war footing even in peacetime."[77] While expansion and changes in organization were going forward, the Red Army was being completely reequipped and again reequipped, as the dividends of industrialization became available; to all intents and purposes it became an entirely new army. The Red Air Force was being built up at a rapid pace and, despite the gloomy comments of some foreign critics, was becoming a formidable weapon. Enormous reserves of munitions, ordnance, and materiel were accumulated. By legislation of 1939-1940 the officer corps was professionalized, a hierarchy of rank was created (including the formerly detested titles of general, admiral, and field marshal), pay and salaries were increased, improved housing and other perquisites were made available, especially to officers. Discipline was tightened and compulsory salutes restored. Every opportunity was given younger officers to earn promotion.[78] As a result of the poor showing made by the Red Army in the early phases of the war with Finland, Marshal Timoshenko was charged with the responsibility of bringing the armed forces to a higher state of efficiency. This he succeeded in accomplishing by the strictest enforcement of the foregoing and other reforms, so that by the time the Germans launched their attack on the U.S.S.R. in June, 1941, the Red Army had acquired a high degree of effectiveness.

Another momentous change in Soviet policy indicated how far Stalin was willing to go in the interest of military efficiency. The question of division of authority between officers of the Red Army and the political commissars had been a subject of bitter debate and complaint ever since the civil war. During the years 1935-1940 the authority of the commissars was under constant scrutiny but was not seriously curtailed. It was something of a surprise, therefore, when, on August 12, 1940, the post of commissar was abolished. It was restored after the German invasion but was again abolished in October, 1942.

[76] For the "party line" on Suvorov, Bagration, Kutuzov, and Alexander I see Eugene Tarlé, *Napoleon's Invasion of Russia, 1812* (English translation, New York, 1942).
[77] Quoted in White, *op. cit.*, p. 360.
[78] See Werner, *op. cit.*, Chaps. II and III; White, *ibid.*, Chap. X.

The commissar system was important partly because of the emphasis it placed upon the revolutionary role of the army and partly because it constituted a curb on the suspect officer class. For these reasons the end of an institution of such prestige is significant to an understanding of the increasing power of the army as an independent body.

One of the striking features of the Russo-German war has been the high quality of staff work in the Red Army. This has been the more remarkable because it had been freely prophesied by persons outside the Soviet Union that the great purge of 1937, which removed Marshal Tukhachevsky and others from the rolls, would disrupt the high command. Such might well have been the case had it not been that the progress of the Five Year Plans, the mobilization of the whole population for war, and the drastic changes made in the composition and organization of the Red Army gathered increasing momentum after 1937 and offset any consequences, psychological and otherwise, of the purge. Furthermore, the new chief of staff Shaposhnikov assured a certain continuity of policy and strategy. The true reasons for the execution of Marshal Tukhachevsky and other high-ranking officers are too clouded in mystery and controversy to warrant further treatment in this volume.

VI

Stalin's political strategy during the years 1933-1941 did not conform to the classical communist dogma. From the moment Hitler came to power the world was in a state of war; in part it was the latent "white war" of the Nazis; in part it was active hostilities in Ethiopia, China, Spain, Austria, and elsewhere. Only in Spain did Stalin follow Lenin's formula of turning international war into civil war, and there he was reacting as much to Hitler's and Mussolini's tactics as to Lenin's theories. Elsewhere he seems to have viewed the international scene in terms of *Realpolitik*, with the U.S.S.R. acting as a great military power rather than as a revolutionary state. The goal of his diplomacy —ably executed by Litvinov—was defense of the Soviet Union through collective security. Guided by Moscow, Communist parties in other countries participated in a common front with bourgeois groups for action against aggressors. In accordance with Lenin's advice, obstructive policies were not carried out for their own sake; on the contrary, the Communists in France and elsewhere not only supported, but took the lead in, rearmament. It was only after Munich had left the U.S.S.R. completely isolated that Stalin used international communism in support of the agreement with Hitler and as a weapon against the "imperialist war" of 1939. But the ruthless suppression of communism in Germany and Italy left Stalin with no illusions concerning the possibility of using it as a "fifth column" for aid to the Soviet Union after the German invasion of 1941.

Indeed one of the most momentous facts of our time is that it was not Stalin and Soviet Russia, but Hitler and Nazi Germany which turned the international war of 1939 into a civil war, according to Lenin's revolutionary formulas. It was fascism, not communism, which produced Quislings and Lavals and Mosleys. The Communists, after the Nazi aggression against the U.S.S.R., turned "patriotic," just as the socialists had turned "patriotic" in 1914 and had thereby earned Lenin's contempt.

Before the German successes in Poland, the Red Army had evolved a fairly coherent military strategy. As has been seen, it was a predominantly offensive strategy which had been outlined by Tukhachevsky, although both he and Shaposhnikov had visualized the possibility that a European war would be on so vast a scale and of such a degree of intensity that it would become a war of attrition in which Russia might have to resort to the defensive for a time.[79] A doctrine for such a contingency had therefore been worked out with some care and had been incorporated in the new Field Service Regulations of 1936. It was based upon the concept of defense in depth. Resistance to an invader was not to be based upon fortifications and position but was to be elastic and founded upon maneuver. Modern weapons, especially the tank, it was pointed out, could be used by an army on the strategic defensive as well as by an army on the attack. In fact, the Regulations of 1936 placed great emphasis upon the impor-tance—indeed the imperative necessity—of close integration of all arms in both offensive and defensive operations. This applied especially to aviation; and although the Red Air Force was coordinate with the army, it had not developed the theory of independent air power as advocated by Douhet and others, but was closer to the Luftwaffe's role of cooperation with ground troops.

Of great value in the tactical execution of the strategical defensive was the emphasis which the Field Service Regulations of 1936 put upon the individual soldier and the junior officer. "All sensible initiative of subordinates must be encouraged through all possible means," said the Regulations, "and must be exploited by the commander in the general interest of [success in] battle. Sensible initiative is based upon an understanding of the commander's inten-tions."[80] The problem, then, was to relate the purposes of the high command to the company commander and his men. Stalin himself understood the diffi-culty, although he suggested no definitive solution: "We leaders see things, events, people, from one angle only, so to speak from above. Therefore our horizon is more or less limited. The masses, on the other hand, see things, events, people from another angle, that is to say from below, and their horizon also is limited to a certain extent. To find a right solution of problems we must combine these experiences. Only then will management be correct."[81] It was undoubtedly a keen appreciation of the importance of individual initiative that led to so high a standard of competence of the Red soldier and hence to such great effectiveness in the defensive operations of 1941-1943.

[79] For an admirable statement of Soviet strategical concepts before 1939 see Werner, *op. cit.*, Chaps. IV, XI, XII. Concerning Tukhachevsky see above.
[80] *Field Service Regulations of the Red Army*, 1936, Chap. I, "General Principles." I am indebted to the United States Army for an opportunity to consult this document. Trans-lation is by V. T. Bill.
[81] Quoted by Marshal Tukhachevsky in "On the New Field Service Regulations" in *The Bolshevist*, May-August, 1937. Translation by V. T. Bill.

It was fortunate for the U.S.S.R. that, despite almost idolatrous worship of the offensive, it had made adequate preparation for defense and, particularly, defense in depth. For the German conquest of Poland in 1939 and the collapse of France the next spring required a complete reexamination of all former military doctrine. A characteristic of Soviet attitude in domestic affairs had been adaptability, a willingness to change the party line (while seeming to keep it the same), and an absorbing interest in attempting the new and untried. This flexibility, easily applied to the military sphere, has stood Stalin in good stead. Readiness, indeed eagerness, to alter existing plans in the face of new conditions has been an outstanding virtue of the Red Army and its personnel during the stresses and strains of 1941 and 1942. Even the junior officers seem to have understood the necessity for constant adaptability to the unprecedented problems which arose in meeting the German assault.[82]

After the collapse of Poland, the U.S.S.R. was faced with a new situation as regards the security of its western borders. A first step in a new strategy of defense was the extension of existing frontiers, so that adjoining territories might be used as a buffer against the first impact of a German attack. In September, 1939, the Red Army marched into Poland and obtained by agreement with Hitler a substantial block of territory with a natural boundary in the Bug River. In March, 1940, came the attack on Finland, resulting in a "rectification" of the Karelian frontier and the acquisition of defensive positions in the Gulf of Finland. The following June Bessarabia was "ceded" by Rumania, and in August, 1940, the three Baltic republics were occupied. The result of all this territorial expansion was that the Soviet Union acquired a cushion against the initial speed and surprise which were inherent in blitzkrieg. In other words, the zone of defense was substantially deepened.

The German campaign in France in the spring of 1940, which was not very different from the campaign of 1939 in Poland, provided the Red Army with a blueprint of the attack against them which was to come a year later. The Germans, the Russians reasoned, would depend upon surprise and speed, aerial assault upon communications and services of supply, mobile warfare aimed at encirclement and annihilation—the most gigantic Cannae in all history. Hitler was determined to try what Falkenhayn, Seeckt, Leeb and others had always thought could not be done—to deliver a knockout blow to Russia within a relatively brief time. The Russians were reasonably sure that unlike the Low Countries they could not be overrun and that unlike Poland they could not be paralyzed by aerial assault. But they knew that they had a prodigious task on their hands in meeting an invasion of such tremendous scale and intensity. It is doubtful, however, that they could have imagined even vaguely the purgatory through which they were to pass before, in the summer of 1943, they could seize the initiative.

What the Russians had to do was fairly obvious. They had to keep the Red Army intact, "in being," at all costs. They had to avoid encirclement as far as possible; such units as could not escape were to resist to the last. They must

<hr>

[82] Leonard Engel, "The Red Officer Corps," in the *Infantry Journal*, LII (1943), 18-24.

trade space for time—that is to say, they must bring about a protracted war by compelling the Germans to punch deep into Soviet territory without obtaining a decision. But the territory which the *Wehrmacht* acquired must be made virtually useless by wholesale devastation and rendered insecure by incessant guerrilla warfare. The resulting warfare of attrition and extended lines sooner or later would give the Red Army the great opportunity for which it had been trained and indoctrinated ever since the civil war—the opportunity to destroy the enemy by an offensive. "According to the [new] Soviet concept, blitzkrieg comes at the *end* of the war, not at the beginning."[83]

In evolving a strategy of retreat for 1941 the Red Army was completely unaffected by the defeatism of Weygand and Pétain, but rather was adopting the policy of active defense which had been ably advanced by Field Marshal Ritter von Leeb of Hitler's army.[84] The facts of geography and the force of historical tradition must have been almost equally persuasive. Space and cold and rain and mud have always stood in the way of the would-be conqueror of Russia—natural barriers perhaps even more formidable, under conditions of mechanized war, than rivers or mountain ranges. When the storm broke over the Soviet Union in June, 1941, the minds of people everywhere traveled back to 1812, the name of Napoleon was on the lips of millions. It is not unlikely that Kutuzov, the quaint and gallant figure of Borodino, stood in spirit at Stalin's elbow during the awful months June to December, 1941. Some of Stalin's own public utterances indicate that such was the case.[85]

A word needs to be said, too, about the exiled, discredited, and murdered Trotsky. He had always warned against a dogmatic view of strategy which sought to be all things for all occasions. He had advocated adaptability and elasticity as being more suited to a revolutionary society and more in accord with sound military principles. Time has vindicated Trotsky's judgment.

But Stalin is a titan in his own right. It took a heart of oak, nerves of steel, and veins of ice to assume the responsibilities which were involved in the Great Retreat. The stature of Marshal Stalin may be measured by the fact that his decisions were military decisions, not decisions of prestige for himself and his regime. There must have been doubts in many minds as to whether any dictatorship could stand up under the long series of blows which threatened to pulverize the Soviet Union during 1941 and 1942. But not once did Stalin, like Hitler, distort the fundamental truths of the situation or subordinate the goal of ultimate military victory to the momentary demands of popular morale. There is something awe-inspiring in Stalin's broadcast to his people of July 3, 1941,

[83] Max Werner, *Attack Can Win in '43* (Boston, 1943), p. 35.

[84] *Die Abwehr* (Berlin, 1938). English translation by S. T. Possony and D. Vilfroy, *Defense* (Harrisburg, Pa., 1943). There is, of course, another thesis—that the Russians never abandoned their offensive strategy and were badly beaten in attempting to carry it out. This seems to me less probable than the belief that they reverted to the classic strategy of defense, as in 1812.

[85] *Cf.*, for example, the speech of November 7, 1941. Kutuzov, of course, abandoned Moscow, whereas Stalin was determined to hold it at all costs. But on the basis of broader strategy the comparison is probably sound.

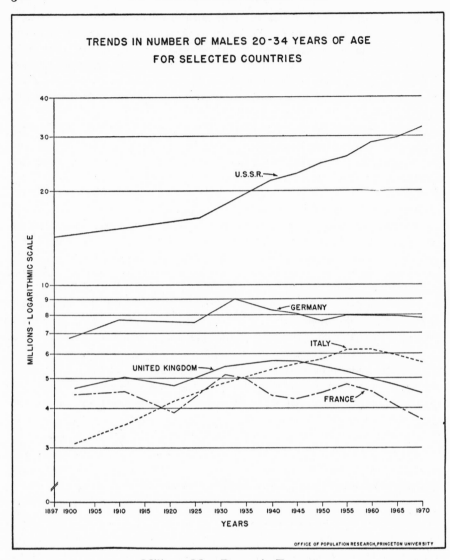

Military Man Power in Europe

exhorting them to scorch the earth and to fight as a nation of guerrillas. It was magnificent, it was terrifying—and it was war.

"In case of a forced retreat of Red Army units," he said, "all rolling stock must be evacuated; to the enemy must not be left a single engine, a single railway car, not a single pound of grain or a gallon of fuel.

"Collective farmers must drive off their cattle and turn over their grain to the safekeeping of State authorities for transportation to the rear. All valuable property including nonferrous metals, grain and fuel which cannot be withdrawn, must without fail be destroyed.

"In areas occupied by the enemy, guerilla units, mounted and on foot, must be formed, diversionist groups must be organized to combat enemy troops, to foment guerilla warfare everywhere, to blow up bridges, roads, damage telephone and telegraph lines and to set fire to forests, stores and transports.

"In occupied regions conditions must be made unbearable for the enemy and all his accomplices. They must be hounded and annihilated at every step and all their measures frustrated."

Under the guidance of the Kremlin and under the leadership of brilliant young generals who won recognition in the inexorable tests of combat, the Red Army proceeded to carry out its long-range war plan.[86] Active defense as conducted by the Soviet forces meant, in the words of the Soviet analyst Professor Minz, "fighting for every inch of territory, holding on to every village and town as long as possible to gain time, bleeding the enemy as much as possible, inflicting the greatest possible losses upon him, wearing down his forces and launching frequent and impetuous counter-attacks."[87] Marshal Timoshenko's chief of staff, General Sokolovsky, described these tactics by the picturesque and illuminating term "blitzgrinding." Active defense required an encircled unit, large or small, to continue to give unrelenting battle to the invader—what the Germans, in frustration, called "senseless resistance." It also included carefully organized guerrilla warfare, of which the Russians have been past masters throughout their history—a type of warfare to which modern armies are particularly susceptible because of the complicated character of their weapons, equipment, communications, and supply. It meant devoted and courageous sacrifice by tens of millions of people in every walk of life. The U.S.S.R., in fact, has come nearer the goal of the nation in arms than any other nation in history.

But throughout the long ordeal of defensive war, the goal of the Red Army has always been an eventual offensive which would bring victory through annihilation of the enemy's forces in the field. The offensives of the winters of 1942 and 1943 succeeded partly because the Russians were prepared for active operations in subzero temperatures and the Germans were not. But there is little doubt that the Russian summer campaign of 1943 will demonstrate, as Stalin has said, that the ability of the Red Army to take the initiative is not confined to the months December-March.

VII

It would be a rash man who would forecast Russian military strategy in Europe and the Far East after the German invader has been crushed. But the foundation of that strategy almost certainly will be the strategic security of

[86] The best account in English is by Max Werner, *The Great Offensive* (New York, 1942), Chap. X.

[87] Quoted by C. L. Sulzberger in a dispatch to the *New York Times*, June 4, 1943. During the month of June Mr. Sulzberger wrote a remarkable series of articles for the *Times* on the Red Army. See also his article "Now It Is the Blitz-Grinders' Turn," in the *New York Times Magazine*, August 1, 1943, pp. 7, 26.

the U.S.S.R. As Mr. Churchill has said with characteristic insight and eloquence, Russian policy "is a riddle wrapped in a mystery inside an enigma; but perhaps there is a key. That key is Russian national interest."[88]

Time alone will tell whether Stalin will be judged in the future as more Russian than Peter the Great or more Communist than Lenin. But there are straws in the wind by which the student of politics may chart his course. And chart a course every statesman must, for the existence of Russian military power of the present magnitude is an entirely new factor in European and world politics.[89] Russia has all the elements of fundamental strength, all the essential factors of the war potential. She has a vast territorial domain, superabundant natural resources, a vital and growing population,[90] an industrialized society, and a dynamic political system. Other nations may wane, but the U.S.S.R. and the U.S.A. will emerge from the present struggle in overwhelming strength. It is even probable that before the twentieth century has run its course, the Soviet Union will be the most powerful nation in the world. How she uses her vast power is portentous for the future of all mankind. The Moscow and Teheran conferences have given us reason to hope that Russian power may be a stabilizing influence in peace, as it has been a determining influence in war.

[88] Broadcast of October 1, 1939.
[89] Werner, *Military Strength of the Powers*, Chaps. I and II.
[90] See Frank W. Notestein, "Some Implications of Population Change for Post-War Europe," in *Proceedings of the American Philosophical Society*, Vol. 87 (1943), pp. 165-174.

CHAPTER 15. Maginot and Liddell Hart: The Doctrine of Defense

BY IRVING M. GIBSON

IN 1939 the French nation went to war with a spirit different from that at the beginning of the World War. In September 1914, when the Germans were marching toward Paris, the old fire of French patriotism, burning fiercely as in the days of Malplaquet, of 1793 and 1870, produced the élan which stopped the invader at the Marne. In 1940 there was little enthusiasm. The French armies after the first disasters caused by the sudden German attack could not be roused to the courage of despair, and retreats could not produce the rebound of counter offensives. They ended in surrender.

This fatalistic spirit and lack of enthusiasm can be traced back mainly to two causes: the violent political quarrels which often involved the army and the great loss of life suffered by the French in the World War. There were many other factors, for instance, the reduction of the time of service to an inadequate period, which helped to create a trend of thought culminating in the passive, almost lethargic attitude of the French at the beginning of the war. But it was the political and military factors which were mainly responsible for the fall of France.[1]

When the Third Republic emerged from the war of 1870-1871 and began to rebuild its defenses, the discussions in the legislature on the reorganization of the army were from the outset dominated by political considerations. The principle of the nation-in-arms, on which the Prussian army had been built, proved beyond doubt that French security in the future could not be guaranteed unless the same system were adopted. On the other hand, the Commune of Paris had frightened Thiers and the bourgeoisie so much that they refused to accept the new principle in its entirety, because a short-service army, presupposed by it, did not seem sufficiently reliable. They wanted a large police force, and thus the defense of *la patrie* and the defense of property produced divergent armament policies.

Finally a compromise was reached in 1873 by dividing the annual contingent into two halves, one serving for five years, the other, the so-called *deuxième portion*, serving for six months. After 1873 politics could no longer be eliminated from French military legislation. The division of the army into two parts, national militia and highly efficient professionals, merely reflected the division of the nation into republicans and royalists, later radicals and conservatives—in short, into left and right.

The political struggle reached its height during the Dreyfus affair and its

---

[1] This evidence was furnished by the Riom trial conducted from February 19 to April 3, 1942, where astonishing revelations were made about conditions in the army, particularly about political intrigue in the highest places, civilian and military alike, undermining morale and discipline. There are no official publications available on the Riom trial, the dispatches of representatives of American newspapers being the principal source of our information.

aftermath, when the army was purged of many strong personalities suspected of harboring reactionary designs.

"The Republic of 1875 lived in the fear, inherited from December 2, 1851, of 'coup d'état generals.' It hoped that it rooted out the brood after the Dreyfus affair. In fact it had succeeded only too well."[2]

The term of service in 1905 was reduced to two years. In 1913 came a reaction regarding the length of time of service but it was temporary and forced by external circumstances. German military legislation early that year increased the effectives of the Reich's peacetime establishment to more than 800,000 men. To balance the German increase the French government had to return to the three-year service, because the low birthrate in France was capable of furnishing the equivalent of only half of Germany's annual contingent of troops. This reaction, however, spurred the left to new efforts. In the 1914 elections the socialists registered a large increase in their vote and, flushed with triumph, they came to the Chamber of Deputies vowing to put an end to the "folly of armaments."[3] On November 11, 1918, they picked up the thread of pacifism which they had been obliged to drop when the World War broke out. They demanded a return to the military law of 1873 and the uniform extension to the whole army of the short service of its *deuxième portion*. Their principal source of inspiration was the book of their great leader Jean Jaurès, *L'armée nouvelle*, published shortly before the outbreak of the war, which demanded the organization of a national militia of eight months' service.[4] As in 1913, however, external influences again defeated the plan to establish a short-service army. The conflict with Germany over reparations and the Ruhr crisis played into the hands of the right and its great expert on military questions, Colonel Fabry, the rapporteur of the government's project.[5] The contest between the two views was long and bitter. Finally a compromise solution was accepted which reduced the term of service to eighteen months.[6]

In the elections of 1924 the *cartel gauche* emerged victorious and the Herriot government immediately proceeded to revise the law of the previous year. It was decided not only to reduce the term of service but to reorganize the entire military system of France, carrying out the principle of the nation-in-arms to the last letter. For this purpose organic laws were prepared envisaging the mobilization of all the resources of the nation in case she should have to fight

[2] A. Géraud ("Pertinax"), "Gamelin," *Foreign Affairs*, XIX (1942), 318. Géraud's remark that the Republic "succeeded only too well" refers to the fact that there was a dearth of talent and initiative in the French army in 1939, producing only weak men like Gamelin.

[3] L. Souchon, *Feue l'armée française* (Paris, 1929), p. 72.

[4] Some went even farther, demanding complete disarmament (Souchon, *op. cit.*, p. 71) or reduction of the service to three months with a liberal admixture of long-service professionals. See Lieutenant Colonel E. Mayer, a military writer of recognized authority, in *La guerre d'hier et l'armée demain* (Paris, 1921), *passim*.

[5] General Duval, "La crise de notre organisation militaire," *Revue de Paris*, VII (1926), 759, says that it was due to Fabry's adamant attitude on the question of the term of service that the *deuxième portion* of the law of 1873 was not restored as early as 1921. See also "B.A.R.," *L'armée nouvelle et le service d'une an* (Paris, 1921).

[6] The text of the law was published in a reprint under the title, *Projet de loi ... sur le recrutement de l'armée présenté par M. Maginot, ministère de la guerre* (Paris, 1923).

for her existence again. After so much political agitation and counter agitation, it was impossible to keep the discussions within the bounds of purely technical considerations, and in order to understand the basic concepts of the organic laws of 1927-1928, we have to consider them in the light of the long political rivalry in army matters between left and right since 1873 and to remember that principles of national defense were often subordinated to political ideologies.[7]

The background of this legislation was a general *malaise militaire* which gripped the army and the nation as well. It filled the military literature of the period and became the common topic of the daily press.[8] The diagnosis of this malady is complicated and must be examined with great care. An immediate cause was the inflation of the French franc which automatically reduced the pay of the officers and men to less than half. While civil employees of the government could obtain increases through collective bargaining, the army officers became the forgotten men. When, in 1926, Poincaré was called upon to save the franc, the emphasis upon universal economy hit the army again, and the number of officers was reduced by five thousand, mostly in the higher ranks. Chances for promotion diminished and the army ceased to attract the younger generation. Military academies and officers' training schools turned out poor material, the type which had been eliminated in the competition for more promising careers.[9]

Besides financial difficulty there was a widespread faith in the efficiency of international treaties of arbitration and in the League of Nations which at that very time, the Locarno period, reached the height of its prestige. In such an atmosphere it was difficult to persuade the legislators to vote large appropriations for defense when solemn pacts heralded the outlawry of war and all men paid at least lip service to the form, if not exactly the spirit, of international conciliation.[10]

All of these circumstances furnished the advocates of a short-service militia with additional arguments. A further reduction of the term of service had, indeed, become a foregone conclusion and even the higher army officers resigned themselves to the inevitable.[11] But their resentment was nevertheless great, because they could not dissociate the program of the left concerning the technicalities of army organization from ulterior political motives. They scented

---

[7] An excellent review of the long political struggle around military legislation in France is found in S. C. Davis, *The French War Machine* (London, 1937).

[8] See Lieutenant Colonel Reboul, "Le malaise de l'armée," *Revue des Deux Mondes,* (March 15, 1925), pp. 378-398.

[9] General Debeney (*Sur la sécurité militaire de la France,* Paris, 1930), discusses the budgetary problem at great length, especially in the chapter "Politique de matériel, politique des effectives," and strongly warns against further economies. He says that because personal items in the military expenditures amount to 68 per cent, whereas material expenditures only to 32 per cent, a continuous temptation exists to reexamine the former and discover a gold mine in it for further economies. This chapter is an excellent analysis of the whole military organization.

[10] Lieutenant Colonel Reboul, *loc. cit.*, comes to the interesting conclusion that the army had no purpose and no aim, in contrast to the pre-1914 period when it had an inexhaustible inspiration in the idea of the recovery of Alsace Lorraine.

[11] General Duval, who often acted as the official spokesman of the army and had great influence in the army committee of the Chamber through his friend Fabry, furnished a complete program based on the one-year service. See the article quoted above.

discrimination, antimilitarism and deliberate attempts further to reduce their prestige. Since Clemenceau's famous decree of June 16, 1907 which gave preference to civil dignitaries over the military chiefs during state celebrations,[12] they were in a sulky and angry mood, always seeing the ghost of Jaurès and Clemenceau everywhere. They were the ones, they complained, who had saved France in the war and the reward was discrimination! It was the nation-in-arms which saved the Republic, was the retort of the left, and they would quote Clemenceau's famous dictum that war was too important to entrust it to the generals.

Long pent-up animosities easily find outlets in mud-slinging and degenerate into standing political feuds, especially so in France where conservative, rightist, even feudal ideologies among the higher officers' corps were long traditions and therefore the age-old target of radical agitation. The *malaise militaire* was caused by deep organic troubles, but its symptoms were political squabbles, journalistic denunciations and the standing feud between the generals, who were accused of fascist inclinations, and the radical leaders, who were despised by the former as antimilitarists, ignorant meddlers, and enemies of France.[13]

Under such circumstances, it should not be surprising that the birth pangs of the organic military legislation lasted so long and caused so much excitement and vituperation in the entire nation. The reorganization of the country's military system furnished the opportunity for a battle royal between two principles as old as war itself: a national militia of free citizens numbering millions and a small elite force of long-service professional or semiprofessional soldiers. French history is a great source of inspiration for contenders of both principles. On the one hand, there were the *levée en masse* of the Convention which brought so much glory to revolutionary France and the citizen army of Gambetta which so hopelessly succumbed before the Germans; on the other, one could point to Napoleon's professionals who carried everything before them and the praetorian guard of Napoleon III who ended so ingloriously at Sedan. There were, moreover, the standing examples of the British and Swiss military systems representing the two extremes in the argument. After 1924 the debates on the question were endless. There were counter proposals by Fabry, Daladier and others which merely increased the already manifold aspects of the question.[14]

The main factor, which practically determined the outcome of the discussions,

---

[12] Cf. Davis, *op. cit.*, p. 80.

[13] There is sharp criticism of the "politicians" in General H. Mordacq's works: *Les leçons de 1914 et la prochaine guerre* (Paris, 1934), *Faut-il changer le régime* (Paris, 1936), and *La défense nationale en danger* (Paris, 1938). A violent enemy of the left, his attitude is very characteristic of the fascist-inclined French officers so much derided at Riom by the defense.

[14] A brilliant and impassioned discussion of the numerous issues involved may be found in the article of an anonymous writer, "XXX," "Notre reorganisation militaire," *Revue Politique et Parlementaire* (September 10, 1926), pp. 371-406. The government proposal, with a long *exposée des motifs* including the more important counter proposals, was published in a separate reprint entitled *Rapport fait au nom de la commission de l'armée chargée d'examiner le projet de loi relatif au recrutement de l'armée*, ed. M. Paul-Bernier (Paris, 1927). The part of this publication entitled "Conditions préalables," is especially important because it contains the fundamental principles of the new army organization.

was the deep-seated desire to approximate the ideal of a national militia as closely as possible. The result was the reduction of the term of service to one year, which naturally carried in its wake the reduction of the peacetime effectives. To offset this, a large professional element had to be added to the yearly contingent of 240,000 recruits so that a peacetime army of 400,000 men resulted, not counting the colonial forces. On the surface it seems paradoxical that the left, which had always shown an instinctive dislike for professional soldiers, should have made concessions sacrificing its long-sought-for ideal of a citizen army based on large reserves and trained by a small active force. The amalgamation of the principle of the nation-in-arms and the long-service professional idea was, however, recognized as an inevitable necessity because of the general belief that the German Reichswehr of 100,000 mercenaries could at any time be increased by the 150,000 state police and brought by secret organizations to a grand total of 400,000. A further powerful argument was that this force could be hurled against the French frontier at a moment's notice in accordance with the military theories elaborated by General H. von Seeckt, the commander in chief of the Reichswehr and the originator of the idea of lightning war or *attaque brusquée*.[15]

History repeated itself. In 1913 German armaments determined French military legislation and forced the three-year service upon the country. In 1927-1928 conditions in Germany, or the interpretation of such conditions by the French military leaders, vitiated the program of the left, which was intended to place the French army on almost the same basis as the Swiss national militia. The principal argument of the military men, always presupposing a German blitz invasion, was that the army should maintain a permanent strong frontier guard, the *couverture*.[16] While such a force would keep the enemy at bay at the beginning of hostilities, the complicated process of mobilization could go on undisturbed in the hinterland. Mobilization takes time, especially if the peace effectives are low. The old system elaborated by the Germans before 1866 and then copied by every military power after 1870, maintained large effectives in peacetime which were merely supplemented and brought to war footing by the inclusion of reserves during mobilization. The new French legislation rejected this practice and in that rejection lies its fundamental character.[17] Whereas formerly training and instruction, mobilization and the guarding of the frontier were all prepared and taken care of by the peacetime army, the new system produced three different organizations: a permanent *couverture*, another permanent element to train and instruct the annual contingent, and a third per-

[15] This was the official view based on the reports of the head of the Allied Military Control Commission in Germany, General Nollet, whose influence upon French military legislation was stated during the opening debates on the Maginot line. See Chambre des Députés, *Débats*, CXXXIX² (1929), 1723. General Nollet later published his views in *Une expérience de désarmament; cinque ans de contrôle militaire en Allemagne* (Paris, 1932).

[16] See General Debeney, "Le problème de la couverture," *Revue des Deux Mondes*, 8th per., XXXVI (1938), 268-294.

[17] See the comprehensive technical discussion in General Brindel, "La nouvelle organisation militaire," *Revue des Deux Mondes*, 7th per., LI (1929), 481-502.

manent staff to prepare mobilization and maintain a framework which would be filled in time of war by the masses of trained reserves. All three permanent elements consisted of professionals of different character and status.

In view of these basic elements of its military organization, France had no army in peacetime in the old sense of the word. She had only a permanent frontier guard and 240,000 recruits under training. As soon as one contingent was well trained, it would be sent back to civilian life. In the old prewar system it was the trained contingent of the previous year, kept under the colors for an additional year or two, which formed the real peacetime army. To the chagrin of the generals, this practice was entirely abolished and the French army henceforth was to exist only as a reserve, if we use the old vocabulary. In reality it was a new conception, a nation-in-arms created in conformity with geographical limitations and other circumstances beyond the control of France. This was the system with which France went to war in 1939, although the two-year service was reintroduced in 1935 to offset the effect of the rapid fall of the birthrate during the World War years.[18]

The critics of the law said that the French army was broken into three distinct parts, or four, since the colonial army received an independent organization with two years' service; that the establishment of so many distinct units with separate functions, organizations, and status would be the breeding ground for rivalries and intrigues which would aggravate the *malaise* caused by politics and ideological differences; that in one year no real military spirit could be instilled into the men and that thereby the army lost its striking power; that by insisting so much on the defensive character of the military organization the spirit of initiative, aggressiveness, and determination to carry the war into the enemy's country would die down; that such an army would eventually be unable to withstand the shock of lightninglike thrust; and that Seeckt's blitz army could force its way into the very heart of the country before the nation-in-arms could get organized and begin to function.[19]

---

[18] For this cf. Marshal Pétain, "La sécurité de la France au cours des années creuses," *Revue des Deux Mondes*, 8th per., XXVI (1935), 1-20. This was a feature article especially recommended by the magazine. It urged the reintroduction of the two-year service because *attaque brusquée* was more than ever possible. The birthrate had depleted the effectives so much, said the Marshal, that some divisions had only twelve companies!

[19] The amount of literature on General Seeckt and *attaque brusquée* and the references to it in the speeches of the Chamber and the Senate are amazing. In fact a well-recognized English military analyst, Major E. W. Sheppard, "Two Generals, One Doctrine," *Army Quarterly*, XLI (1940), 105-118, comes to the conclusion that De Gaulle elaborated his *armée de métier* to cope with this danger. General Debeney, *op. cit.*, devotes a long discussion to the question of how to deal with sudden attack. In another book, *La guerre et les hommes: reflexions d'après guerre* (Paris, 1937), in a chapter on the *couverture*, he goes into an even more detailed examination of this danger. So do generals Maurin in *L'armée moderne* (Paris, 1938), and Chauvineau, *Une invasion, est-elle encore possible?* (Paris, 1939), both of which had a profound influence upon French military thought. General Seeckt's few speeches and articles dealing with blitzkrieg were published in *Gedanken eines Soldaten* (Berlin, 1929; English translation *Thoughts of a Soldier*, London, 1930). The fears of *attaque brusquée* were aggravated by the spread of the doctrines of the Italian general Giulio Douhet on air war or, as the Germans put it, *Schrecklichkeitskrieg*, or "frightfulness war." It was expertly summed up by General A. Vauthier, *La doctrine de guerre de Général Douhet* (Paris, 1935), with a preface by Marshal Pétain, and widely read in the French army.

There was much in these arguments that could not easily be refuted. Lack of offensive spirit did not worry the French because that was the general attitude of the nation. What they wanted was security from attack. They had it as long as a French army of occupation stayed in the Rhineland. When it had to be withdrawn because of international agreements France lost a very valuable strategic advantage, a *couverture par excellence*. After the evacuation of the Rhineland a new *couverture* had to be established on the French frontier and in consequence security seemed to be jeopardized. For with the Rhineland under occupation France could gain time and put her nation-in-arms system on a war footing even though a German lightning attack might have initial success. Only German territory would be lost should the *couverture* on the Rhine be forced back. But after 1929 this advantage was gone and in that very year the huge appropriations asked by the government since 1926 for the fortification of the northeast frontier were voted. This is how the Maginot Line was born.[20]

The Maginot Line had its origin in the *couverture* idea, unknown in other armies. The tradition of permanent frontier fortifications was older and stronger in France than in other countries and could be traced back to Vauban.[21] After 1870 it received a new stimulus from Serré de Rivière's system, of which Verdun of World War fame was an important bastion. The very fact that during the battle of Verdun the concrete of the forts of Vaux and Douaumont could withstand the worst German bombardment had much to do with the final acceptance of the plan to erect permanent and very elaborate fortifications facing the German frontier. There were long discussions prior to the voting of the appropriations by parliament in 1929.[22] In fact the organic laws of 1927-1928 and "the fortification on the northeast frontier," as the Maginot Line was originally called, were discussed simultaneously and Paul Painlevé, minister of war at this period, was responsible for both. The credit is due to Maginot only in so far as the huge sums were voted by the Chamber at the time when he was minister of war. Maginot merely carried out the plans, prepared to the minutest detail and handed over to him by his predecessor.[23]

There is great significance in the fact that the Maginot Line was designed and prepared by Painlevé. It was Painlevé, speaking as prime minister during the disastrous Nivelle offensive of 1917, who said to the Chamber: "There will be no more offensives."[24] Painlevé was soon replaced by Clemenceau who did believe in offensives and won the war—but with British and American help.

[20] It was exactly for the above reasons that General Debeney's *Sur la sécurité militaire de la France* was published in 1930.

[21] L. Montigny, "Les systèmes fortifiées dans la défense de la France depuis 300 ans," *Revue Militaire Francaise*, LVII-LVIII (September-December 1935, in four installments).

[22] There was also much disagreement in the Supreme War Council, according to the minister of war, Paul Painlevé, in a speech on March 4, 1927 in the Chamber. See Chambre des Députés, *Débats*, CXXXI, 484. See also the *Exposée des motifs* attached to the bill in *Documents*, CXVI, 367.

[23] See P. Belperron, *Maginot of the Line* (London, 1940), who on p. 83 says that E. Herriot proposed in 1936 that the line should be called the Painlevé line.

[24] Cf. Souchon, *op. cit.*, p. 168. The attacks against him and their refutation are summed up in P. Painlevé, *De la science à la défense nationale* (Paris, 1931), p. 122.

When the Anglo-American armies returned home, France reverted to the passive attitude which ultimately led to her downfall in 1940. Its most visible manifestation was the Maginot Line, the true symbol of the deep aversion the French nation felt for offensive warfare and best summed up by General Maurin who, as minister of war, made the following declaration in the Chamber on March 3, 1935:

"How could it be believed that we are still thinking of the offensive when we have spent billions to establish a fortified barrier? Should we be such fools as to go beyond that barrier, seeking some God knows what kind of military adventure? This I say to show you the thought of the government. Because the government, that is at least as far as my person is concerned, knows the war plans perfectly well."[25]

The Maginot Line mentality would not be complete without the Verdun legend woven around Marshal Pétain's figure. At Verdun France was saved, and the enemy received a disastrous moral blow. *All this was done by defense,* which proved to be unconquerable, and Pétain became its living symbol though he himself was not an absolute believer in this principle. But popular legends are not based on accurate estimates and scientific investigations.[26] The French mind created a glorious and at the same time a very comfortable illusion, which portrayed the World War as a heroic defensive effort which brought ultimate victory.

The Maginot Line mentality gradually emasculated French foreign policy and made the rise of Hitler, the remilitarization of the Rhineland and all the rest of German aggressive acts possible.[27] Thus we may establish a causal relationship between the nation-in-arms principle driven *à outrance* in 1927-1928, the *couverture* mentality resulting in the Maginot Line, and the Verdun legend glorifying the defensive. The sum total of it all was a passive attitude which took hold of the French national mind, giving it a sense of false security and in its ultimate effect creating an atmosphere in which the dry rot could spread in all branches of the French army, ending in what De Gaulle described as "l'obscure sentiment d'impuissance."[28]

[25] Paul Reynaud, *Le probléme militaire française* (Paris, 1937), p. 27. This little book is an invaluable source of information. A very good summary of the Maginot Line myth is in H. Pol, *Suicide of a Democracy* (New York, 1940), pp. 106-110.

[26] This is well expressed by J. M. Bourget, "La légende de Maréchal Pétain," *Revue de Paris,* January 1, 1931, pp. 57-70. The legend is cruel, the author says, because it distorted Pétain's historic significance by completely disregarding the role he had played in the second battle of Verdun and the campaign in the Riff.

[27] See Lieutenant Colonel Jean Fabry, "La stratégie générale affaire de gouvernment," *Revue Militaire Générale,* I (1937), 387-390, which discusses the government's feeble policy and points out that the type of army organized in France did not allow any bold diplomatic action at the time when Hitler remilitarized the Rhineland.

[28] In his admirable memorandum submitted to the French high command on January 26, 1940, De Gaulle predicted the piercing of the Maginot Line by the Germans and urged the formation of armored divisions. See the French text of this memorandum with English translation in *National Review,* CXV (London, 1940), 393-405. The quotation is on p. 400.

De Gaulle's entire military career in France, his agitation for an armée de métier, his fanatical belief in the superiority of the offensive were only the most brilliant examples of a long series of arguments in French military literature trying to uphold the traditions of the offensive. The merit of having been the first in the post war period to criticize severely the passivity of the military doctrine belongs to Lt. Col. E. Alléhaut (*Le guerre n'est pas*

With the Maginot Line as the background we can easily trace the origin of the military doctrine of a war of limited liability, based on defense, with which France went to war in 1939 and to which she clung even after the campaign in Poland had given a striking demonstration of the power of attack. We must bear in mind that the French army was nothing but the nation-in-arms and that the French nation was thoroughly saturated with the Maginot Line mentality. The army would naturally reflect these illusions, making it incapable of elaborating a doctrine implying initiative and vigorous action. A defensive doctrine pushed so far would naturally create sluggish mentality, slow movements, and a clumsy system of supply and coordination. This is all clearly reflected in the military doctrine worked out after the World War. Russian opinion as early as 1932 summed up the situation as follows: "Most of the French equipment is obsolete and cumbrous, the troop units are slow in movement and in manoeuvre, the calculations of the high command are too pedantic and in general the offensive power of the army is insufficient."[29]

The French military doctrine was based on the magic word fire power, a mere variation of the views based on the defensive theory. Fire power built everything around the concept of "shelter," the essential principle behind the Maginot Line. Fire power produced mainly by artillery would establish a protective cover over the army just as the heavily fortified border in the northeast would protect the nation. The origin of this faith in fire power is to be sought partly in the terrible shock caused by the heavy losses of the first great battles in 1914 and partly in the Verdun legend.

In 1914 the French army had no heavy artillery attached to the field forces, whereas heavy cannon was rumbling behind the German troops in long columns always ready to give support to the lighter field artillery. The devastating effect of the heavy barrage of lead thrown into French ranks by the Germans was the first lesson in fire power. The French had to pay an exorbitant price for it in the form of human lives. Lacking machines, they had to throw men into the gaps torn by the German artillery.[30] When the Germans renewed their large scale offensives on the western front in 1916 at Verdun, it was the newly created French heavy artillery which beat them back and saved the country. That was the other lesson so well remembered, so often and so solemnly commemorated, so much glorified that the great tank battles of 1918 were almost forgotten.

Fire power became a fetish,[31] to which every prospective innovation was

---

une industrie, Nancy, 1925), one of the great prophets of offensive warfare. As general, Alléhaut was the only high ranking French officer who immediately recognized De Gaulle, gave him full credit but at the same time pointed very clearly to his faults. See his *Etre prêts* (Paris, 1935), pp. 167-179. So great was the impression created by Gen. Alléhaut's book in Germany, that in reply to it General Leeb wrote his famous *Die Abwehr* (Berlin, 1938). For a precise evaluation of De Gaulle's role, cf. also H. A. DeWeerd, "De Gaulle As a Soldier," *Yale Review*, XXXII (1943), 760-776.

[29] *Christian Science Monitor* in its April 8, 1942 issue cites this summary from the Russian military periodical *Voyna i Revolucia* in connection with the Riom trial.

[30] This is dramatically described by Paul Bénazet, chairman of the committee for air warfare of the senate, in his book *Défense nationale, notre sécurité* (Paris, 1938), pp. 1-15.

[31] See Ph. Barrès, *Charles De Gaulle* (New York, 1941), pp. 21-23. He discusses De Gaulle's years in the war college in 1924-1926 and describes General Moyrand lecturing on

automatically subordinated. Army aircraft merely became an adjunct of artillery, tanks could not operate outside the visible range of fire power,[32] and improved means of transportation were good only for bringing more and more ammunition and other material to feed the Moloch of fire power.

Fire power could destroy the enemy's entire defensive apparatus and then and only then would the French infantry move to the attack. This meant a clumsy system of transportation encumbered with material. Movement and surprise were forgotten, and the French doctrine hardened into a rigid code of pseudoscientific calculations as to how to prepare every square yard, lure the enemy into it and destroy him. For "fire kills" was the mystic motto of Marshal Pétain, and his word became dogma.[33] The dogmatized defensive theory based on fire power was finally reduced to a credo and published by General Chauvineau in a book which obtained great popularity, because Marshal Pétain recommended it in a notorious foreword[34] and because it summed up the Maginot mentality, the defense doctrine, and helped to confirm Frenchmen in their cherished illusion about a comfortable war in which fire power would consume the legions of the enemy and protect the French army in its impregnable fortifications.

What the fire power doctrine and the Maginot Line mentality really did to the French army was to petrify the tactical and strategic lessons gained from the World War, kill imagination and initiative, and nullify the efforts of those who believed in the offensive and tried to make something out of the only positive element in the French military doctrine: the idea of elastic defense.[35] However, counterattack, no matter how carefully prepared, requires vigor and offensive

the principle of the "section de terrain." They studied the terrain very minutely, dividing it into squares and assigning the necessary fire power to each square. An essential element of this theory was that the enemy would be lured into certain sections of the terrain where his destruction would be well arranged in advance. Barrès comes to the conclusion that the Maginot Line emerged out of these theories.

[32] This was officially stated by General M. Sciard, tank inspector, during the Riom trial. See New York *Sun*, March 20, 1942. The assumption was that since tanks were tied down to artillery protection they should not be supplied with more than five hours' fuel. In consequence they were often captured by the Germans because they had exhausted their fuel.

[33] H. Pol, *op. cit.*, p. 188. On the "fire kills" theory see Daladier's speech on February 2, 1937, in Chambre des Députés, *Débats*, CLXI, 291, which shows that Daladier himself fully endorsed it.

[34] Much was made of this foreword at the Riom trial because the defense turned it against the prosecution and incriminated Pétain himself. It was read by Daladier on March 18: "Direct action of aerial forces in battle is a moot question, for troops engaged in combat on land are disposed to receive blows and return them. It is by individual action over the rear that aviation activity is exercised most efficiently." *New York Times*, March 10, 1942. It meant the rejection of the air arm as a major weapon, a step also advocated by General Chauvineau. His book is a classical summary of the passive French defense doctrine pushed to the very extreme.

[35] The *Revue Militaire Générale* published an article on tanks in its September 1937 issue entitled, "Considérations sur l'offensive." The author, Major Krebs, reviewing the new official *Instructions sur l'emploi tactique des grandes unitées*, points to its advocacy of the attack whenever the enemy gives a favorable opportunity. That there was a good deal of offensive element in the French doctrine is further proved by the review of the *Instructions* in the July 1938 issue of the German tank magazine *Kraftfahrkampftruppe* under the title: "Französische Grundsätze über Verwendung mechanisierter und motorisierter Einheiten und die Abwehr dagegen." The columnist of the *Militärwochenblatt* on tanks, Colonel Braun, compares in the October 22, 1937 issue the German and French tank doctrines with very favorable conclusions for the French.

spirit, essentially French characteristics and suited to their temperament, but at a very low ebb then, because of the influence of the Maginot Line mentality. The brilliant innovators like De Gaulle, the air expert Rougeron, Generals Velpry and Doumenc and their spokesman in parliament, Paul Reynaud and his group, hurled their arguments in vain against this Chinese wall of prejudice. The attack remained taboo, for as Chauvineau's credo said, attack required a threefold superiority and even more in certain cases, with a corresponding higher rate of casualties.[86]

There were to be no new holocausts in the next war if war must come. Let the enemy pile up his dead in front of the French positions. When time for counterattack comes the French army will reap an easy harvest. This narrow-minded, petty, bourgeoislike speculation about a cheap war and an easy victory led to the ultimate ruin. On the eve of the second World War the French, let us hope only temporarily, ceased to be *la grande nation*, because with their Maginot Line mentality they tried to escape the heavy responsibilities resting on the shoulders of great nations.

II

After the World War the British army was quickly liquidated and only skeleton forces were left. Public opinion and the government readily accepted the policy of reducing armaments to the bare minimum which then was carried out consistently in all three services, particularly in the army. In Great Britain the first line of defense has always been considered the navy. Since the World War the air force has come next and the army, "the Cinderella of the forces," occupied last place.

This was exactly the reverse of the French system, and for purely geographical reasons. Whereas in France the nation-in-arms has always been in the forefront of discussions, in Britain this principle was considered completely out of harmony with the age-old system of long-service professionals. Conscription was established for the first time during the World War, but after the Armistice of 1918 it was quickly discarded. Not even the Haldane type of army was considered necessary, and a gradual return if not to the letter, at least to the spirit, of the old Cardwell system took place.[87]

After the World War there was no likelihood of employing the British standing army in other than colonial operations. In Europe the Peace Treaty, the League of Nations, and the armies of France and her allies would deal with any emergency. On the other hand, in distant colonies the tasks of the British garrisons would consist merely of putting down local outbreaks or guarding against the incursions of desert or mountain tribes. A few battalions, brigades at the most, would suffice for operations of this nature, and armament heavier than small infantry weapons and light field artillery would be too cumbersome in tropical jungles or high mountains. Why, then, maintain at home

[86] General Chauvineau, *op. cit.*, pp. 131-132. The idea of elastic defense is evident even in Chauvineau's book but he fails to elaborate it properly.

[87] Major E. W. Sheppard, *A Short History of the British Army* (London, 1940), pp. 373-375.

a complicated organization like the Haldane system with heavy armament? The temptations of the old Cardwell system were too great to resist. According to its traditions the army at home would merely instruct and train new recruits for the imperial police force which was to guarantee *Pax Britannica*. Therefore, the return to the original conception of Cardwell's world-wide constabulary seemed logical.

The army clung to this principle even after the rapid rearmament of Germany raised the specter of a new European conflict. The White Paper of March 3, 1936—the first serious armament proposal submitted to parliament since the armistice—said that, inasmuch as the army had been reduced by twenty-one battalions below the 1914 effectives:

"His Majesty's government proposes to raise four new battalions of infantry, which will to some extent mitigate the present difficulties of policing duties which our imperial responsibilities have placed upon us."[38]

The return to the Cardwell conception of army organization meant the doom of the tank corps, the pride of the B.E.F. at the end of the World War. The veterans of the great tank actions of 1918 departed, and new men had to do as well as they could with the old models. Development of new design and construction was reduced to a skeleton and lack of appropriations prevented all progress. We read:

"In 1931 a medium tank of superior design was issued, but the great depression and pacifist agitation on top of it prevented large-scale production. When this was finally decided in 1936 the tank proved to be out of date. There was debate and debate . . . and the tank has yet to reach the men."[39]

These conditions are much in line with the French development, but whereas France in the 'twenties had no military writer who could establish a tank doctrine, Britain produced a brilliant expert and a school of thought around him. Major General J. F. C. Fuller was chief of staff of the Royal Tank Corps in 1918 and after the armistice took upon himself the task of waging a one-man crusade in favor of mechanization. Despite the brilliant record of the British tank corps in the World War, his difficulties were overwhelming; but Fuller, endowed with great literary skill and constructive imagination, was able to interest ever larger and larger circles in his program. His most signal success was the conversion in 1925 of Captain B. H. Liddell Hart, at that time already recognized as the leading military writer in Britain.[40]

[38] *Accounts and Papers*, Cmd. 5107, "Statement Relating to Defense" (London, 1936), p. 10.

[39] Major E. W. Sheppard, *Tanks in the Next War* (London, 1938), pp. 77-80. Major General J. Fuller goes much further in his criticism, declaring in his *Army in My Time* (London, 1935), pp. 188-189, that in 1933 the army had reached its lowest tactical level since the end of the World War and added that the first tank brigade was born in the 1934 army estimates when tank-mindedness had at last taken root. Cf. also Major General Ironside, "Land Warfare," in the *Study of War* (London and New York, 1927), pp. 116-147.

[40] In 1923 General Fuller published his *The Reformation of War*, which shows strong influences of Mitchellism and Douhetism. In his book, he became the advocate of the air arm as the new arbiter of the battlefields and prophesied the disappearance of the traditional soldier, relegating infantry only to garrison duties. His favorite weapon, the tank, received its due share of praise. The highly imaginative author even saw tankodromes

Liddell Hart was also a veteran of the World War, having served as captain of infantry. In 1918 he was entrusted with the task of revising the British field service regulations for the infantry, in accordance with the latest German shock troop tactics, which quickly fell into disuse and then complete oblivion. It proved easy for General Fuller to proselytize among such open-minded officers as Liddell Hart who, after his conversion, soon equaled his teacher in exposing the backward conditions in the army. As a result of their efforts some reforms were introduced under Lord Milne but rather slowly and cautiously because, as Liddell Hart said, the Chief of the Imperial General Staff was an old man and a Scot, two reasons for a somewhat over-circumspect policy of progress involving heavy expenditures.[41]

In the Royal Air Force the situation up to the early 'thirties reflects the same mentality and the same conditions as prevailed in the Royal Tank Corps despite the fact that, as an independent branch of the armed services, the air force was second only to the navy. Its higher status, however, was recognized merely by the fact that separate legislation in 1923 had provided for a minimum standard to be maintained. Even this low limit, however, was later discarded.[42] As in the case of the tank, so also in the case of the plane, new design, construction, and organization were sacrificed to considerations of pacifism, disarmament, and politics, and it was a long time before the British people were roused from their dreams of collective security and the universal application of arbitration in international disputes by Hitler's merciless policy of aggression and cynical disregard for treaties.[43]

In comparison with the French attitude, it must be noted that in Britain political animosities were, on the whole, absent in armament and military legislation, though there were isolated criticisms that class discriminations con-

carrying tanks on sea which would be deposited on invasion beaches, an idea ridiculed at the time. This book was followed by *The Foundations of the Science of War* (London, 1926). To understand the background of these two fundamental works of the military school of thought named after him, one should read his memoirs and reflections written after his retirement *The Army in My Time* (London, 1935), and *Memoirs of an Unconventional Soldier* (London, 1936). The motto of this last volume, printed on the title page, is a quotation from Heraclitus: "Asses would rather have refuse than gold," an indication of the outspoken nature of these volumes.

[41] See his "The New British Doctrine of Mechanized War," *English Review*, XLIX (1929), 608-701, and *The Remaking of Modern Armies* (London, 1927), written as a *magnum opus* after his full conversion to mechanization. Concerning the administration of Lord Milne, cf. "Seven Years: the Regime of Field Marshal Milne," *English Review*, LVI (1933). In another article, "Grave Deficiencies of the Army," *English Review*, LVI (Feb. 1933), 147-151, Liddell Hart points out that mechanization was slowed down too greatly since there were only four tank battalions to 132 infantry battalions. Mobility was absent and there was a great deficiency in artillery.

[42] *Accounts and Papers*, "Memorandum by the Secretary of State for Air to Accompany the Estimates" (London, January 11, 1934), and "Statement Relating to Defense Issued in Connection with the House of Commons Debate" (London, March 11, 1935).

[43] The best work to show the influence of the disarmament conference upon English rearmament is J. F. Kennedy's *Why England Slept* (New York, 1940). He refers to a passage in the above quoted "Memorandum": "These estimates, therefore, in broad outline are the outcome of our desire to pursue disarmament and to study economy on the one hand; on the other of our reluctant conviction that the policy of postponement of the 1923 program cannot be continued." See also R. A. Chaput, *Disarmament in British Foreign Policy* (London, 1935), and A. Wolfers, *Britain and France between Two Wars* (New York, 1940).

tinued to be prevalent in the army.[44] However, the situation in Great Britain could in no way be compared to the parliamentary feuds in France which made the army a football in the political arena.

In Britain the delay in readjustment to modern conditions in armament was due to other causes, though in certain respects the gross neglect of mechanization can be traced to a military mentality identical with that of France.[45] First, the great depression worked havoc with British finances, the effects of which were immediate, whereas in France it appeared late and in a mild form. The second Labor government, with implicit faith in world peace and disarmament, deliberately neglected even the minimum standards in the Royal Air Force, and the army fared no better. The Nationalist government which succeeded the Labor administration could not disregard the deep-seated pacifist ideology of the British public and, as the principal sponsor of the disarmament conference at Geneva, it could not disavow its principles by preaching pacifism abroad and expending huge sums for rearmament at home. Baldwin's government, burdened with such handicaps, could do nothing but wait and play for time until the psychological atmosphere should become more favorable. But the elections of 1935 were conducted in the sign of *pax*, and regard for the ballot box made it imperative for Baldwin to leave problems of rearmament in the background despite his memorable pronouncement on July 20, 1934, that the eastern frontier of Britain was the Rhine.[46]

The first real effort at rearmament took place in 1936 for two reasons: because the escalator clause of the London naval disarmament conference became operative in that year and because of Hitler's denunciation of the Locarno treaties and the remilitarization of the Rhineland. But, even then, the navy came first, and a special board of investigation was appointed to inquire into the merits of the bomber-versus-battleship controversy which had filled the pages of naval and military periodicals ever since General William Mitchell's sinking of the "unsinkable" *Ostfriesland* off the Virginia Capes by a few heavy bombs dropped on and near it from the air. Needless to say, the special board of investigation, just like similar boards appointed in the United States by Presidents Coolidge, Hoover and Roosevelt, came to the same conclusion: that available evidence would not warrant any major departure from existing technical principles in naval warfare. At any rate, the verdict was that as long as other powers were building battleships, Britain had to follow suit, a very comfortable excuse of the rigid and dogmatic military mind for the avoidance of responsibilities. The credo of fire power had just as tight a hold on the

[44] See L. Clive, *The People's Army*, with an introduction by Major C. R. Attlee (London, 1938), and articles on armament in the *Labor Monthly* of April and May 1937. See also Lord J. M. Strabolgi, *New Wars, New Weapons* (London, 1930), a vicious criticism, by the Laborite lord, of traditionalism and of the caste spirit in the armed services. Lord Strabolgi later wrote an article, "What's Wrong with the British Army," *Collier's*, August 22, 1942, renewing his accusations concerning clique rule in the army. For this article he was mercilessly taken to task on October 1, 1942 in the House of Lords by Lord Lovat, a hero of the Commandos. *New York Times*, October 2, 1942.

[45] There is a particularly extreme but well documented criticism in J. R. Kennedy, *Modern War and Defense Reconstruction* (London, 1936).

[46] J. F. Kennedy, *op. cit.*, p. 72.

minds of British admirals as it had on the French generals. The only difference was in application: the French applied it to the land, the British to the sea. Moreover, because of the fact that naval traditions were overwhelmingly predominant in British military thought, and influenced all the other services, the effect upon modernization was just as baneful as the similar mentality in the French army.[47]

Compared with other branches of the services the air force received attention second only to that accorded to battleships when rearmament started after 1936. Designs for the Spitfire and Hurricane models were worked out and improved in this period, even though their manufacture on a large scale was still delayed.[48] The tank corps, on the other hand, could make no headway. There was a decidedly inimical attitude in the army toward mechanization on the Fuller pattern, and this foremost expert in tank warfare had been sent into retirement without any attention being paid to his brilliant suggestions. In fact, Fuller's fate bears a remarkable similarity to that of De Gaulle. In 1937, the year of great rearmament activity in the British army, Fuller, already in retirement for five years, wrote his *Lectures on Field Service Regulations III: Operations between Mechanized Forces*,[49] which was published in Britain in an edition of five hundred copies but distributed in the Russian and German armies by the thousands. Though the Germans did not copy it, finding De Gaulle's system more efficient, the Russians apparently considered it better adapted to their tactical conditions. Concerning Fuller's *Lectures on F. S. R. III*, one of his disciples wrote the following:

"It is, I believe, the most farsighted military manual or commentary that has been written by anyone intended to inform Englishmen as to the nature of future war between mechanized armies. It failed of its purpose because Englishmen would not read it, though 30,000 copies were published for the German

[47] The findings of the special board of investigation were published in *Cmd. 5301* and treated at great length in the *Royal United Service Institution Quarterly* (henceforth quoted as *"RUSI"*), LXXXII (1937), 409-414. Rear Admiral H. G. Thursfield, "Battleships," *National Review*, CXV (1940), 531-543, discussing the background of the board of investigation, says that there were so many official declarations concerning the restrictions, even the complete abolition of the battleship, that the public really began to believe them. The board, says the same author in "Sea and Air," *National Review*, CXVII (1941), 285-294, after sitting for four months, could not publish most of the findings because among them were some of the most cherished secrets of the navy and the air force. In this connection, the revelations of B. Acworth, one of the navy's spokesmen and author of several works, are most interesting. In *Britain in Danger* (London, 1937), he declared that in these questions the navy had an answer for the public and that, acknowledging the tremendous threat of airplanes, they were taking steps to increase the security of ships. But among themselves the navy men, in order to prevent money being transferred from the navy to the air force, agreed that it would be advisable to make reductions in the space and weight of ships from 16 to 20 per cent, using the savings for the construction of aircraft and antiaircraft guns. A violent critic of the government in the ship-versus-plane dispute was J. L. Garvin. In his "Arms, Money and Muddle," *Observer*, March 3, 1929, he reveals and condemns the inter-departmental intrigue for allotment of the appropriations for defense which stood in the way of modernization. The article is extensively quoted in P. R. C. Groves, *Behind the Smoke Screen* (London, 1934), pp. 238-245. Garvin calls the naval policy fossilized because the 1929 estimates allotted 57 per cent for the navy, 31 per cent for the army and only 12 per cent for the air force.

[48] Even more perfect models were designed in 1937 by Sydney Camm, originator of the two previous models, but not manufactured. See *New York Times*, October 3, 1942, p. 5.

[49] London, 1937.

army and the work was also widely circulated through the military forces of the USSR. Had the proportion of democratic and totalitarian interests in *F. S. R. III* been the reverse of these results, it is conceivable that the present war would not have happened."[50]

For reasons which are still obscure and will be cleared up only when the records of Mr. Hore-Belisha's administration of the War Office are made accessible for investigation, the British army rejected the theory of mechanization in favor of motorization, laying emphasis on light armored machine gun carriers. The striking power of these carriers was small, but this was offset by their great mobility. For the heavy tanks of the German panzer divisions, they were mere match boxes, but they could get away fast, and thus the way to Dunkirk was well paved.

The biggest problem, however, concerned not the technical equipment but the size of the army which should be sent to Europe in case a general conflagration should again break out. This problem was of general interest, and there was no question as to the attitude of the vast majority of Englishmen. They were against the very idea of a large army, and this opposition continued until after Munich, when the government, in the spring of 1939, finally decided to introduce conscription, in a very limited sense. The origin of this attitude goes back to the World War when Britain sent a vast force to the continent and left 600,000 dead in the military graveyards of France. In all discussions as to the size of the future expeditionary force reference was made to the frightful losses suffered in the campaigns of the fall of 1917 and Passchendaele became a mournful symbol. During the passionate rearmament discussions of 1936, the Royal United Service Institution announced that the size of the British expeditionary forces would be the subject of its annual gold medal essay. The reward was given to Captain J. C. Slessor for suggesting a small but highly trained elite mechanized force; the author and the journal itself made explicit reference to the fact that the British public was totally averse to the idea of a new mass army. This "gold medal army" became very popular and received the endorsement of many military writers, a characteristic sign of the trend of strategic thought.[51]

[50] S. L. A. Marshall, *Armies on Wheels* (New York, 1941), p. ii. See also Fuller's own comments on the reception of his *F. S. R. III* in his *Memoirs of an Unconventional Soldier*, pp. 490-494.

[51] *RUSI*, LXXXII (1937), 463-484. Captain Slessor received the full endorsement of the famous writer, Major E. W. Sheppard ("Does Defense Mean Defeat?" *RUSI*, LXXXIII [1938], 291-298), and from Capt. H. M. Curteis ("The Doctrine of Limited Liability," *RUSI*, LXXXIII [1938], 695-701). His severest critic was V. W. Germains ("The Army in War," *National Review*, CXII [1939], 759-767). Captain Slessor was professor at the General Staff School at Camberly and his lectures on air doctrine in 1931-1934 were published in *Air Power and Armies* (London, 1936). He advocated a pliable military doctrine as regards cooperation between the air force and ground troops. His view closely approximates the methods of blitzkrieg, and if the RAF had acted on these principles in 1940 in France the German drive might have taken a different course. But Slessor was disregarded, as were many others. He further elaborated these views in his gold medal essay. On the reaction of the British public to the question of sending a large expeditionary force to France, see C. Q. Martel, "Mechanization," *RUSI*, LXXXII (1937), 280-302. This is a reprint of lecture delivered by Colonel Martel to a distinguished audience of army officers, which included Major General A. P. Wavell, who supported Martel's views.

On October 25, 26, and 27, 1937, Liddell Hart, as military correspondent of the *Times*, published three articles which suggested that Britain definitely accept the theory of limited liability in her military obligations and return to her traditional policy of blockade and economic warfare, for which she was eminently suited by her mighty navy and the unlimited resources of her empire. With respect to the continent, he favored a strictly defensive strategy because it was more suitable to the British temperament, and because in view of the greater superiority of defense over attack, it would bring better results in the long run. Only a small expeditionary force should be sent to France and, since the Maginot Line with its French garrison would hold the enemy, the British force should be kept in the rear as a strategic reserve of high mobility.[52]

These three articles immediately became the object of an endless controversy with reverberations abroad. A reply was sent to the *Times* by the French General Baratier protesting, in a dignified but determined manner, "against that tendency with which England has so often been reproached, i.e., to throw the main weight of the war upon her allies. It cannot be refuted too strongly. If ever Germany attacks our country, she will wage a totalitarian war, in the course of which France and England, in order not to succumb, will have to throw into the balance all their resources on land, on sea and in the air."[53]

There were angry protests from British writers as well. The *Army Quarterly* rejected the defense doctrine immediately in an editorial and opened its pages to critics of Liddell Hart's views.[54]

One of the ablest among these was General H. Rowan-Robinson, a member of the Fuller school, who had published a number of works[55] which demonstrated a well balanced judgment and a broad view which was not restricted to the shores of the United Kingdom but extended to problems of imperial defense. He clearly foresaw the menace of the airplane and urged a vast rearmament program to secure the lines of communication between Hong Kong and Gibraltar. As a former artillery man he put his faith in fire power plus high mobility, that is, in the tank. His active mind, for a long time in harmony with Fuller's ideas, rejected with indignation Liddell Hart's suggestions for a passive war and the avoidance of the attack. Such views, he said, were neither military nor British. He reminded Liddell Hart that the Empire had not been

[52] He later summed up these views in *The Defense of Britain* (New York, 1939), a comprehensive work stressing the doctrine of limited liability which was to a larger extent responsible for Dunkirk.

[53] *Le Temps*, January 4, 1938, republished in *Army Quarterly*, XXXVI (1938), 123-127.

[54] *Army Quarterly* (henceforth quoted as "*A.Q.*"), XXV (1938), 200-202. General Baratier's article was not the only one that took notice of the *Times* articles outside of England. The American *Infantry Journal* reviewed them in vol. XLV (1938), 174-175, and in vol. XLVII (1939), 98-112, the well known American military writer, Colonel A. Q. Phillips wrote an article, "Attack and Defense," strongly advocating the attack. For further comments of the foreign military press cf. the *Command and General Staff School Quarterly* and its reviews of foreign military periodicals (Leavenworth, Kansas, 1938). The *Times* published a letter on October 29 from Sir Frederick Morris who declared that "we will have to send an army to our ally and we will not be free agents in choosing our tactics."

[55] The most remarkable being *Imperial Defense: a Problem in Four Dimensions* (London, 1938).

built with limited liabilities. However, Rowan-Robinson, as a disciple of full mechanization, emphasized quality at the expense of quantity.[56]

The *Army Quarterly* published also an article by Major E. W. Sheppard,[57] the conclusion of which was that Britain cannot wage an offensive war because of inadequate resources:

"In fact, the offensive always has been and is certainly more than ever today so difficult to carry through cheaply and successfully that it can be undertaken only in reliance on some definite factor or factors of superiority on the side of the assailant. The exact nature of those factors will vary of course with the particular case but if no factor of superiority of any kind be present then only an irresponsible madman would venture to assume the offensive, since to do so would merely be to condemn his command to defeat and himself to disgrace."[58]

Perhaps the ablest critic Britain produced was V. W. Germains who had contributed several articles to various magazines prior to the controversy on defense and attack. His book, *The "Mechanization" of War*[59] was perhaps the best commentary on the World War and especially on the tank battles of Cambrai and Amiens, favorite subjects of postwar military literature in Britain. Germains was the advocate of an all-out continental war with a new British expeditionary force of sixty divisions equipped with everything modern war requires. He deplored all the one-sided ideas of the so-called "scientific school" and chose for his motto the opinion of General Malin Craig, chief of staff of the United States army, who in a report on the Spanish civil war said, "The decisive factor is still infantry; the new arms can aid the man on foot but cannot replace him," which he called the nutshell of the problem.[60]

[56] "Defense or Attack," *A.Q.*, XXXV (1938), 277-291. Long before the controversy provoked by the *Times* article began, Major General H. L. Prichard ("The New Army," *The Nineteenth Century and After*, CXXI [1937], 796-804), rejected the limited liability theory, declaring, like Sir Frederick Morris in the *Times*, that the size of the British expeditionary force would be determined by other than purely British factors.
[57] "Defense or Attack," XXXVI (1938), 38-50.
[58] *Ibid.*, p. 41. This is one of the most representative expositions of the doctrine of defense. Why such writers never assumed that Adolf Hitler might have every disposition to act as such a madman with an instrument, the newly created German army, which might have success, is beyond comprehension. The *A.Q.* in its editorial comment on this article questioned whether the issue was so clear and advocated a well rounded training in defensive and offensive tactics. In the same volume Colonel E. G. Hume, "Attack or Defense," pp. 338-342, comes to the same conclusion. Major Sheppard also wrote in *RUSI*, LXXXIV (1938), 291-298, replying to Colonel R. H. Beadon, one of Liddell Hart's severest critics. Colonel Beadon ("Defense or Defeat," *RUSI*, LXXXIII [1938], 58-68) sharply condemned the *Times* articles and their limited liability idea. One of the ablest contributors to this debate was Captain H. M. Curteis. His article "The Doctrine of Limited Liability," *RUSI*, LXXXIII (1938), 695-701, was a reply to Sheppard but in reality exposed the fallacies of Hart's doctrine. He warned that to leave France alone might have disastrous consequences and might create a situation in which the Low Countries could easily be lost by a sudden attack. In such an eventuality, a small English army would never be able to reconquer the Low Countries, since "the defense is strong on both sides!" He urged a different strategic frame of mind and, as a compromise, suggested the gold medal army.
[59] London, 1927.
[60] "The Quart Measure and the Pint Pot," *National Review*, CX (1938), 453-465, was his principal contribution to the attack-defense controversy. In another article, "The Army in War," *National Review*, CXII (1939), 759-767, he attacked the "faddists," particularly Fuller who "lost all sense of balance," and Groves, "the fanatic of air power," and by implication Liddell Hart, because they misled the young generation of officers in the army

V. W. Germains was the only military writer of note who saw the situation with a prophetic vision: "For more than a decade the British public was trained to put faith in every conceivable means of winning wars save by fighting battles and beating the enemy."[61] He never ceased to demand conscription and the organization of a large army. In Parliament he was supported by Sir Edward Grigg, but their efforts remained as unsuccessful as those of Germains' counterpart in France, Senator Bénazet. Germains considered the army to be the hardest hitting arm of the three services, the air force being the eyes, ears, and nervous sensory organs. Instead of the government organizing a large force by conscription, the army was treated as the Cinderella of the forces and the happy hunting ground for faddists, exactly like the navy in France.[62]

The true spokesman of the British mind was Liddell Hart. He did not abandon Fuller's theories and still remained the advocate of the tank. He would not, however, use the new armored divisions of the British army in France but instead would keep them at home as a main strategic reserve with an eye on the Low Countries and the Near East. After Munich, when the probability of another world war was greater than ever, he improved on the theory first published in the October 1937 articles of the *Times* and declared that except for some technical troops no expeditionary force should be sent to France. For once a British army, no matter how small, should be employed on the front in an attack and the attack should be beaten back—a likely eventuality in view of the strength of the defense—new units would be sent to France, and the generals would try to attack again and again, until Britain had an army of a million in France with corresponding casualty lists. At the most, for the sake of bolstering French morale, three armored divisions should be sent over, but only on condition that they would not be employed in an offensive campaign.[63] These divisions should serve as a mobile reserve to be used in counterattacks, in close cooperation with the air force. His assumption was that, if war should start on the continent, it would come with lightninglike suddenness. The Germans, he said, had three armored divisions and two more were in the course of organization. They might breach the Maginot Line, but such mobile defense formations could easily pinch off their spearheads and stop the gap.

These views, based on the experiences of the Spanish front, notably the battle of Guadalajara, were excellent and remind one of De Gaulle's advocacy of a similar force and tactics as set forth in his famous memorandum.[64] In fact, if such divisions with the proper air force supplement had been used they could have disturbed German operations and perhaps intervened with decisive results in the May days of 1940. But, as Liddell Hart himself says, the British

with such talk as "the British traditional method," and "the increased power given by modern weapons to the defensive."

[61] "Military Lessons of the War," *Contemporary Review*, CLVIII (1940), 146-155.
[62] "Some Problems of Imperial Strategy," *National Review*, CX (1938), 738-749.
[63] "The Defense of the Empire," *Fortnightly Review*, new series, CXLIII (1938), 20-31.
[64] *National Review*, CXV (London, 1940), 393-405.

had only one armored division and that was wasted by the French at the Somme.[65]

What was the reason for these grievous neglects? Liddell Hart wrote a self-vindicatory pamphlet after the disasters of 1940 in which he said: "After Mr. Hore-Belisha became war minister, officials who might have come in contact with him were sternly warned that they must not suggest that any basic reorganization of the army or any increase in mechanized forces was possible."[66]

The *Army Quarterly*, reviewing *Dynamic Defense* rejects this self-defense and puts the blame on Liddell Hart who was "at Hore-Belisha's elbow."[67] On the other hand, Germains, who has no good word for either Liddell Hart or Hore-Belisha, said during the Munich crisis: "The real nature of the mentality ruling the roost at the war office in the present year of grace is an apotheosis of the mentality of the learned military goose. . . ."[68]

Apparently Liddell Hart himself could not escape from the confusion. Under the effect of the Nazi blitzkrieg in Poland which demonstrated with overwhelming proof the efficacy of attack, Liddell Hart, upon the request of some government officials, wrote a memorandum on September 9, 1939.[69] In it he still harped on the superior strength of the defense, although he qualified this by a statement that defense was superior to attack "where there was no room for mobility." He went so far in clinging to his unfortunate doctrine as to suggest that the government make a declaration renouncing military attack as a means of combating aggression and, in this way, avoid ridicule because of the complete inactivity of the allied armies. He suggested also that concentration on moral and economic weapons might bring the internal collapse of the enemy. Thus, Britain would concentrate on the enemy's weak flank and refuse to expose her own, that is, to expose it *by an offensive*. Such a spirit, such a belief in defense *über Alles*, was bound to have a harmful influence upon the whole conduct of the war, and it inevitably brought to England sweat, blood and tears.

Liddell Hart's policy was not followed to this extent. At the beginning of

[65] B. H. Liddell Hart, *Dynamic Defense* (London, 1941), pp. 30-35.

[66] *Ibid.*, p. 36. He adds that the most brilliant tactician in mechanized warfare, General Hobart, was sent to Egypt after Munich in order to get him out of the way. Apparently Hore-Belisha was trying to overcome a Chinese wall which was too strong for him. In a later book, *The Current of War* (London, 1941), Liddell Hart explains the riddle in a chapter entitled "Hore-Belisha," and sums it up by saying (p. 147): "The root of his offence has been that he wrought changes in an institution which instinctively resented change."

[67] *A.Q.*, XLI (1941), 380.

[68] "To Be or Not to Be," *National Review*, CXI (1938), 349. In another article he refers to Hore-Belisha's attitude toward Field Marshal Sir Cyril Deverell's speech of October 24, 1937 at Guild Hall which advocated a mass army and for which the C.I.G.S. was dismissed by the Secretary for War. ("The Army in War," *National Review*, CXII [1939], 759-767.) Germains' other accusation was that there existed a great uncertainty concerning military doctrine, and he quotes an article of the London *Times*: "The country would welcome any measure for national defense, however drastic, once convinced of its absolute necessity." ("To Be or Not to Be," *loc. cit.*, p. 350.)

[69] This memorandum is contained in his *The Current of War*, which includes other interesting articles written by him.

the war a small British expeditionary force appeared in France and remained as a strategic reserve far behind the Maginot Line, as was suggested by Liddell Hart in his attack-defense articles in the *Times*. When the great German drive began on May 10, 1940, the British army rushed to contest the passage of the Germans through the Liége gap, paying no attention to the Ardennes forest where the Germans actually prepared the break-through at Sedan. It was Liddell Hart's view that the ruggedness of the terrain and the dense forests of the Ardennes country would make it unsuitable for large-scale operations.[70] This was also the French view and apparently that of the Belgians as well.

There was lack of proper equipment in the British army, as there was uncertainty about the proper doctrine. But that was not the principal mistake. The road to Dunkirk was prepared by the doctrine of limited liability based on a purely defensive type of warfare, which had been so mercilessly criticized by many in Britain during the attack-defense controversy. But it would be most unjust to blame Liddell Hart alone since he was not the author of it. He had merely summed it up and, since he was the foremost military publicist of his country, his name was identified with it. The doctrine of the defense was the product of a great complexity of causes, and it was accepted officially in France as well as in Britain.

"Presumably, the present government has decided that in the event of our having to take part in another major war we shall not send an expeditionary force either to the Continent of Europe or elsewhere. We are to revert, it seems, to what was our policy in the past: to give assistance to our allies by means of our sea power and in the air."[71]

In summing up it is evident that the defense doctrine of the two great western democracies of Europe was not the product of a few men, timid politicians or narrow-minded experts, but the result of the trend in national thought arising from a superior civilization which turned horror stricken away from the holocaust of war. Democratic institutions produced the pacifistic influences and the general reluctance of the British and French to appeal to the arbitrament of war in the event of international disputes. Moreover, the overwhelming victory at the end of the World War which resulted in Germany's unconditional surrender left behind a well-justified pride and self-confidence and an

---

[70] *The Defense of Britain* (London, 1939), p. 381, and the whole chapter on "The Method of Defense."

[71] *A.Q.*, XXXLV (1937), 3, an editorial interpreting the speech of Secretary of War Sir Thomas Inskip in connection with the debate on the military loan in which Prime Minister Baldwin also participated. The editorial condemned this policy. Yet it remained, with slight modifications, the official attitude of the Chamberlain administration until the fall of France. In France, the *Revue Militaire Générale*, a semiofficial organ of the French army, reviewing the controversy on attack and defense between Liddell Hart and Rowan-Robinson, took the side of the former in his isolationist view and condemned the *A.Q.* for its criticism of Liddell Hart. See *Revue Militaire Générale*, III (1938), 501-509. Pierre Cot, air minister of France during the crisis years of 1936-1938, in his *L'armée de l'air 1936-38* (Paris, 1939), pp. 106-108, seems to bear out the same idea by saying that the favoritism shown the air arm in France would destroy the main basis of the mighty Franco-English alliance. England cannot send, he says, more than a few divisions to France!

unshakable belief in the invincibility of the defensive. For four years the World War demonstrated how unconquerable the defense was. After the war, with new methods and improved materials, it was believed that the defense would become increasingly more effective. Behind the Maginot Line would stand the inexhaustible resources of the two mightiest colonial empires the world had ever seen. It is not really surprising, therefore, that the people of Britain and France looked with complacency on the future.

The victory psychology fostered a belief in the invincibility of the armed forces, and traditionalism turned into rigid formula. The prestige of the army became a serious obstacle to free discussion. At the same time the deep aversion by public opinion to mass murder of the World War type would admit no consideration of offensive ideas. Financial troubles, social disequilibrium, and pacifistic views all created an atmosphere in which only the theoretical discussions of a few experts were possible on other than the universally accepted dogmas. But even if changes were advocated, the will to act was lacking. Democratic institutions are based on the will of the masses. Mass, however, if it acquires momentum, is extremely slow in changing its direction, a delay that is sometimes fatal. Herein lies the weakness of democracy when faced with a resolute and cunning enemy. Not even the eloquence of Demosthenes was sufficient to rouse the Athenians to the superhuman effort necessary to save the liberties of Greece from the designs of Philip of Macedon. In consequence, ancient Attica, like modern France, was trampled underfoot by the conqueror's armies.

On the other hand, if the transformation does take place in due time, the mass momentum produced by the will to victory in a democratic nation has deeper and more abundant sources because moral energies spring from conviction and not from compulsion. Reviewing the war on October 12, 1942, Winston Churchill summed all this up very eloquently:

"But, after all, the explanation is not so difficult. When peaceful people like the British and Americans, who are very careless in peacetime about their defense; carefree, unsuspecting nations and people who have never known defeat; improvident nations, I will say reckless nations, who despised military art and thought war so wicked that it never could happen again—when nations like this are set upon by highly organized and heavily armed conspirators who have been planning in secret over years on end, exalting war as the highest form of human effort, glorifying slaughter and aggression and prepared and trained to the last boundary which science and discipline permit, it is natural that the peaceful and improvident should suffer terribly, and the wicked, scheming aggressors should have their run of savage exultation.

"That does not end the story. It is only the first chapter. If the great peaceful democracies could survive the first few years of the aggressor's attack, another chapter would have to be written. It is that chapter which we shall come to in due time. It will ever be to the glory of these islands and this empire that while we stood alone for one whole year we gained time for the good cause to arm, to organize slowly and to bring the conjoined, united, irresistible forces of outraged civilisation to bear upon the criminals."[72]

Herein lies the inherent strength of a democracy, which may lose many battles but wins the last. It has often been said that the Republic of Rome conquered the world against her own will, because she was attacked, surprised, and often defeated by able and cunning military leaders. In this modern Punic war we have witnessed the battle of Cannae, but the battle of Zama has yet to come.

[72] *New York Times*, October 12, 1942.

CHAPTER 16. Haushofer: The Geopoliticians

BY DERWENT WHITTLESEY

GEOPOLITICS is a creature of militarism and a tool of war. As its name implies, it is the offshoot of both geography and political science, although it was originated and largely developed by geographers. Sometimes it is made to appear as a twin of political geography, but it is much younger, having grown up in the generation that comprises the two world wars.

I

Long before geopolitics emerged and received its name, geography occupied an increasingly important place in the life of central European peoples, especially the Germans. Indeed it took shape, as a modern science, in Germanic Europe, where it was part of the stream of philosophic and scientific thought usually traced from Immanuel Kant.[1]

A formative period—in which geography, while remaining mainly descriptive, was systematized—covered the first half of the nineteenth century, contemporary with critical political and military changes that transformed Germanic Europe from a neighborhood of petty dynastic states to an association of vigorous nationalistic states. In the last half of the century, geography put forth the branches which gave it the form it has today. In these decades Germany was industrialized and unified and emerged as a great power. The new German empire derived its strength from a twofold base: the political authority traditional in the agrarian east, and the economic force built upon the mineral wealth and the overseas commerce of the maritime west.

Geography, perhaps more than other subjects, is colored by and in turn assimilates its surroundings. It is therefore natural that German geographers who worked amid the ferment of German political unification and expansion should have taken a particular and absorbing interest in political geography. They made it one of several new branches of a rapidly growing discipline.

As the nineteenth century drew to a close, rising German interest in sea power ranged alongside age-old German preoccupation with land power. At this juncture, the first formulation of political geography as a scientific study was made by the geographer, Friedrich Ratzel. In his *Politische Geographie*[2] he posed and discussed most of the major topics which have since been accepted as essentials of the subject. Among them, he inevitably considered war as an important aspect of both politics and geography. This topic is mentioned briefly at several points in his volume, and in the second edition (1903) the

---

[1] For these origins see Richard Hartshorne, *The Nature of Geography* (Lancaster, Pa., 1939), chap. 2.

[2] Friedrich Ratzel, *Politische Geographie* (Munich, Berlin, 1897; 3rd ed., 1923). See particularly the analytical table of contents. Ratzel's systematic study of political geography has nothing in common with the conventional "general geography" of political units, to be found in some more recent books labeled political geography.

word "war" appears in a subtitle, along with "state" and "communication."[3] This and all other topics are presented with objectivity, as befits a work of scholarship. His scientific temper required him to treat political phenomena in their total geographic setting, without reference to any particular nation.

At the end of his life Ratzel allowed himself to be swept off the narrow path of scientific geography by the nationwide surge of enthusiasm for German naval power.[4] This appears, however, to have been a momentary aberration with him. It calls our attention to the danger of confusing political geography with political fealty, but it does not make Ratzel a geopolitician. His departure from scientific discipline is far from the method of the geopoliticians, who glory in manipulating their findings to serve the presumed interests of the German state.

Although Ratzel was not a geopolitician, he nevertheless did contribute signally to geopolitics, because its creators borrowed from him certain concepts fundamental to their system.

Perhaps the most powerful of these directives was his opinion that the state is in some sense an organism. This concept has appeared in several branches of science and presumably derives from Darwinism and other nineteenth-century advances in biology.[5] Ratzel believed the resemblance between geographic and political structures on the one hand and biological organisms on the other to be a useful analogy. Although in practice he overworked it, he was aware of its limitations. The geopoliticians later failed to grasp the vital distinction between the political and biological worlds and based their system on the assumption that the state *is* an organism.

Also basic in Ratzel's analysis is the concept of the "space" (*Raum*) occupied by political groups, which is the subject of the fifth and final part of his treatise. This idea was taken up by the geopoliticians and rechristened "living space" (*Lebensraum*). Their treatment, however, presupposed the right of a state to *Lebensraum* whereas Ratzel was concerned rather with studying the space relationships of states.

Ratzel profoundly influenced the geographic thinking, not only of central Europe, but also of the English-speaking world, France, and places farther afield.[6] He is universally recognized as the founder of political geography. He is also the progenitor of geopolitics though not its founder.

Although it marks a critical distinction between dispassionate scholarship and selective indoctrination, the borderline between political geography and geopolitics is not always easy to see. Certain studies made in the transition period between the publication of Ratzel's *Politische Geographie* and the First World War cannot with assurance be allocated to one or the other category.

[3] Friedrich Ratzel, *Politische Geographie oder die Geographie der Staaten, des Verkehres und des Krieges* (Munich, Berlin, 1903).

[4] Friedrich Ratzel, "Flottenfrage und Weltfrage," *Kleine Schriften* (Munich, Berlin, 1906).

[5] H. F. Raup, "Trends in the Development of Geographic Botany," *Annals of the Association of American Geographers*, vol. 32 (1942), pp. 327, 341-343.

[6] Hartshorne, *op. cit.*, pp. 297-298.

At the time Ratzel was revising and republishing his treatise, the English geographer Halford J. Mackinder expounded the shifting power of contemporary states. He ranged himself in opposition to Mahan's then popular views by calling attention to the rise of rivals threatening British supremacy on the world's oceans. "Other empires have had their day, and so may that of Britain. . . . The European phase of history is passing away, as have passed the Fluviatile and Mediterranean phases."[7] Not long after, he amplified his meaning in a memorable paper read before the Royal Geographical Society. In this address he recognized, for the first time and with prophetic insight, that the end of the period of world-wide explorations signalized the establishment of "a closed political system" on the earth, and that mechanical transport was altering the relative strength of land power and sea power. He pointed out the possibility of "the oversetting of the balance of power in favor of the pivot state." He exemplified his thesis in a world map, supported by argument posing the concept of a Eurasian Heartland power and its potential threat to sea power.[8] Fifteen years later, he expanded his argument and elaborated his predictions in *Democratic Ideals and Reality*.[9] His stated object was "to measure the relative significance of the great features of our globe as tested by the events of history, including the history of the last four years, and then to consider how we may best adjust our ideals of freedom to these lasting realities of our earthly home."[10]

War had brought the findings of the political geographer into the arena of practical politics. That, by one definition, is geopolitics. In the meantime the original statement of the Heartland concept, as set forth by Mackinder in 1904, had been seized upon by the German geopoliticians, who adapted it to serve the interests of Germany. In that form it became one of the major systems for world organization as proposed by them.

On the eve of the First World War, the British geographer, James Fairgrieve, wrote a book which bore the significant title, *Geography and World Power*.[11] Most of the volume is a sketch of the political geography of the major political areas of the earth, past and present. In the final chapter the author makes "an attempt to discover what are the possibilities of change or further advance."[12] Written independently at the same time and published only a few months earlier was a study of the eight contemporary great powers by Rudolf Kjellén, a Swedish political scientist. It was translated into German before the end of 1914.[13] It seems more than coincidence that these books were written just before the great powers went to war on a scale unknown in previ-

[7] Halford J. Mackinder, *Britain and the British Seas* (London, 1902; reprinted with additional notes, 1906), chap. 20, p. 350.

[8] Halford J. Mackinder, "The Geographical Pivot of History," *Geographical Journal*, vol. 23 (1904), pp. 421-444.

[9] Halford J. Mackinder, *Democratic Ideals and Reality* (New York, 1919; republished 1942).

[10] *Ibid.* (1942 edition), p. 4.

[11] James Fairgrieve, *Geography and World Power* (London, 1915; New York, 1917; republished with addenda, 1942).

[12] *Ibid.* (1917 edition), p. 330.

[13] Rudolf Kjellén, *Die Grossmächte der Gegenwart* (Leipzig and Berlin, 1914).

ous history. Rather they appear to have been responses to an awakening interest in the natural environment as the foundation of national power, in the new era of "a closed political system" on the earth.

Two years later Kjellén produced his system of politics, *The State as an Organism*.[14] The author recognized five subdivisions of politics, one of which he named geopolitics. As defined, the term appears to have been intended as a synonym for political geography.[15] Kjellén was progressively affected by the war and published other works in which he used geopolitics to advocate the German cause. All his work discloses the enthusiastic convert to natural science, unrestrained by the rigorous technique of science. For instance, he treats the state as an organism, adopting Ratzel's analogy as if it had been a fundamental principle or a natural law. He not only gave geopolitics a name, he also advocated space concepts later embedded in Nazi policy, especially autarky. Whereas Mackinder suggested using geography as an aid to statecraft, Kjellén formulated a system of statecraft founded on geography.

By the close of the First World War the mating of geography and politics had hatched geopolitics. It was fledged in Germany, where defeat and rearmament engrossed the population between that war and its successor. Both the terms and the ideas of Kjellén were eagerly adopted by a group of German geographers who desired to use their knowledge of the earth to aid in restoring Germany to the rank of a great power.

II

Intense interest in military affairs was nothing new in Germany. The states of eastern Germany began as bulwarks against non-German hordes pressing westward from the heart of Eurasia. They continued as spearheads in the reflex push of German settlement eastward—the *Drang nach Osten*, celebrated by German historians. Centuries of vigilant military preparedness created a militaristic state and ultimately a militaristic habit of mind among the people, at first in Prussia and later in the German empire as a whole.

A conservative outlook on politics and devotion to monarchy are associated with agrarian society. These traits characterized the life of all Europe during the formative Middle Ages when the soil was the one natural resource people knew how to exploit with any degree of thoroughness. The commercial, industrial, and agricultural revolutions that modernized the economic and political structure of western Europe left Bavaria and the Prussian lands beyond the Elbe almost untouched.

These states lacked the minerals that give rise to large-scale mining and manufacturing—except for the far corner of Silesia, exploited too late to affect political thinking. Prussia is scarcely less landlocked than Bavaria, for the Baltic is an inland sea which, in modern times has not fostered a maritime outlook. Hence the east German states felt only indirectly the rise

[14] Rudolf Kjellén, *Staten som Lifsform* (Stockholm, 1916); translated into German under the title *Der Staat als Lebensform* (Leipzig, 1917).
[15] *Ibid.*, p. 43 in the German edition of 1924.

of ocean trade. Their soils proved unresponsive to fertilizer and never encouraged, much less enforced, the breakup of large estates handed down by inheritance from the Middle Ages. Extensive agriculture remained the foundation of economic life, political forms and personnel continued without alteration, and medieval ideals of society held sway over the public mind.

Western Germany, in contrast, was during the nineteenth century swept by the application of science to efficient manufacturing, transportation, and farming. There the soils are amenable to increased production through scientific management. In the northwest were found the minerals on which the German empire subsequently built its industrial leadership. Also, only in the northwest did a beckoning seacoast encourage ocean trade. Way for this technologic transformation had been paved by revolutionary ideas spread throughout western Germany in the wake of Napoleon's conquests. Prussia, on the other hand, had vigorously resisted these ideas as enemy propaganda.

It is immensely significant that Prussia, ultrareactionary in social structure and frame of mind, succeeded in making itself the ruler of Germany during the half century after Napoleon's downfall. Thereby the militaristic Prussian state, controlled by feudal landholders, fell heir to regions in process of rapid industrialization and alive to the accompanying change in social outlook. The resulting conflict produced a compromise. The west was granted protective tariffs on manufactures, government fostering of overseas trade, and a big navy. The east obtained protection for agriculture, a big army, and continuance of its ruling caste of large landholders as the dominant element in the new national government.

Between its unification in 1871 and the outbreak of the First World War, Germany passed through two stages. First came two decades devoted to consolidating the political and economic rearrangements of the preceding period. Then followed a quarter century of rising aggressiveness expressed in demands for eastward overland expansion, and for aggrandizement in the Atlantic sphere and elsewhere throughout the world by means of sea power and overseas colonies. These ambitions were vociferated by the Pan-German League, the Navy League, and the "saber-rattling" German emperor. They may have been no more than surface eddies but they were significant because they disclosed powerful undercurrents of German life which eventually undermined the political structure of occidental society.

Most obvious of these currents was the rise of Germany to the status of a world power and the consequent upsetting of the "balance of power" among European states. Less apparent, but equally critical for the future of Europe, was the unabated dichotomy between eastern Germany and western Germany in economic and social structure and in customs and outlook upon life.[16] Almost unobserved was the penetration of the militaristic habit of mind into all parts of Germany. All three of these fundamental forces in German life reappeared after the First World War.

[16] Derwent Whittlesey, *The Earth and the State* (New York, 1939), chap. 7.

For a while Germany was powerless to rearrange the newly drawn map of Europe, but powerful interests in Germany never ceased to make blueprints of territorial revisions for future use. The success of Hitler's foreign policy, so portentously set forth in *Mein Kampf*, can be explained in large part by his avowed revisionist and expansionist purposes.

Whatever revolutionary force the socialists may have had following the German collapse in 1918 appears to have dwindled after the government was taken over by the moderate parties of the Weimar coalition in 1919. This trend to the right became even more marked after French occupation of the Ruhr (1923) when, for the first time in the history of the young republic, heavy industry and big business acquired predominant influence in the government. The development of an aggressive foreign policy thereafter was still more unequivocal. In 1925, the election of Hindenburg as president of the republic and symbol of the state was a signal that the landholders of eastern Prussia had resumed direction of the national government.

The old general was also the symbol of militarism, which for some time had been spreading from its entrenched position in Prussia to the rest of Germany. Before the end of 1919 the general staff had been reinstalled and was functioning in violation of the Versailles Treaty. It was openly and formally reestablished in 1935 along with the restoration of the conscript national army. Every step toward rearmament taken by the Nazi party was hailed with enthusiasm not alone in Prussia but throughout Germany. Meanwhile the national economy was reorganized in order to equip an army that should regain for Germany its position as the leading power of continental Europe.[17]

As warfare has come to engross an increasing proportion of the belligerent populations until now it is "total," so military geography has grown far beyond tactical problems of varying terrain and strategic plans for military operations. It now comprises, in addition, surveys of the natural resources of all possible belligerents and useful neutrals, planning for the location of factories which produce war materiel or can be converted to war production, and stimulation of scientific discoveries and inventions calculated to improve the utilization of national resources. Activities of this nature may be described as political strategy. It has recently taken a place alongside military strategy.[18] Germany was the first nation to comprehend the value of political strategy as an adjunct to war and to recognize that it was rooted in geography. In Germany, geographic explorations and surveys were habitually supported by state subventions. It was

[17] A. T. Lauterbach, *Economics in Uniform* (Princeton, 1943), chaps. 3-4.

[18] If all the resources of the nation are to be directed toward the making of war, they must be controlled by the overall *political* authority of the country, rather than exclusively by its *military agent*; hence the term used here, "political strategy." Use of the same term also appears in a study of modern warfare from the viewpoint of political science by Edward Mead Earle, "Political and Military Strategy for the United States," an address before the Academy of Political Science at its annual meeting on "The Defense of the United States," New York, November 13, 1940: *Proceedings of the Academy of Political Science*, vol. XIX (1941), pp. 2-9. The author advances the name "Grand Strategy" to designate the integration of the policies and armaments of the nation (p. 7). Note also the title *The Axis Grand Strategy: Blueprints for the Total War* (New York, 1942).

but a short step to support by the state of geographic contributions to political strategy.

Geopolitics was groomed to bring geography to the service of a militarized Germany. Its functions were to collect geographic information, to orient it to serve the purposes of the government, and to present some of it to the public in the form of propaganda.

III

The portentous period of German history initiated by political unification coincides with the life of Karl Haushofer, who has made himself the *alter ego* of German geopolitics. A movement is often associated with a personality but such complete identity as exists in this case suggests a unique parallel between the man and his calling.[19]

Haushofer was born in Munich, the capital of Bavaria, on the eve of the Franco-Prussian War. Bavaria was a predominantly agrarian state, largely because it possessed no mineral resources. Haushofer was a member of an ancient landowning family which, two or three generations before, had turned to the professions and the middle-class life of the city. Before the war of 1914-1918 Bavaria was separatist, Catholic, and on the whole liberal, as contrasted with the nationalistic, Protestant, and reactionary tendencies of the Prussian ruling classes of Berlin. But after the collapse of 1918 Bavaria became a hotbed of monarchism, reaction, and antirepublicanism; above all, it was the capital of Hitler's National Socialist movement and the home of other chauvinist groups. It was in this postwar atmosphere that Haushofer rose to national prominence outside purely military circles and enjoyed the opportunity to make his influence felt on the Nazi movement.

Upon completing his schooling at the local gymnasium in 1887, he embarked upon a military career in which he early demonstrated proficiency. During these formative years, he developed a serious interest in geography. He "seized all available opportunities to learn to know Europe through travels."[20] From the outset of his professional career he was repeatedly assigned to general-staff duty with the Bavarian army and to teaching at the general staff college. By the time he was forty, he had proved his interest and proficiency in warfare, geography, and academic instruction.

At the beginning of 1909 he went to the orient. "I obtained, in connection with a furlough for further travels in India and southeastern Asia, an order for service study of the Imperial Japanese Army for two years."[21] He traveled to the Far East by sea, stopping in India and elsewhere. Upon arrival in Tokyo, he not only shouldered his general-staff duties of studying the Japanese army but also made it his business to learn the Japanese language. A part of his tour of duty was spent in travels in East Asia. He undertook systematic field studies

---

[19] For details see the appended chronology of Karl Haushofer, p. 410.

[20] Karl Haushofer, "Der deutsche Anteil an der geographischen Erschlieszung Japans und des subjapanischen Erdraums, und deren Förderung durch den Einflusz von Krieg und Wehrpolitik." *Mitteilungen der geographischen Gesellschaft in München* (Munich, 1914), vol. 9, p. 111.

[21] *Ibid.*, pp. 111-112.

in middle and southern Japan, in Korea, Manchuria, and North China. The remainder of his time was spent with different arms of the Japanese military service. In the summer of 1910 he returned home by way of Siberia.

Upon arriving in Germany he returned to the general staff college to lecture on his experiences abroad. After another tour of duty with troops as battalion commander, "a severe illness contracted in foreign service" was followed by a "long leave."[22] This leave lasted from 1912 until the outbreak of war recalled him to active service.

In those two years he published two books on Japan. To the first he gave the title, *Dai Nihon*,[23] which can be translated either "Great Japan" or "Greater Japan." The analogy with "Great[er] Germany," at that period the slogan of the Pan-Germanists, was probably intentional. The book was the first of a long series of works in which Haushofer has presented Japanese development and Japanese ambitions in a favorable light.

The second book was presented to the philosophical faculty of the University of Munich in the form of a dissertation for the Ph.D. degree. The degree was conferred in 1914 in geography, geology, and history, "with highest distinction." The long title of the book goes far to describe it: *The German Share in the Geographical Opening-Up of Japan and the Sub-Japanese Earth Space, and Its Advancement through the Influence of War and Defense Politics.*[24]

Haushofer aimed these books at different audiences. *Dai Nihon* was prepared for the general public and touched only the fringes of military geography. The dissertation was written from the conviction that only "a firm scientific summarizing of my frequently interrupted cultural progress in geography would permit me to extract from my practical experiences in the foreign service their full value."[25]

In these first volumes, Haushofer's overlapping interests in military matters and geography were coming into a common focus. In dedicating his dissertation to Ratzel, he quotes that master's commendation of the geographic work done by "travelers and sea voyagers, priests, soldiers, and traders." The traveler and soldier, Haushofer, begins his scientific study of Japanese geography with a quotation from Heraclitus: "War is father of all things." He goes on with an illuminating statement of his intellectual convictions: "To a soldier who obviously comes to the scientific workfield of geography from a militaro-geographic approach, it is but a step to wish to demonstrate the worth of Heraclitus's axiom also for this field, as a justification of his admission to it."[26] The military approach to geographical study he never abandoned in later years. On the contrary, four years of experience in war confirmed him in it.

To be sure, military geography is no invention of Haushofer's. Probably it is as old as armies. Both the tactics and the strategy of warfare are conditioned at every point by natural environment. Military geography of this sort has

[22] *Ibid.*, p. 112.
[23] Karl Haushofer, *Dai Nihon: Betrachtungen über Gross Japans Wehrkraft, Weltstellung, und Zukunft* (Berlin, 1913).
[24] For the German title see note 20.   [25] *Ibid.*, p. 112.
[26] *Ibid.*, p. 1.

always been taught in military schools. Long before Haushofer's day army officers in the German-speaking world had also valued geography as a window on the world. The great Moltke studied under Karl Ritter, one of the two nineteenth-century founders of modern geography. After he became chief of staff, if not before, the army stressed geography, particularly for general-staff officers. Perhaps even more important was Ritter's influence on Roon, the famous Prussian minister of war and close friend of Moltke. At the suggestion of Ritter, Roon wrote a well known textbook on political geography which was widely used in Prussian officer schools as well as throughout the Prussian secondary school system.[27]

Haushofer, with unusual opportunities to study geographic problems of tactics and strategy, might have applied his knowledge narrowly to his duties as a field artillery officer. But his military career synchronized with the elaboration of general-staff functions, and his frequent tours of staff duty gave him both experience and personal connections which carried over into the postwar period of his life.

IV

By the terms of the armistice and the peace treaties at the close of the First World War, the German army was reduced to little more than a police force and its general staff was abolished. Haushofer, like many other officers, returned from the front a general officer, only to be retired from active service. The militaristic pulse of German life was merely driven beneath the surface by fiat of the victors. As is now well known, the general staff continued nevertheless to function under General von Seeckt, who used whatever assistance he could find to nourish the militaristic ideals and objectives so long grounded in the German government and in German thought. Among his associates was Oskar von Niedermeyer, who had been a colleague of Haushofer's on the Bavarian general staff and who also had worked abroad, chiefly in Russia, on a mission similar to that of Haushofer in the orient.

More than ever before, the work of an alert army staff after 1919 entailed studies geographic in nature and world-wide in scope. The mechanization of warfare, and its increasing dependence upon the productive capacity of the homeland, extended the service rendered by geography from problems of tactics and supply on the fighting front to long-range logistics and the nation-wide manufacture of complicated equipment out of resources drawn from the four corners of the earth.

For obvious reasons no general staff publishes accounts of its activities. Bits of evidence which have come to light make it clear that the German general staff devotes much attention to thoroughgoing geographic studies of all potential sources of supply for German forces, as well as of all imaginable combat areas. With the increasing scale of German ambitions, these studies have come

[27] *Grundzüge der Erd-, Völker- und Staatenkunde* (1832). A second, equally well received work by Roon, though of a more elementary character, was *Die Anfangsgründe der Erd-, Völker- und Staatenkunde*, published in 1834. For further details concerning Roon see the article in *Allgemeine deutsche Biographie*.

to comprise the whole world. A good deal of this work has been undertaken, not by the general staff itself, but by other agencies with which it is closely articulated. Several of them are geographic in character.

Among those who worked to revive the German army, Karl Haushofer filled a special niche. Ostensibly he merely shifted once more from his career as an army officer to his career as a geographer. An appointment in 1919 as lecturer on geography in the University of Munich gave him a connection upon which he built up geopolitics both as an intellectual concept and as a method of statecraft.[28] His academic advancement was paralleled by his leadership in geographic societies and organizations. These, while in appearance scholarly undertakings initiated privately or by the university, were among the instruments used by the general staff both during its *sub rosa* period and after it had been openly restored.

It was natural for Haushofer to return to his own city and university. It was perhaps more than a coincidence that Munich should also be the center of the intrigues which resulted in the Beer Hall *Putsch* and the Nazi party. Bavaria had remained faithful to the agrarian and monarchist tradition of medieval Germany because its lack of mineral resources had inhibited the rise of powerful industrial magnates. Its preservation of semiautonomy in the German federation and its remoteness from Berlin made it a handy place for conspiracies against the established government of the nation; but it could serve equally well as a nonsuspect base for activities intended by that government to revive German military power contrary to the treaties it had signed. Retention of its own general staff had made the Bavarian army a leader in geographic activities up to the First World War, as is illustrated by the career of Haushofer.

Other remoter connections helped to create a favorable intellectual climate for geopolitics in Munich. The young Ratzel had taught geography there. Oswald Spengler, with his philosophic notions of the destiny of different parts of the earth, was still living and working there. Ludendorff, exponent of total war and military dictator of Germany during the final months of the First World War, retired to Munich and became associated with its elements of extreme reaction, including the Beer Hall *Putsch* of Hitler's Nazi party. And Bavaria was the home of the Wagners, early adherents of Nazism and exponents of the Nordic-German tradition celebrated in Wagnerian opera.

It has often been remarked that many non-Germans have advocated the Prussian viewpoint more vigorously than the Prussians themselves. Munich has been the chief center outside Prussia where Germans have enthusiastically championed Prussianism, including its all-pervading militarism.

After returning from the war, Haushofer associated himself with a number of geographers and a few political scientists and publicists in the founding of geopolitics. Most of these men were his peers in training and position. At the end of a decade many of them had quietly fallen away from the movement. They were replaced by Haushofer's disciples and pupils—mainly younger men

[28] The steps in his postwar career are given in the appended chronology, p. 410.

who slavishly copied their master. They adopted his views and even imitated his opaque literary style.

They have been less successful, however, in equaling his habitual avoidance of commitment. Hence their more direct statements frequently help to elucidate the objectives and philosophy of geopolitics. Because all German writings on the subject are cast in the mold set by Haushofer, no attempt is made in this study to assign credit to particular authors. Instead, their work has been drawn upon as a single reservoir.[29]

This reservoir brims with a flood of words but from it issue no limpid streams of ideas. Repeated dipping into it brings up a mixture of theories, programs for action, and propaganda.[30] When analyzed they are found to contain matter drawn from several natural and social sciences, ranging from geology to psychology. The most fundamental items in the mélange are concepts about "space" because the claims made for geopolitics as a guide to statecraft are based upon geography.

An objective search for the space concepts used by the geopoliticians has to date uncovered only five.[31] They form a recurrent *leitmotiv* in these prolix and repetitious writings. When stripped of the ambiguities with which they are generally surrounded, they are seen to be presented in different versions, often mutually contradictory. All this is the antithesis of the spirit and method of science which the geopoliticians look upon as their guiding star, for natural science sets as its goal impartial search, accurate observation, lucid generalization, and precise formulation. Two of the five are applications of geographic theory, two others are proposals for world organization, and the fifth is a facilitating device.

Autarky. Autarky, as used by the geopoliticians, is the ideal of national self-containment in the economic sense. It assumes that every political unit ought to produce everything it requires. Thus the state will be in economic balance and independent of the products of foreign parts of the world.

It is self-evident that no area less than the entire earth can contain all the products useful to man. Autarky, in the literal sense, can therefore not be attained short of world unity. Existing states fall far short of the ideal. Even the largest comprise only a few of the major climates which, along with soil and drainage, determine capacity to yield agricultural and forest products. The distribution of minerals is so erratic that even if a state were to include every major type of climate it would have no assurance of being adequately provided

[29] For quotations from particular authors see Andreas Dorpalen, *The World of General Haushofer* (New York, 1942), and Derwent Whittlesey, *German Strategy of World Conquest* (New York, 1942).

[30] Werner J. Cahnman, "Methods of Geopolitics," *Social Forces*, vol. 21 (1942), pp. 147-154, lists the kinds of materials utilized.

[31] Months before I knew of the project for the present volume I made an attempt to sort out the geography residual in geopolitics in the hope of discovering factual material that could be subjected to scientific criticism. My findings surprised me. I used them in addressing several audiences of diverse character with no thought of publishing them. I now present them without change, except for phraseology.

with any particular mineral, to say nothing of a balance of all the minerals needed by a modern industrial nation.

If Germany comes as near autarky as any industrial state it is only because it makes a practice of importing huge stockpiles of the commodities most needed to supplement its large and varied internal production. With such accumulations on hand, it still has to live on short rations. Autarky at this price is dearly bought. Moreover, it is specious because it vanishes when stockpiles are used up.

The Nazi government cannot have adopted autarky, independent of excessive stockpiles, as an attainable goal within existing national boundaries, although many Germans may have been led to believe so by artful propaganda. Rather it must have intended to achieve other and practical ends. If the German people could be made to strive toward autarky, their self-denial would immediately release both internal and imported supplies for some special purpose of state. This turned out to be preparation for war, as was openly proclaimed in the later 1930's, notably in the slogan: "Guns for butter."

With preparedness expedited, war could be precipitated so much the sooner. The power to pick the moment to fight gives the aggressor a tremendous advantage, particularly if its potential enemies are not equally eager to make war or have not realized their danger. Once a country is at war, the continued low rate of civilian consumption is an aid to winning a quick victory.

A state embarking upon this program will hope to be immune to attack from any enemy while reorganizing its internal resources and building up stockpiles. Then, when its military strength is at the peak, it can threaten war, and if necessary fight. With luck it may hope to make conquests. If then it can convert these into perpetual domains, it will have annexed territory and natural resources which will make it more nearly self-sufficient, that is to say autarkic, than it was before the war.

Without a plan for war and the enticing promises of conquest, no government could obtain popular backing for a program of self-denial to gain autarky. The accomplishment of real autarky within the boundaries of any existing national state would be merely a painful method of reducing the national level of living. There is ample proof that the short rations of Nazi Germany were adopted only to expedite re-creation of the German military establishment.

If all states were to follow the German lead, and adopt autarky as its working ideal, no one of them could realize its ambition. Either they would have to fight eternally, or they would have to isolate themselves within their frontiers and reduce their living to the measure of their internal productivity. Peaceful economic interdependence, as practiced among nations in recent decades, is incompatible with autarky.

Lebensraum. *Lebensraum* (living space) is understood by the geopoliticians to be the *right* of a nation to ample room for its population. In addition to mere area, *Lebensraum* takes account of all the natural and human resources to be found in any area claimed by a state as its rightful living space.

The claim to this right rests upon a fact and a theory. The fact: the differential increase in population growth among the nations. The theory: the state is an organism subject to biological laws. A corollary of this theory is that a youthful, growing state must expand.

In German geopolitics only the great powers are envisaged as growing; small states are seen as doomed to extinction. It is further recognized that other great powers may sooner or later dwindle, but the possibility that Germany might decay or die is never considered.

The German application of the theory of *Lebensraum* is vitiated by fallacies. First, a state is not a biological organism although it may appear superficially analogous to such an organism. Therefore, there is no "natural law" to decree that a state must either grow or decay. Second, the area which Germany has traditionally claimed as its *Lebensraum* is the part of Europe to the eastward of Germany, where the population is largely Slavic. Among these Slavs and others the birth rate has long been higher than in Germany. According to the theory, their need of increased territory is therefore greater than Germany's. The illogic of the geopoliticians' position is also conversely demonstrated. Until the fortune of war dropped France unexpectedly into the German lap, they never claimed a *Lebensraum* west of the Franco-German linguistic line. Yet France is the one European territory bordering on Germany where the birth rate has been considerably lower than in Germany for many decades. This merely proves that the *Lebensraum* theory has been made to fit the traditional policy of German expansion—toward the east rather than toward the west.

Attempts to gain *Lebensraum* involve either war or the effectual threat of war, as the events since 1938 have amply proved. It can only be concluded that the concept of *Lebensraum*, although geographic in character, is in its operation a politico-military device. It appears to provide scientific support for the greed for conquest inherent in many governments.

Panregions. As the ambitions of German chauvinists have risen, it is natural that geographers among them, viz., the geopoliticians, should have staked out particular claims to parts of the earth their nation might hope to conquer and annex.

It has long been the fashion in Germany to demand that the nation's political area be extended to include all people of German speech, whatever their history. Definition of this area varies. Sometimes the Dutch and Flemish languages are assumed to be German. Ever since the heyday of Pan-Germanism (1890-1918) claims have run far beyond the boundary of uninterrupted German speech. A Germanic "culture area" and a Germanic "trade area" have been mapped as part of Germany's nature-given domain. However, the territory earmarked for this anticipated expansion lies to the east of Germany. Claim to it is based on the German settlements of earlier centuries scattered as linguistic enclaves in a region of Slavic and other non-Germanic languages, and on the use of German as a *lingua franca* for commercial purposes throughout the area.

The territorial aspirations of the geopoliticians are still more grandiose. As substitutes for the ideal but at present unattainable autarkic world dominion,

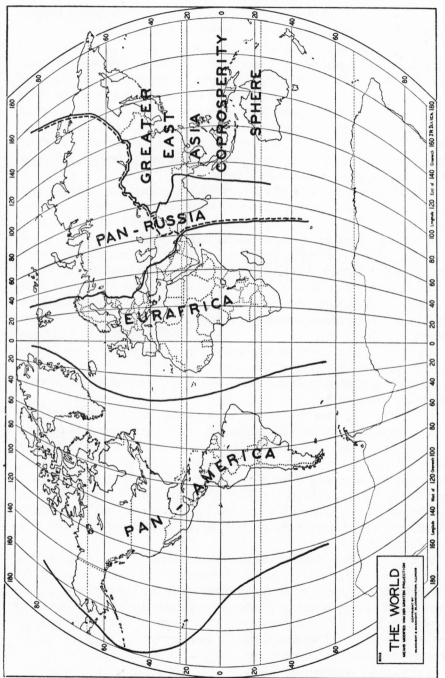

FIG. I. A Construction of Principal Panregions, based on writings (not maps) of the geopoliticians. No specific frontiers for the regions have ever been proposed. They are here assumed to be existing political boundaries, except in eastern Siberia. An alternative subdivision into three regions, allotting India to Japan and Russia to Eurafrica, is suggested by the dotted line.

they propose a political aggregation of the earth into three (or sometimes four) "panregions." Each of these vast areas is supercontinental. (Figure 1.)

Each unit combines middle and low latitudes, thus providing the diversity of agricultural and forest products which only the gamut of climates can yield. Being so large, each of these units is sure to incorporate a wide variety of mineral resources. However, since minerals are distributed without relation to present-day climates and populations, there is no assurance that the proposed panregions would possess equal shares and it is certain that each area would lack a few vital minerals. Each of the proposed panregions would, however, have ample contact with the ocean and therefore would be in a position to import the few items not produced at home.

The panconcept grows out of a recognition that present-day rapid transportation and communication have destroyed independence of action for nearly all small states and some fairly large ones. It is optimistically assumed that the time has come when areas larger than the largest existing nations can be economically unified. It concludes that they should therefore be politically unified. It fails to take account of the fact that water transportation often links lands on opposite shores of an ocean more readily and effectively than land transportation connects parts of a continental mass severed by deserts, highlands, or forests.[32] These assumptions and oversights underline the fact that the panregion is an invention of land-minded people, and specifically of denizens of Europe, the one continent lacking formidable barriers created by nature's harsher aspects—deserts, great plateaus, polar ice.

Each of the proposed political units would have at its head one of the existing great powers.

1. Pan-America is the most obvious of the panregions, because it is set apart from other lands by wide oceans. The United States is assumed to be at the controls. The geopoliticians greatly admire the Monroe Doctrine as the earliest of the panideas and one which the United States has skillfully kept functioning for more than a century. They concentrate entirely upon its political aspects and its aggressive manifestations. They ignore the cultural elements connoted by the word "Pan-America," despite the fact that the present generation of North Americans takes for granted that the essence of the concept is cooperation for the mutual benefit.

2. Pan-Asia comprises the eastern Asiatic mainland, Australia, and the islands between. Japan is its presumed master. The Japanese name for this panregion is "Greater East Asia Co-Prosperity Sphere." Japan is the only country outside Germany whose geographers have turned eagerly to geopolitics. The personal influence of Haushofer appears to have opened their eyes to his application of geography. They have found it a useful prop to the plans for conquest already formulated by the militaristic rulers of the nation.

The German geopoliticians tacitly assume, and sometimes flatly state, that both Pan-America and Pan-Asia are temporary expedients, permitted by Ger-

[32] Eugene Staley, "The Myth of the Continents," *Foreign Affairs*, XIX (1941), pp. 481-494.

many until it shall have integrated its own panregion. In due time these territories are to be swallowed up by an expanded Germany.

3. The panarea to be dominated by Germany combines Europe and Africa. Not only are the lesser countries of Europe merged in "Eurafrica," but also the great powers, France and Italy. The status of insular Great Britain and the vast land area of the Soviet Union are recognized to create difficulties.

One solution to the problem of the Soviet Union is to permit it to set up a fourth panarea along with India. This proposed region does not quite reach the equator and therefore lacks certain of the climates which all the others possess. An alternative suggested is that Germany make itself master of Russia while Japan takes India.

Whether the number of political units envisioned is three or four is a minor and theoretical issue. More critical is the actual existence of several panregions. These the geopoliticians grudgingly recognize but do not admit their validity. They are the British Empire and in a minor way the other established overseas empires. These political aggregations of territory cut across all the proposed panregions and therefore must be obliterated before their successors can be set up. The geopoliticians brush aside the lesser colonial structures as easy prey but they acknowledge the British Empire to be a bar in their path.

It is self-evident that none of the panregions contemplated can be formed without recourse to war. States in the middle latitudes to be obliterated include great powers which will not submit to conquest without a struggle, as the present war proves. Military domination of the low-latitude areas is easier but politically it is no more feasible because it involves large colonial holdings of middle-latitude states. Seizure would embroil those powers as certainly as an attack on their home territory.

Other practical obstacles to the organization of panregions arise from their geographic nature when analyzed as land units. In each case the dominant middle-latitude area is separated from its low-latitude dependency by a formidable natural barrier to overland contact. The two Americas are in effective contact only by sea and air. Their first overland connecting highway is far from completed, being interrupted by mountains and jungles at several points. Southern South America is further cut off from its northern fringe by the Amazon forest and marshland, coupled with the high, rugged, and dry Andean Plateau. Japan can control its Co-Prosperity Sphere only by maintaining naval strength requisite to link the numerous islands and land strength sufficient to dominate the eastern part of the Asiatic continent. Russia and India are separated from each other by almost impassable mountains and by cold plateaus and desert lowlands.

Eurafrica is twice broken. The Mediterranean Sea will compel a dominant Germany to maintain sufficient sea power to close its outlets and to maintain ferry service across it. This has been done by land power in the past and could perhaps be repeated. The vast waste of the Sahara, however, remains a barrier unsurpassed on earth except by the polar ice caps. To circumvent it, the power

which controls Eurafrica must be master of the ocean no less than the overlords of Pan-America and Pan-Asia.

Every attempt to set up a panregion would result in war, as is being proved
in two cases right now. If established by conquest, only the Russia-India combination could be maintained without dominant sea power. The existence of
more than one such power is incompatible with the unity of the world ocean.
Hence the proposed panregions would be no more stable politically than the
present great powers.

In spite of difficulties in the way, the panidea is not wholly visionary. The
considerable success of both Japan and Germany in conquering territory they
have staked out is a first step toward assimilation, however ephemeral these
conquests may prove to be.

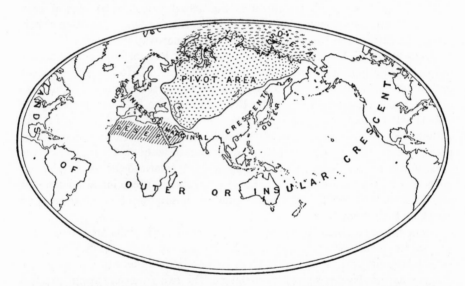

Fig. 2. The Heartland and World Island vis-à-vis the Insular Continents, according to Mackinder. Mackinder, H. J., "The Geographical Pivot of History," *Geographical Journal*
XXIII (1904), 435. (Courtesy Royal Geographical Society.) This is the map used by the
geopoliticians. The "Pivot Area" is the unit which the author later renamed the "Heartland." The "World Island" is the double continent Eurasia-Africa. Compare maps and
diagrams in Mackinder, H. J., *Democratic Ideals and Reality* (New York, 1942), especially
67, 69, 78-79, and 105.

Land Power vs. *Seapower.* A somewhat different approach to the integration
of the earth has been taken over by the geopoliticians from Mackinder. It recognizes the land mass Eurasia-Africa as by far the largest, most populous,
and richest of all possible land combinations. (Figure 2.) Merely because of its
size, it is inevitably the center of gravity of human life. It may be thought of
as the principal island in the world ocean. The other continents lie around the
periphery of this "world island" as a ring of smaller islands.

The heart of the world island is a vast region cut off from the oceans.
Throughout most of this area the rivers either flow into the ice-blocked Arctic

Ocean or lose themselves in inland seas and saltpans. Nowhere do navigable streams make contact with navigable oceans. This "Heartland" is an ample base for land power, potentially the greatest on earth.

Fringing the Heartland on the west, south, and east lies a crescent of marginal lands with access to the oceans. All of them are to some degree maritime. They are separated from each other by mountains, deserts, or seas. The two large insular groups, the British Isles and the Japanese Archipelago, are major sea powers. Beyond this inner crescent the world ocean is interrupted by an outer crescent of continental "islands"—the Americas, Black Africa, and Australia. Their interests are twofold, divided between their considerable expanses of land and their access to oceans. At present they swing in the orbit of sea power, but they might be overwhelmed by a Heartland state which had added sea power to land power by mastering the inner crescent.

The Heartland concept to this point is borrowed intact from Mackinder. The geopoliticians apply it to the political expansion of Germany in the following argument. The Heartland is nearly coterminous with Soviet Russia. Germany, likewise a significant land power, shares with Russia the interior of Eurasia. In contrast to Russia, it also has access to the sea and therefore the capacity to make itself a sea power. If only it can become dominant in a partnership with Russia, it might control first the Heartland, then the inner crescent, including British and Japanese sea power, and finally, the outer crescent of lesser continents.

The only conceivable method of achieving German expansion according to this formula is by war. The geopoliticians' expressed hope that Germany and Russia might peacefully cooperate to attain a partnership that Germany would dominate rose high with the treaty of 1939 only to be dashed to the ground by Hitler's attack on the Soviet Union in 1941.

The current war of six continents and all the oceans is evidence that the earth's political structure is unready for such stresses as the Heartland theory thrusts upon it. In actuality adjacent states tend to seek alliances, not with their immediate neighbors, but with the states next beyond. This results in alternate linkage across Eurasia, with Germany and Russia in opposed camps. Combinations of land power with sea power are often made. Great powers with interests outside of Europe throw their weight into the resulting balance independently of either Germany or Russia—nations with which they may have little contact.

Too far from present realities to be put into effect just now, the Heartland proposal might be rejected without consideration but for its hold on the minds of the German people. To them it appears to be a feasible way of satisfying their ambitions.

Frontiers. Whatever the territorial pattern proposed for a new political order on the earth, and whatever the philosophic concepts adduced to justify conquest, there must also be a practical method of moving toward the goal. This method is supplied by the geopoliticians' interpretation of political boundaries. Americans and others who have naively supposed that a frontier represents a fixed margin of the state will be surprised to learn that to Germans who follow the geopoliticians such a boundary is merely a temporary halt of a nation in its march toward world domination. To them a newly established boundary offers a breathing spell until the nation can again gird for further conquest. Then, at the desired moment the existence of the new frontier may be made to serve as a fresh cause for war.

History has abundantly proved that a frontier is, of all political devices, the one most readily made the occasion for war. Border incidents have accidentally or intentionally touched off many a conflict. The geopoliticians have gone farther. They claim that the nation has a right to "natural frontiers." These may lie beyond its existing political borders. If so, they invite aggression. Implicit in the term is the idea of a physical barrier. Most frontiers described as "natural" are in no sense barriers today, whatever they may once have been. Therefore, once achieved, frontiers marked out by nature can as readily be made an excuse for further wars of conquest as "artificial" boundaries. Any frontier, even a considerable barrier, is unstable if it separates nations having widely different power potentials. The lust for *Lebensraum* or autarky makes of a weak neighbor a tempting morsel to a great power, especially if it possesses rich or reciprocal resources, or occupies a strategic location.

It is by no means an accident that Europe is the classic continent of conquest, with the most numerous and longest frontiers.[33] It is also the continent with the least wasteland and the most valuable natural resources in proportion to its area. The diversity of its terrain induced the rise of many different political units, segregated by natural barriers. As these separate units grew and came in conflict with their neighbors, they reduced or destroyed the intervening barriers. On making close contact they met as rivals or enemies. Fierce struggles for tiny bits of territory ensued. The boundaries of Europe are the results of this chain of events. The geopoliticians chime with history in viewing them as a standing excuse for war and the one most readily available.

V

The space concepts on which geopoliticians have built their system are presented by them as scientific postulates. This gloss of respectability covers an interlarding of fact with fallacy which perverts whatever geography they contain. Even if the reasoning were scientifically sound, a government adopting it as a program would be led irresistibly into war. To convert the theories of space into geographic actualities would entail rearrangements of the political map certain to meet bitter resistance, as the current war abundantly proves. From its inception geopolitics has been inextricably bound up with war. It grew up in a militaristic country and its development has been directed by a military man.

Geopolitics, then, is the product of military thought. To what degree has it in turn permeated military thought?

Its service to military preparedness in Germany is of two sorts. It intensified

[33] S. Whittemore Boggs, *International Boundaries* (New York, 1940), pp. 13-15.

the methodical gathering and organizing of information about all parts of the earth—a habit already well established among German geographers. It channeled its findings directly into the military establishment and, in the garbled form of propaganda assuming the guise of applied geography, to the people of Germany and other nations.

The military establishment, through its general staff, had long been in the habit of gathering systematically and digesting information about the earth. The geopoliticians, in their larger function of engaging in applied science, were the first to extend this fact-finding to serve the needs of the government to lay a course for its political strategy. This entails geographic research in the broadest sense, by no means confined to geopolitics or even to political geography. The research groups assembled by Haushofer in his Institute of Geopolitics at the University of Munich and elsewhere, were doing in expectation of war precisely what geographers in the United States government agencies, including the War Department, are now belatedly doing for the American armed forces in consequence of war. The realization of total war has moved geography into a key position among techniques and viewpoints serving the interests of every nation.[34] The geopoliticians, desiring and foreseeing war, urged and obtained support for geographic fact-finding agencies years before war came.

Connection between Haushofer and the German general staff has never been proved by documentary evidence. Nevertheless, competent foreign observers, especially those long resident in Germany, are convinced that it existed. Both military attachés and businessmen have given evidence of their convictions.[35]

The geopoliticians, in their second and narrower function, have confirmed the German people in its militaristic frame of mind. Their space concepts are set up as practical objectives for the German government but these can be realized only by going to war. This the geopoliticians know. They justify their proposals by asserting that war is the natural state of man. They extol its effects and the effects of military preparedness on the nation. They present a recipe for the prosecution of war by geographic means.[36]

All this has bolstered the nation's faith in its destiny as a "master race" with a mission to conquer and rule. That attitude grew out of German history and the mysticism of the German temperament, and was prevalent long before the day of geopolitics. But in a scientific age, blind faith clutches at natural science for support. Geopolitics provides the desired rationalization for German racism. The space concepts adduced deal incontrovertibly with the material earth. Moreover, they are couched in semiscientific lingo, impressive to the layman. The formulas of geopolitics have gained for it not only the adherence of

[34] Derwent Whittlesey, "The Role of Geography in Twentieth Century War," J. D. Clarkson and T. C. Cochran, eds., *War as a Social Institution* (New York, 1941), pp. 78-87.
[35] For the most part orally; among them Eric Archdeacon, an American banker, in a broadcast *Address before the Canadian Club of Montreal, February 8, 1943*. Mimeographed Transcript, pp. 6-7. Fact-finding agencies that appear to have worked with or for the general staff are named in Whittlesey, *German Strategy, op. cit.*, pp. 107-109.
[36] See especially Ewald Banse, *Raum und Volk im Weltkriege* (Oldenburg, 1932). Translated as *Germany Prepares for War*. English edition, 1934 (New York, 1941).

the public, but also the official sanction of the rulers. Connection between geo-politics and the Nazi party has been questioned by both Germans and non-Germans. Nevertheless, study of events and of internal evidence has caused unbiased students and many of the doubters to conclude that its influence has been large.

Rudolf Hess, at the outbreak of the present war officially named Number 3 man in the Nazi government, was an aide-de-camp to Haushofer, and one of the most devoted of his early disciples in geopolitics. While Hitler was in deten-tion near Munich after the Beer Hall *Putsch*, Haushofer frequently visited him. At that time Hess was acting as amanuensis in the writing of *Mein Kampf*. A few of the more striking passages from *Mein Kampf* will indicate that several of the concepts and some of the program of geopolitics appear to have influenced Hitler profoundly.

Discussing the relationship of *Lebensraum* and a nation's military potential, Hitler said:

"The size of a people's living area includes an essential factor for the deter-mination of its outward security. The greater the amount of room a people has at its disposal, the greater is also its natural protection; because military vic-tories over nations crowded in small territories have always been reached more quickly and more easily, especially more effectively and more completely, than in the cases of States which are territorially greater in size. The size of the State territory, therefore, gives a certain protection against frivolous attacks, as success may be gained only after long and severe fighting and, therefore, the risk of an impertinent surprise attack, except for quite unusual reasons, will appear too great. In the greatness of the State territory, therefore, lies a reason for the easier preservation of a nation's liberty and independence, whereas, in the reverse case, the smallness of such a formation simply invites seizure."[37]

Continuing in much the same sense he said later in the book:

"*The foreign policy of a folkish State is charged with guaranteeing the exist-ence on this planet of the race embraced by the State, by establishing between the number and growth of the population, on the one hand, and the size and value of the soil and territory, on the other hand, a viable, natural relation-ship. . . .*

"*Only a sufficiently extensive area on this globe guarantees a nation freedom of existence.*

". . . *the area of a State has also another, military-political significance than as a direct source of nourishment of a people.* When a people has secured its nourishment for itself by virtue of the extent of its soil and territory, it is nevertheless necessary to think also of securing the territory in hand. This depends on the State's general power-political force and strength which is to no small extent conditioned by geo-military considerations."[38]

In these last quoted sentences the use of the term "geo-military considera-

---

[37] Adolf Hitler, *Mein Kampf* (New York, Reynal & Hitchcock edition, 1939), p. 177. Quotations reproduced here through the courtesy of Houghton Mifflin Company, owners of the basic copyright.
[38] *Ibid.*, pp. 935-936. All italics in the original.

tions" makes the existence of a direct influence of Haushofer on Hitler a not unreasonable assumption.

Particularly significant are the following quotations which neatly blend geopolitics and National Socialist *Weltanschauung*. They prove definitely that for Hitler the racial and language problem is only one among several at least equally important factors which ought to determine the extent of the dynamic frontiers of Germany.

"The demand for the re-establishment of the frontiers of the year 1914 is political nonsense of such a degree and consequences as to look like a crime. Entirely aside from the fact that the frontiers of the Reich in the year 1914 were everything but logical. For they were, in reality, neither complete with respect to the inclusion of people of German nationality, nor intelligent with respect to geo-military appropriateness. They were not the outcome of considered political action, but momentary frontiers of a political struggle in no way concluded, indeed, partly the result of the play of chance. With equal justice, and in many cases with greater justice, one could select any other milestone of German history to proclaim the re-establishment of the relations then prevailing as the goal of foreign-policy activity.[39]

"We National Socialists, however, must go further: *the right to soil and territory can become a duty if decline seems to be in store for a great nation unless it extends its territory.* Even more especially if what is involved is not some little negro people or other, but the German mother of all life, which has given its cultural picture to the contemporary world. *Germany will be either a world power or will not be at all.* To be a world power, however, it requires that size which nowadays gives its necessary importance to such a power, and which gives life to its citizens."[40]

It is quite probable that history will show an intimate relationship between the expansionism of Nazi foreign policy and the ideas of Haushofer and other geopoliticians.

In addition to these facts, the rise of Haushofer and other geopoliticians in the Nazi world, particularly after 1933, is too striking to be accepted as sheer coincidence.[41]

The army itself cannot have remained unaffected by a movement so admirably designed to serve its purposes. Within the ring of German culture, then, it must be admitted that geopolitics is an integral part of official and military thought and public militaristic opinion.

Attempts to domesticate it in other countries have not succeeded except in Japan. That nation was militaristic long before the day of geopolitics, but Haushofer's paternal interest in Japan has been reciprocated. His books have been translated into Japanese, and geography is actively sponsored by the state-supported universities. In adopting geopolitics, the Japanese have adapted it to suit their government's program of conquest.

[39] *Ibid.*, pp. 944-945.   [40] *Ibid.*, p. 950.
[41] For positions held by some of these men see Whittlesey, *German Strategy, op. cit.*, pp. 74-76.

In all other nations geopolitics was disregarded until the present war forced everybody to give it attention. The word has now become familiar and the history and meaning of the movement in Germany has been made the subject of a number of books and articles.[42] Most people look upon it as a *Frankenstein*. Some endorsers of the term "geopolitics" have tried to remodel it for use by one or another of the United Nations by adopting its name and fact-finding procedure. In other words they would gather all the sound geographic information needed by the government and call it geopolitics. Others have attempted to formulate an American geopolitics by drawing upon geography to support a political program for the United States' participation in world affairs. This invariably turns out to be a design for the practice of American power politics.

The alternative usage serves to differentiate geopolitics from geography. Knowledge of world geography is one of the securest foundations for an intelligent national policy in war and peace because it records and arrays material facts about the earth which must be taken account of in a realistic national program. Geopolitics in the narrow Haushofer sense can lead only to war, no matter what nation adopts it. However disguised, it is a recipe for only one brew—power politics and aggression.

CHRONOLOGY OF KARL HAUSHOFER, ARMY OFFICER AND GEOPOLITICIAN

1887—Finished his studies at the Royal Maximilian Gymnasium.

1887—Cadet in the First Royal [Bavarian] Field Artillery Regiment. All of Haushofer's military service was with the Bavarian army.

1889—Left the Military Academy (Kriegsschule) with the mark "highest distinction." Commissioned Lieutenant.

1890-1892—Ordered to the Artillery and Engineering School. Returned to regiment in the earlier place of service with "Qualification for Special Service."

October 1, 1895—Entered the General Staff College (Kriegsakademie) on passing the entrance examination.

October 1, 1898—Completed the General Staff College with "Qualification for the General Staff and for Teaching." Promoted to the grade of First Lieutenant.

1898-1899—On the Adjutant's staff of the First Field Artillery Brigade.

November, 1899-1901—Employment in diverse places, chiefly in a two-year tour on the General Staff.

September, 1901—Captain and Battery Chief, serving with troops.

Early 1904—On the General Staff once more.

October, 1905—Teacher of the new military history at the Military Academy.

January, 1907—General Staff officer of the Third Division in the exercise of General Staff troops.

Winter 1908-1909—Visit to Japan.

Summer 1910—Journey home.

[42] See annotated bibliography, appendix.

October, 1910—Teaching in the General Staff College, specifically to give account of experiences in the Orient.

Summer 1911—Promoted to major and battalion commander in Eleventh Field Artillery Regiment.

1912—Retired from active service in the army.

1914—Ph.D. in geography, geology, and history, "With Highest Distinction," Royal Bavarian University of Munich. Recalled to active service at outbreak of First World War.

1919—Retired as Brigadier General. Lecturer on geography, University of Munich.

1921—Honorary Professor of Geography, University of Munich.

1924—Editor-in-Chief of the *Zeitschrift für Geopolitik,* a publication sponsored by the University of Munich.

1924—Senator in the German Academy.

1933—Professor of Geopolitics and Dean of the Faculty of Sciences, University of Munich.

1933—Director of the Institut für Geopolitik of the University of Munich.

1934-1937—President of the German Academy.

1938—President of the People's Organization for Germans Living Abroad.

SECTION V

Sea and Air War

CHAPTER 17. Mahan: Evangelist of Sea Power

BY MARGARET TUTTLE SPROUT

NO OTHER single person has so directly and profoundly influenced the theory of sea power and naval strategy as Alfred Thayer Mahan. He precipitated and guided a long-pending revolution in American naval policy, provided a theoretical foundation for Britain's determination to remain the dominant sea power, and gave impetus to German naval development under William II and Admiral Tirpitz. In one way or another his writings affected the character of naval thought in France, Italy, Russia, Japan, and lesser powers. He was a historian of distinction and, at the same time, a propagandist for the late nineteenth century revival of imperialism. By direct influence and through the political power of his friends, Theodore Roosevelt and Henry Cabot Lodge, he played a leading role in persuading the United States to pursue a larger destiny overseas during the opening years of the twentieth century.

Mahan's epoch-making book *The Influence of Sea Power upon History, 1660-1783*, published in 1890, appeared at a uniquely propitious time. The following decade was crowded with international events of great moment in naval history: the decision of Germany to construct a modern fleet, the rise of the Japanese navy, the Spanish-American War and the consequent emergence of the United States as a world power. Furthermore, naval architecture and naval technology were then passing through the later stages of the industrial revolution: sails had given way to steam, wooden hulls to ironsides and armor plate, smoothbores to rifled guns. New weapons were looming on the horizon, and specialized types of naval vessels were being designed for specialized naval functions.

But naval thought lagged behind naval technology, especially in the United States.[1] Indeed American naval doctrine in the 1880's showed little advance over revolutionary and post-revolutionary days. We still seemed to be obsessed with the twin theories of coastal defense and commerce raiding. When, as in 1776 and 1812, we were faced with vastly superior naval forces and could accept fleet action only at dire peril, there was both wisdom and caution in a purely defensive strategy. But our naval officers and statesmen seemed to forget that when occasion permitted—that is, when we engaged a weaker naval power—we had gained command of the sea and throttled enemy commerce. Under Jefferson we carried offensive war to the Mediterranean and blockaded the ports of the Barbary pirates. In the Mexican War we controlled the waters of the Gulf and thereby enabled an American army to land on Mexican shores. The Union's command of the overseas and river communications of the South

[1] For detailed discussion of American naval policy see H. & M. Sprout, *The Rise of American Naval Power* (rev. ed., Princeton, 1942). On naval technology see Bernard Brodie, *Sea Power in the Machine Age* (Princeton, 1941).

undoubtedly was a major factor in the downfall of the Confederacy. Despite these significant experiences, however, the popular conception of the navy at the time of the Spanish war was that it was designed for defense of the American coast.

Coast defense and commerce raiding—the *guerre de course,* as the French called the latter—as theories of naval power seriously hindered the development of American naval strategy and naval technology. But conservatism could not long be controlling in such matters, especially as during the 1880's a new navy was in course of construction and a new consciousness was abroad in the land that the navy was an important instrument of national policy. Advances in technology were natural enough in an industrial nation like the United States, and the writings and teachings of Mahan provided the bases for a new naval strategy.

I

In 1884 Admiral Stephen B. Luce, president of the recently established Naval War College at Newport, Rhode Island, invited Captain Mahan to become lecturer on naval history and tactics, thus providing him with a platform from which he was to rise to world fame.[2] By training and temperament Mahan was well qualified for the post at the Naval War College, which he accepted with eagerness.

Alfred Thayer Mahan was the son of Dennis Hart Mahan, a professor at the United States Military Academy who had shown great interest in the art of war and had written extensively on military engineering. The younger Mahan therefore grew up not only in the military atmosphere of West Point but in the distinctly intellectual atmosphere of his own home. He entered Columbia College in New York but, much against his father's wishes, transferred to the Naval Academy at Annapolis, from which he graduated with the class of 1859. After serving in the Union Navy during the Civil War, he was sent in 1867 on a two-year cruise in Asiatic waters. It was during this particular tour of duty that he began the systematic study of history which was to be a lifelong habit. He returned from the Far East by way of Europe, which he saw not merely as a tourist but as a student of commerce and naval affairs. Beginning in November, 1872, Commander Mahan served three active years on the South Atlantic station. Later, he made determined efforts to reform the Boston Navy Yard, without much success.[3]

In 1878 Mahan's first published work, an essay on "Naval Education for Officers and Men," won third prize in a competition of the United States Naval Institute. By this time scholarship was his constant avocation, and he habitually employed his leisure reading the best military literature of the time, including the foreign professional journals. Napier's *Peninsular War* aroused his interest because of its emphasis upon "the military sequences of cause and effect," a

[2] Capt. W. D. Puleston, *The Life and Work of Captain Alfred Thayer Mahan* (New Haven, 1939), chap. XI .

[3] For Mahan's early life and career, see *ibid.,* chaps. I-X.

matter to which he gave some attention in his first book, *The Gulf and Inland Waters*, published in 1883.[4] At about this time, also, Mahan began to give serious thought to the influence of the projected isthmian canal on the international position of the United States. It is of interest to note that during the earlier years of his life Mahan was an outspoken anti-imperialist. Thus, while in command of a sloop off the west coast of South America in 1883-1884, he was thankful that he did not have to land sailors to protect American interests during the Peruvian revolution.[5] While he was stationed off Callao in 1884, he received the invitation to lecture at the War College, and a new life was opened up to him.

During the following two years Mahan prepared assiduously for his new duties, principally by acquiring the historical background which he considered essential to his work. He began with Mommsen's *History of Rome*. Mommsen's discussion of Hannibal's reasons for selecting the overland rather than the sea route to Italy led Mahan to wonder what the result of Hannibal's campaign might have been had he made the opposite choice. Pondering this question, he concluded that "control of the sea was an historic factor which had never been systematically appreciated and expounded."[6] As a result, Mahan proceeded to make a systematic study of the naval and military history of the seventeenth and eighteenth centuries, which provided ample material for analysis of the factors which affect the rise and decline of nations. It also presented opportunities for studying "the analogy between land and naval warfare" from which he hoped to derive a theory of naval tactics. For his lectures at the War College he decided to examine "the general history of Europe and America with particular reference to the effect of sea power."[7] In this manner Mahan prepared and wrote his three works: *The Influence of Sea Power upon History, 1660-1783*, which appeared in 1890, *The Influence of Sea Power upon the French Revolution and Empire, 1793-1812* (1892), and *Sea Power in its Relation to the War of 1812* (1905). These volumes, considered together, are an entity in which successive maritime events are strung upon a slender "thread of general history," with frequent digressions into "questions of naval policy, strategy, and tactics."[8] To these monumental studies, Mahan added a steady stream of articles and books dealing with naval strategy and international affairs.[9]

Mahan's principal books and essays are history written with the focus of interest on sea power. His discussions of national policy, naval policy, naval strategy, and tactics therefore are not presented separately but are interwoven,

[4] See *ibid.*, pp. 59-65.
[5] For his early views, see *ibid.*, chap. XI; and A. T. Mahan, "The Growth of our National Feeling," in *World's Work*, III (February 1902), pp. 1763, 1764.
[6] For Mahan's preparation, and the origins of his ideas, see Puleston, *Mahan*, pp. 66-73, 74-80, 93 ff. Quotation is on p. 69.
[7] *Ibid.*, p. 75. See also A. T. Mahan, *The Influence of Sea Power upon History, 1660-1783* (1890), p. iii.
[8] *Ibid.*, p. vi.
[9] See bibliographical essay in appendix. For an even more complete list of the volumes of essays, with subtitles, see Puleston, *op. cit.*, pp. 359-364.

largely at random, into a chronological narrative. A systematic appraisal of Mahan's place in the history of military thought requires the disentangling of many separate strands and their rearrangement in orderly manner. From such reappraisal and rearrangement the following emerge as Mahan's great contributions to modern strategy: first, he developed a philosophy of sea power which won recognition and acceptance far outside professional naval circles and found its way into the councils of state throughout the world; second, he formulated a new theory of naval strategy; finally, he was a critical student of naval tactics.

Naval strategy and sea power were conditioned, in his judgment, by certain fundamental natural phenomena (such as a nation's insular or continental situation) and by national policies related to navies, the merchant marine, and overseas bases. Naval tactics, on the other hand, are concerned with operations after the beginning of actual combat. Tactics, being the art of using weapons which have been forged by man, may change as weapons themselves change. But the principles of naval strategy, having a broader foundation, "remain, as though laid upon a rock" and operate in time of peace as well as in time of war.[10]

This clear distinction between strategy and tactics was one of the things which raised Mahan above the level of earlier writers, who had considered naval forces as little more than a branch of the military establishment, designed principally to protect merchant shipping and to help repel invasion. Mahan's studies convinced him that sea power, conceived on a broader scale, would constitute for the United States, as it had constituted for Great Britain, an instrument of policy serving to enhance the nation's power and prestige.[11]

The main theme of the three volumes dealing with the influence of sea power upon history is, then, the supreme importance of sea power in the shaping of national destinies. Mommsen, whose work had first aroused Mahan's interest in military history, had demonstrated that the outcome of the Punic Wars was due to "naval power." But Mahan's interest in sea power was much broader, as is indicated by his choice of the seventeenth and eighteenth centuries—an era of mercantilistic imperialism during which sea power was transcendent—as a period for special study, a kind of laboratory of naval strategy.[12]

II

From his basic hypothesis that sea power was vital to national growth, prosperity, and security, Mahan proceeded to an examination of the elements of sea power, naming six fundamental factors which affect its development: geographical position, physical conformation, extent of territory, population, national character, governmental institutions.[13] His conclusions may be roughly paraphrased in the following manner.

[10] *Influence of Sea Power upon History*, pp. 8, 20, 23, 28, 88-89.
[11] *Ibid.*, chap. I, especially p. 2, and pp. 287-288.
[12] See *ibid.*, pp. iii-vi, and Puleston, *Mahan*, pp. 70 ff. For mercantilistic influence, see H. & M. Sprout, *Toward a New Order of Sea Power* (rev. ed. 1943), pp. 9 ff.
[13] *Influence of Sea Power upon History*, chap. I.

Geographical position as a factor in sea power is best appreciated by examination of the insular position of Great Britain as compared with her chief rivals of the seventeenth and eighteenth centuries, France and Holland. The security of the homeland relieved the British government of the necessity or temptation of maintaining and using a large army, with its attendant drain on the national wealth. The British Isles were near enough to the continent of Europe to be within striking distance of potential enemies, but far enough away to be relatively safe from invasion. Operating from its strategically located home base, the British fleet could be concentrated and yet used simultaneously for defense or for the blockading of continental ports. France, on the contrary, had to divide her navy between the Atlantic and Mediterranean coastlines. The almost unique geographical situation of the British Isles, furthermore, made it feasible for Britain to control the shipping lanes to and from northern Europe. By the acquisition of important islands and other strategic bases like Gibraltar, Britain also was in a position to maintain a large measure of control in the Mediterranean, which has "played a greater part in the history of the world, both in a commercial and a military point of view, than any other sheet of water of the same size."[14]

Physical configuration of the national domain determines in large measure the disposition of a people to seek and achieve sea power. The character of the coastline governs accessibility to the sea; good harbors imply potential strength; the character of the soil may win people away from the sea or drive them to it for a livelihood. The Dutch were driven to the sea, but their almost complete dependence upon it was a source of weakness. The fertility of French soil made it unnecessary for the French to turn to the sea unless they so desired. Insular or peninsular nations like Britain, Spain, and Italy must of necessity be strong upon the sea if their pretensions to power are to be made effective. To any nation with a coastline the sea is a frontier, and national power will largely be determined by the manner in which it extends that frontier.

Extent of territory may be a weakness rather than a strength, depending upon the degree to which the land itself is supported by population, resources, and other factors of power. If vast stretches of territory are cut up by rivers and estuaries, the latter will constitute an additional source of weakness. The South during the Civil War is cited by Mahan as an instance of a nation with too much land in proportion to its population and resources, and too much coastline and too many inland waterways in relation to its inherent strength.

Size and character of population must both be considered in the measurement of sea power. A seagoing nation like England must not only have substantial numbers of men, but must have a large proportion of them engaged directly or indirectly in maritime occupations. A nation's peacetime commerce is an index of its "staying power" in naval war. There must be a large reserve, among the population as a whole, of those skills which are essential to the maintenance of ships both in time of peace and in time of war. England, for example, was not only a seagoing nation, but a shipbuilding and trading nation,

[14] *Ibid.*, p. 33.

and hence had the human and technical resources so essential to success in naval war. The existing "shield of defensive power," however, must always be maintained. "If time be . . . a supreme factor in war, it behooves countries whose genius is essentially not military, whose people, like all free people, object to pay for large military establishments, to see to it that they are at least strong enough to gain the time necessary to turn the spirit and capacity of their subjects into the new activities which war calls for. If the existing force by land or sea is strong enough so to hold out, even though at a disadvantage, the country may rely upon its natural resources and strength coming into play. . . ." In this manner Mahan emphasized the lesson which democratic peoples so often need to be taught, that potential power and actual power are not the same and must be kept in rational balance.[15]

National character and aptitudes are an essential factor in the success of a seafaring people. The desire to trade, and the ability to produce the commodities which enter into trade, together constitute "the national characteristic most important to the development of sea power. Granting it and a good seaboard, it is not likely that the dangers of the sea, or any aversion to it, will deter a people from seeking wealth by the paths of ocean commerce." If a people have an aptitude and liking for commercial pursuits, they are almost certain to develop an extensive peacetime commerce, which is one of the very first prerequisites to sea power. It was the union of a large maritime commerce and a great naval establishment which made Britain the predominant sea power of the world. Closely related to commercial pursuits is, of course, the planting of colonies which, when firmly bound to the mother country, offer markets and "nurseries for commerce and shipping." It is not surprising, therefore, that Britain was the foremost colonizing power as well as the foremost commercial and naval power.

The character of government is of vital importance in the achievement of sea power. The "most brilliant successes" have ensued when a government has intelligently and persistently fostered and directed a national interest in, and an aptitude for, the sea. British policy since the reign of James I has been determined to assert and maintain colonial, commercial, and naval supremacy and to adopt all measures necessary thereto. This adherence to a single line of policy was easier, Mahan believed, because the government of Britain lay in the hands of a single class—the landed aristocracy.[16] He expressed some doubt concerning its continuance under the more democratic government of his own day, for he believed that democracies were unwilling to pay the price of continued naval power and had not the foresight to ensure adequate military preparedness. The French under Colbert had attempted to become a great sea power, but the policy did not long survive Colbert's tenure of office and was, in any event, inadequately supported by commerce and a prosperous colonial empire.

[15] *Ibid.*, p. 48.

[16] More recent research would cast doubt upon Mahan's historical generalizations concerning the influence of the landed aristocracy and would place more emphasis upon the rising commercial classes.

The efficiency, intelligence, and determination of a government will be determining factors in the development of sea power. The government controls the size of the navy, the quality of the naval establishment, the capacity of the naval organization to expand quickly in time of war, the spirit of its men, and its effectiveness in combat. Furthermore, the strategical doctrines of a government may well be crucial in relation to the actual power of the nation on the seas. The French government, for example, had for years insisted that its admirals keep the sea as long as possible, while avoiding action which might entail the loss of ships—a strategy which prevented the French navy from taking decisive action and which precluded the very possibility of conclusive victories over the British fleet. (As Mr. Kiralfy points out in Chapter 19, considerations of the same sort seem to have influenced Japanese naval thought and may prove to be a factor of major importance in the present war.)

In analyzing the factors which enter into sea power Mahan came to view imperialism through spectacles somewhat different from those of his earlier days. Then, as we have seen, he was an anti-imperialist. Now he began to see the relation between colonies and sea power. In establishing colonies a naval power "won a foothold in a foreign land, seeking a new outlet for what it had to sell, a new sphere for its shipping, more employment for its people, more comfort and wealth for itself.

"The needs of commerce, however, were not all provided for when safety had been secured at the far end of the road. The voyages were long and dangerous, the seas often beset with enemies. . . . Thus arose the demand for stations along the road, like the Cape of Good Hope, St. Helena and Mauritius, not primarily for trade, but for defence and war; the demand for the possession of posts like Gibraltar, Malta, Louisburg, at the entrance of the Gulf of St. Lawrence—posts whose value was chiefly strategic, though not necessarily wholly so. Colonies and colonial posts were sometimes commercial, sometimes military in their character," and no territorial acquisition could be judged without keeping these basic facts in mind.[17] The government's decision as to whether and where to secure naval bases would have a vital bearing on the nation's power on the seas. The ships of a nation which, like the United States, were without adequate overseas bases were like "land birds, unable to fly far from their shores."

It is clear from Mahan's analysis of the factors conditioning sea power that Britain's predominant position rested not only on the greater material strength and the superior strategic doctrines of the British navy but also on the control of the "narrow seas." These narrow seas, which play so large a role in modern naval history, may be defined, roughly, as those bodies of water—such as the English Channel, the straits of Gibraltar, the Sicilian narrows, the Dardanelles and Bosphorus—which may be controlled with relative ease from either shore. Britain had succeeded in acquiring a substantial number of the outposts of sea power which, in combination with her battle fleet, gave her virtually undisputed control after Trafalgar of the eastern Atlantic and the Mediterranean.

17 *Ibid.*, pp. 27-28.

And as there were no great naval powers outside Europe in 1890, when Mahan published the first of his books on sea power, control of European waters meant control of all the oceans of the world. It was only with the rise of non-European powers that Britain's world-wide command of the seas was threatened. Throughout the nineteenth century, however, British naval supremacy was such that the principal sea routes of the world were, in effect, the internal communications of the British Empire.[18]

III

Mahan doubted whether Britain would continue indefinitely to maintain her position as the world's greatest sea power. The broad basis of that power, he wrote, "still remains in a great trade, large mechanical industries, and an extensive colonial system." But, "whether a democratic government will have the foresight, the keen sensitiveness to national position and credit, the willingness to insure its prosperity by adequate outpouring of money in times of peace, all of which are necessary for military preparation, is yet an open question."[19] Actually, however, the alterations in the naval balance of power which were to come during the twentieth century were the result of developments over which Great Britain had little control. They were the rise of new naval powers, the vastly increased strength of land power vis-à-vis sea power in Europe, and certain technological developments which made sea blockade a less deadly weapon than it had previously been.[20] It is doubtful if Mahan grasped the significance of these world-wide changes or, indeed, if anyone could have appraised them except in the perspective of a later time.

The rise of Japanese naval power undermined England's strategic dominance in Europe as well as in the Far East. Through one of the ironies of history, Englishmen themselves contributed materially to this result. British shipyards in the 1880's and 1890's built one warship after another for Japan. And British naval officers were loaned to the Mikado's government to teach the elements of naval science and administration. It could be argued, of course, that someone else would have built the ships and given the advice if England had refused. It could also be argued that Great Britain needed a counterpoise to Russian imperialism which was at that time encroaching on British preserves in Asia. But such reasoning does not alter the fact that a modern Japanese fleet in Asiatic waters fundamentally altered the strategic situation to the disadvantage of Great Britain; that British squadrons guarding the English Channel, the North Sea, and the Mediterranean no longer *ipso facto* dominated the sea communications of the Far East.

Meanwhile, parallel developments were taking place in the Western Hemi-

[18] See summary in Sprout, *Toward a New Order of Sea Power*, pp. 3-28. Compare Mahan, *Influence of Sea Power upon the French Revolution and Empire, 1793-1812* (1892), I, pp. 10, 15 ff., and II, p. 106; "Considerations Governing the Disposition of Navies" in *The National Review*, XXXIX (July 1902), pp. 701, 709-711; *The Interest of America in International Conditions* (1910), pp. 5 ff., 61-62; and *Naval Strategy* (1911), p. 177.
[19] Mahan, *Influence of Sea Power upon History*, p. 67 and *Naval Strategy*, p. 110.
[20] For a general summary see Sprout, *Toward a New Order of Sea Power*, chap. 2. In the present chapter I have borrowed a number of sentences from this work.

sphere. Prior to the Civil War, the United States had both a navy and a naval policy. But neither affected the main currents of world politics in any large or continuing manner. Even within the Western Hemisphere, England rather than the United States was the dominant naval power. After a brief growth induced by the Civil War the American navy passed into a prolonged eclipse. Reconstruction commenced in the early 'eighties, and by 1890 was acquiring some momentum. Mahan's writings, in conjunction with other influences, accelerated the pace and changed the direction of American naval development. By 1898, the Navy of the United States had evolved from a handful of commerce-raiding cruisers into a rapidly growing fleet of first-class battleships. Control of Europe's narrow seas no longer assured naval dominance in the New World. Only by progressively strengthening its overseas squadrons could the British Admiralty have preserved even a semblance of its former primacy in American and Far Eastern waters. And whatever the desires and inclinations of British naval authorities, developments nearer home soon rendered such a course practically impossible.

Acceleration of the naval building pace in Europe, especially the very rapid growth of the German navy after 1900, threatened England's historic dominance in European waters. So instead of strengthening its overseas squadrons, the British government had progressively to deplete them in order to maintain a safe margin of superiority in the narrow seas and eastern Atlantic.

The implications of all this are clearer in retrospect, of course, than they were in prospect. With increasing difficulty the British government did manage to keep a margin of naval superiority that seemed to assure its hold on the sea approaches to Europe. There was a fair presumption that Great Britain could still cut off its continental enemies from their overseas colonies and from the foodstuffs and raw materials of the Western Hemisphere and the Far East. However, the ability to maintain such a blockade would thereafter depend not only on Britain's naval dominance in European waters but also on the attitude and policy of the transoceanic naval powers, Japan and the United States. British statesmen could still exercise a large, often a decisive, influence on world events, through commerce, finance, diplomacy, and propaganda. But they had irretrievably lost the ultimate sanction of superior force in the Western Hemisphere and in the Far East. Great Britain's world-wide command of the seas had vanished, and with it the historic balance wheel of the vast, intricate, and smoothly running machinery of that advantageous world economic community and quasi-political order which British sea power had fostered and supported during the preceding century.

It is probable that Mahan himself accelerated the forces which undermined England's world-wide command of the seas. His interpretation of history, linking sea power with national greatness and imperialism with sea power, stimulated expansionist impulses already stirring in Europe, in the Far East, and in America. His gospel of sea power strengthened the trend of political and economic events which were already encouraging the growth of navies, and these navies in turn fostered and supported the new imperialism which even further quickened the pace of naval construction. In proportion as other navies increased in power, England's margin of supremacy declined.

Not only was England's strategical position vis-à-vis other sea powers gradually deteriorating, but scientific and mechanical developments were taking place which lessened still further the political and military importance of sea power. One of these developments was the phenomenal improvement in railway and road transportation on the continent of Europe. Another, not to be fully appreciated until the rise of Nazi Germany, was the manufacture of synthetic substitutes for strategic raw materials which formerly had to be brought to Europe from overseas.

British influence at its peak owed much to the primitive state of overland transport on the continent. A very high proportion of European commercial traffic moved by water, along rivers and canals, through coastal waters, or upon the high seas. Goods sent from northwestern to southern Germany, for example, might normally go by ship from the northern ports, through the English Channel, around Gibraltar, through the Dardanelles, and up the Danube to their destination. The construction of an elaborate and efficient system of railway transport, and subsequently of motor roads, altered this state of affairs to a certain extent and had strategic as well as commercial repercussions. Mahan recognized these new developments but contended that "transit in large quantities and for great distances" was "decisively more easy and copious" by sea than by land, and that this was the reason why command of the sea was so important. Sea routes, he insisted, were still the "inner lines of communication" which gave decisive military advantage.[21] In an article written in 1907 Mahan admitted that "numerous alternatives to sea transport" had become available. But to arguments that the "former efficacy" of sea transport could "no longer be predicated," he replied that "for obvious reasons of cheapness and facility, water transport still "maintained its ascendancy." It might grow relatively less important but "unless we succeed in exploiting the air, water remains, and must always remain, the great medium of transportation."[22]

In 1907, despite the fact that automotive transport was still in its infancy, the shape of things to come was already discernible. In this as in other technical matters, Mahan remained conservative. Of course, he never dreamed of the mobility which land forces were to achieve in the Second World War. This new mobility of land power not only deprived sea power of its "inner lines of communication," but threatened the security of the land bases without which sea power cannot exist.

It would have been even more difficult in the early years of the century to foresee the tremendous scientific advances which would progressively relieve great continental countries of their dependence upon products brought from distant lands. Production of motor fuels from coal, manufacture of synthetic rubber, and the other miracles of applied science still lay far in the future.

---

[21] *Problem of Asia*, pp. 124 ff. For Friedrich List on the strategic significance of railways to Germany and to Britain's position in the Near East see Chap. 6.

[22] *Some Neglected Aspects of War* (1907), pp. 174 ff., and *The Problem of Asia* (1900), pp. 125-126.

Nevertheless, their coming was one day to deprive sea blockade of some of its potency and was to prolong the time necessary to achieve its maximum effectiveness. It is conceivable that at some future time air transportation may so completely supersede sea transport that the value of sea power in international politics may be still further jeopardized. Certainly the seventeenth and eighteenth centuries, upon which Mahan based his theories, constituted the heyday of sea power.

Mahan believed that sea power in all its ramifications was the royal road to national wealth and prestige for all countries capable of its development. France, he pointed out, had a most favorable situation for the development of commerce and naval power. The position of France was stronger than that of any other European nation for operations against England, but the French chose to be primarily a land, rather than a sea, power. Germany suffered a severe handicap because all its sea-borne commerce had to pass through either the North Sea or the English Channel, almost literally under the guns of the British navy.[23] Furthermore, it is improbable, as Mahan pointed out, that any European nation with a land frontier to defend against powerful neighbors could ever safely divert from its army enough of its human and material resources to win primacy at sea.

IV

As an American naval officer, Mahan naturally emphasized the value of sea power to the United States and the steps necessary to secure it. At the time he wrote *The Influence of Sea Power upon History*, however, he was unwilling to accept the program of territorial expansion which seemed to be inherent in his theories of sea power, and his views of an American naval program must be considered moderate.[24]

With respect to the defense of our seaports, he declared, there is "practical unanimity in theory and entire indifference in practice" that a navy is necessary. The principle that free ships make free goods seemed to make it safe for the United States to trust its commerce to neutral flags in time of war. But this rule did not hold, he pointed out, when the ship was bound toward a blockaded port. To break a blockade, or to avoid having our own ports blockaded, the United States needed a naval force strong enough to drive off the blockading forces. Events of the Civil War indicated that the great length of the American coastline would not, as some persons had argued, make such a blockade impossible. Contrary to early American theory, the enemy must be kept "not only out of our ports, but far away from our coasts." The "influence of our government should make itself felt," Mahan concluded, "to build up for the nation a navy which, if not capable of reaching distant countries, shall at least be able to keep clear the chief approaches to its own."[25]

Mahan saw clearly that the United States possessed the elements conducive

---

[23] *Interest of America in International Conditions*, pp. 53 ff., 192, 195 f.
[24] See note 6 above; *Influence of Sea Power upon History*, pp. 83-86.
[25] *Ibid.*, pp. 84-88.

SEA AND AIR WAR

to the growth of sea power to a lesser degree than England. In the first place,

to the growth of sea power to a lesser degree than England. In the first place, the geographical position of the United States was comparable to that of England only in the single feature of insularity. He believed that the opening of a trans-isthmian canal at Panama would alter the relation of the United States to the Caribbean so that it would resemble that of England to the Channel, or of England to the Mediterranean. With proper military preparations, the United States could exercise dominant sea power in this area. He failed, however, to point out one critical dissimilarity. No other great power fronts on the Caribbean. Hence American control of that waterway could give the United States no such leverage on other great powers as England exercised through control of the Channel and the Mediterranean.[26]

In the matter of physical conformation the United States possessed elements both of strength and of weakness. Numerous and deep harbors were dangerous if not properly defended. The "extent, delightfulness, and richness of the land" which tended to keep the French people from the sea, Mahan saw "reproduced" in this country. The situation had been different when the United States comprised only a fringe of settled land along the Atlantic Coast. The center of power now lay in the interior. But when "the day comes that shipping again pays, when the three sea frontiers find that they are not only militarily weak, but poorer for lack of national shipping, their united efforts may avail to lay again the foundations of our sea power." Until that time, Mahan believed, those who understand the importance of sea power may "mourn that their own country is being led" like France "into the same neglect of that instrument."[27]

Except for Alaska, the coastline of the United States presented "few points specially weak from their saliency" and all parts of the coast could be easily reached from the interior, either by water or by rail. The nation had no distant colonies essential to its existence. Thus with our "boundless" resources we could easily "live off by ourselves indefinitely" unless a new "commercial route through the Isthmus" should give us the "rude awakening of those who have abandoned their share in the common birthright of all people, the sea."[28] Construction of the isthmian canal would give every position in the Caribbean an "enhanced commercial and military value." The canal itself would become "a strategic centre of the most vital importance," and the nation which ruled the sea approaches to the canal would control the canal itself. Without more military and naval power than the United States possessed in 1890, Mahan feared that the opening of the canal would be "nothing but a disaster" to this country.[29]

From the point of view of commerce and the carrying trade, Mahan declared that the position of the United States was unique. "Facing the older worlds of the East and West," our shores are "washed by the oceans which touch the one

[26] *Ibid.*, pp. 33-34; also *Interest of America in Sea Power* (1897), pp. 104, 110, 124, 277; *Arms and Arbitration* (1912), p. 180.
[27] *Influence of Sea Power upon History*, pp. 38-39.
[28] *Ibid.*, p. 42; *Problem of Asia*, p. 198.
[29] *Interest of America in Sea Power*, pp. 12-13; *Problem of Asia*, pp. 182-184.

or the other, but which are common" to the United States alone. This position had advantages in defense also, for the "remoteness of the chief naval and military nations from our shores" would make naval operations against us difficult. The "jealousies of the European family of states"—that is to say, the balance of power—would further limit the ability of European powers to send forces against our shores.[30]

The extent of our territory, Mahan argued, might be the same source of weakness against a stronger power that it was for the South against the North in the Civil War. Not only did the Confederacy have no navy, but its population was not interested in the sea and was not "proportioned to the extent of the sea-coast which it had to defend." The United States must be able to exert its strength not only on one long coast, but on two.[31]

Mahan pointed out that the United States was like Holland in that the people would not spend money for their own defense unless danger actually stared them in the face. Consequently we had no "shield of defensive power" behind which we might develop our reserves of strength. Our seafaring population was far from adequate for possible needs, and foundations for such a class, Mahan believed, could "be laid only in a large commerce" under the American flag.[32] Mahan saw no reason to doubt that his compatriots possessed aptitudes for commerce, for self-government, and for independence similar to those of the English. He thought that, if "legislative hindrances" could be removed and "more remunerative fields of enterprise be filled up, sea power would soon begin to develop." The "instinct for commerce," the love of "bold enterprise in the pursuit of gain," and a "keen scent for the trails that lead to it" all existed in the American people.[33]

As a democracy, it seemed to Mahan that the United States was at a disadvantage. Democratic governments tended to lack foresight and willingness to keep up their military expenditures in peacetime.[34]

The nation was also handicapped by a lack of colonies. In 1890 we had neither colonies nor naval stations. In Mahan's view the United States then possessed only the first of the three great links of sea power: internal development and production, peaceful shipping, and colonies. "In the present condition of the navy," he continued, an attempt to blockade the ports of this country would not entail a greater effort than has been made before by great maritime nations. "The people of the United States would not starve, but they may suffer grievously." We must, therefore, have a sizable naval force to keep the enemy forces away from our coasts. Mahan regretted, at the time he was writing, the lack of a motive to stimulate American naval development, but

[30] *Interest of America in Sea Power*, pp. 6, 110, 181-182; "Current Fallacies upon Naval Subjects" in *Harper's Monthly Magazine*, XCVII (June 1898), pp. 45, 49; *Naval Strategy*, pp. 18-19.
[31] *Influence of Sea Power upon History*, p. 43; *Problem of Asia*, pp. 181, 198-199; also "The Panama Canal and Distribution of the Fleet" in *North American Review*, CC (September 1914), pp. 407 ff.
[32] *Influence of Sea Power upon History*, p. 49.
[33] *Ibid.*, pp. 57-58.   [34] *Ibid.*, p. 67.

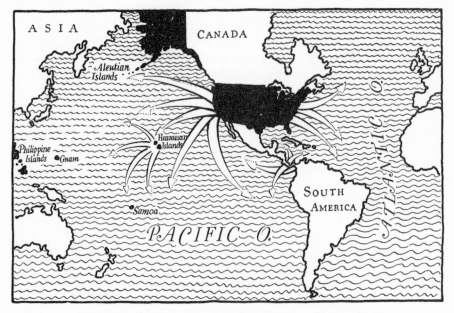

Geography and American Sea Power, 1898-1922

he suggested that such a motive might eventually be found in the opening of a trans-isthmian canal.[35]

Although the American navy was weak, and the position of the United States was not comparable to that of England in the essential elements of sea power, Mahan felt that the United States could make itself a great naval power, if not, indeed, the greatest of all naval powers. If the nation would adopt a policy of consistently enlarging and strengthening its navy, of acquiring suitable naval bases and overseas colonies, and of building up our merchant marine, our position upon the sea would be assured. Our central position, our relative security from attack, combined with our great industrial development, might offset our lack of other advantages.

Just as time and technology were qualifying the significance of sea power as an instrument in world politics, so they were shifting the importance of some of the factors which Mahan considered to be basic elements of sea power. Mahan saw that steam, by making ships independent of the uncertainties of wind, had already in some measure lessened the protection afforded by insularity.[36] Steam had also enhanced the importance of naval stations by circumscribing the fleet's radius of action. But that was only the beginning. Submarines, and later airplanes, were gradually to undermine the protective value of insularity. Mahan did not live to see German submarines set up a counterblockade of England which actually threatened during 1917 to starve the British

[35] *Ibid.*, p. 86; *Interest of America in Sea Power*, pp. 199-200.
[36] *Interest of America in International Conditions*, pp. 62-63; "Blockade in Relation to Naval Strategy" in *U.S. Naval Inst. Proc.*, XXI, pp. 851-866.

into surrender. Nor did he see the aerial blitz of 1940 against the industries and the ports which were the backbone of British naval power. The traditional command of the sea by great surface fleets could no longer guarantee the security of the British island base.

Furthermore the battle fleet itself was vulnerable to attack by these new weapons. Elaborate precautions must be taken at all times to protect the fleet from enemy submarines. Narrow seas or coastal waters dominated by land-based enemy aircraft became exceedingly precarious, if not untenable, for warships. The new weapons, in short, circumscribed more closely the area in which a fleet might exert its power.

Mahan reiterated many times the fact that secure bases constituted the necessary foundation for power upon the sea. He could not have foreseen that mechanized warfare was going to challenge the existence of sea power by attacking its vulnerable bases, especially from the air. The extent of a nation's territory acquired an importance far beyond that perceived by Mahan. Depth could be converted into defense, not only against the new mechanized land warfare, but against attack by any sort of aircraft. Naval bases without depth and without large sources of supply—bases like Singapore and Hongkong—could not be held by sea power in the face of determined mechanized attack by land.

In the years since 1914 warfare has become increasingly a conflict of technologies. Not only is the extent and availability of a nation's resources a critical factor in creating the new weapons, but the technical ability of a people is also essential. A government's ability to choose the middle path between undue conservatism and too hasty experimentation in technical matters may spell the difference between victory and defeat.

Most of the basic changes affecting sea power have operated, on the whole, to the advantage of the United States. In view of the development of the long-range bomber, the American home base and American war industry are less vulnerable to attack than those of any other major power. In these and in other respects Mahan's six elements of sea power need qualification, shifts in emphasis, and some reappraisal. But it is clear that his basic ideas still are sound. Position, physical conformation (including natural resources and climate), the character of the people and of the government are still the "principal conditions affecting the sea power of nations."

V

If Mahan's first objective was to determine the influence of sea power upon the destiny of nations, his second was to derive from a study of naval warfare certain fundamental and immutable principles of naval strategy, comparable to the principles of land warfare formulated by Jomini. Although the basic theme of the three works dealing with the influence of sea power upon history was the political importance of sea power, these volumes also contain, scattered here and there, discussions of many elements of Mahan's theory of naval strategy and defense. These ideas were more completely developed in his later books and essays.

Mahan had been reading European history, especially French history, for over a year and a half when he began studying Jomini,[37] especially his *History of the Campaigns of the Revolution and Empire*, and *The Summary of the Art of War*. Mahan frequently acknowledged his debt to the Swiss writer, for he learned from Jomini to view the events of naval history as illustrations of "living principles." He discovered from Jomini that there was no sharp distinction "between diplomatic and military considerations," and from the same source he derived the method of critical analysis of campaigns and battles.

Jomini had formulated principles of war[38] which were built around fundamental ideas of position, lines, communications, and concentration of force. Mahan attempted to discover analogous principles underlying naval tactics and strategy, and he found many of Jomini's concepts equally applicable to naval warfare, although others required "modification or limitation." The principles which Mahan thus formulated became the foundation of a system of naval strategy that was to affect the plans and policies of all the leading navies.

A central position, Mahan perceived, afforded the same defensive and offensive advantages upon the sea as on the land. Such a position gave "interior lines, shorter lines, by which to attack." An interior line was, in fact, an "extension of a central position," or "a series of central positions connected with one another." The possessor of such lines could concentrate his forces on any one of several fronts more quickly than his enemy and hence utilize his forces more effectively.[39] Suez, for example, was an interior line as compared with the Cape of Good Hope, Panama as contrasted with the Strait of Magellan, and the Kiel Canal as compared with the Skagerrak.

The strategical value of a position, Mahan reasoned, depends not only on its relation to strategic lines, but also upon its intrinsic strength, and upon the resources of the place itself or of the surrounding country. A position such as Dover or Gibraltar which is near sea routes, or close to a crossing of several sea routes, is likely to be a central position. The value of such strategic positions is enhanced by the fact that the sea lanes in these places become very narrow, and because many ships must pass through them. The military strength of a particular site may be increased by proper fortifications but if all materiel has to be brought from a distance that position will still be inferior to another "having a rich and developed friendly region behind it."[40] Gibraltar, for example, is under a disadvantage in this respect.

In Mahan's system, the term "communications" refers to the "lines of movement between the force and its sources of supply." Communications, he wrote, "are the most important single element in strategy, political or military." The "eminence of sea power" lies "in its control over them." "The power, therefore, to insure these communications to one's self, and to interrupt them for an

---

[37] For the influence of Jomini on Mahan's work, see Puleston, *Mahan*, pp. 61, 69, 77 ff., 287, 295 ff.

[38] *Ibid.*, pp. 78-79. See also Chap. 4, above, for a detailed account of Jomini's work.

[39] *Naval Strategy*, pp. 31 ff.

[40] *Ibid.*, pp. 130 ff., 163. See also *Influence of Sea Power upon the French Revolution*, I, pp. 110, 184; *Interest of America in Sea Power*, pp. 41-42; "Considerations Governing the Disposition of Navies" in *The National Review*, XXXIX (July 1902), pp. 707-709.

adversary, affects the very root of a nation's vigor. . . ." "This is the prerogative of the sea powers; and this chiefly—if not, indeed, this alone—they have
to set off against the disadvantage of position and of numbers. . . ." The longer
the communications, the greater the benefit conferred by sea power. A central
position which affords protection to one's communications affords great advantage. Such was the position of France in her war against Spain and Austria. It
was also the position of England against France. The English navy could
blockade the French coast and at the same time cover British interests "from
the Baltic to Egypt."[41]

Concentration of force, Mahan emphasized, is a fundamental principle of
both sea and land warfare. The value of a central position lies in the fact that
it facilitates, indeed encourages, concentration of one's forces. Should a fleet
be confronted with two enemies, the proper course is to go after one first and
destroy it, then if possible seek out the other.[42] It was this principle which
President Theodore Roosevelt had in mind when he urged his successor never
under any circumstances to divide the American fleet between the Pacific and
Atlantic Oceans.

Mahan was convinced from his historical studies, that "battles of the past"—
sea as well as land—had "succeeded or failed according as they were fought in
conformity with the principles of war."[43] The supremacy of the British navy
was attributable in no small degree to the superior British naval strategy. Over
a long period of years the British had discovered that certain types of naval
operations were more successful than others. In the second Anglo-Dutch War,
Mahan points out, the fleet of Charles II met defeat because it was divided to
meet both the Dutch and their French allies at the same time. This war, thanks
to the parsimony of the English king, had been waged mainly by preying on
enemy commerce, and it ended only after the Dutch occupied the mouth of the
Thames—quite unlike the outcome of the campaigns of fifteen years earlier in
which Cromwell's powerful fleets of ships-of-the-line had shut Dutch merchantmen within their own ports.[44]

Between 1689 and 1698, Mahan recalls, the French sent great fleets to dispute the command of the sea with the British, who suffered greatly as a result.
During the War of the Spanish Succession (1702-1712), however, French
fleets were "practically withdrawn" from the ocean and the number of French
commerce-raiding cruisers greatly increased. But despite the loss of hundreds
of merchantmen, British commerce prospered increasingly while French merchantmen all but disappeared from the seas.[45] The *guerre de course*, therefore,
was an altogether unsatisfactory substitute for fleet action.

British naval operations against France in the Seven Years War (1756-1763)
show how British strategy continued to develop. Mahan points out that in this

[41] *Problem of Asia*, pp. 124 ff.; *Influence of Sea Power upon the French Revolution and
Empire*, I, pp. 95-96, 184 and II, p. 106.
[42] *Naval Strategy*, pp. 45 ff.
[43] *Influence of Sea Power upon History*, p. 9.
[44] *Ibid.*, pp. 118, 125, 131-133; *Naval Strategy*, p. 74.
[45] *Influence of Sea Power upon History*, pp. 133 ff.

conflict, for the first time, the British navy undertook a close blockade of Brest, to prevent either great fleets or small squadrons from getting out without fighting. This blockade helped to neutralize the enemy's only offensive weapon, his battle fleet, and to keep the French in a "state of constant inferiority in the practical handling of their ships." The British navy attacked the French coasts with small flying squadrons for the purpose of keeping French land forces divided. They stationed a fleet in the Mediterranean near Gibraltar to prevent the French Toulon fleet from getting into the Atlantic to unite with the other French forces. With the French ships thus bottled up, the British sent expeditions to seize the French colonies in the West Indies. French commerce was annihilated; English trade prospered.[46] By the end of this war, Mahan concludes, the English government had come to realize that control of the sea was the secret of prosperity and success. Through that means the "kingdom of Great Britain had become the British Empire."[47]

The French, on the other hand, failing to see the dangers of losing control of the sea, in this war and in later conflicts kept their fleets in port as much as possible. When circumstances forced the fleet out to sea, the basic French objective remained to save the ships and avoid action if possible. They considered it more important to capture than to destroy British ships. If forced to fight, French admirals habitually chose the lee-gage because it imposed on the enemy "the necessity of attacking with all consequent risks," and usually enabled the French to "cripple the enemy as he approached." The British ordinarily chose the weather-gage which enabled them to steer for their opponents. Their "steady policy was to assail and destroy their enemy."[48]

The differing British and French concepts of naval strategy, Mahan argues, reflected divergent views as to the "true end of naval war. If it is merely to assure one or more positions ashore, the navy becomes simply a branch of the army for a particular occasion, and subordinates its action accordingly." This was in general the French view, notwithstanding the fact that the leading French tactician of the day, Bigot de Morogues, first director of the French Académie de Marine, had declared that "at sea there is no field of battle to be held, nor places to be won." History leaves little doubt that to the French naval warfare was a "war of posts," and the "action of the fleets" was "subordinate[d] to the attack and defence of the posts."[49]

If, on the other hand, Mahan observes, the "true end" of naval forces "is to preponderate over the enemy's navy and so control the sea, then the enemy's ships and fleets are the true objects to be assailed on all occasions." This represents the British view. Their fleets attempted to "break up the enemy's power on the sea, cutting off his communications with the rest of his possessions, drying up the sources of his wealth in his commerce, and making possible a closure of his ports. . . ."[50]

[46] Ibid., pp. 296 ff.
[47] Ibid., p. 291; also Sea Power in its Relations to the War of 1812, I, pp. 9 ff.
[48] Influence of Sea Power upon History, pp. 289 ff. For explanation of the terms weather-gage and lee-gage, see ibid., pp. 5-6, 5 note.
[49] Ibid., pp. 288-289, 533, 538 ff.   [50] Loc. cit.

The French concept of naval warfare as merely a branch of land warfare, and their preoccupation with expensive military campaigns on the continent, would seem to explain the French predilection for cruiser or commerce-raiding operations. This type of operation, the *guerre de course*, not only was cheaper, but also offered a method of attack on the most obvious manifestation of British power and wealth—merchant shipping. Mistaking the branch for the root, the French did not see that as a means of crushing the enemy, commerce-raiding was a dangerous delusion which presented itself, as Mahan put it, "in the fascinating garb of cheapness to the representatives of a people."[51] Doubtless the popularity of privateering, in France as in America, reinforced the faith of the government in this type of warfare.

The history of the conflicts between France and England, as recounted by Mahan, provide many more examples of their differing naval strategies and tactics. Even during the wars of the French Revolution, which constitute the "great and conspicuous instance of commerce-destroying, carried on over a long series of years, with a vigor and thoroughness never surpassed," that type of naval operation was not successful in bringing the British to terms. Instead, British control of the sea became more and more complete. Access to French ports became more and more dangerous. The cost of living rose in France, and sufferings on the continent became increasingly acute. Through it all British trade expanded, and British naval power increased.[52]

American independence, Mahan points out, was won at a moment when the British gave up their usual policy of blockading the French fleet in Brest, in favor of trying to protect their scattered colonies by dividing their forces. The French and their allies failed to seize the opportunity to secure control of the English Channel, but they did wield a temporary control of American waters, which resulted in the Franco-American capture of Cornwallis at Yorktown.[53]

The essence of Mahan's strategic doctrine was the necessity of controlling the sea which, he believed, could be done only by a concentration of force able to drive enemy naval and merchant ships from the seas. "It is not," Mahan concluded, "the taking of individual ships or convoys, be they few or many, that strikes down the money power of a nation; it is the possession of that overbearing power on the sea which drives the enemy's flag from it, or allows it to appear only as a fugitive; and which, by controlling the great common, closes the highways by which commerce moves to and fro from the enemy's shores. This overbearing power can only be exercised by great navies."[54]

If "the great end of a war fleet" was "not to chase, nor to fly, but to control the seas," the dominant characteristic required of such a fleet was not speed, but "power of offensive action." It was of no use "to get there first unless, when the enemy in turn arrives, you have . . . the greater force." This principle, Mahan believed, was even more true on the sea than on the land, because on the

[51] *Ibid.*, p. 539.
[52] *Influence of Sea Power upon the French Revolution*, II, pp. 222-223, 227, 288, 343.
[53] See discussion in *Influence of Sea Power upon History*, chap. XIV, especially pp. 529 ff., 534 ff.
[54] *Influence of Sea Power upon History*, p. 138.

former lack of gun power could not be balanced, as it could sometimes be on land, by judicious choice of ground. Speed was useful, but not at the cost of gun power. The heavy ships-of-the-line—in later days, the battleship or capital ship—therefore constituted the backbone of fleet strength.[55]

Should a nation be so unfortunate as to have an inferior fleet, Mahan suggested that its most useful disposition would be to shut it up within an impregnable port and so impose upon the enemy the duty of constant guard to prevent its escape. This is substantially the course followed by the German fleet in World War I, and by the Italian fleet in World War II. The existence of such an inferior fleet, often called a "fleet in being," compels the enemy to restrict his operations until it shall be destroyed. Mahan discussed this concept of "fleet in being" at some length in an essay on *Lessons of the War with Spain* (1899), and concluded that the value of such a fleet has "been much overstated." The superior force will in the end run the inferior to earth.[56]

What was the proper role of the navy of the United States in the light of this capital-ship-command-of-the-sea doctrine? Mahan suggested that the prevalent American view that our navy was "for defense only" was widely misinterpreted because of confusion between "defense" in the political sense and "defense" in the military sense. "A navy for defense only, in the *political* sense," explains Mahan, "means a navy that will only be used in case we are forced into war; a navy for defense only, in the *military* sense, means a navy that can only await attack and defend its own, leaving the enemy at ease as regards his own interests, and at liberty to choose his own time and manner of fighting."[57]

Mahan's distinction between the strategic offensive and the political offensive is fundamental to any intelligent program of military preparedness. But Americans in 1890 were still interpreting a navy for defense only in the narrow military terms of preventing the sacking and shelling or occupation of the harbors up and down the coast. The navy, Mahan contended with vigor, is not "the proper instrument" for "coast defense in the narrow sense of the expression, which limits it to the defense of ports." The "passive defense" of our shores is properly the work of the army; and, if the navy undertakes such defense, it is merely using its trained men in garrisons when they could be better used elsewhere. Furthermore, "if the defense of ports, many in number, be attributed to the navy," the naval forces will be so divided that its real strength will be lost. The defensive strength of sea powers should depend on fortifications, which are not the job of the navy but in which the navy is interested because secure bases are a necessary foundation for naval power.[58]

"If, instead of a navy 'for defense only,' " Mahan continued, "there be one so large that the enemy must send a great many ships across the Atlantic, if he sends any, then the question whether he can spare so great a number is very

[55] *Lessons of the War with Spain* (1899), pp. 81 ff.

[56] *Ibid.*, pp. 75 ff.

[57] "Current Fallacies upon Naval Subjects" in *Harper's New Monthly Magazine*, XCVII (June 1898), pp. 44-45.

[58] *Naval Strategy*, pp. 145 ff., quoted in Allan Westcott, *Mahan on Naval Warfare* (1941), pp. 71 ff.; *Influence of Sea Power upon History*, p. 87 note.

serious, considering the ever-critical condition of European politics." If we had twenty battleships, no European nation except England "could afford to send over here twenty-five battleships, which would be the very fewest needed, seeing the distance of their operations from home. . . ." Britain "equally cannot afford the hostility of a nation having twenty battleships, and with whom her points of difference are as inconsequential to her as they are with us." Such a navy would be defensive only in the sense that its existence protects the country from invasion, because it commands the seas.[59]

Even while Mahan was still writing critical minds were beginning to inquire about the effect of new technical developments upon Mahan's theory of command of the sea. In 1895 the Royal United Service Institution asked Mahan to express an opinion on the question whether the close blockade, upon which the naval strategy of the past had depended, could be maintained under conditions of steam, steel, and torpedoes. Mahan's answer was "yes." Just as wind had limited the movements of sailing vessels, modern fleets were "extremely hampered . . . by the very elements to which they owe much of their power." Torpedoes might be dangerous to the blockading force but they could be used also against the ships trying to escape. These new conditions had "simply widened the question, not changed its nature."[60] Sixteen years later the same question arose in connection with the submarine, the improved automotive torpedo, and wireless telegraphy. Mahan's answer was the same. The submarine and the new torpedo, he said, would place a "far greater strain on the blockaders, and compel them to keep at a much greater distance," but the principles of strategy will remain unchanged.[61]

When the efficacy of blockade came to be tested in the First World War, British control of coaling, docking, and other maritime facilities enabled them to devise a new system of enforcement. Neutral shipowners, applying to British commercial agents with proof that the ship's cargoes conformed to the laws of contraband, were issued certificates known as navicerts. These papers entitled the ships to immunity from capture or destruction, and to use of coaling and docking facilities without which they could not sail at all. This same system was followed in World War II. In any case, the blockade was maintained at so "much greater distance" as no longer to be a close, or coastal, blockade at all.

A real challenge to the capital-ship theory arose from the vulnerability of capital ships to attack by submarines and aircraft. The experimental bombing of several surrendered German ships after the First World War started a controversy which still rages on the subject of planes vs. battleships. At least two conclusions may be stated with certainty. Narrow seas within range of enemy land-based aircraft seem to have become untenable for capital ships except for short periods and with strong air protection. Upon the open sea the battle fleet must have both antisubmarine and antiaircraft weapons in great strength. No other form of military or naval power has yet appeared which can completely

---

[59] "Current Fallacies," p. 45; and Allan Westcott, op. cit., p. 72.
[60] "Blockade in Relation to Naval Strategy" in U.S. Naval Inst. Proc., XXI (November 1895), pp. 851-866, esp. p. 857.
[61] Naval Strategy, pp. 2 ff.

take the place of the battle fleet with its battleships and battle cruisers, although, it is true, the present war in the Pacific has brought the development of a new strategy involving the use of aircraft-carrier task forces of great speed and striking power in operations involving long distances.[62]

VI

Few persons leave so deep an imprint on world events as that left by Mahan, and fewer still live to see so full a realization of their life's work. When Mahan died in December 1914 the impact of his writings had been felt in every admiralty; his views had profoundly affected civilian thinking and public policy in America, in Europe, and even in the Far East.

Mahan's influence on American naval policy became evident even before the appearance of *The Influence of Sea Power upon History* in 1890. His lectures were first delivered at the War College in 1886, and in the years before 1890 many naval officers and public men, including Theodore Roosevelt, became familiar with his work.[63] The Secretary of the Navy, B. F. Tracy, may well have read or heard some of his lectures, for his annual report for December, 1889 bears the unmistakable imprint of Mahan's ideas. Although defense, not conquest, was the object of American policy, Tracy maintained that we required a "fighting force." Unarmored cruisers did not constitute such a force. What we needed was twenty "armored battleships" with which "to raise blockades" and to "beat off the enemy's fleet on its approach. . . ." We must have a fleet capable of diverting a hostile force from our coast "by threatening his own, for a war, though defensive in principle, may be conducted most effectively by being offensive in its operations."[64] Closely following Tracy's report came a still more revolutionary document, the report of the so-called Policy Board, a committee of six naval officers appointed by Tracy to study the naval requirements of the United States. This board, taking a broad view of its commission, outlined a program in terms of the large national destiny envisaged in Mahan's interpretation of history. Although admitting that we had no colonies, that our overseas commerce was mainly carried in foreign vessels, that our manufactures were competing with other nations in but few markets, that Great Britain was our only potential enemy, the board nevertheless advised building more than two hundred modern warships of all classes. Its members stated their belief that we were about to enter upon a period of commercial competition and expansion, including development of our own carrying trade, and that the opening of an isthmian canal would constitute a source of danger. The board specifically recommended a number of battleships with short cruising radius for coast defense, and a fleet of battleships with long cruising range for offensive operations.[65]

The report raised a storm of protest, both in and out of Congress, but it is

[62] For a defense of the capital ship as against air power see Bernard Brodie, *A Layman's Guide to Naval Strategy* (Princeton, 1942), chap. VIII.
[63] Puleston, *Mahan*, chap. XIII.
[64] Sprout, *Rise of American Naval Power*, pp. 207-209.
[65] *Ibid.*, pp. 209-211.

significant that the House Naval Affairs Committee provided in the next naval bill for what were termed, enigmatically, "three sea-going, coast-line battle-ships" with "heaviest armor and most powerful ordnance. . . ." With the passage of the Naval Act of 1890, Congress was clearly moving in the direction of the naval policy implicit in Mahan's historical analysis of sea power.[66] And while Congress was debating the naval needs of the country, the Navy Department was struggling to grasp and to apply the new strategy of naval defense, with the result that by 1897 the North Atlantic Squadron was developing into a fighting force which might realistically be called a fighting fleet.[67]

Congress continued to authorize battleships, in addition to other men of war. In November, 1893, Secretary of the Navy Herbert, inspired by Mahan's second large work, *The Influence of Sea Power upon the French Revolution and Empire, 1793-1812*, endorsed the capital-ship theory of naval defense in terms even stronger than those used by his predecessor. Indeed Herbert went further to justify the navy as an instrument of power with which to promote national interest abroad and to put teeth into diplomacy generally, even in time of peace. Other factors were pushing in the same direction. The naval battles of the Sino-Japanese War (1894-1895) were widely interpreted as proving the fighting value of capital ships. The Venezuelan boundary dispute with Great Britain provided impetus for naval preparations. The beginning of the crisis which led to the war with Spain offered effective propaganda material.

Most important of all was the steadily increasing vogue of Mahan himself. The captain's name was not even mentioned in the naval debate of 1890. There is no indication that either senators or representatives had heard of his lectures or of his book. By 1895, however, Mahan's name and ideas were well known, frequently cited, and widely if not universally endorsed in congressional circles. In the naval debates of 1895 and 1896 a substantial number of senators and representatives, for the first time, displayed a fair understanding of the strategic theory implicit in all American naval legislation since 1890. The general endorsement of this theory squarely aligned the national legislature with the political executive and the service, both now thoroughly committed to the policy of seizing indisputable command of the sea throughout a wide zone extending outward from our continental seaboards. By 1897 the day was not far distant when the American navy, though still numerically inferior to those of several European powers, would hold command of all the sea approaches to the continental United States.[68]

The war with Spain was a historic milestone in the development of American naval thought, as it was in the rise of American naval power. The conflict was interpreted as proving the validity of the strategical principles which Mahan had reiterated in his many books and articles. The defeat of the feeble Spanish force in Manila Bay is of interest less as an example of naval strategy than for its political results. But in the Caribbean the strategic situation clearly depended upon naval command of the sea. In order to free Cuba from Spanish rule—the avowed political objective of the war—all Spanish forces must be

[66] *Ibid.*, pp. 211-213.   [67] *Ibid.*, p. 217.   [68] *Ibid.*, pp. 217-222.

driven from Cuba and Cuban waters. A naval blockade of the island might eventually starve out the garrison; military invasion might hasten its collapse. American military authorities, however, were willing to risk invasion of Cuba only after establishing command of the Caribbean and its adjoining waters. Such control could be effected only by destroying or immobilizing the naval forces of Spain, which obviously would do their best to maintain Spanish communications with Cuba.[69]

Viewed in retrospect, one of the most illuminating results of the war was a startling exihibition of public ignorance of the principles of naval strategy, as preached by Mahan. The civilian public in general and the daily press in particular succumbed to panic over improbable rumors of prospective naval assaults on our seaboard, and recklessly demanded protection in the form of warships for each and every coastal city. Such a division of our forces, which Mahan, cabling from Europe, "emphatically disapproved," might well have produced grave or disastrous consequences.[70] The Navy Department for the most part, however, resisted public demands for protection for the Atlantic seaports, and the fleet sailed for the Caribbean, where it first blockaded and then destroyed the Spanish squadron. Naval power exerted at some distance from our own shores determined the outcome at every stage, thus giving a death blow to the coast-defense theory of naval strategy.

If the war with Spain confirmed Mahan's theory of naval strategy, it also launched the United States on the expansionist career then advocated by Mahan. Naval primacy in the Caribbean had become a settled American naval objective long before the war with Spain, so that the annexation of Puerto Rico and the occupation of Cuba did little to enlarge or alter our defense problem in that region. But the war and the annexations which followed did accentuate our need for an isthmian canal, and emphasized the necessity of commanding the approaches to such a canal.[71]

The new insular possessions of the United States in the Pacific profoundly altered the strategical situation of the country. Mahan had long regarded the annexation of the Hawaiian Islands as a military necessity for the security of our Pacific Coast. Acquisition of the Philippines added another argument for annexing Hawaii, and was perhaps the decisive factor in bringing this about during the summer of 1898. Acquisition of part of the Samoan archipelago in 1899 completed the considerable list of new possessions. For better or for worse the United States was launched on a program of overseas expansion which constituted, according to Mahan, one of the three important links in the chain of sea power. These new outposts sustained the naval power which Mahan believed was essential to the support of American diplomacy in the Far East, but in turn the outposts themselves had to be defended. Although the strategical problems involved in their defense were difficult, they were not insoluble. But there were powerful political and emotional forces standing in

---

[69] *Ibid.*, pp. 223-232.   [70] *Ibid.*, pp. 234-237; Puleston, *Mahan*, p. 187.
[71] Sprout, *Rise of American Naval Power*, pp. 237-241. It took the *Oregon* sixty-eight days to steam from San Francisco to Key West by way of the Strait of Magellan, nearly 13,000 nautical miles.

the way of carrying out fully the military plans prepared by the Navy.[72] Although the United States had actually embarked upon the road to sea power advocated by Mahan, the people as a whole were not ready to pay the price necessary to carry out the rest of the program, or even to safeguard what they had acquired. The United States drifted along with a comfortable feeling of security until on a Sunday morning in December, 1941 a rain of bombs falling on Pearl Harbor plunged the nation into war in the Pacific.

Although the United States never put fully into practice the broader aspects of Mahan's philosophy of sea power, his doctrine of naval strategy and his belief in the necessity of preponderant naval power were eventually accepted by his own navy and by the country at large. Mahan's acquaintance with Theodore Roosevelt developed into a close and lasting friendship. Roosevelt became thoroughly conversant with Mahan's ideas, and found them singularly compatible with his own. As Assistant Secretary of the Navy just before the outbreak of the War with Spain, Roosevelt had had much to do with preparations for that conflict. The assassination of McKinley in September 1901 made Roosevelt President of the United States, thereby altering the whole course of American naval development, for Mahan's philosophy of sea power entered the White House in the person of Theodore Roosevelt.[73]

In the years that followed, Theodore Roosevelt dominated both the naval and foreign policies of the nation.[74] He had come to the presidency thoroughly committed to the building of an isthmian canal, which Mahan had been urging for years. The initiation of this undertaking was one of the notable achievements of Roosevelt's administration. Heeding Mahan's warnings that such a canal would be a strategic asset only if the United States navy held indisputable command over its approaches, he launched a campaign for the increase of our navy. Although as a matter of fact the growing German navy and the British navy tied each other to European waters, Roosevelt was absolutely convinced that the Germans would one day start trouble somewhere in this hemisphere. Only a fighting fleet second to none save Britain's, he argued, could forestall or resist such aggression and guarantee the security of the Panama Canal. Furthermore, Roosevelt utilized a crisis with Japan to harp on the need of stronger forces for the Pacific. Year after year Roosevelt continued his drive for a larger navy and by 1905 he had obtained impressive results. Ten first-class battleships, four armored cruisers and seventeen other vessels of different classes had been authorized by Congress. In the latter years of his presidency he secured four more capital ships and twenty destroyers.

Whole-hearted adoption of Mahan's strategic doctrines required more than a large number of ships. These ships, and the officers and crews who manned them, had to be organized and drilled into an efficient fighting machine. This was as great a task, if a less spectacular one, than securing the ships. The world cruise of the Atlantic Fleet which Roosevelt carried out in the face of

---

[72] *Ibid.*, pp. 241-245.

[73] *Ibid.*, pp. 249-250; Puleston, *Mahan*, chap. XXVII; *ibid.*, index: Theodore Roosevelt.

[74] For Roosevelt's impact on American naval policy, see Sprout, *op. cit.*, chap. XV.

congressional opposition may be said to have marked the debut of the United States as one of the world's great naval powers.

For five years after Roosevelt left the White House the navy continued a slow, uncertain growth. But the first year of the First World War foreshadowed an upward surge of navalism which carried the American naval program far beyond the standard of "a navy second only to Great Britain" which had guided policy since 1901. Had Mahan lived another year, until December, 1915, he could have heard President Wilson asking Congress for a navy "equal to the most powerful" in the world. The Naval Act of 1916 actually authorized for the United States "a navy second to none."[75]

VII

Mahan's countrymen were not slow to give him the acclaim his work deserved but, as his biographer points out, "it was in England that Mahan achieved his greatest immediate popularity." *The Influence of Sea Power upon History* appeared in England at the precise psychological moment to win the greatest attention and approval.[76]

During the 'eighties a new imperialism was born from the intensive competition among European nations for markets and raw materials. Expanding British interests in overseas trade and shipping, in loans, concessions, and spheres of interest, inevitably came in conflict with similar interests of other nations. This clash produced a general trend toward naval rearmament. To the British, who considered their own navy a "vital necessity" but those of other noninsular nations as "mere luxuries," these growing navies "could only be intended for eventual aggression against [the British] themselves."[77] This jealousy of naval power other than their own is understandable if one considers the complete dependence of Britain upon sea power for her very existence as a free nation.

In 1889 the government presented to Parliament a naval expansion plan based on the principle of a navy equal to those of any two other European nations—the "two power standard." Mahan's book appeared at the right moment to provide clear and irrefutable arguments to justify this program. It also provided welcome ammunition with which to defeat the demands of British army officers and others for elaborate and costly fortifications along the entire coast of Britain.[78]

During the previous years the British public had begun to take a renewed interest in the navy, but Mahan's two books appearing in 1890 and in 1892 were the most important single factor in making the whole nation navy-minded. It is not difficult to understand the appeal of Mahan's ideas to the British people. He had perceived beneath the maze of events in English naval and political history the basic principles which had made Britain mistress of the seas, and he

[75] *Ibid.*, chaps. XVI-XVIII.   [76] Puleston, *Mahan*, p. 110.
[77] A. J. Marder, *The Anatomy of British Sea Power* (1940), chap. II, especially pp. 10-11, 13; also Puleston, *Mahan*, chap. XVI.
[78] Marder, *op. cit.*, p. 25; R. H. Heindel, *The American Impact on Great Britain* (1940), p. 117; Puleston, *Mahan*, pp. 106-107.

stated these principles in a form even laymen could understand. With an ancient tradition of sea power behind them, with a naval race already in progress, and a new era of imperialism in the making, it is no wonder that Englishmen at once eulogized his work.[79]

Mahan visited England several times in the course of a "ceremonial cruise" of European waters during 1893-1895 and was acclaimed in an almost unprecedented manner. He was "dined by the queen and by the prime minister, awarded honorary degrees by Oxford and Cambridge, and entertained as a guest of honor by the Royal Navy Club—the first foreigner to receive this honor." The staid *Times* of London eulogized him as one who had done for naval history what Copernicus had done for astronomy. A naval critic compared him with Priestley: "Sea power, of course, has influenced the world in all ages. So also has oxygen. Yet just as oxygen, but for Priestley, might have remained until this day an indefinite and undetected factor, so might sea power but for Mahan." His views were widely disseminated in the press, in the quarterlies, through the professional journals, and by such influential societies as the Royal United Service Institution. One Englishman, perhaps with a note of regret, said that it took "a Yankee to wake up this generation of Englishmen to the meaning and importance of sea power." But the navy, the government, the universities, and the general public were glad to pay him the supreme compliment of being the only foreigner who had the right "to offer them counsel upon the conduct of their affairs." Even Gladstone, the arch opponent of armaments, said Mahan's work on the French Revolution and the Empire was the "book of the age." None of these compliments was undeserved, for Mahan put the term sea power into the contemporary vocabulary of Englishmen and did more than any other single person to stimulate the interest of Britons in their own navy. For the first time, wrote Lord Sydenham of Combe, "we had a philosophy of sea power built upon history." It might be added, too, that Mahan contributed to the growing feeling on both sides of the Atlantic that control of the world's ocean highways was a matter which encouraged and, indeed, made almost imperative effective Anglo-American understanding and cooperation.[80]

But Mahan's influence in Europe was by no means limited to Britain. In 1890 the world's second naval power was France, where Mahan's first book was promptly translated and came to the attention of French naval officers. Captain Darrieus, a former professor of strategy and naval tactics at the French Naval War College, was greatly impressed with Mahan's criticisms of French naval policy. In a book of his own covering much the same period of history as Mahan, he reiterated the faults of French naval strategy which Mahan had pointed out.[81] Another French naval officer, Admiral Raoul Castex of the French War College, declared that Mahan's two ideas, first the paramount importance of command of the sea, and second, the necessity of

[79] Marder, *op. cit.*, pp. 25, 45-48; Heindel, *op. cit.*, pp. 117-118.
[80] Heindel, *op. cit.*, p. 117; Marder, *op. cit.*, pp. 45-47; Puleston, chap. XXIII.
[81] Puleston, *Mahan*, pp. 107, 326.

organized force, would alone "entitle him to the consideration of posterity."
Castex did not like Mahan's style, but he believed that it did not in any way
detract "from the priceless value of Mahan's work which was truly creative
in the field of strategic theory."[82] Nevertheless, Mahan's ideas never evoked
in France the wide acclaim which they enjoyed in Britain, in the United States,
and in Germany. After all it could hardly be expected, as Mr. Ropp shows in
the next chapter, that the French would receive with enthusiasm a philosophy
which was so at variance with their historic policies.

Mahan's writings had a practical influence on German policy second only to
their influence in the United States. As in England and in America, *The Influence of Sea Power upon History* appeared at a critical moment. Emperor William II had recently dismissed the aged Bismarck largely because of the latter's
obstinate insistence that Germany should remain a continental power, and the
nation under the young kaiser launched itself on an avowed policy of imperialistic expansion overseas. The newly founded German navy was an essential part of the new policy, and although small, the imperial fleet was growing.[83]

Sea power was not for Germany the result of spontaneous and natural
processes as it had been in England. Interest in a navy was from the first
artificially aroused and stimulated. The *Reichsmarineamt* and, later, the Navy
League (*Flottenverein*) undertook a systematic campaign to popularize naval
power.[84] The leaders in this effort to enlist public interest at once realized the
value of Mahan's writings for their purposes and the Navy itself took a hand
in translating them into German. Ernst von Halle, one of the leaders of the
movement to popularize sea power, himself wrote a book applying Mahan's
doctrines to the events of German history. The propagandists presented the
need for naval power to the German people as a business proposition, using
Mahan's contention that the development of profitable economic interests overseas was dependent upon the possession of sea power.[85]

Germany, in contrast to France and Britain, had practically no naval history
at all, and hence few naval traditions. German officers were, therefore, more
than willing to accept ready made the doctrines of strategy which Mahan had
formulated. Mahan's picture of sea power as something which derives from
a combination of colonies, overseas trade, and naval power had an almost
intoxicating effect on a people just awakening to an interest in colonies and
foreign commerce. The message which the German kaiser sent to a friend in
May, 1894 is typical of the enthusiastic reception Mahan's ideas found in
Germany. "I am just now not reading but devouring Captain Mahan's book,"
he wrote, "and am trying to learn it by heart. It is a first-class work and classical

[82] *Ibid.*, p. 332.
[83] Marder, *op. cit.*, pp. 288 ff.; A. Vagts, *Deutschland und die Vereinigten Staaten in der
Weltpolitik* (1935), p. 652, note 1. The author wishes here to express appreciation for the
assistance of Dr. Felix Gilbert in summarizing German source material.
[84] Concerning the *Flottenverein*, with its hundreds of thousands of members, see Chap.
13, especially note 4.
[85] E. Kehr, *Schlachtflottenbau und Parteipolitik 1894-1901* (Berlin, 1930), pp. 38, 45, 101-
110. E. von Halle, *Die Seemacht in der deutschen Geschichte* (Leipzig, 1907). See especially
p. 6 on "das epochemachende Werk des Amerikaners A. T. Mahan."

in all points. It is on board all my ships and constantly quoted by my captains and officers."[86]

The guiding spirit of German naval policy was Alfred von Tirpitz. From 1897 until the First World War, Tirpitz was entirely responsible for the material as well as the administrative and ideological organization of the German navy. Tirpitz was familiar with the writings of Mahan and always spoke well of them, and the effect of the American officer's ideas on Tirpitz's program and policies is evident.[87] Tirpitz maintained that only a world power, meaning a power with interests around the globe, could be considered a Great Power. No mere land power could, therefore, be a Great Power; possession of sea power was necessary. The acid test of a nation's possession of sea power was not a large fleet of cruisers for use in a *guerre de course* against enemy commerce but the existence of a fleet of battleships. It is interesting that Tirpitz's first act as secretary of the navy was to replace his predecessor's program for increasing the number of German cruisers with proposals for building a fleet of battleships.

Tirpitz was really identifying sea power with battleship strength—a position which might conceivably have resulted from too casual perusal of Mahan's writings, for it was certainly not altogether consistent with Mahan's views. Tirpitz also put special emphasis on "the political importance of a navy." A navy will increase the alliance value of a state, he maintained. It will serve to gain influence and to win friends. He further explained in his memoirs that "only a fleet which represented alliance value to other Great Powers, in other words, a competent battle fleet, could put into the hands of our diplomats the tool which, if used to good purpose, could supplement our power on the continent."[88] This alliance-value idea was the basis of Tirpitz's famous "risk" theory, the essence of which was as follows: Since "the German navy could not be made strong enough for a reasonable chance of victory against every opponent, it should be made so strong that its destruction would cost even the strongest sea power such heavy losses, endangering its supremacy vis-à-vis third navies, that the mere thought of that risk would act as a deterrent against an attack."[89] Tirpitz hoped that instead of risking her control of the seas by fighting the German navy, Britain would make concessions and come to an understanding with Germany. But events of the First World War did not substantiate Tirpitz's theory. The English neither came to an agreement with the Germans nor were deterred from attacking the German fleet. Germany's battleships were finally bottled up in German waters and the Allied fleets once more demonstrated the real value of command of the sea.

[86] Vagts, *op. cit.*, p. 652, note 1. The Kaiser's statement is quoted in Puleston's *Mahan*, p. 159.

[87] For Tirpitz's views on the value of a navy to Germany, see Tirpitz's draft of a speech in the Reichstag, March 1896, published by H. Hallman, *Krügerdepesche und Flottenfrage* (Stuttgart, 1927), p. 85. For Mahan's influence on his ideas of the "political justification of the battle fleet" see Hallman, *Der Weg zum deutschen Schlachtflottenbau* (Stuttgart, 1933), p. 128.

[88] *My Memoirs* (Eng. trans. 1919), I, p. 79.

[89] H. Rosinski, "German Theories of Sea Warfare," *Brassey's Naval Annual, 1940*, p. 89.

The influence of Mahan on German naval thought before 1914 is undeniable, but there is good evidence that it was a misunderstood Mahan which the Germans adopted.[90] Mahan's writings contain many discussions of the disadvantages under which geography placed the German nation with relation to British sea power. Had the Germans understood the real essence of sea power they inevitably must have realized that only a naval force large enough to defeat the British fleet could ever transform Germany into a sea power in Mahan's sense of the word. Tirpitz appears not to have comprehended the basic geographical limitations on the German position with relation to the control of the sea.[91] Furthermore, Tirpitz failed to heed Mahan's warning that a nation cannot hope to be a great land power and a great sea power as well.

In the postwar years German naval theorists apparently realized that faulty naval strategy was responsible for the failure of the German navy to play a more useful role in the war. Tirpitz's prestige remained so great that none dared attack him openly, but there was a definite trend in the direction of a new naval theory presumably better fitted to the German geographical situation. The pendulum swung to the other extreme and in recent years the Germans have turned their attention to a new *guerre de course* waged with submarines, aircraft, and surface raiders, but in essence similar to that which French naval strategists followed in the past.[92]

No estimate of the influence of Mahan on military thought could be complete without mention of the part played by Mahan's theories in the development of German *Geopolitik*.[93] This new German approach to statecraft comprises a theory of state power and growth built on expanding land power, roughly analogous to Mahan's philosophy of growing sea power. According to Robert Strausz-Hupé, in his recent book explaining German geopolitics, Mahan was "one of the several Anglo-Saxon thinkers whose influence is most clearly noticeable throughout Haushofer's own teachings—in spite of, or perhaps because of, the fact that Haushofer's own doctrine of land power is the most extreme negation of Mahan's theories."[94]

A large part of geopolitical writing and thinking is concerned with the conflicts between land power and sea power. The exponents of land power were eager to find the "weak spots in Britain's strategic setting." Hence they turned eagerly to the works of the British geographer, Sir Halford Mackinder, who had pointed out the "sensitive regions in the anatomy of British sea power."[95] In developing his theory of the Eurasian Heartland Mackinder critically reexamined the Mahan doctrine of sea power. He restated certain qualifications of the sea power theory which Mahan had pointed out but which his followers had frequently ignored. It was Britain's position and the location of her adversaries

---

[90] See Rosinski, "Mahan and the Present War," *Brassey's Naval Annual, 1941*, pp. 9-11.
[91] See Vagts, *op. cit.*, p. 1524.
[92] See the analysis of German post-war naval theory by Rosinski, "German Theories of Sea Warfare" in *Brassey's Naval Annual, 1940*, pp. 88-101.
[93] On geopolitics, see Chap. 16.
[94] R. Strausz-Hupé, *Geopolitics: The Struggle for Space and Power* (New York, 1942), p. 246.
[95] *Ibid.*, p. 252; also p. 53.

which gave the British fleet such a powerful leverage on world affairs. Insularity alone gave no "indefeasible title to marine sovereignty." Mackinder also raised the vital question of the relation of the bases of sea power to land power.[96]

German geopolitics studied the history of sea power, Strausz-Hupé suggests, "only to be able to conclude categorically that the day of island empires was drawing to a close and that in the future landpower would be in the ascendant." Haushofer and his followers apprehended the changes in strategic geography produced by the railroad and later by mechanized land warfare much sooner than the exponents of sea power. They put their fingers on the weaknesses of Hongkong and Singapore years before war demonstrated these weaknesses to the world. They realized, if the rest of the world did not, that the "world political potential of seapower had been in full retreat before the rapidly increasing potential of landpower, long before the first submarine had plunged below the surface and the first plane had taken to the air."[97]

German geopolitical writers have frequently expressed their admiration for Mahan, whose global philosophy was built on a scale more grandiose and more audacious than any European expansionist theories of his day. Haushofer considered Mahan a great geopolitical thinker, a "seer" who had set "the United States on the path to greatness" and who had taught American statesmen to "think in terms of world power and greater space."[98]

It is one of the strange quirks of history that the American naval officer whose doctrines became the guiding principles of the world's leading sea powers should have inadvertently provided inspiration for the creation of an antithetical theory of land power. Had the grand strategy of Haushofer and Hitler succeeded, it would have spelled the doom of sea power as Mahan understood it. And as the modern world has been largely predicated upon British and more recently upon Anglo-American control of the seas, the results of a Nazi victory would have been revolutionary far beyond their best hopes and our worst fears.

[96] *Ibid.*, pp. 53 ff., 249 ff. Mackinder's views on Britain's position are summarized in his *Britain and the British Seas* (2nd ed. 1930), chap. XX, and especially p. 358.
[97] Strausz-Hupé, *op. cit.*, pp. 253-254, 255-264, especially p. 261.
[98] *Ibid.*, pp. 243-246.

CHAPTER 18. Continental Doctrines of Sea Power

BY THEODORE ROPP

D ESPITE the general acceptance of Admiral Mahan's theories in both the British and American navies, and with all due regard to his influence on German naval thought, his great works never entirely displaced the various continental schools of naval strategy. Although continental naval theorists have usually acknowledged their debt to him, he wrote little on their crucial naval problem—the means of countering British sea power with numerically inferior forces. Mahan's works were, in fact, intended partly to answer the arguments of the most original of these continental schools of thought, the *Jeune Ecole* of the French navy of the early 1880's. The *Jeune Ecole's* interest in commerce destroying has its background in a long series of unsuccessful French attempts to defeat England by the *guerre de course* in the wars of the eighteenth century,[1] but their basic ideas came from the wars of 1854 to 1871, the first to test the new weapons of the industrial era. The defeat of France in 1870-1871 by a state with hardly any navy placed naval theorists on the defensive. In both France and Germany it was considered as confirmation of the traditional views of military men that sea power was a matter concerning trade and colonies and that, at best, it had only a secondary role to play in the relations of the great European states. Naval personnel and naval ordnance had manned the forts of Paris after Sedan, but the navy as such had done little to avert defeat, and its blockade of the German coast had been a complete failure. Neutral steamers had entered the blockaded ports as easily as they had run the Union blockade of the Confederacy, while the French ironclads were able to maintain the blockade only by coaling under the lee of neutral Heligoland. The mines and small coast defense vessels which had proved so effective in the Crimean and American wars had sealed the mouths of the German rivers against naval attack. Landing operations, which had played an important part in the American Civil War, had been next to impossible in the face of the Prussian army's excellent railway system and its effective use for the movement of troop reserves.

As long as military men prepared for wars as short as those of 1866 or 1870, sea power played little part in European military thinking, and the navy remained a stepchild of continental strategy. Neglected in France, pampered in William II's Germany, the navy was in both instances regarded by orthodox military men in these countries as a stepchild. On the whole the *Jeune Ecole* agreed with this analysis. They knew that defeated France could not afford to build a navy as large as England's; and they agreed that a war of revenge on Germany would be fought primarily on land. A few French thinkers suggested hitting the Triple Alliance by a land and sea attack through Italy but

[1] Raoul Castex, *Synthèse de la guerre sous-marine de Pontchartrain à Tirpitz* (Paris, 1920).

this idea did not win wide support either in the 1880's or in 1939. A blockade of the traditional type was nearly impossible. In a short war neither France nor Germany would be seriously affected by blockade, even if one could be imposed; and the small sea-borne expeditionary forces of the Crimean period —60,000 men in the original attack—would have little effect on an up-to-date military power. Practically all of the naval operations of the mid-nineteenth century were coastal forays against little or no naval opposition and the strategists of this period had almost forgotten the importance of organized fleet action. It was then widely assumed that the day of great naval battles was past and that an inferior fleet would confine itself to remaining "in being." The strategic purpose of the small German fleet of the 1880's was coast defense; the sailing cruisers which protected German interests abroad could neither fight nor run from a modern enemy. Field Marshal Sir Henry Wilson's well known views of the uselessness of British sea power in an Anglo-French war with Germany reflected the orthodox continental military opinion of his period.[2]

To reinforce these general conclusions from the wars of the 1860's, the next decades saw the appearance of the high speed torpedo boat, the forerunner of the present-day destroyer. This weapon was the answer to the continental navy's financial problem, and was immediately recognized as invaluable for mobile coast defense. The torpedo boat crystallized the ideas of a whole generation of French naval thinkers, and the *Jeune Ecole's* leader, Admiral Théophile Aube, made it one of the foundations of his naval theory.[3] His specific tactical suggestions were as premature as Mahan's were belated, but his works attracted a host of enthusiastic and uncritical followers.[4] At the height of the torpedo boat craze, Germany, Austria-Hungary, and Russia abandoned their battleship programs, and the British admiralty felt obliged to apologize to parliament for finishing those already under construction.

As far as we can now tell from the accounts of peacetime maneuvers, the torpedo boat would have proved itself a formidable coastal weapon. The *Jeune Ecole's* hopes of terrorizing the coastal population of Italy or England by using torpedo craft to lob high explosives and incendiaries into their cities probably would have proved more illusory. Complete disregard of international law and naive estimates of the damage to be done by such indiscriminate attacks were similar to the arguments of many of the early air enthusiasts. But the torpedo boat was the first short range weapon to challenge the rule of the battleship; and, although the "Davids" have not yet driven the "Goliaths" from the seas, the latter's freedom of action has been appreciably curtailed. Military powers were commencing to extend their sovereignty to cover the seas around them. The French built special torpedo craft for operations between Corsica and North Africa and began a modern naval base system on both shores of the western Mediterranean.

[2] G. E. Tyler, *The British Army and the Continent, 1900-1914* (London, 1938), ch. VI.
[3] His most important work is the series of articles collected in *A terre et à bord, notes d'un marin* (Paris, 1884).
[4] Perhaps the best known was the journalist Gabriel Charmes, *La reforme de la marine* (Paris, 1886).

Italian naval theory in this period reflected this same general trend. Their huge battleships were primarily seagoing coast defenders, to break up French bombardments or landing operations. Their fast, lightly armored ships were not intended to win the command of the western Mediterranean from the more heavily armored French; they were a mobile fleet-in-being to render the French command of the sea tenuous. Outside the western Mediterranean the Italians spent little money and maintained no naval bases, but they hoped that in this single area they could neutralize French superiority.[5]

In spite of pretensions to world-wide diplomatic interests after 1898, the United States, too, confined its naval efforts to hemispheric defense. Within this area we have gradually become paramount through our overwhelming control of its naval bases. In the same way the Japanese have won the local command of the seas of East Asia.

In every case this process of regional control by other powers has weakened Britain's world maritime supremacy, and it is interesting therefore that Britain's first major naval allies were America, France, and Japan. As early as the Russo-Japanese war, some French thinkers felt that Indo-China could not be defended against a serious Japanese attack and the English began to discount the importance of Hongkong at about the same time.

These considerations were all, however, primarily defensive; there still remained the other problem of putting pressure on British sea power, to break the colonial and maritime monopoly which she had built and was still extending. There were a number of theoretical solutions, the first of which dates from the introduction of the steamship in the 1840's. The French dream was that all the delays and failures which had wrecked French invasion schemes for a century would end; French military power could be brought to bear directly on the enemy. Steam had "bridged the Channel," steamships would not have to wait for wind and tide, and the British could no longer blockade them in their ports. Although this trend of thought was less prominent in the 1880's than it had been thirty years before, the Japanese invasion of China in 1894 reopened the whole question. It was a favorite subject for French and German staff officers about the time that the Boer War was revealing the utter ineptitude of the pre-Haldane British army. The effects of this threat to British military theory are naturally outside the scope of this chapter, but many thinkers on both sides of the channel have long been concerned with it. At the very least, French strategists predicted, the threat of invasion would pin a large part of the British field army to the British Isles and cut down British pressure on the French colonies.

Admiral Aube, however, was only mildly interested in such invasion schemes; his highest card was the revival of the *guerre de course* against English merchant shipping. The sensational successes of the Confederate commerce destroyers had revived the traditional French enthusiasm for this type of warfare. Though some of Aube's arguments date back to the days of Louis XIV, there

---

[5] Their most important theorist was the prolific writer Domenico Bonamico. Typical is *Il problema marittimo dell' Italia* (Turin, 1881).

were two new factors which promised success where all France's previous efforts had been failures. For the first time, England's whole economic structure depended on foreign food and raw materials. For the first time, as the blockade runners of the Civil War had also proved, England would be unable to confine the raiders to their bases. With some qualifications, these two factors are still the basis of modern commercial warfare. Britain is absolutely dependent on sea-borne supplies, and British sea power must give direct convoy and patrol protection to every ton of shipping moving into the islands. The efforts of the Royal Air Force, supplemented by the Eighth United States Air Force, to reestablish indirect protection by bombing U-boat bases and factories is one indication of the seriousness of this aspect of the problem.

The repeal of the Corn Laws and the disastrous cotton famine of the 1860's had made England's dependence of sea-borne food and raw materials perfectly clear. Aube's generation was the first to appreciate the real gravity to Britain of the problem of protecting her overseas supply lines. Mahan's and Tirpitz' appeals for the acquisition of colonies and foreign trade clouded this particular issue in both Germany and America. Nobody who advocated building up a great foreign commerce could be expected to stress the difficulty of protecting it. Only the *Jeune Ecole*, who assumed that France was self-sufficient, and their British contemporaries, who were faced with the problem of protecting an existing economic organism, have added much to the study of the *guerre de course*.

Much of the damage suffered by American shipping during the Civil War was due to an owners' panic; a similar panic might disrupt the whole of British economic life. The first proposals for national insurance against war shipping losses were supposed to give the government far too much power over business and many shipowners felt that their only salvation would be mass transfer to neutral flags. The economic disorganization attending such a panic was one of the *Jeune Ecole's* main strategic purposes. In view of the Union government's complete inactivity in a similar crisis and the French government's feeble economic policies during the siege of Paris, it is no wonder that many people expected the British merchant flag to disappear from the seas in wartime. Their views were reinforced by the enormous American claims for indirect damages in the *Alabama* case, while Britain's acceptance of the *Alabama* award seemed to show her fear of an American attack on her commerce. The *guerre de course* has always been the best known portion of the *Jeune Ecole's* strategic theory. Aube predicted that even torpedo craft would prey on merchant shipping in the Mediterranean and the Channel, in complete disregard of the rules of cruiser warfare. For nearly twenty years the *Jeune Ecole's* coast defense and cruiser combination was very important in continental naval theory. Its influence was greatest in France and the United States, the ancestral homes of the privateer and the monitor; but even the Russians made careful studies of the routes of British shipping and built the pirates *Rurik* and *Rossia* to match our fast *Columbia* and *Minneapolis*, or the French *Chateaurenault* and *Guichen*.

One of Mahan's main purposes was the refutation of the *Jeune Ecole's* heresies, by appealing from the limited experiences of the wars of 1854 to 1870, to the wider history of the sailing period. Like many another great writer he served different functions in different countries. In America and Germany he was one of the great prophets of maritime and colonial imperialism. In France he restored the sound military principles of Jomini and Clausewitz, which the *Jeune Ecole* had completely forgotten. His insistence on battle and the importance of the organized force was a necessary corrective to the *Jeune Ecole's* extravagant enthusiasm for the latest technical wrinkles and their hopes of gaining a cheap victory by attacking only nonmilitary objectives. Their doctrine of "Shamelessly attack the weak, shamelessly fly from the strong," would have ended in complete military demoralization. Mahan, on the other hand, pointed to the offensive-defensive of the great Dutch sea fighters as the true policy for an inferior fleet. They had never ignored the English fleet but had sought to destroy portions of it by well-timed strategic concentration. Steam had greatly increased the chances of well-planned offensive actions of this type. Even if commerce destruction was the primary objective, Mahan's followers pointed out that it would have to be supported by attacks on the British battle force. Mahan's proof that, historically, commerce destroying had never been successful failed to deal with the *Jeune Ecole's* contention that conditions were now quite different. They simply remained convinced that technical changes do alter the principles of military and naval strategy.[6]

Bismarck's means of putting pressure on England's superior sea power were primarily diplomatic, a method seriously neglected by many naval authorities. It does not bolster theories of the world supremacy of sea power to show the British Empire of the 1880's—with a navy roughly equal to those of most of the rest of the world put together—giving ground to Bismarck in purely colonial matters. Even Mahan, naturally enough for an American of the 1880's, failed to appreciate adequately the connection between diplomatic and naval policy and the European balance of power. His remarks on diplomacy are too largely concerned with the mere acquisition of naval bases. He could see the results of the partition of China and he was anxious that America acquire equally good positions while the division of the world was going on. But, like most Americans of his day, he did not completely realize the importance of the forces working behind the scenes in the great European capitals. Our fears of German aggression in the Caribbean and certain aspects of our open door diplomacy show this same underestimation of the workings of the European state system. The founding fathers were more experienced, and so is our own generation, but both have had far more constant contact with European diplomacy than the Americans of Mahan's time. The importance of British sea power in checking continental military aggression in the New World is a commonplace of the 1940's, as it was of the 1820's. But the notion of the

[6] His most important French followers were Gabriel Darrieus, *War on the Sea* (Annapolis, 1909) and Réne Daveluy, *L'esprit de la guerre navale* (3 vols., Paris, 1909-1910). The leading Italian naval strategist was Giovanni Sechi, *Elementi di arte militare marittima* (2 vols., Leghorn, 1903-1906).

British navy as our own first line of defense scarcely figured in Mahan's elaborate arguments for Anglo-American friendship.

It must not be forgotten, however, that Bismarck's method of diplomatic pressure can be successful only for minor objectives, while Britain is threatened in other quarters by more dangerous enemies. His relatively modest colonial demands had to be balanced against the major threats of France and Russia in Asia and the Mediterranean. His diplomatic blackmail was successful as long as England regarded France and Russia as her most dangerous maritime and colonial rivals. It inevitably failed when English opinion began to think of Germany as her greatest potential enemy. This was the basic error of Tirpitz' famous *Riskogedanke*. His theory was formulated in a period when England was isolated between the Dual and Triple Alliances. His "risk" fleet needed only to be strong enough to hold the naval balance of power between England and the Franco-Russian combination. Its whole psychology reflects the kaiser's shining armor diplomacy.

When William's demands had driven England into colonial agreements with Russia and France, the German navy did not abandon a program which had been fitted to this specific diplomatic situation. The German fleet was technically close to perfection; its weakness lay in its high command. From the abolition of the German naval staff to give William personal control of naval policy (the exact condition which the general staff of the army had been established to prevent), the German navy was a royal toy. In 1914 William could think of no better use for it than to hold it in reserve for the peace conference, and more blackmail. The navy had no plans for an invasion of England, for a serious attack on English commerce, or even for the positive domination of the Baltic Sea. It did not even try to hinder the landing of British troops in France, though the right wing of the German army was the key to the whole Schlieffen plan. The world's second fleet spent four years defending the German coast. It fulfilled that function well enough—as well as the French fleet had done it in 1870, as well as the little German fleet of the 1880's would have done. The High Seas Fleet's fate at Scapa Flow in 1919 was peculiarly appropriate. Tirpitz had added battleship to battleship for exactly twenty years. At the end, the great pile of material which was designed to blackmail Britain into colonial concessions proved to be just so much scrap metal.

In the 1880's, the *guerre de course* had failed to win many advocates in Germany, partly because of her lack of foreign bases and the distance of her home bases from the main English trade routes. The same arguments were also used against it in Italy. In the German case this certainly delayed the adoption of commerce-destroying submarines, while the Italians show comparatively little interest even yet in this type of operation. When the submarine campaign did come, it seems to have been entirely improvised, a factor which must be remembered in any criticism of the British failure to meet it.[7] The Nazi campaigns in Norway, Crete, and Africa are an odd contrast to the High

[7] There seems to be no positive reason for doubting the official account of Admiral Spindler, *Der Handelskrieg mit U-Booten* (Berlin, 1932), vol. I, p. 8.

Seas Fleet's Scandinavian policy of 1914 to 1918. In its late discovery of commerce destruction by submarine and its failure to develop a more vigorous Baltic policy, the old German navy clearly revealed its uncertain appreciation of the basic ideas of the *Jeune Ecole*. Whether a true *Jeune Ecole* navy would have been less obnoxious to the English than the building of the High Seas Fleet proved to be is, of course, an open question. At least the British in 1935, as part of a policy of appeasement, were willing to sign an international agreement accepting the former type of fleet.[8]

The final method of putting pressure on England by-passes the need for sea power completely, through development and control of the strategic land areas. Mahan himself saw that the most significant factor in his America had been the rise of a self-sufficient industrial empire. But to him this process was only a beginning and he looked to the day when the America of big business would export population instead of importing it. His colonial ideas may seem to have been an anachronism in the years when American business men were developing our characteristic methods of continental exploitation. It is equally ironical that the German founder of *Geopolitik*, Friedrich Ratzel, saw the sea as the prime source of national greatness, while the British geographer Mackinder was pointing out the importance of the nineteenth century's developments in land transportation. But before 1914 even the Russians continued their painful push to the sea, to two disastrous wars and the total collapse of the old monarchy. By 1914 Mahan was as popular on the continent of Europe as in England, America, and Japan. The German navy's enthusiastic acceptance of his ideas was all that was needed; for the Italian, Russian, and French fleets were by this time actual or potential allies of England.

In France the *Jeune Ecole* had declined in influence until it was a mere military sect, repeating its dogmas with little effect on French naval policy. Neglect of the most elementary military principles had proved their undoing. In the ridicule which now greeted their tactical proposals, many people had already forgotten that Aube and his followers had been the first people to analyze many of the salient features of modern naval war. The *Jeune Ecole* had become involved in politics with the usual disastrous results. The Radical minister Camille Pelletan had wrecked both the fleet and the *Jeune Ecole* by trying to purge the navy of many supporters of the opposing doctrine. The defeat of France's ally in the Russo-Japanese war was widely interpreted as proving the correctness of Mahan's ideas. Few writers, except Mahan himself, took the trouble to examine the unorthodox elements in Japanese strategy. The mine and torpedo did not, of course, come up to the exaggerated hopes of the faithful though they established themselves as weapons of considerable importance. Russian attacks on Japanese commerce had been, on the whole, unsuccessful; the Russian fleet's passive belief in the importance of the fleet-in-being was as disastrous as the same belief was later to prove in Germany. It

---

[8] One of the bitterest *Jeune Ecole* critics of Tirpitz' policies is Captain Lothar Persius. Typical of his work is *Warum die Flotte versagte* (Leipzig, 1925).

was also as great a violation of Mahan's own principle of an active offensive-defensive for a temporarily inferior force.[9]

This decline in the vitality of the *Jeune Ecole* was largely responsible for a general failure to foresee the strategic possibilities and the portentous future of the submarine, the weapon which was the hope of Britain's enemies from Fulton to Dupuy de Lome. The competition which eventually produced the first successful French submarine had been started by Aube himself. But when this key weapon actually appeared, it caused very little speculation concerning naval strategy. Even the French regarded it as a sort of diving torpedo boat for coast defense. The *Jeune Ecole* had been completely discredited by the violence of their previous claims. (Any student of Douhet will notice a similar parallel. With his perfect fighter bomber almost in sight, his theories have been so discredited by the antics of his journalistic followers, that his bombers have had a hard time getting assigned to the strategic tasks he outlined for them.)

The most original naval thinkers of the period immediately before the war were the Englishmen, Fred T. Jane and Julian Corbett.[10] Like certain other English writers,[11] they took the *Jeune Ecole's* contentions seriously. Both Jane and Corbett stressed the importance of the torpedo boat and, later, the destroyer in limiting the action of a superior fleet and both men agreed with Aube that, because of these newer types of ship, close blockade was probably impossible. Corbett's careful studies of the military and diplomatic sides of the great wars of the sailing period gave a much fuller picture than those of Mahan and there was a general revival of interest in combined military and naval operations. Corbett felt that the indirect protection of commerce could best be effected by capturing the enemy's bases; unless this was done—and in this he agreed with Aube—the protection of British commerce would be difficult indeed. Germany, fortunately, could be blockaded indirectly, out of the range of German torpedo operations. The navy, which publicly proclaimed its frontiers to be the coasts of the enemy, began and won the war in stations as far away from enemy torpedo strength as possible. Its respect for the torpedo was, in fact, so great that it hampered tactical action. The Germans, who should have studied the *Jeune Ecole's* doctrines most carefully, contemptuously rejected them as a sign of weakness, and naively assumed that the superior force would appear where the inferior navy preferred to have them. The realistic and scientific German commanders were an odd contrast indeed to the conservative brass hats of the British admiralty.

British naval theory since the war of 1914-1918 has been primarily concerned with tactical matters—the Grand Fleet's fear of torpedo attack at Jutland, and the high command's failure to deal promptly with the submarine menace. American theorists have followed the British pattern, though our criticisms of British naval leadership naturally have been much more severe.

[9] Réne Daveluy, *La lutte pour l'empire de la mer* (Paris, 1906).
[10] Jane's most important work was *Heresies of Sea Power* (London, 1906). Corbett's revision of Mahan is *Some Principles of Maritime Strategy* (London, 1911).
[11] Among them the brilliant brother of Sir John Colomb, Admiral P. H. Colomb.

Like the British, we have accepted both Mahan and Corbett. Our coasts-of-the-enemy theory is now tempered with a sound respect for both the mine and the torpedo, as well as, more recently, land-based aircraft. The most involved British strategical discussions have been largely historical in nature.[12]

Like the French, the British before 1939 seemed reluctant to consider the tremendous new dangers facing the empire. Their intellectual preparation for meeting the air danger was, on the whole, inferior to that with which they met the torpedo in 1914. No European naval strategist has added much to the discussion of air and sea power; this whole development has been largely focused on the special problems of the Pacific. The Germans, as could have been expected, have carefully explored the causes of the High Seas Fleet's tragic failure. Too late, their discussions have repeated the French pattern of the 1880's. Too late, many Germans, echoing the war ministry's criticism of the period just before 1914, have attacked the kaiser's whole naval philosophy. *Mein Kampf's* doctrines of an empire to be won in the east on land show a belated recognition of some of the basic fallacies of the imperial navy. The unsuccessful attempt of the west to suppress the Russian Revolution immediately after the last war, and the vast development of Soviet power since that time, have proved, at least to many geopoliticians, that sea power is again on the wane and land power in the ascendancy. Tirpitz' attempt to proportion the navy to the state's foreign trade is now seen to be completely absurd. No one would dream of computing the size of the military air force in terms of commercial air traffic.

Some Germans have belatedly adopted the ideas of the *Jeune Ecole* but they have not added much to the *Jeune Ecole's* theory. Perhaps the most brilliant suggestion was Admiral Wegener's spring to the north through Norway,[13] but even this goes back to the days of Tromp and Ruyter, who tried to sneak Dutch convoys around by way of Bergen. Some German writers favored large scale surface raiding in connection with submarine attacks but the actual raids of the German pocket battleships and heavy cruisers have accomplished very little. Only in convoy actions off the Norwegian coast have the Germans used such forces successfully against surface shipping. We naturally know little of German plans for a land invasion of England but the speed and success of their drives in Norway and Crete show that they might well have considered something similar for England. From the outbreak of the war, the German command has certainly avoided the kaiser's indecision; the story of their actions so far clearly sustains Mahan's contention that even an inferior fleet can cause great damage by an active offensive-defensive.

On the whole, however, the most original continental works on naval theory in the postwar period were produced in France and Italy rather than in Germany. Both powers had been tremendously impressed by the naval warfare in the English Channel and the Adriatic, and the damage which could be inflicted

---

[12] Even the works of the able Sir Herbert Richmond have approached most strategical problems from this standpoint.

[13] *Die Seestrategie des Weltkrieges* (Berlin, 1929).

by the skillful use of even an inferior navy's light forces. The Mediterranean was more suited to the lightning clashes of special task forces than to the methodical maneuvers of entire fleets. In the Mediterranean, as in the contested areas of the South Pacific, both France and Italy expected to be able to use the sea for essential military and commercial convoys; neither expected to win the complete command of the entire area. For this infighting, both powers developed great numbers of shore based aircraft, heavily gunned cruisers, and heavy battleships. The carrier and the big flying boat were both unnecessary and the capital ship for such conditions needed to be as nearly unsinkable as possible. The Italian fleet of the 1880's was the direct ancestor of that of the 1930's; its aim was still to interfere with a superior sea power's use of the Mediterranean. As late as the spring of 1943 the Italian efforts at more pretentious tasks had failed; but their fleet had been a constant threat to British military and commercial convoys in the Mediterranean. With the help of shore based aircraft, it had protected the vulnerable Italian peninsula against serious landing operations. In the light of Italy's hopeless strategic situation, geographically, this was a considerable achievement. Mussolini's fleet had failed to win the command of the Mediterranean but it had constantly hampered British movements in that area. Sixty years ago, Italian strategists would hardly have dreamed of anything more.[14]

Certainly the most interesting of modern continental naval strategists is the French Admiral Raoul Castex, whose five volumes on naval strategy[15] are the climax of a long literary career. Like his predecessors Daveluy and Darrieus, Castex represents perhaps the best synthesis of Mahan and the *Jeune Ecole*. In his definition of sea power and the military principles on which it must rest, Castex avoids the basic heresies of the *Jeune Ecole*. He is well aware of the importance of the command of the sea and of the need for an organized force to obtain it if possible. Wars are not to be won easily or cheaply but only by fighting to destroy the organized force of the enemy. In the fundamentals of military science, Castex agrees almost completely with Mahan. In his analysis of the conditions of modern naval strategy, however, he agrees rather with the *Jeune Ecole*. To Castex, as to Aube and Douhet, strategy varies with the material conditions of warfare. To him, as to Aube, England's position seems to be gradually deteriorating, and sea power will play a lesser role in the future than it did in the sailing ship era. Like other French and Italian theorists, Castex stresses the weakening of offensive sea power in both time and space: the end of continuous blockade has made it less effective in time; submarine and air attack have limited its action in space by denying access to certain coastal waters. Sea power today, as Jane contended thirty years ago, is increasingly dependent on the security of land bases.

Castex' first volumes were concerned with the hopelessness of trying to

[14] The best known modern Italian works are Romeo Bernotti, *La guerra marittima, studio critico sull' impiego dei mezzi nella guerra mondiale* (Florence, 1923); G. Fioravanzo, *La guerra sul mare e la guerra integrale* (2 vols., Turin, 1930-1931); and Oscar di Giamberardino, *L'arte della guerra in mare* (2 vols., Rome, 1937).

[15] *Théories stratégiques* (2nd ed., 5 vols., Paris, 1937).

defend Indo-China against Japan and he took the lead several years ago in contending that not only that colony, but Syria and perhaps Madagascar, should be abandoned in the face of a serious attack. This inability of the great oceanic powers to defend their outlying possessions is, again, a new limitation on sea power. At the Washington Conference, Castex created a sensation by a memorandum approving the German submarine campaign and he and other French strategists continue to consider it the most dangerous possible threat to England. Though most French naval men were not as outspoken as Castex on this subject, the *Jeune Ecole's* theories of the *guerre de course* still have strong support in the French navy. The building of the giant submarine cruiser *Surcouf* was a clear indication of this continuing trend. In general, the Italians are less enthusiastic about the possibilities of U-boat warfare than the French or Germans. For this, as for almost every other type of naval warfare, their strategic position is difficult, if not hopeless. Finally, Castex had been as greatly impressed as the Germans by the fact that land power is escaping the pressure of sea power almost entirely in areas like Russia and China. One volume of his strategic compilation is devoted completely to studying duels of land and sea power; to him, the parallels between the age of Stalin and the age of Genghis Khan are obvious. This may have been partly a reflection of France's traditional hopes in the Russian alliance, for at least one French admiral, before the days of Darlan, exhibited a bust of Stalin as the future savior of France. He may not have been a poor prophet!

CHAPTER 19. Japanese Naval Strategy

BY ALEXANDER KIRALFY

THE subject of Japanese naval thought is virtually unexplored territory. It is a field of inquiry wherein special geographic factors and the heavy hand of national policy have left a far deeper imprint than has been the case elsewhere. Accordingly it is necessary to depart from the usual treatment of a naval or military topic to the extent of considering many phases of Japanese naval theory which could be taken for granted or entirely disregarded when inquiring into the doctrine held by other powers.

Such an approach to this subject is particularly important when it is realized that Japanese naval thought differs so radically from that of the western world that Japanese ideas of sea power cannot very well be stated in occidental terminology. It is because non-Japanese inquiries into this subject have adhered to western ideas and formulae that the true essence of Tokyo's strategy and tactics has been frequently misinterpreted. Investigators in the United States and in Europe were naturally prone to attribute to the Japanese intentions and designs which were accepted as sound in their own countries. Some of these ideas, however, either never entered the Japanese mind at all, or they were pondered and set aside as being inapplicable to the Far East.

It follows that a recourse to works of non-Japanese thinkers cannot very well give a picture of Japanese naval thought. Rather will it present a western appreciation of Nipponese naval activities, and highly commendable studies of this type exist. This means that Japanese methods of sea fighting have been approved or disapproved depending upon whether or not they have adhered to western doctrine. The criticism is generally of the man rather than of the principle. Japan is customarily accepted as an orthodox naval power of the American or European type, and her admirals are expected to act in an orthodox manner. A Japanese admiral is blamed for having adopted the very course of action dictated to him by his government which, in turn, had an eye to peculiar eastern circumstances.

The distinctions between Japanese and western thinking were further obscured by the friendly treatment usually accorded Tokyo by occidental writers. It had become almost a general rule to qualify an item of criticism with the observation that the action under examination was probably justified by the special conditions then obtaining. A cycle was thus set in motion wherein the failure of a Japanese admiral to prevent the escape of a hostile fleet because he had been acting as an escort for his troopships is explained away on the grounds that he had been on convoy duty and hence could not have prevented the escape.

In thus assigning western theories to Tokyo, western nations began to rely upon the assumption that a Japanese fleet would always act in the conventional manner, that it would, for instance, always seek a decisive engagement at the first opportunity when, in fact, that was, as often as not, the very thing a Japanese admiral might not desire. Had western authorities scrutinized Japanese theory rather than Japanese practice, Japanese rules rather than the followers of the rules, more light might have been thrown upon the subject. Admittedly such a course would have been a difficult one to follow in the United States and Great Britain. The peoples of these countries, averse to maintaining large armies, had come to rely completely upon their naval establishments and to hope that potential enemies could be defeated at sea. Any line of reasoning tending to show that Japan might have to be met in full strength on distant land fronts would have been highly unacceptable. Yet that very conclusion becomes inevitable when the normal factors associated with sea power are studied and the divergency between Japanese theories and those of the west is realized.

The normal factors in western ideas of sea power are deducible from a study of the naval theories of Great Britain and the United States, on the one hand, and of the lesser European naval powers on the other. As regards Anglo-American doctrine, decisive naval battle and pressure of the blockade held first place. But in the case, first of France, and then of Germany, interest centered upon commerce destruction and the whittling down of hostile fleets by secondary action, by the employment of such special weapons as the torpedo and bomb, and by reliance upon tactics that would lead to the destruction of detached parts of a numerically superior enemy force.

Although western methods of naval warfare involved the strategy and tactics of superior and inferior forces respectively, general all-embracing theories of sea power were widely discussed throughout Europe and America. A vast literature sprang up wherein the superior naval powers studied defense against the *guerre de course* while their rivals discoursed at length upon the evolutions incidental to decisive fleet action. The outbreak of war accordingly set in motion forces that were already well understood. In 1914 and 1939 British cruiser squadrons took up their predetermined positions in anticipation of German attacks upon British commerce. In both wars British battleship squadrons endeavored to bring the enemy fleet to action. As the crisis in the Pacific loomed, the United States battle force was dispatched to Hawaii and then ordered to be in readiness for eventualities. On each occasion the Allies imposed a blockade and took steps to destroy commerce raiders.

It would seem, on first analysis, that these broad theories were likewise relevant to Japanese naval activities. In its wars with China in 1894-1895 and with Russia in 1904-1905, Japan gave evidence of seeking fleet action. On these occasions as well as in connection with the undeclared war against China, Japan showed that she understood the significance of the blockade. Yet in the momentous struggle which began on December 7, 1941, the Japanese not only avoided fleet action, as such, but appear to have done very little with respect to a naval blockade or a counterblockade. However, if Nipponese activities in 1941-1942 are scrutinized in conjunction with earlier actions, a number of underlying principles may be glimpsed which acquire basic significance when considered against the background of Japanese history and national policy. The great obstacle to this line of reasoning, though not an insurmountable one, is the fact that one may search in vain for a carefully formulated theory of sea power written by a reputable Japanese author which would substantiate the principles thus visualized.

Inasmuch as naval thought stems largely from records and analyses of naval activities, the lack of such material in Japan seems worthy of comment.

The reasons for the lack of literature are psychological and political. From a military standpoint the Japanese mind may be described as being subjective rather than objective. In peacetime, an American writer can impassionately discuss a war in the Pacific just as a British student can compose a disquisition upon command of the Mediterranean; and either can discuss at length imaginary campaigns wherein France is opposed to Italy or Germany to Russia. The Japanese, on the contrary, lack interest in waters which do not directly concern them. Whereas the western student will proceed along purely academic lines, concentrating upon the naval factors alone, the Japanese find difficulty in eliminating the national-political approach. As a rule they have been unable to discuss Guam without stating or implying that it was a threat to their country which must be removed.

Politically, the publication of any serious writings which would disclose the theories accepted by Japan has been considered contrary to Japanese policy. It will be shown below that Nipponese naval planning has been concrete, directed against specific enemies and territories, and hence aggressive. The intended victims were accustomed to formal indoctrination, to a high respect for the authoritative printed word. The appearance of lengthy treatises clearly setting forth the methods to be employed by Tokyo on and after December 7, 1941, would have aroused an alarm among the democracies far greater than would have been expected from a piecing together of actual Japanese activities. As the United States and Great Britain were peaceful nations with no secret aggressive designs, there were no naval theories which their nationals could not safely reveal in print, and hence there was no censorship. The tendency of the Japanese to limit their thinking to their own part of the world, and the prohibition of most writings which would reveal their true thoughts, led to a marked paucity of literature.

This condition was aggravated by the fact that the Japanese, surprisingly enough, were interested in naval matters only during periods of hostility. This was stressed by a competent Japanese authority in 1933 in correspondence with the author. Consequently the publication of works on naval subjects was commercially unattractive. Most formal treatises, as well as many official documents, are evidently considered restricted or secret. The writer was officially informed, for instance, that no original Japanese work has been published on the Battle of Jutland, which has been formally treated not only in English and German, but at some length in French and Italian. The publication of the Japanese counterpart of *Jane's Fighting Ships* does not appear to have been contemplated before 1935, after which it was brought out in sections but seemingly with little success.[1] Articles on naval matters in scientific magazines have been usually short and superficial. A number of pictorial books have appeared, but they cannot be said to represent a contribution to naval thought.

In Nipponese naval magazines the same reticence likewise is obvious. One such periodical, *The Navy*, began publication shortly after the close of the Russo-Japanese war, only to be suspended a few years later. It was revived in 1926, but passed out of existence shortly thereafter. The leading periodical, *Sea and Air*, is much more elementary in content than the Proceedings of the United States Naval Institute, to cite a typical American publication. Its treatment is in the accepted formal manner. It has had, for example, special "cruiser" and "destroyer" numbers wherein the ships of the world's navies of these categories have been analyzed from the standpoint of speed, gunpower and so forth.[2] In so far as can be determined it did not deal with the Battle of Jutland until 1939 when it printed, in installments, a condensed translation of Lieutenant Commander Frost's treatise published at Annapolis in 1936.[3]

In the field of popular naval writings there are, besides somewhat superficial articles, the so-called sensational books and pieces. Such works are common to other countries and generally receive little serious consideration. Literature of this kind, when published in the United States and Great Britain has no political importance. In the latter part of the nineteenth century, for instance, a host of articles printed in London portrayed a coming Anglo-Russian war, yet not only did such expectancies fail to materialize, but in 1914 and 1941 Briton and Russian fought against a common enemy.

Quite patently, sensational literature in Japan could not run counter to the government's political and military plans, or make disclosures of fact or theory. Furthermore it had to run substantially along orthodox or western lines. Some examples of this literature may be cited to show these tendencies. In Ishimaru's *Japan Must Fight Britain* the issues are decided in orthodox naval combats in the East Indies wherein the aerial arm is conspicuous by its absence.[4] Not only is the military weapon played down and the airplane almost entirely overlooked in Kinoaki Matsuo's *How Japan Plans to Win*, but the whittling down of the United States battle fleet by the submarine in particular is emphasized, the remnants of that body being shattered in conventional battle.[5] Such works would furnish little new material to those who had read such a British book as

[1] "1935 Naval Pocket Annual," Tokyo, to be published in six parts. Part 1, Japanese Navy, publication delayed. Though Part 2 was to deal with the British Navy, the first handbook advertised early in 1936 was Part 3, United States Navy.

[2] *Sea and Air*, Tokyo, Destroyer Number, September 1935; *ibid.*, Light Cruiser Number, November 1935.

[3] *Sea and Air*, Tokyo, January-November 1939, covered the account to the opening of the battle cruiser action.

[4] Lieutenant Commander Tota Ishimaru, I.J.N., *Japan Must Fight Britain*. Translated by Instructor-Captain G. V. Rayment, C.B.E., R.N. (ret'd) (New York, 1936).

[5] Kinoaki Matsuo, Japanese Naval Intelligence, *How Japan Plans to Win* (Boston, 1942).

Mr. Hector Bywater's *Great Pacific War*[6] or, in the United States, Denlinger and Gary's work on the same topic.[7]

A final category of Japanese naval literature is represented by confidential writings. There can be little question but that the Nipponese *have* devoted considerable attention to the Battle of Jutland, but such discussions are undoubtedly kept under lock and key—as is true in the case of such special staff treatises in Washington and London. This also appears to be the case with the type of formal writing openly published in the western world. In the United States and Great Britain the Battle of Jutland is reexamined for its objective lessons, lessons which, with understandable qualifications, would be applicable to any and every naval battle in the future. The Japanese would be more interested in studying lessons for their own future use, a fact which gives a confidential character to all such writings.

It is therefore apparent that there could not have been a Japanese Mahan, Corbett, Castex, Groos or Manfroni—except, of course, of the "unsung" kind. The Japanese translated and studied Mahan, but did not publish any significant works of their own on sea power. They were not so much interested in discussing what hypothetical fleets could do under hypothetical conditions as in finding out what their ships could do under *planned* conditions, and this was obviously not a fit subject for general dissemination.

The inquirer into Japanese naval thinking is hampered by more than the lack of authoritative native treatises. Such writings as are available to him present special problems because of the difficulty of the language in which they are written, a fact of which even well educated Japanese have complained. While a children's book on the navy is printed in the simple katakana characters and an "elementary" naval magazine appears in Sino-Japanese ideograms with hiragana annotations, the periodical *Sea and Air* is largely lacking in such helpful phonetic symbols. The booklets issued to reserve officers, and presumably to those on the active list, contain obscure forms of the native syllabary, in one case a sixth variation having been used. Furthermore, the language is so involved that translation is difficult, and in some cases the peculiarities defy rendition into a foreign language in a manner that would preserve the sense of the original author. The translator is therefore compelled to resort to paraphrase and circumlocution and may be tempted to omit portions of compound verbs and even phrases when these appear to be redundant. Those unacquainted with the amphibious nature of Nipponese operations are therefore not unlikely to select a naval term when a military implication is intended. Ordinarily, and when translating factual information, the above complexities may not act as a deterrent. But deficiencies of this nature are serious stumbling blocks when one is peering into Japanese *thought*.

One important case may be given by way of illustration. In the French translation of the formal oral report given by Admiral Togo at the conclusion of the

[6] Hector C. Bywater, *The Great Pacific War* (Boston and New York, 1932).

[7] Sutherland Denlinger and Charles B. Gary (Lieutenant Commander, USNR), *War in the Pacific* (New York, 1936).

Russo-Japanese war the words "principal enemy forces" are employed.[8] In view of the context, these words seem rather clearly to imply "land forces" as much as naval units. In a translation of the same speech made by Japanese writers into English an important phrase in the French version is omitted and the words "principal enemy forces" is translated "main forces of the enemy *squadron*," the obvious redundancy suggesting that the word "squadron" was added in the translating.[9] In the official German translation the term "principal squadron" is employed, though the aforementioned omitted phrase reappears.[10] This speech will be reconsidered later in connection with the hostilities to which it referred.

There remain for consideration articles written by Japanese naval officers for the western technical press. When these are not strictly orthodox, and hence not representative of Nipponese theory, they are political and self-serving. To take an example of the first kind, an officer writing in 1921 on the subject of the Battle of Jutland discussed it on the basis of the "traditional" mission of the British navy.[11] His criticisms of the opposing commanders for breaking off action at nightfall and failing to press to a decisive victory are the very criticisms western writers have leveled on occasion at Admirals Ito and Togo of the Japanese navy.

In the political and self-serving field a few examples may be selected at random, written during the period when Japan refused to extend the treaties of naval limitation. Commander S. Takagi in the 1935 issue of *Brassey's Naval Annual* stressed his country's adherence to the Washington Treaty *despite* the development of aircraft, the very factor which was so much to Japan's benefit. In the November 1934 issue of *Current History* Captain G. Sekine pointed out that the modernization of capital ships and the "remarkable advance in aviation" had weighed heavily in favor of a force attacking Japan, though events were to prove that the opposite was the case. In *Foreign Affairs* for January 1935, Admiral K. Nomura claimed that the 5:5:3 ratios were highly repugnant to the Japanese people, an argument of doubtful validity.

All of these factors conspired to make Japanese naval thought virtually a closed book to the western world. The question therefore arises as to how it would be possible to discover the theories followed by Tokyo. It is believed that this can best be done by considering the "spirit" of Japan, the geographic characteristics of the Far East, and the manner in which Japanese naval strategy and tactics have adapted themselves to these factors. By this procedure we should be able to discover a number of constants, a number of courses of action always followed under given conditions, and these can properly be accepted as the fundamentals of Japanese naval thought.

[8] *Revue Maritime*, Paris, CCI (April-June 1914), p. 339.
[9] Vice-Admiral Viscount Nagayo Ogasawara, *Life of Admiral Togo*. Translation J. and T. Inouye (Tokyo, 1934), p. 391.
[10] *Der japanische-russische Seekrieg 1904/5, Amtliche Darstellung des japanischen Admiralstabe* (Berlin, 1911).
[11] Lieutenant Commander Ichira Sato, I.J.N., *Brassey's Naval Annual* (London, 1921-1922), pp. 77-84.

II

Japanese thought has always been closely related to historical-ideological factors. In this regard the statements of Professor Katsuro Hara of the Imperial University of Kyoto are most pertinent.[12] Military art, he wrote, was the only one that had been practiced in Japan since antiquity: it was the most honorable of all, and practiced at the expense of other attainments. Japan adopted the military instruments of the west and put them to use in a field already well ploughed. According to him, "whether or not her evolution pleases or displeases other nations, Japan cannot fix limits to the progress she must yet make, for, where there is no progress, there must be stagnation." In 1926 the word "progress" might have had a purely abstract significance, despite the context, but when it is considered in the light of Japanese history it is impossible to dissociate it from territorial aggrandizement by military means.

Such aggrandizement would naturally radiate from Japan through the Far East. The propinquity of Japan to the Asiatic continent has frequently been compared with the position of Britain with respect to Europe. Except from the strictly geographic standpoint this is not a proper parallel. The British Isles faced a militarily powerful Europe and possessed an empire, a series of naval bases, and an ocean trade that called for, supported and serviced a mighty navy. Japan faced no comparable potential enemies, had no distant dominions or possessions and could at no time have protected its overseas trade because of the lack of far-flung naval stations and of sufficiently powerful fleets to station at those bases. Having no designs on the neighboring continent, modern Britain developed a strategically defensive sea power. Little more was demanded of it than the ability to defeat an enemy on the high seas. The same observations naturally apply to the United States.

The military weakness of Asia and Australasia, the immense land and sea distances separating this region from important western military and naval concentrations, the manner in which the Japanese islands commanded the western Pacific littoral and the numerous closely spaced islands leading from Japan to the south—all these factors favored the success of maritime aggressions. The bellicose spirit of the samurai, and the militarily attractive features of his surroundings, dictated that a navy was to be used for troop transport or troop escort and not as a high seas fighting force. It was, in fact, the "troopship" which brought the Japanese to Japan.

The Japanese came by sea to the home islands they now occupy. After centuries of bitter fighting they exterminated the aborigines or drove them into the northernmost islands. In these operations the newcomers made ample use of their warships, or more properly of their transports. While the principal invading armies moved by land, important fleets skirted the coasts where they landed strong forces on the flanks of the enemy in precisely the same way that the Japanese squadrons functioned during the early months of the Second World War.

[12] *Histoire du Japan* (Paris, 1926).

It was thus that Jimmu-Tenno—referred to as the first emperor of Japan— (660-585 B.C.) moved from Kyushu, the most southerly of the main Japanese islands, through the Inland Sea to a point near which Osaka now stands. For hundreds of years the Inland Sea was a Japanese military thoroughfare. In the numerous wars between the leading Japanese clans the ship was the favorite means of locomotion, and land was the element selected for battle. The sea fight—notably that fought in A.D. 1186 between the 700-ship fleet of the Minamotos and the 500 vessels of the Taira clan at Dan-no-ura, near Shimonseki—was a rare occurrence.

As a result of this employment of shipping the Japanese became impressed with the concept of a vessel as a means of transport rather than of fighting. A subsidiary influence was the close relationship perceived between the land and sea. The transports generally coasted and their principal function otherwise was to embark or disembark land forces. The naval tactics formulated during the medieval days of Japan clearly show this influence. The terms assigned to the various dispositions to be assumed by a fleet frequently included a relationship to land. One such term signified that the fleet was drawn up in a certain order with respect to a nearby mountain, or to a pair of mountain peaks. Another indicated that the fleet was stretched across the mouth of a bay—drawn up either in a line or in crescent formation, depending upon the terminology used.[13]

As fighting at sea then differed little if any from fighting on land, and as the Japanese were customarily on the offensive against land objectives, fleet and squadron formations appear to have been designed with a view to cooperating with and protecting expeditionary forces. A tactical disposition astride a bay would seem to imply that a fleet was protecting an embarkation or disembarkation under way in the sheltered waters of the inlet against hostile naval interference. To the basic theory of the "transport-fleet" would be added the important theory of the "protective squadron." These theories were put to successful use about the year 1200 when the Japanese crossed over to Korea and conquered that land.

In the years 1274 and 1281 the attempted invasions of Japan by the Mongol Kublai Khan profoundly affected Japanese naval thought. The Tsushima Islands were seized by the enemy, many of whose troops landed on Kyushu. While battles took place on the beaches the Japanese ships, smaller and far less numerous than those of the Mongols, were repeatedly defeated despite their reckless attacks and exhibitions of ingenuity.[14] Their nimbleness, however, brought them some success and impressed upon the Japanese the value of speed. The enemy, on the other hand, contrived to protect himself by lashing his vessels together in fortresslike groups, a defensive improvisation which was not lost on the Nipponese mind. It will be recalled that Kublai Khan's armada was destroyed by a violent typhoon.

[13] Captain Viscount Ogasawara, "Historical Essay on the Japanese Navy." Translated in *Revue Maritime*, Paris, CC (January-March 1914), pp. 96-97.
[14] Nakaba Yamada, *Ghenko, the Mongol Invasion of Japan* (New York, 1916), p. 185.

The Japanese then turned to piracy on a large scale. In the period 1400-1600 they ravaged the coasts of the Far East from Korea (whence they had been driven by the Koreans and Chinese many years previously) to the island of Singapore, the shores of Sumatra, Borneo, and the Philippines. At a number of points the pirates—whose depredations were organized under the control of Japanese feudal chieftains—established beachheads which were the forerunners of those which have been established by the Japan of today.

In the year 1587 Toyotomi Hideyoshi, hailed by the Japanese as the founder of their navy, proceeded by sea from the main island to Kyushu with 150,000 men to put down a Satsuma clan rebellion in the latter island. He then turned north, and again by means of sea power subdued a refractory prince. In 1592 he set out to reconquer Korea. Organizing an army aggregating 300,000 men, he called upon the princes whose dominions were on the seaboard to prepare a large fleet. Hideyoshi himself supervised the construction of some of the larger ships, including the *kikan*, or flagship, *Nihon Maru*. This vessel and its attendant craft were transports rather than fighting ships. The mind that built these "square-enders" was thinking more in terms of a military headquarters on the coast of Korea than of a battle squadron intended for action in the Straits of Tsushima or the Yellow Sea.

The manner in which the headquarters squadron was organized and moored suggests considerable attention to detail, the absorption of lessons from the past, and a larger amount of caution than has generally been attributed to the Japanese. The *Nihon Maru* and its attendant ships anchored behind a defensive line consisting of a large number of small vessels linked to each other by cables and anchored at both extremities. This was obviously a flexible adaptation of the Mongol arrangement. The ships in the center of the protective cordon were fitted with projectile-throwing devices. Beyond this defensive bulwark fast scouting ships kept a lookout for the enemy. The flagship itself was surrounded by smaller warships while various transports and supply ships, including water-carriers and horse transports, formed the rear.[15] Measures had been adopted to "armor" the squadrons against the clumsy muskets of the day by means of layers of bamboo.

In the campaign against Korea the two main Japanese armies landed in the south and quickly seized the capital. Cruising squadrons, protecting the communications met and drove off a number of enemy squadrons attempting to interfere with the operations. A third Japanese expeditionary force was dispatched to outflank the Koreans from a point on the west coast, and another movement, through the Yellow Sea and against China herself, was contemplated. At this critical moment a Korean admiral, Yi-sun, appeared aboard an "armored" *kopukson* or "turtle-back" dreadnaught leading a vast concourse of ships which may have included other units of the same novel type. The invading warships and transports moving up the west coast were annihilated; the Japanese protective squadrons were shattered. Cut off from Japan, Hide-

[15] *The Navy* (Tokyo, April 1926), p. 5.

yoshi's armies fell back to the coast. But for the intrigue in the Korean court, which possibly had been instigated by the invaders, and which resulted in the dismissal of Yi-sun, the Japanese would have been driven out of the country altogether.

In this campaign the Japanese made no attempt to secure command of the seas as a preliminary to invasion. Judging from their future actions the lesson derived from this bitter experience by the Japanese was not that they should secure command of the sea, but simply that their transports and warships should be assured better protection from a material as well as from an organizational standpoint. Hideyoshi's admirals might have attempted to destroy the Korean fleets at the start of the war. But this would have meant long delays, during which Korea and China could prepare to repel the invaders. To secure this respite it would only have been necessary for the Koreans to avoid battle by retiring among their islands and anchoring behind bars. The utilization of coastal features and of the state of the tide were well-known factors in the tactics of the day.[16] Furthermore every ship searching for the Korean fleet meant a ship that could not act as a transport; it meant a ship that would not be available to defend the transports should they be attacked by an enemy detachment. The political situation at the time was complicated by nonmilitary circumstances but one might speculate as to what might have happened if the rivalry in the Japanese camp had not interfered with operations. The Koreans might have been compelled to sue for peace before the "turtle back" could have changed the fortunes of war.

Recalling the fate of the Mongols who had been abandoned on the islets between Kyushu and Tsushima by loss of the command of the sea, Hideyoshi cannot have been unmindful of this factor. He apparently had to choose between clearing the seas and running the risk of meeting an enemy who had become impregnable on land, or of accepting risks at sea in order to guarantee the success of his military enterprises on the continent. The Japanese had underestimated their enemy at sea and had suffered a severe check at the hands of a new and dangerous weapon. Thereafter they were to keep a close eye on their actual and potential enemies, as well as to explore to the full the possibilities of new weapons.

In a few years the essential soundness of the Japanese theories was to be proved. In 1597 Hideyoshi repeated his attempt against Korea on whose coasts he had contrived to maintain beachheads, a significant term in Nipponese thinking. The reduced Korean fleet had almost rotted away at its anchorages whereas Hideyoshi had constructed a new fleet. A naval battle occurred which was a complete victory for the Japanese, yet it was a fruitless triumph. Since the first invasion the Koreans, now ably assisted by the Chinese, had so fortified their country that Hideyoshi's forces, which had formerly advanced with ease and rapidity, were slowed down to a mile by mile advance and brought to a final stop at the capital, Seoul. In one case the Nipponese land forces participated in a naval engagement which was fought practically on the beach, a precursor

---

[16] Ogasawara, *Revue Maritime*, Paris, CC, 96.

of the simultaneous struggles in all elements which characterized the aggression against the western powers in 1941-1942. The Japanese military position rapidly deteriorated and became hopeless when Yi-sun, recalled to command the Korean fleet, inflicted a defeat of secondary importance upon the invading armada. The Nipponese land forces, suffering from attrition, promptly took advantage of the setback of their naval forces by evacuating Korea. In the course of the withdrawal Yi-sun fell upon the troopships, inflicting heavy losses upon them, although at the cost of his own life.

Hideyoshi, primarily a landsman, while attributing the disaster to the discomfiture of his fleet rather than of his armies, revealed the amphibious nature of Japanese strategy by stating that his armies of land and sea had not properly supported *each other*. This bitter lesson taught the Japanese, not that they should revise their naval theories, but that no time should be wasted or measures overlooked to guarantee success on land. But for the invention of the "turtle-back" the first invasion of Korea must have been successful even though the northern shores of the Yellow Sea had been lined with unbeaten Korean and Chinese warships. Hideyoshi had struck when the political and military situation on the mainland was especially tempting, precisely as it was to be on December 7, 1941.

Subsequently Japan's rulers determined upon absolute isolation from the outer world. The prohibition of intercourse with other nations was implemented by an ordinance forbidding the construction of vessels exceeding seventy-five feet in length. Japan remained a hermit nation until opened to foreign relations by the visits of Commodore Perry in 1853-1854. In the interim naval and military establishments practically vanished, the former having degenerated into a species of coast guard bent upon enforcing isolation. While westerners have speculated as to what Japan might have done in the absence of this withdrawal from the world—and with numerous powerful fleets of seventy-fours—the fact remains that naval thought slept in Nippon for over two hundred years.

Japan retired from international society at a time when galleys were plying the Mediterranean and before the clumsy galleon had developed into the ship-of-the-line and frigate that were to hold sway during the seventeenth, eighteenth and part of the nineteenth centuries. Japan awoke in the days of steam and of iron ships. The awakening was punctuated by trouble with foreign powers and the bombardment of Kagoshima and Shimonoseki. The Japanese lost little time adapting the new instruments of the west to their special needs. The old, dormant naval theories were strengthened rather than weakened by the new developments. If, in the past, the Japanese did not go out of their way to risk the loss of warships, they must now have realized that any costly victory would leave them weak in the presence of powers which might profit thereby. Fighting ships were no longer numbered by the hundreds—Hideyoshi had sailed for Korea with seven hundred. A half dozen battleships or cruisers, a score of torpedo boats, now were imposing forces.

From this point on Japanese naval theorizing can be judged, not only from the strategic distribution of squadrons and their special employment, but also from the involved tactical formations they adopted. On the latter point one is compelled to proceed from the general to the particular, from the strategy to the tactics. Such are the complexities of even the simplest naval engagement that it is often unwise to select individual details of the action and attempt to adduce broad theories therefrom. Heavy seas and poor visibility interfere with the best laid plans, as does a shortage of ammunition. An admiral might break off action temporarily not only to save his ships from punishment but simply to readjust his line or to reappraise the situation. Where the enemy course and formation are confusing, a commanding officer may believe that he is pressing forward when he is actually widening the range. Any movement made under the press of combat might be actuated by these or many other factors without implying that it formed part of a deep-laid scheme. However, when such tactics reappear almost automatically during a number of battles, and when they fit into a general political-strategic picture, they acquire the stature of formal doctrine. It is on this basis that the operations of the navies of modern Japan will be briefly considered.

While the modern Japanese navy adhered to the old principles, it adopted special means to avail itself of the latest weapons as well as to adjust itself to new conditions. The vital dependence of overseas expeditionary forces on their communications made it necessary for Japanese admirals to watch hostile naval forces more closely than Hideyoshi had done. On the other hand, the factor of speed was eagerly adopted by Tokyo because it fitted so perfectly into a set of theories that had become traditional. To the British navy engine speed meant an ability to force an enemy to accept battle. While American technical opinion long favored armor rather than speed in capital ships, adherence to the doctrine of speed as well as armor has found favor in recent years. To the Japanese, however, speed meant an ability to fight at long range or to break off action; it also spelled an ability to force an enemy to conform to Nipponese "flanking" movements—a favorite Nipponese movement in all elements, including, in a sense, the air. Fast ships were also particularly well adapted to rapid fire, a method of gunnery at which the Japanese have always been adept.

Speed made it possible for the Japanese to translate their strategic thinking into the domain of tactics, of battle practice. It meant that the Japanese could strike fast, retire if hard pressed, and limit their risks as the occasion required. In the field of strategy it had been deemed necessary to inflict only that amount of damage which could be counted upon to keep an enemy away from "troopship" waters. Tactically the Japanese objective was so to injure a portion of a hostile fleet as to compel it to turn away from these forbidden waters. Damage of a greater degree was to be inflicted only at a considerable reduction in accepted risk.

In the Sino-Japanese war of 1894-1895 there was displayed that close cooperation between Nipponese armies of the land and sea, the lack of which Hideyoshi had so greatly deplored. The Japanese knew that the Chinese fleet of Admiral Ting did not constitute the menace it represented on paper. Because of inefficiency, if nothing worse, the Chinese heavy ammunition was both insufficient and, in part, defective. The Nipponese rate of fire was four to five times that of their opponents. High explosive shells, which dealt shattering though not fatal blows, wherever they struck, were preferred by the Japanese to armor piercing projectiles which had to strike vital spots in order to be fully effective. The Japanese were presumably more anxious to stop an enemy in the early stages of an engagement than to expend the additional time required to sink him. The longer a battle lasted, the greater was the opportunity accorded the enemy in which to damage or destroy Japanese vessels.

The war began with the torpedoing of the Chinese transport *Kowshing* and the overwhelming of a number of small Chinese warships. Admiral Ting undertook a sortie, but failed to locate his rival Admiral Ito, who was busy escorting troopships. Alarmed by the progress the enemy was making in Korea, and fearing that troops moved by land would arrive too late, the Chinese dispatched a fleet of transports across the Yellow Sea escorted by Admiral Ting's squadrons. The Japanese failed to prevent either this transfer of troops or the return of the empty transports. They did, however, engage the Chinese fleet and at one time had it practically surrounded.

This battle has been well described in a few words by Captain G. Darrieus of the French Navy in his *War on the Sea*:[17]

"Admiral Ito . . . anxious to make up for the disadvantage of the insufficient protection of his ships, and at the same time wishing to profit by his undoubted superiority in speed and gunnery, while he kept his two columns at a distance from the enemy always greater than three thousand meters, followed a very gradually changing course . . . so as little by little to outflank the right of the Chinese squadron. That wing . . . was almost immediately crushed, and the main force of Chinese battleships, turning two points to starboard to come to their assistance, destroyed all regularity in their formation; the fire of some ships became masked by others, and the battle was lost to the Chinese."

Admiral Ito, several of whose ships had received some injury, did not push his advantage. Though a Japanese officer has expressed the opinion that another hour of fighting would have meant the capture of the two biggest units in the Chinese fleet, the latter, severely handled, extricated itself and returned to its base. The Japanese commander reported that as "my Flying Squadron was separated by a great distance from my Main Squadron and considering that sunset was *approaching*, I discontinued the action."[18]

A contemporary authority observed that "Admiral Ito has been blamed for not having destroyed the whole Chinese Fleet."[19] A later critic remarked, "The victory was thus far from Nelson's ideal, annihilation of the enemy."[20] Darrieus admitted that it was "incomplete."[21] These, of course, are western opinions. From the standpoint of Japanese thinking the Chinese, without significant dam-

---

[17] Page 105. This work was translated and published by the U.S. Naval Institute (Annapolis, 1908).
[18] *Brassey's Naval Annual* (London, 1895), p. 107.
[19] Vladimir, *The China-Japan War* (Kansas City, 1905), p. 126.
[20] H. W. Wilson, *Battleships in Action* (London, n.d.), I, p. 105.
[21] *War on the Sea*, p. 106.

age to their opponents, had been prevented from interfering with the movement of troopships across the Yellow Sea. Admiral Ito proceeded to escort the transports heading for Port Arthur and to participate in the bombardment of that Chinese fortress, which was quickly seized.

The principal final action of the war was the reduction of the Chinese base of Wei-hai-Wei, an operation that could have served as a model for the Japanese assault upon Java early in 1942 and for their operations elsewhere in the Indies thereafter. Troops and marines were landed on the flanks of the Chinese positions, which were kept under steady naval bombardment. After several costly failures Japanese destroyers broke into the landlocked harbor to inflict severe damage on Admiral Ting's ships anchored therein, a large unit being sunk by shellfire from a Chinese fort which had been captured by the invaders.

The amount of naval pressure employed by the Japanese in this war was little more than was necessary to guarantee the unimpeded progress of their land forces. To have attempted more would have appeared unwise—as being unnecessary and hazardous, because a ship lost in an unnecessary battle or prolongation of a battle was a ship that could not be risked in the course of a necessary battle, one upon which the fate of the troopships might depend.

The essential differences between this thinking and the orthodox theories of the west could hardly be better shown than by comparing Japan's "minimum-risk" sea fights and employment of naval forces against land objectives with some comments of Britain's Admiral Nelson. In 1796 the latter wrote that he could not "help being more than commonly displeased" at the prospect of using British ships to cooperate with land forces. He stated that his task was to "hunt" for enemy ships "and if I find them in *any* place where there is a *probability* of attacking them, you may depend they shall be either taken or *destroyed* at the *risk of my Squadron*."[22]

The outbreak of war with Russia in 1904 found Viscount Ito, the admiral, in full charge of all operations. The sea forces were under Admiral Togo. Though, six years previously, an American army had not been allowed to proceed to Cuba until the Spanish fleet had been located and destroyed, the Japanese lost no time in dispatching transports to the enemy coast. In the American case, delay meant the elimination of risk to the transports without any benefit accruing to the enemy. Admiral Ito, however, could not afford to give the Russians time in which to grow stronger. A strong but poorly indoctrinated Russian battleship force was at Port Arthur and a small cruiser squadron rode at anchor in Vladivostok waters. In time the Czar could send his Baltic Squadron to the Far East. It seems patent that the Japanese could not afford a victory over the Port Arthur fleet of a type that would leave them too weak to deal with Russian reinforcements. It is also obvious that the seizure of the Russian-dominated coasts in the Far East would automatically eliminate any naval threat to Japan. None the less the Russian squadrons in eastern waters were more dangerous than those of the Chinese had been and so would require

[22] Sir Nicholas Harris Nicolas, *The Dispatches and Letters of Vice Admiral Lord Viscount Nelson* (London, 1846), VII, pp. lx, lviii.

stronger treatment. It should be emphasized, however, that Admiral Togo acted defensively—to protect his country's armies—rather than offensively, with an eye to the enemy naval forces alone.

The latter observation is substantiated by the opening sentences of the official Japanese report dealing with the early naval activities of the war. No such report would have been written in connection with the actions of a Nelson or a Perry, a Drake, a Paul Jones or a Farragut.

"The Japanese combined fleet had successfully made a *threatening* movement against the enemy's fleet, causing them to *abandon* their departure from Port Arthur. . . . Commander-in-Chief Togo, in order to aid the *army's* operations . . . issued orders . . . to the Saien divisions . . . to divert and threaten the enemy in order to aid the third army."[23]

By causing the enemy to "abandon their departure from Port Arthur" a naval battle was avoided, but the transports leaving Japan were safeguarded. The above quotation is particularly significant when read in conjunction with the *French* translation of Admiral Togo's speech previously referred to:

"As soon as hostilities began, and based on your majesty's orders, I considered the topography of the sea and *land* and the location of the *land fighting* and concluded that the principal enemy force was on the side of Port Arthur, so I busied myself in that direction and not towards Vladivostok."[24]

Before a formal declaration of war Admiral Togo steamed in the direction of Port Arthur, dispatching ahead a force of destroyers which found the Russian fleet at anchor and proceeded to torpedo two battleships and a cruiser without, however, sinking them. Togo himself, though a stirring signal was run up on the halyards, advanced cautiously, overcautiously it is insisted in the west, where it is generally conceded that a decisive attack would have resulted in the annihilation of the enemy. When it became apparent that the defender's land batteries might inflict appreciable damage on the attacking ships, the latter were withdrawn. This was the "threatening movement" quoted above.

Thus far in the war Japanese landings had been confined to southern Korea. When the time came to send an army to western Korea strenuous attempts were made to prevent the Russian fleet from leaving Port Arthur—by sinking blockships in the channel. This operation, however, was unsuccessful. Admiral Togo next embarked on an amphibious move by landing marines at a point near the narrow isthmus separating the peninsula on which Port Arthur stands from the mainland to the north. The beachhead was opposite the Elliot Islands in which the Japanese had established an advanced naval base. The islands were joined together by antitorpedo boat booms and the approaches to the landing place were mined. Thanks to the arrival of a Japanese transport fleet the Russian fortress and naval base was cut off from the rest of the Czar's forces in Manchuria. The Japanese furthermore obtained a base whence attacks could be launched against the main enemy armies from the west as well as from Korea

---

[23] *United States Naval Institute Proceedings* (September-October 1914), p. 1283.
[24] *Revue Maritime*, CCI (April-June 1914), p. 339.

to the east. Thus was demonstrated the truism of Japanese naval thought expressed by Vice Admiral Viscount Nagayo Ogasawara:

"As to the tactics of Jimmu Tenno, the making of long detours by sea in order to crush an enemy from the rear, it is identical with that of our modern fleets."[25]

The Japanese who had landed opposite the Elliot Islands marched upon the narrow isthmus against which a small squadron of gunboats was sent. When these failed to arrive on time the land attack was delayed until the next day. Such was the relationship between Tokyo's land and sea operations.

While the major Japanese land movements emanated from a number of points along the Korean-Manchurian coast and were directed northward toward Mukden and ultimate victory, the amphibious moves mentioned above, though conducted on land in so far as the drive on Port Arthur was concerned, were essentially naval as to objective. Blockaded on both land and sea, the Russian base could not be expected to hold out indefinitely. It was the knowledge that the Baltic Fleet was being dispatched to the Far East which caused the attackers to press the siege. Their goal was not so much the fortress as the fleet it sheltered. The Japanese could better afford expending thousands of lives and much military material to secure their objective than to submit their warships to danger. Lieutenant General Sir Ian Hamilton, attached as British observer to the staff of the Japanese First Army, has recorded:

"I had a long talk today with a staff officer, mainly about Port Arthur and the effects of the eleven-inch howitzers. It appears they have hit the ships in harbour several times, and that once General Nogi can sink them some of his army will probably be brought up here."[26]

As the "here" referred to an inland point more than two hundred miles to the north, it can be seen that the Japanese felt they could relax their efforts against Port Arthur once the Russian fleet had been destroyed. In short the objective would have been accomplished though the fortress had not fallen.

Before the latter event occurred, however, the Russian fleet made a sortie the consequences of which were remarkably like the Battle of the Yalu between Admirals Ito and Ting. It was the intention of the Russians to reach Vladivostok and the aim of the Japanese to prevent this "escape." In both cases the underlying thought was military. Tokyo could not look upon a siege of Vladivostok without grave misgivings. The region was highly defensible and Nipponese officers are understood to have shrunk from the task of assaulting the position. Furthermore it had little intrinsic value to Japan. Not only was it commanded by the Japanese, whose islands almost enclosed the Vladivostok salient, but far richer prizes lay to the west—Korea, Manchuria—and eventually China herself. The juncture of the Czar's Port Arthur and Vladivostok naval forces was a matter of little consequence, infinitely less significant than would be a joining of hostile German and Italian naval forces to Britain in the

[25] Ogasawara, *Revue Maritime*, Paris, CXCVIII (July-September 1913), p. 260.
[26] *A Staff Officer's Scrap-Book during the Russo-Japanese War* (New York, 1907), II, 270-271.

present war. Despite occasional Japanese statements to the contrary—and these statements themselves contradict sounder expressions of Nipponese views—the Russian Port Arthur fleet was to be kept at its base so that it could be captured by land forces if it could not be whittled down by a series of engagements at sea wherein the Japanese could accept a minimum amount of risk.

The tactics as well as the strategy of the opposing forces clearly show the Russian disregard of all factors except the basic one of reaching Vladivostok. Their antagonists were actually more interested in thwarting this Russian *plan* than in destroying the Czar's *fleet*. Close action plus rapid fire could be expected to accomplish the latter aim. Although the most efficient long-range gunnery labored under disadvantages, the Japanese overcame this fact by a compromise. They adopted the policy of long-range tactics as a basic formula, to be modified by occasional hit-and-run attacks wherein a maximum of fire was developed.

In the Battle of the Yellow Sea, August 10, 1904, which followed the Russian sortie, Admiral Togo inclined more to long than to short range. As in the case of Admiral Ting's defeat, the Russian line was thrown into confusion, but little advantage was taken of this fact by the enemy. Again the approach of night, and the presence of Russian destroyers, led Togo to break off action. One Russian battleship, badly damaged and much confused, safely reached Kiaochau, the others reentered Port Arthur. "Togo, unlike Nelson, shrank from continuing the action until his enemy was destroyed." Admiral Togo himself reported that the engagement "dealt a severe blow to the enemy's *plans*." Success had been achieved, for the enemy had been prevented from reaching Vladivostok.

Though not a single enemy ship had been sent to the bottom, when the news of this battle reached the land fronts there was great rejoicing. Sir Ian Hamilton wrote: "Headquarters are overjoyed and Sugiura says they expect that Port Arthur will fall in three or four days, and that the fleet will become the spoil of the conquerors."[27] Not only did the Japanese navy refrain from *annihilating an enemy for the sake of annihilation*, but it did not insist that an enemy force was its natural objective simply because it floated. The hostile formation had to move beyond the reach of the Nipponese land arm in order to become the legitimate objective of their naval arm. The advent of the airplane was to bring special significance to this point of view.

While the Battle of the Yellow Sea was in progress the three Russian cruisers stationed at Vladivostok hurried southward in the hope of joining the Port Arthur fleet on its northward move. In the Sea of Japan a few miles to the north of the Straits of Tsushima the squadron ran into a superior force of enemy cruisers under Admiral Kanimura. History repeated itself. The fighting was at long range, and when Kanimura discovered that one Russian ship was stopped and that the other two, badly damaged, were heading back for Vladivostok, he broke off action. British official history has criticized both Togo and Kanimura for breaking the "cardinal rule of warfare that once battle

[27] *Ibid.*, II, 9.

is accepted no effort should be spared to make the result decisive."[28] This, as should now be patent, was not a cardinal rule in the Japanese system of naval warfare.

When the Russian ships had been driven back to their harbors, the Japanese Army proceeded against Port Arthur with great energy. The key to the whole situation was 203-Metre Hill which dominated the harbor, thus permitting observation of the shelling of the fleet it sheltered. After a series of furious infantry assaults had been beaten back with a loss of some ten thousand men, the hill was captured and equipped with a *naval* observation post. Now that the Japanese siege guns could be accurately directed against the Russian ships it was little more than a matter of hours before they were destroyed.

After the fall of Port Arthur the Japanese overhauled their navy, mined Vladivostok against the egress of the few ships it contained, and awaited the arrival of the Russian Baltic fleet—officially known as the Second Pacific Squadron—which was commanded by Admiral Rojdestvensky. The Port Arthur fleet no longer existed, a decisive land campaign had been fought and won, and the journey-worn Baltic fleet could not constitute a serious menace as regards the final outcome of the war. Nelson had pursued the French fleet of Villeneuve across the Atlantic to the West Indies and thence back to Trafalgar. Admiral Togo showed not the slightest inclination to anticipate the Russians by steaming into the Indian Ocean, and there is little doubt that, had Admiral Rojdestvensky changed his mind when in the vicinity of Shanghai, no attempt would have been made to prevent his return to Europe. The Japanese fleet was a dynamic barrier rather than a javelin. In prior operations the enemy fleet had been driven back to its base, and the base had been captured. This time the enemy was to be kept away from his base. As much damage was to be inflicted upon the enemy as would frustrate his plan without weakening the attackers. That this damage resulted in annihilation was due far more to Russian mistakes than to Japanese planning. The latter, having correctly appraised the true strength of the Russians and having provided an economical means of defeating them, was essentially sound—in the eastern rather than in the western sense.

When Togo steamed out to meet the enemy he called for an annihilating victory. This was an exhortation to his men rather than a statement of his objective. The ineffective attempt against the Port Arthur fleet at the opening of the war had been accompanied by a Nelsonian signal. In his post-war address, already referred to, Admiral Togo naturally discussed the decisive battle with Rojdestvensky. His words have been variously translated. The following is the French—and "long" version:

"I had assembled all my forces in the Strait of Korea. I counted upon *harassing* the enemy, but the battle fought so courageously by our men, thanks to the divine aid, obtained the great result."[29]

While Togo was a modest man and intended to praise his subordinates, this

[28] Quoted by H. W. Wilson, *supra*, p. 224.
[29] *Revue Maritime*, CCI, 340.

might have been done in a number of ways without recourse to a word or sentence construction which could in any manner imply the idea of harassing rather than destroying. His activities during the engagement seem quite plainly to show that the "great results" grew out of the phenomena of the battle rather than from the original plan. It must be admitted, however, that a fleet as poorly led and badly trained as that of Admiral Rojdestvensky could hardly hope to survive a defeat in water hundreds of miles away from the nearest friendly or neutral harbor.

Two earlier statements made by Admiral Togo which may be taken as representing part of his mental outlook at the time of the Battle of Tsushima warrant mention.

"In a battle the most important thing is caution. It has often happened in the past that there have been matters for regret after a battle."[30]

This observation can be readily associated with the phases of long-range gunfire and interrupted action which characterized the operations of Admirals Ito, Togo and Kanimura in the past. The other statement was:

"If an unarmored vessel was to silence an enemy warship's fire with its fiercer fire, it would practically be the same as if the vessel were the better armored of the two."[31]

Besides being a commentary upon Japanese interest in rapid fire and gun power that was, shortly before the battle, "the subject of many lectures to the British Admiralty," this remark is particularly applicable to the Battle of Tsushima. At that engagement the Russians had a decided superiority in capital ships. The Japanese filled out their line with heavy cruisers, though one authority has described these as being "battleships in disguise."[32] The Japanese possessed a decided advantage in speed.

The essential difference between the Battle of Tsushima and the engagements of the Yalu and Yellow Sea was that in the latter cases the enemy was being chased, in the former he was merely being repulsed. Chasing tactics had involved the rapid crushing of part of the hostile fleet, with resultant confusion and retreat, pursuit at long range, abandonment of heavy ship action, continuation of the chase by means of destroyers. At Tsushima the Japanese appear to have employed the chasing tactics "in series," each composed of a "run-in" followed by a lengthening of the range during which they tended to place themselves in a better position across the Russian line of advance. Action was broken off at night, which was devoted to destroyer and torpedo boat attacks. In the meantime the Japanese had retired northward before the Russians in order to take up a position between the remnants of the Russian fleet and Vladivostok—there to receive the surrender of what remained of the Czar's forces. Under western theory a fleet that had seized the initiative and secured a decided superiority of fire should press the advantage to complete victory. As repeatedly exemplified, the Japanese followed a different line of reasoning.

[30] Ogasawara, *Life of Togo*, p. 326. [31] *Ibid.*, p. 327.
[32] Archibald S. Hurd, *Naval Efficiency* (London, 1902), p. 106.

Accurately informed as to the enemy's dispositions Admiral Togo[33] left his base in southern Korea and steamed on an easterly course, passing to the north of the Straits of Tsushima. In this way he could deploy in any direction in order to keep himself between the Russians and Vladivostok. The Japanese commander in chief led the First Division, consisting of the heaviest ships in the Nipponese navy. Admiral Kanimura, commanding the Second Division— composed of armored cruisers—was given a good deal of latitude in his movements. As the battle progressed it became apparent that Togo's was the squadron which blocked the Russian advance, Kanimura's the harassing squadron which was permitted to stray away from the sea lane to Vladivostok in order to fall upon the Russian flank. The Third Division, which had for some time scouted to the west of and on a parallel course to the enemy, subsequently concentrated upon the smaller vessels and auxiliaries constituting the rear of the Russian fleet. It thus became the counterpart of Ito's Flying Squadron, a "herder" as much as a "fighter." At the Yalu, because of the proximity of the Chinese base, it had contributed to the failure to destroy the enemy. Because the Russians lacked a base to the rear, such was not the effect at Tsushima.

Shortly after 12 o'clock noon Rojdestvensky, who was passing through the eastern Strait of Tsushima in a northeasterly direction turned more toward the north. Togo, pressing ahead on a southeasterly course, had to turn sharply to the west to prevent the enemy from passing behind him and thus reaching Vladivostok. Between 1:20 and 1:40 p.m. he headed directly for the Baltic fleet, perceived its disposition—an extremely faulty one—and held away to the northwest in order to execute a turning movement which would bring his fleet back across the line of march of the enemy, on an easterly course. This evolution was of a type long deemed perilous, and hence reflects Togo's slight regard for Russian gunnery. The new course and Japanese marksmanship forced Rojdestvensky to swerve to the east.

The Japanese shooting was concentrated upon the van of the enemy fleet. The range was closed, cautiously. In less than thirty minutes of firing—between 2:15 and 2:45 p.m.—interrupted by smoke blowing down the range, the battle was virtually decided. The *Suvorov*, Rojdestvensky's flagship, was forced out of the formation and the *Ossliabya*, fatally injured, was to sink in a matter of minutes. As Togo had by now passed to the east of the Russians, the latter once more pushed north for Vladivostok and again the Japanese turned away to reform on a westerly course to thwart the enemy's plan. The Baltic fleet fell away to the southeast and then to the south, but the Japanese held on to the west, rapidly opening the range. Shortly after 3:24 p.m., "the enemy changed course and, as the range increased, were soon lost in the smoke and haze."[34]

In the next phase of the battle Togo executed another turn to the east, a repetition of the shuttling back and forth across the enemy's route to Vladi-

[33] The account of the battle is largely based on the translation of the official Japanese General Account in the July-August 1914 issue of the United States Naval Institute Proceedings.

[34] *U.S. Naval Institute Proceedings* (July-August 1914), pp. 970-971.

vostok. Both fleets were heading roughly to the east, the Russians being to the south of their opponents. At 4:15 p.m. the Russians are described as being in a "desperate condition." The situation at 4:35 p.m. is interesting from the standpoint of Japanese naval thought, though conditions which are not clear reduced the significance of the brief operation to be described. According to chart no. 8 of the Japanese official account, "The enemy changed course to the *south*, range unknown" (i.e. away from the Japanese). Togo headed *north*. The text of the account reads: "The enemy turning more and more to starboard (i.e. to the *south*), there was some doubt whether they would cross astern of the Japanese ships or escape to northward. In order to *command their course* (i.e. block the path to Vladivostok) the First Division at 4:35 p.m. . . . stood to the *northward*."

None the less, shortly after this change of direction Togo moved south in search of the enemy, two of whose heavy units were discovered at 5 p.m. but not attacked with vigor. At 5:30 p.m. the Japanese turned north, and a "run-in" occurred in which a concentrated fire was poured into the head of the enemy line. The battleship *Alexander III* suffered the fate of the *Ossliabya*, and the *Borodino* blew up. The day action ended in this manner shortly after 7 p.m., when the flagship *Suvorov*, repeatedly the target of Japanese heavy guns and torpedoes, finally sank.

Admiral Togo broke off action for the night, cruising to a point farther north ready to repeat his tactics the next morning. He made the following report: "The combined fleet today met and gave battle to the enemy's fleet . . . sinking at least four of their ships and inflicting serious damage on the rest."[35] This was not annihilation. Had Togo closed the range more frequently, risking the loss of a number of his battleships or armored cruisers, had he changed his plan from one of barring the route to Vladivostok to one of destroying the major portion, if not all, of Rojdestvensky's fleet, the outcome of the day's fighting must have been more impressive. The Japanese commander in chief, however, was willing to turn over the sea to his destroyers and torpedo boats for the night, and to resume his cautious fencing and rapier-like thrusts on the morrow. During the night the Russian armada was assaulted from every direction, not only by the destroyers that had accompanied the Japanese fleet, but by torpedo boats held in reserve in the Tsushima Islands for that purpose. Two Russian dreadnaughts and an armored cruiser were torpedoed and sunk. What remained of the Baltic Fleet, confronted by Togo the following morning, surrendered—less several ships most of which temporarily escaped, three reaching Manila. A cruiser and three destroyers actually got through to Vladivostok.

In their own waters, and at a time when the torpedo did not constitute a daytime menace, American and British admirals would have insisted on complete victory during the main engagement. But Togo had kept the Russians out of Vladivostok—incidentally destroying their fleet—at a cost of three torpedo boats. He then divided his squadrons among "various bays and cruising zones"

[35] Captain (Kichitaro) Togo, *Naval Battles of the Russo-Japanese War* (Tokyo, 1907), p. 82.

to cover the flow of reinforcements to the mainland, including the northeastern Korean approaches to Vladivostok. Peace was signed before it was necessary to operate in the latter direction. Togo, Kanimura, Ito, Hideyoshi and Jimmu Tenno, all appear to have followed the same general type of naval thinking.

III

Between the close of the above hostilities and the First World War the Japanese expressed their naval theories through warship construction. Speed rather than armor was called for and care was taken to make ships withstand the impact of the torpedo. Rapid firing was insisted upon and director control was introduced, permitting the instantaneous discharge of all turret guns. The turrets themselves were fitted with both hydraulic and electric power to reduce the probability of their being immobilized.[36] In later years the Japanese were to be much criticized for overgunning their warships of all types, notably cruisers and destroyers. As the rapidity of fire of the individual gun reached its limit, they evidently tried to increase the rapidity and volume of fire by adding more guns to their ships. Their strategic and tactical thinking also remained unchanged, and, as ever, the general political plan was all-important. Japan's war against Germany in 1914 gave fruitful evidence of the nature of Japanese naval strategy. At the outbreak of the First World War a western thinker would have conjectured that Japan, as a major Pacific Ocean sea power, would have grasped at the opportunity of seeking out and destroying the German Far Eastern squadron of Graf von Spee, both as a matter of national pride and as an obvious, elemental naval move. But, according to demonstrated Japanese theory, it was natural to assume that the Japanese high command would pay scant attention to the German warships, provided they did not interfere with land operations. The capture of the German base at Tsingtau by land assault supported by naval bombardment, and the seizure of the numerous German islands in the Pacific Ocean were events that were to be expected, judging by Japanese naval precedent rather than prevailing western concepts of sea power.

To draw a number of parallels, the hazards incidental to the presence of a German squadron in these waters would be eliminated by the seizure of Tsingtau, just as the reduction of Wei-hai-Wei and Port Arthur had forestalled similar dangers. Landings could be effected in the German colonies, as had been done on Chinese and Russian-dominated coasts. There would be no greater desire to undertake far-reaching cruises in search of the Kaiser's armored cruisers than there had been on the part of Hideyoshi, Ito or Togo.

On the other hand, there was a new and modifying factor which had to be taken into account—Japan's alliance with Great Britain. When war broke out von Spee was already out in the Pacific. Judging from the official German naval history[37] the Japanese must have had a fair idea as to his whereabouts. Tokyo unquestionably had considerable respect for German naval science and

---

[36] "Notes on Japanese Navy," *Revue Maritime*, Paris (July-September 1913).
[37] *Der Kreuzerkrieg in den ausländischen Gewässern*, I.

cannot have relished the idea of risking important units of its fleet in a distant action. It is possible that Japan's delayed entrance into the war was caused in part by the desire to locate von Spee's squadron and the direction in which he was heading.

When Japan finally entered the war, two squadrons were immediately assigned to nearby waters, one to safeguard the troopships moving toward Kiao Chau, the other to blockade Tsingtau and participate in active operations against the fortress. It was not until September and October that squadrons were dispatched to search for von Spee. In one case, and possibly in a number of instances, one of the squadrons weakened itself by sending a detachment to seize the Pacific island of Jaluit. This may have been part of a broader strategy or it may have been reluctance to come to grips with the *Gneisenau* and *Scharnhorst*. The latter theory is strengthened by the fact that von Spee picked up a wireless message issuing from close at hand, signed by the battle cruiser *Kongo*.[38] It was *en clair*, addressed to the Japanese consul at Honolulu, and apprising him of the fact that this vessel (or probably another ship using its name) was steaming for Hawaii. A Japanese officer, in an account of his country's share in the war, stated that the Japanese drove the Germans across the Pacific to the Falkland Islands, where they were destroyed by the British.[39] As von Spee was never sighted during his somewhat slow movement across the Pacific, the driving process can have been accomplished only by the extremely injudicious use of the wireless. The Japanese writer in question failed to note that before the Battle of the Falklands the German squadron had met and overwhelmed a weaker British force. This was a disaster the Japanese seem to have taken pains to avoid. As distinguished from the cold planning of the staff officer, this subordinate Japanese writer seems to have gone to much trouble to explain why his nation's fleet had not been more "active." Refraining from risking losses against an enemy squadron that had removed itself from the zone of military operations was strictly orthodox in the Nipponese sense.

Tsingtau was shelled from the sea and successfully assaulted on the land side by a Japanese army. As usual the navy furnished much assistance. Included in the Japanese fleet were ships designated as coast defense vessels. Of these the official British historian of the naval history of the war has observed: ". . . if we may judge from the use to which they were put it was coastal attack rather than coastal defence for which they were intended."[40] These words were, in effect, a commentary upon the entire field of Japan's naval thought.

The mandate which Tokyo acquired over the former German islands in the Pacific provided its navy with a considerable number of natural "aircraft carriers." Despite Japan's obligation to the League of Nations and the Great Powers, these carriers were promptly and efficiently incorporated into the Imperial naval establishment. A new significance was given to the medieval

[38] *Ibid.*, pp. 125-126.
[39] Commander G. Nakashima, I.J.N., *Brassey's Naval Annual* (1919), p. 65.
[40] Sir Julian S. Corbett, *Naval Operations* (London, 1920), I, 291.

relationship between the man-of-war and the bay or mountain. Little space need be devoted to a consideration of Japanese naval theory as expounded by the delegates to the Washington conference on the limitation of naval armaments, inasmuch as they were usually of the orthodox, western type, and hence self-serving. Only when anxiety was displayed with regard to aircraft carriers did Tokyo give an inkling as to its modernized theories and intentions.

At the Conference, Admiral Baron Kato emphasized Japan's "defensive" purposes, describing them ambiguously as being "necessitated by the Far Eastern situation."[41] He also took care to dismiss lightly and to postpone into some future period considerations of warplane limitations.[42] Article XIX of the treaty, hastily agreed to, provided that "no increases shall be made in the coast defences of the territories and possessions" in specified regions in the Pacific area. Had the western powers attempted to place anything stronger than nominal air forces in their Far Eastern bases Japan would have undoubtedly called attention to Article XIX on the theory that a bomber was a new "weapon" or piece of "equipment." Had the other signatories protested this interpretation, Tokyo might have alleged a material "change of circumstances" and called for a new conference. Spared a naval race, Japan was able to concentrate her resources upon other military preparations and to profit from the peaceful intentions of nations who had no desire to deal in technicalities. The powers who might have defeated Japan in 1922, or compelled her otherwise to renounce aggression, understandably drifted from thoughts of naval disarmament into an attitude of general disarmament.

The Washington Treaty of Naval Limitations appears to have been, therefore, the equivalent of a major naval victory for Japan. It was a guarantee that the susceptibility of potentially hostile ships in regard to torpedo and mine damage could not be reduced, and as the significance of the air arm became fully appreciated, the treaty was assurance that any enemy fleet approaching Japanese "troopship waters" would do so at extreme peril. The airplane came to fit perfectly into the Japanese conception of rapidity of fire and minimum of risk. It was the ideal substitute, under favorable conditions, for the quick-firing gun of the Sino-Japanese War and for the "run-in" movements in the encounters with Russian fleets and squadrons. It was hoped to be an up-to-date version of the "turtle-back" device.

In the undeclared war with China proper which began in 1937 the Japanese made considerable use of the troopship and occasional use of the new naval weapon, the aircraft carrier. Modern warships were not risked in the enterprise, ancient craft being employed to cover landings. The principal reliance was upon the army and land-based aircraft. As already intimated, Japanese writings and utterances were in the defensive, conventional manner. On some occasions the significances of Nipponese statements might have given a hint of things to come, though the defensive clothing of the language was difficult to pierce. An instance was afforded early in 1937 when Vice Admiral Yonai,

[41] Conference on the Limitation of Armaments (Washington, 1922), p. 140.
[42] Ibid., p. 796.

Minister of the Navy, pointed out that "the policy of the navy is to destroy enemy air bases before enemy planes can reach the Japanese mainland."[43] By mentioning the Japanese mainland, attention was directed away from the objectives of a campaign of military aggression.

It will be recalled that Hideyoshi had contrived—by diplomatic means short of war—to maintain valuable beachheads in Korea, and that the Japanese had always made it a point to remove the threat of naval attack by seizing the bases from which it could spring. By securing Chinese ports and coastal airfields or air base sites, the Japanese not only denied their use to the western powers which were vitally interested in the Far East, but secured for themselves bases close to intended fields of action. Though this involved hostilities with China, it was "short of war" in so far as the other nations were concerned. As the Japanese navy was also susceptible to the latest weapons, the newly acquired coastal positions provided it with improved methods of defense. The plea of the necessities of war with China furnished Tokyo with the pretext of entering French Indo-China. This, in turn, supplied the Japanese with a continuous chain of airfields and naval bases from Korea to the borders of Thailand, where intrigue advanced the Nipponese positions to the frontier of Malaya.

The theories pursued by Premier Tojo differed in no essential respect from those followed by his predecessors. Fleet action was generally to be avoided, certainly under circumstances involving that appreciable risk which American and British admirals had always willingly accepted. In 1893 a Japanese emperor had warned that "if one mistake is made in matters of national defense, its consequences may be felt for a century."[44] This basic thought, applied to naval warfare, was as fresh in 1941 as it had been in 1893 or centuries earlier. More expendable means—troopships, armies and air squadrons—were to bear the brunt of attack. Some portion of the fleet was to be used, as usual, against land objectives, but the main battle force was to be withheld as a last resort, and then to engage only under highly favorable circumstances. The only essential change was geographic. In the past Japan had faced west. Once established along the western littoral of the Pacific, Tokyo directed its attention to the east and south, to the islands of the Philippines, Indonesia and Malaya.

War came to the Pacific area on December 7, 1941, by a repetition of the Port Arthur type of peacetime raid. As at Port Arthur the assailants failed to press their attack home by actually invading Hawaii. They may have feared that the arrival of United States aircraft carriers would cut them off from Japan, or elected to husband their own vessels of this type for employment elsewhere. As the principal objective in Port Arthur had been the Russian fleet, so the United States Pacific fleet was the objective in Hawaii. To the extent that Tokyo believed the latter had been put out of action, the desirability of the actual seizure of the islands was diminished. Midway Island was, in a sense, and from an aerial viewpoint, the modern counterpart of 203-Meter Hill at Port Arthur. From safe bases in the mandated groups the Japanese

[43] Associated Press dispatch, Tokyo, February 27, 1937.
[44] Ogasawara, *Revue Maritime*, CC (January-March 1914), p. 123.

could send bombers to a forward and almost equally safe victualing point at Wake Island, whence Pearl Harbor via Midway could be subjected to repeated attack, the bomb in this case replacing the 11-inch howitzer shell. The Japanese therefore seem to have felt justified in sending a powerful maritime force against Midway Island in that shattering type of attack which has characterized Nipponese operations in the past. The attempt was frustrated by the timely arrival of United States aircraft carriers, which presumably evaded or silenced Japanese scouting submarines. The Japanese suffered the rudest shock they had ever experienced since the days of Yi-sun, a greater shock, in fact, because this time no novel weapon was employed by their opponent. It was furthermore discovered by the attackers that they could not break off action in the naval sense so long as their enemy had land or carrier based planes with which to undertake a pursuit.

In the meantime Japanese maritime moves in the Far East had developed rapidly. The initial operations had a distinct naval aspect. It was generally assumed by western theorists that the Japanese would first land in Lingayen Gulf in the Philippines and thence march upon Manila. Instead, they directed their early efforts against northern and eastern Luzon. Within two weeks they were at Davao, on the south coast of that island. By seizing such positions it could be hoped that the passage of United States vessels through or around the Philippine group could be prevented by air and submarine attack.

In strict keeping with Japanese theory was the withdrawal of a small naval force before the guns of the United States Far Eastern fleet off Luzon on the one hand and the successful aerial attack on H.M.S. *Prince of Wales* and *Repulse* off Malaya on the other. This is suggestive of the manner in which Admiral Ito broke off action with the Chinese fleet, only to send destroyers after enemy units off their own base—certain aspects of aerial warfare having been substituted for destroyer action. Tokyo thought of its ships in terms of beachheads seized, not of hostile ships subjected to shellfire. A destroyer heavily laden with marines might take a strategic airfield site which would be more desirable than the sinking of an enemy destroyer, especially as such a sinking would involve the risk of being sunk by the intended victim.

In Malaya the Japanese operations were but an enlarged version of their earlier attacks upon the aborigines of their home islands and upon Korea, Manchuria and the peninsula upon which Port Arthur stands. After escorting troopships to the initial beachheads the Japanese navy proceeded to protect the landing of forces to outflank the defenders already under frontal attack. After establishing naval-air positions around the perimeter of the Philippines, in Malaya and western Borneo, the invaders fanned out into the Netherlands Indies from the north, closing all waters against the arrival of enemy reinforcements. Considerable use was made of aerial scouting to determine whether or not such reinforcements were approaching. Without hesitation the Japanese hurled their troopships against the various objectives, oblivious to losses and disregarding theories which would call for the preliminary establishment of an absolute command, by air and sea, of the beachhead under assault.

Japan's participation in the Second World War brought about a number of changes in that country's naval planning because of the fact that it no longer faced essentially weak enemies. For the distinctly inferior warships of Admirals Ting and Rojdestvensky were substituted the highly efficient vessels of the United States and Great Britain. For the ineffective destroyers of old China and Czarist Russia were substituted United Nations' aircraft, many of which early in the war established decided qualitative superiority over those of Japan. Last but not least Japan was compelled to give much thought to defensive, as distinguished from offensive or defensive-offensive, operations.

Especially in or near enemy-dominated waters the Japanese found it necessary to substitute night for daytime naval attack in order to reduce the aerial risk, a condition that could only be changed by the acquisition of local aerial superiority. Old theories were evidenced in the "run in" attack delivered by Japanese forces against a United Nations cruiser squadron on patrol off Savo Island at the time of the first American landings in the Solomon group. Though successful, the assailants withdrew without pursuing their advantage. A subsequent Nipponese night operation wherein a strong force including capital ships attempted to attack the United States position on Guadalcanal Island, only to be repulsed, showed that night still held the disadvantages feared by both Ito and Togo. The darkness that sheltered the attackers from the aerial arm failed to avail them against the determined action of fast surface ships.

As Japanese theories of naval strategy and tactics are the reflection of geographic and political considerations, it is to be expected that the changes in the latter factors have affected the former. Throughout the course of Nipponese history the navy had been primarily a protective shield for the homeland and for the military transport. The insular territories seized by the Japanese require continuous maritime communications, or provisioning to stand a long siege, supplemented by aerial transportation. The somewhat narrow waterways separating these islands and quasi-islands might be closed to hostile use by a *sufficiency* of air power, both qualitative and quantitative—the latter because of the vast extent of the regions under Japanese dominion. In the final analysis the Imperial navy has been strictly a servicing unit for the Japanese armies. In so far as such servicing might be more economically rendered by other instruments, such as air power, the significance of the sea weapon would diminish as far as the defense of the homeland and conquered territories are concerned, and hence greater risks could be accepted. Japanese sea power would still retain its old significance with respect to coastal maritime moves.

The loss of strategic positions and the presence of strong hostile forces at points near Japan—as well as the fact that Japan's enemies are imbued with an offensive as distinguished from the historic defensive spirit of most of Japan's previous enemies—may well influence Japanese concepts of naval warfare. As no evidence exists of the employment of Japanese naval forces on decisive operations of an oceanic nature, it is not possible to consider the status of Japanese naval thought on that subject. It would represent a revolutionary change in theory from a more or less necessary defensive in the sense previously outlined to an outright offensive, and the latter, in the light of history, would hardly be undertaken so long as the defensive qualities of sea power were required nearer home. Japanese surface ship operations in great strength and far at sea would be some evidence that Tokyo considered that the strategic position on the home front could be adequately defended by airpower plus the strenuous resistance of land forces, the latter being the most expendable commodity in militaristic nations. Where oceanic fighting was involved it is logical to assume that Japanese theory would tend to conform to that of the west in all cases where the enemy was strong at sea. One may speculate, none the less, as to how much tradition, based on the lessons of the past, would continue to color all operations.

The above discussion of Japanese naval thinking as it relates to the Second World War is obviously handicapped by the lack of full and detailed information. For example, beyond the employment of extremely small submarines little can as yet be said about Japanese theories with respect to that weapon generally. While the torpedoing by submarines of damaged warships is suggestive of the torpedo-boat action at Tsushima and elsewhere, it is nothing new in naval warfare and was to be expected in such a vast region as the Pacific.

To summarize the matter briefly, the Japanese navy has been a floating wing of a powerful army occupied with offensive operations in an area militarily far weaker than Europe or North America. It was inevitable that its theories and practices should differ to a considerable degree from those accepted in the United States and Great Britain where the navy was usually a powerful defensive force backed by a relatively small army. It was also inevitable that Japanese thinking should differ from that in France and Germany, where the usual objective was on land and attainable by means of land power alone. There are some significant similarities between Japanese and Italian naval thought because these countries sought to employ both sea power and air power to effect relatively nearby conquests.[45] In the case of Italy the employment and coordination of the various arms were distinctly inferior to Japanese practice.

Japanese naval thinking as explored above is consequently not a revolutionary system comparable with that involved in Napoleon's marshaling of the large and small mass or with the development of the pincer movement. It is simply the direction of politically aggressive thinking to the problem of utilizing sea power in waters and under conditions which differed essentially from those obtaining in the regions where orthodox concepts of sea power were formulated.

[45] Note also the similarity to French naval strategy in the eighteenth century. See Chapter 17 above.

CHAPTER 20. Douhet, Mitchell, Seversky: Theories of Air Warfare

BY EDWARD WARNER

I T IS only in a very limited sense that one can speak with literal accuracy of theories of air power. Obviously any discussion of air power has postulated the existence of flight; but the discussions that have actually taken place have typically postulated not only the existence of aircraft, but also the existence of particular kinds of aircraft, possessing particular characteristics. The ultimate conclusions have depended upon the assumptions adopted with respect to the characteristics of the materiel. Douhet and others have made notable contributions to the theory of the mode of air power's employment; but the great debate of the past twenty years, that has made the names of Douhet and Mitchell household words, has not been concerned with a choice among theories of that type, but with the acceptance or rejection of a fundamental doctrine.

The fundamental doctrine is that the airplane possesses such ubiquity, and such advantages of speed and elevation, as to possess the power of destroying all surface installations and instruments, ashore or afloat, while itself remaining comparatively safe from any effective reprisal from the ground.

If the correctness of that doctrine be conceded, the predominance of the role of air force in the military establishment—and the necessity of planning all campaigns with primary reference to creating the most advantageous conditions for one's own air forces and the most unsatisfactory for the enemy's —obviously and automatically follow. The long-continued controversies over the relation of air force to ground arms in a military organization, and distribution of effort among the various arms, have been incidental to a controversy over the inherent capacity of the airplane, however it may be employed. The difference of opinion between the most zealous adherents of the cult of air power and the most stubborn and extreme of the skeptics has not been a difference about theories of strategical or tactical employment of the available instruments, but about the fundamental power of a particular weapon.

The advocates of the doctrine of the inherent supremacy of air power wrote, on the whole, in terms of prophecy until shortly after the end of the war of 1914-1918. In the month of July 1921, the United States Army Air Corps sank the ex-German battleship *Ostfriesland* in tests off the Virginia Capes, giving the first actual demonstration that gravity-propelled bombs could send a heavily armored vessel to the bottom. Thereafter the prevailing tense of the statement of air power's claims to supremacy changed from the future to the present. But the implications behind airplane supremacy—after a higher degree of technical improvement had been obtained—were well understood before the *Ostfriesland* experiment.

In dealing with aircraft, as with the submarine, the serious students of military art found themselves anticipated by novelists and by the dramatizers of science for popular consumption. Such writers have nothing of present importance to offer to military science; but they are significant as having fully anticipated the doctrine of aerial supremacy already cited as underlying the theories of Douhet and Mitchell, and of their modern successors, Seversky and Ziff. H. G. Wells' *War in the Air* portrayed an aerial invasion leisurely, systematic, and destructive in a degree exceeding even the world's experience of actual destruction during the present war. The Brooklyn Bridge was "broken down" by a single airship in its first passage overhead:

"The City Hall and Court House and the Post Office [of New York] were a heap of blackened ruins after the first day of the enemy's aerial visit."

"So it was that Bert Smallways saw the first fight of the airship and the final fight of . . . the ironclad battleships. . . . Money had to be found for them at any cost,—that was the law of a nation's existence during that strange time, and then cheap things of gas and basketwork made an end of them altogether, smiting out of the sky!"[1]

The first writers of specific studies of military aeronautics, writing between the time when the armies of the great powers began to acquire their first airplanes (1909) and the opening of the first World War, were as apocalyptic in their visions of the future as was the novelist. To cite but a few samples:

"Picture a great capital in the feverish excitement incident upon a declaration of war . . . and then imagine, amidst all this excitement and enthusiasm, a flock of strange bodies appearing suddenly in the sky.

"Airships! What an awful meaning the word would convey to the crowd suddenly struck dumb in their martial rejoicing! Who can describe the horror that would seize their hearts as they helplessly gazed at the strange monsters in the sky? . . . Ere the populace could find words the airships, with a few well-directed shells, would have wrecked the Parliament House, . . ."[2]

"In a critical time before war was declared, an aerial fleet might be massed from 40 to 50 miles from our [British] coast, and on receiving a wireless message could strike within two hours of war being declared! . . . Sheerness, Portsmouth, and Rosyth would all be open to land or sea attack [following air raids], whilst another section of the aerial fleet could make destructive raids on London, the Midlands, . . . and other great commercial ports. . . . The German aerial fleet, by crippling our naval forces at two such points as Sheerness and Portsmouth, would open the way for a German naval raid covering an expeditionary force . . . and it would be the last chapter of the war!"[3]

This was written five years before the first World War began and not many years after the first flight of the Wright brothers. Read in the light of the imperfections and uncertainties of the aircraft of the period, it was bold prophecy. It may have been defective only in marking the author as too far

[1] H. G. Wells, *The War in the Air* (London: George Bell and Sons; New York: The Macmillan Co., 1908), pp. 167-208.
[2] R. P. Hearne, *Aerial Warfare* (London: John Lane, The Bodley Head; New York: John Lane Company, 1909), pp. 136-137.
[3] *Ibid.*, p. 169.

ahead of his time; but it was the sort of prophecy that has laid a great handicap upon more recent writers on the same subject.

A less picturesquely phrased, but more specific and hardly less alarming, example bears a slightly later date:

"A military expert of high repute, speaking of the havoc that a hostile air fleet might work by an attack upon the Thames valley between Hammersmith and Gravesend, has observed: 'This whole 50 miles of concentrated essence of Empire lies at the absolute mercy of an aerial machine, which could plant a dozen incendiary missiles in certain pre-selected spots.' It was only the other day that a famous constructor showed how . . . it would be possible for an enemy to drop a couple of hundred tons of explosive matter upon London. . . . What such an aerial attack as this would mean has been pictured by Lord Montagu of Beaulieu. Suppose London was thus assailed from the air, at the beginning of a war, he says: 'What would the results be? Imagine the Stock Exchange, the chief banks, the great railway stations, and our means of communication destroyed.' Such a blow at the very heart of the Empire, declares Lord Montagu: 'would be like paralyzing the nerves of a strong man with a soporific before he had to fight for his life; the muscular force would remain but the brains would be powerless to direct.' "[4]

I

The advocacy of air power as the predominant instrument of war gained in force when it came from men who had been subjected to the tests of actual warfare. The war of 1914-1918 brought to the fore the two men who remained the leading protagonists for the doctrine of the supremacy of aircraft for a dozen years thereafter, and whose writings played a large part in the evolution from a simple faith in that doctrine to its use as the basis for theories of tactical employment of forces and the selection of objectives.

There were striking parallels between the lives of Giulio Douhet and William Mitchell. Both entered the army in youth, long before the Wright brothers had made their first flight. Both possessed an imagination which made them seek employment in connection with the promotion of other mechanical novelties in the military service before they were attracted to aviation. Both became hotly critical of the military leadership of their time and suffered punishment in military courts for the form of their criticism and the manner of its expression. Both were fluent and appealing writers,[5] Mitchell addressing himself, in his published writings, primarily to persuading the public, while Douhet wrote more specifically for a professional military audience.

Giulio Douhet was born in 1869 and died in 1930. He entered the army as an artillery officer, a product of the regular Italian course for commissioned rank. He became an early advocate of the development of motor transport for the service of armies; he further displayed his scientific bent by conducting re-

[4] Claude Grahame-White and Harry Harper, *The Aeroplane in War* (London: T. Werner Laurie, 1912), pp. 208-209.
[5] See bibliography for titles.

search on gases at low temperatures; and he first wrote on the importance of air power in 1909. In 1915 he had already conceived the image of total war, and of that shattering of civilian morale by air attack which played a large part in the later evolution of his thought. He was also advocating the "destruction of nations" from the air as a military measure. At the end of 1916, after he had sent to a member of the Italian cabinet certain memoranda highly critical of the existing policy of the Italian staff, he was court-martialed and sentenced to a year of imprisonment. The decision of the court-martial was formally repudiated and expunged in 1920. In the meantime he had been recalled to service, in February 1918, and placed at the head of the Central Aeronautical Bureau. He attained the rank of general in 1921, the year from which his first serious writings on air power date. He was designated commissioner of aviation immediately after the fascist march on Rome, but he withdrew from the government to concentrate on the literary advocacy of his views concerning policy.

William Mitchell was born ten years later than Douhet, and lived six years longer than his Italian colleague. He enlisted in the infantry in 1898 and was commissioned in the Signal Corps within a very short time. He served in Alaska as a young officer, where he built a considerable section of the Alaskan telegraph line, and his realization of the immense strategical importance of that territory, and of subarctic territory in general, played a great part in the later development of his thought on military science. He was active in the early application of radio and motor transport in the army, and upon the entry of the United States into the first World War in 1917 he turned to a more recent application of science, and transferred to the air service. He learned to fly in 1916, was sent to Europe as an observer just before America entered the war, and rose steadily to the highest command over American air service operations, which he headed during the closing weeks of the war. After a postwar tour of the major European States to study the status of aviation in allied and enemy countries, he was appointed assistant chief of air service, with the rank of brigadier general, a post which he held from 1921 to 1925. His zealous advocacy of a unified air force, separate from the army and navy, and his vehement criticisms of the policies then being pursued by the war and navy departments finally culminated in a statement which charged the war and navy departments with "incompetence, criminal negligence, and almost treasonable administration of the national defense" and asserted that officers "and agents sent by the War and Navy Departments to Congress have almost always given incomplete, misleading, or false information about aeronautics." Court-martialed in the autumn of 1925, he was found guilty and sentenced to suspension of his rank for five years. He resigned from the army on February 1, 1926. He thereafter devoted much of his time during the remaining ten years of his life to lecturing, writing, and appearing before committees to plead the cause of the unified air force and to denounce the existing direction of both military and civil aeronautics in the United States, as well as American aeronautical research and the American aircraft industry, which he held responsible for retarding technical development.

General Mitchell had been the leader of the campaign for a larger recognition of air power in American military organization for six years before his resignation from the army. His name continued to be, and still remains, the symbol of that campaign.

II

Air power had claimed Douhet's attention as early as 1909. Even at that time he had foreseen its revolutionary significance for military strategy; but his views were first developed in some detail (although still in a very preliminary and limited form) in his first book in 1921.[6] They received a more comprehensive expression, and in respect of general conclusions a final one, in a revised edition of the same work six years later.[7] Douhet was also the author of a long series of articles in various military and aeronautical periodicals, the best known and most often reprinted of this series being a fictional portrayal of an imaginary future war between Germany and France.[8]

Although the "Douhet theory" quickly became the subject of frequent glib reference, it was only after the author's death that translation made his own exposition of it directly available to large audiences outside of Italy. A substantial part of his book was printed in French in 1932;[9] and that in turn was translated into English and put in mimeographed form for officers of the United States Army Air Corps in 1933. It was also published in an abstracted form in a British military periodical.[10] A German translation appeared in 1935; and at the end of 1942 a complete translation of Douhet's principal military writings was first made generally available to an English-reading public.[11] Douhet's own works have been usefully supplemented by a number of critical studies, reviewing not only the content of his own writings but also the course taken by his numerous exchanges with his opponents.[12]

The major assumptions underlying the Douhet theory are that:

(1) Aircraft are instruments of offense of incomparable potentialities, against which no effective defense can be foreseen;

(2) Civilian morale will be shattered by bombardment of centers of population.

Upon that foundation he reared the theory of which the basic elements are:

[6] Giulio Douhet, *Il Dominio dell' Aria; saggio sul' arte della guerra aerea* (Roma: Stab. Poligr. per l'Amministrazione della guerra, 1921).
[7] Douhet, *Il Dominio dell' Aria* (2nd ed.; Roma: Instituto Nazionale Fascista di Cultura, 1927).
[8] Douhet, "La Guerra del' 19—," *Rivista Aeronautica* (March 1930), pp. 409-502.
[9] Douhet, "La Guerre de l'Air," *Les Ailes*, Paris, 1932.
[10] *Royal Air Force Quarterly* (April 1936), p. 152.
[11] Douhet, *The Command of the Air*, as translated by Dino Ferrari (New York: Coward-McCann, Inc., 1942). In all subsequent references to Douhet's publications, so far as they are included in "The Command of the Air," specific reference has been made to this translation rather than to the Italian originals, as it is presumed that it will be available to a larger number of English-speaking readers.
[12] Col. P. Vauthier, *La Doctrine de Guerre du General Douhet* (Paris: Berger-Levrault, 1935); H. de Watteville, "Armies of the Air," *The Nineteenth Century and After* (October 1934), pp. 353-368; N. N. Golovine, "Air Strategy," *Royal Air Force Quarterly* (April 1936), p. 169.

(a) "In order to assure an adequate national defense, it is necessary—and sufficient—to be in a position in case of war to conquer the command of the air."[13]

(b) The primary objectives of aerial attack should not be the military installations, but industries and centers of population remote from the contact of the surface armies.

(c) An enemy air force, in particular, should not be dealt with by combat in the air but primarily by destruction of the ground installations and of the factories from which its supplies of materiel come.

(d) The role of surface forces should be a defensive one, designed to hold a front and to prevent an enemy advance along the surface and in particular an enemy seizure by surface action of one's own communications, industries, and air force establishments, while the development of one's own aerial offensive is proceeding with its paralysis of the enemy's capacity to maintain an army and the enemy people's will to endure.

(e) In the interest of the most economical application of total effort, the use of specialized fighting aircraft for defense against enemy bombers should be foregone. The basic type of air force equipment should be a "battle plane," which conducts bombardment and is at the same time self-defending, or can alternatively be used solely for combat purposes.

Of the correctness of the first of these conclusions there can now be little doubt, if "command of the air" be interpreted in a very strict sense. If a belligerent state is able to attack its enemy from the air at will, and if all defense against such attack has been liquidated, the victory of the state possessing the free use of air power over the state having no such power and no static defense against air attack is inevitable. The conclusion is sound; but difficulty arises in attaining that degree of command of the air, which is a much more difficult process than Douhet had foreseen. Enemy defenses, both aerial and anti-aircraft, cling stubbornly to life and to a degree of effectiveness at least sufficient to impose substantial restrictions on the freedom of use of attacking air power.

It was in his insistence upon the second of his basic assumptions, the preference of the industrial over the military objective, that Douhet most closely anticipated future evolution. The idea was not a new one. Books appearing even before 1914, and foretelling annihilation of industry and of centers of financial control from the air, have already been quoted here; but it was Douhet who first made the industrial objective for air attack the center of a complete body of military doctrine, developed and proclaimed and promoted over a long period of time.

The importance of the industrial objective and the enemy air station as points of attack has grown with the passing of time; but the full extension of Douhet's theory of choice of objectives, based as it was on his assumption of the comparatively fragile quality of civilian morale, has weathered the years less successfully.

[13] *The Command of the Air*, p. 28.

Somewhere, at last, the limit of human endurance can be reached; but the voices of the surviving members of unconquered populations from Chungking to Coventry raise a chorus of denial that it is to be reached with any such ease as Douhet had foreseen. It must be said in Douhet's support that he was considering the consequences of the tremendous concentration of attack against a virtually undefended population which would follow on the attainment of "command of the air"; but, even so, he appears clearly to underestimate the toughness with which bombardment is endured.

"At this point I want to stress one aspect of the problem—namely, that the effect of such aerial offensives upon morale may well have more influence upon the conduct of the war than their material effects. For example, take the center of a large city and imagine what would happen among the civilian population during a single attack by a single bombing unit. For my part, I have no doubt that its impact upon the people would be terrible. . . .

"What could happen to a single city in a single day could also happen to ten, twenty, fifty cities. And, since news travels fast, even without telegraph, telephone, or radio, what, I ask you, would be the effect upon civilians of other cities, not yet stricken but equally subject to bombing attacks? What civil or military authority could keep order, public services functioning, and production going under such a threat? And even if a semblance of order was maintained and some work done, would not the sight of a single enemy plane be enough to stampede the population into panic? In short, normal life would be impossible in this constant nightmare of imminent death and destruction. And if on the second day another ten, twenty, or fifty cities were bombed, who could keep all those lost, panic-stricken people from fleeing to the open countryside to escape this terror from the air?

"A complete breakdown of the social structure cannot but take place in a country subjected to this kind of merciless pounding from the air. The time would soon come when, to put an end to horror and suffering, the people themselves, driven by the instinct of self-preservation, would rise up and demand an end to the war—this before their army and navy had time to mobilize at all!"[14]

Douhet's conception of destruction from the air, whether of industries or of cities, was worked out with mathematical nicety. Tacitly assuming a perfectly uniform bombing pattern, 20 tons of bombs would suffice "for the complete destruction of everything in a circle 500 meters in diameter." This would amount to the attainment of "complete destruction" with an expenditure of 250 tons of bombs per square mile of territory devastated. "What," Douhet asks, "would happen in a large city like London if, in the central part of the city, there are completely destroyed one, two, or four areas of 500 meters in diameter?" He assumes that with a fleet of 1000 bombing airplanes, 50 such areas could be destroyed each day. The notable discrepancy between this prediction and the actual performance of 1940-1941, or between the Douhet

[14] *Command of the Air*, pp. 57-58.

prediction and the weight of bombs that has actually been used by the Royal Air Force in its raids on Germany to secure much smaller effects than Douhet had portrayed, is due mainly to his underestimation of the extent to which the explosive force of bombs dropped among solidly built structures is dispersed in the free air and to an overoptimistic anticipation of the perfection with which the bombs can be laid in a pattern that will avoid any waste by duplicated assaults on areas already disposed of. Where light buildings such as small residences are concerned, the Douhet estimate that a one-ton bomb would destroy (or at least render unfit for immediate further use) an area 375 feet in diameter has been approximately sustained by experience; but the radius of devastation of bombs dropped in city streets among stoutly built structures of brick or concrete, and especially among modern steel-framed buildings, has been much less. Even among light structures, the realization of Douhet's computation would have required that bombs be placed in a uniformly spaced checkerboard.

Douhet was correct in his anticipation of the major role to be played by the incendiary bomb; his prediction of large and effective use of gas bombs has not, up to this time, been borne out in practice. In this connection, it should be noted that Douhet gave little weight to talk of limitations on such weapons. In words which may reasonably be taken to represent the views of the author, a character in Douhet's fictional account of a future war says: "In the face of instinctive self-interest, of national survival, every convention loses its value, every humanitarian sentiment loses its weight. The only principle to be considered is the necessity of killing to avoid being killed."[15]

Although Douhet's writings give surface forces an explicit function in defensively holding a surface line, either behind permanent fortifications or in entrenchments, and although a corresponding defensive role was contemplated for naval forces, the implication is strong that with a proper development in the use of air power the progress of events would be so swift that little opposition would be needed to delay an enemy surface force long enough to keep it from doing any harm before the issue had been otherwise decided. In his prospectus of a future war[16] Douhet pictures France, with four of her cities having been "transformed into flaming masses" by an hour's bombing, clamoring for peace within 36 hours after the first belligerent act. Douhet, more than any other writer, inaugurated the concept of victory by swift obliteration which thereafter rose to an apocalyptic climax of expression in the visions of lay authors,[17] and to a more sober and expert anticipation that the war which began in 1939 would progress to a conclusive determination within a few days' time after fighting had actually started.[18]

Although Douhet had been a technician and a scientist, he showed but little knowledge of the problems of aeronautical engineering. Like many other writ-

[15] *Ibid.*, p. 309.
[16] Douhet, "The War of 19—," as it appears in *The Command of the Air*, pp. 374-389.
[17] See, for example, Stuart Chase, "The Two-Hour War," *Men and Machines* (New York: The Macmillan Co., 1929), p. 307.
[18] See, *inter alia*, Alford J. Williams, quoted in *Time* (October 23, 1939), p. 32.

ers on the subject he saw the airplane as a far cheaper and simpler instrument than it actually is, and one far more readily subject to changes of function by simple shifts of equipment in the field than is actually the case. His supersession of the specialized fighting craft by the "battle plane," which has already been mentioned as one of the features of his theory, was derived partly from his belief that aircraft could be so designed that either bombs or fuel could easily be substituted for offensive or defensive armament. Such a substitution on a really major scale would have been far from easy at the time when he wrote, and has become more and more difficult and, indeed, impracticable with the increasing specialization of type and mechanical refinement since that time. He did not entirely disdain the fighter; but he concluded that its characteristics should be essentially the same as those of the bomber, and that it would be an uneconomical development of force to build machines which had no function except that of combat.

During the years before 1939, Douhet's ideas on the prospective waning of the importance of fighting aircraft were shared by a certain number of officers with practical experience in operations. Their reasons were not, however, the same as Douhet's, and three years of actual experience in war have left the single-seated fighter still clearly in the front rank of importance. The fact that American bombing formations in western Europe were able, in the latter part of 1942 and in 1943, to pass through heavy fighter attacks without suffering prohibitive losses might afford some measure of support to Douhet's beliefs; but the experience gained up to the time of writing this chapter is still inconclusive in that respect, and the most enthusiastic of the bomber's advocates in 1943 would scarcely follow Douhet in proclaiming the fighting plane's futility in general terms.

But Douhet went still farther with his idea of universality of function, and conceived an intermingling of military and civil employments. "If we examine carefully the functional characteristics of bombing and combat planes as I have tried to define them, we can readily see that they are in general almost identical with the functional characteristics of civil aviation. When all is said and done, the bombing plane is essentially a transport plane of medium speed and sufficient radius of action, especially equipped to carry bombs. . . . By mutual understanding between military and civil aviation, civilian planes could be turned into military planes in case of need. This in turn implies that, with the strides being made in civil aviation, an Independent Air Force can rely for many of its needs and much of its equipment upon civilian progress in addition to military progress."[19] He recognized that such machines would be ideal for neither military nor civil purposes, but continued: "It will always be necessary to compromise between the two extremes. War is made with masses, and the masses are composed of averages. An air force has need of aircraft with average characteristics similar to those of civil aircraft."

It should be recorded that after 1927 Douhet's confidence in the direct com-

[19] *The Command of the Air*, pp. 47-48.

bat utility of the civil transport aircraft appeared to be fading, and such machines were conceived in his later writings as a reserve for secondary functions.

The statement that war is made with masses, and the intimation that the use of converted civil aircraft might be necessary in the interest of economy, are somewhat at variance with Douhet's customary excess of optimism on the economy of air operations. Thus:

"An air fleet capable of dropping hundreds and hundreds of tons of bombs is easy to construct.

"These projectiles [bombs] require neither special metals nor very exact work in manufacturing.

"An air force adequate to gain the mastery of the air, especially in the first period of the conflict, only requires limited weapons, a limited personnel, and small financial resources. This force can be organized without attracting the attention of probable adversaries."

In Douhet's picture of the war of the future[20] the Germans are portrayed as having only 1500 bombers. Of this number, only 100 are heavy bombers, that is, bombers substantially equal in weight and power to such heavy bombers of 1942 as the American Fortress and Liberator and the British Lancaster. In the first day of the imaginary war, one-third of that total fleet is described as having been lost, but in the same time the Germans succeed in wiping out most of the French fighter force sent against them, and are left with the future course of events entirely at their own command.

The anticipated loss of a third of the total of the victorious air fleets in the first day of battle is a characteristic note. Douhet was no believer in reserves. All air power should be thrown into the balance at the outset of war. To hold anything back was a symptom of the defensive policy which he abhorred as both more expensive and less effective than bold and unlimited offensives for immediate destruction of the enemy's bases and resources.

In his emphasis upon the value of the offensive, he brushed defensive measures aside without regard to the possibilities that they, too, might undergo technical improvement. In particular, he failed to foresee the extraordinary development of radiolocation of enemy aircraft, and he assumed that for fighters to be able to intercept enemy bombers would be very much a matter of luck. He dismissed antiaircraft fire with the observation that: "The employment of artillery against airplanes resolves itself only into a useless waste of energy and resources." He thought but little better of air defense, and cited as a historical precedent the fact that "every time that an air attack was carried out vigorously in 1915-18, it reached its goal."

The worst of all of Douhet's failures in dealing with technical development, whether judged by the standards of the specifically aeronautical tacticians of his own time or by the accumulated experience of the succeeding fifteen years, was in his belittling of the significance of speed in military aircraft. Perhaps impressed by the naval analogy, he fixed his attention primarily upon strength.

[20] *The Command of the Air*, p. 337.

He does not completely neglect speed; he mentions it frequently as a desirable quality; but over and over again he comes back to the conclusion that it is of very inferior significance as compared with other elements of performance. "Whether bombers or combat planes, they need no more than a medium speed. No emphasis need be placed upon speed; it is of little importance that technical advances may soon produce bombing or combat planes which, while retaining other basic characteristics unaltered, will have a speed of 10 to 20 more miles per hour. . . ."[21] He was right in anticipating the importance of the heavy bomber; he was right, and at a time when he had very few sympathizers, in anticipating the resumption of the use of armor on aircraft; but he was completely wrong, on every showing up to this time, in his comparative indifference to speed, which has in fact remained one of the most vital characteristics of every military type.

In the matter of speed, the error was in the appraisal of values; but in other cases Douhet went astray even to the extent of gross error on technical facts. He believed, for example, that the minimum power required to maintain an airplane in flight would decrease with increase of altitude; that the speed of an airplane would be doubled by so increasing the altitude as to halve the air density[22] (actually the increase of speed resulting from such a change of altitude is not 100 per cent, but something under 25 per cent); and that very large aircraft "probably could not land or take off except on liquid surfaces. We may have to build artificial lakes for their landing."

Misinformed as he was on some of these matters, however, Douhet displayed a caution in technical prophecy that leaves his books virtually free of the exaggerated promises of the existing or immediately prospective technical capacities of aircraft that have disfigured a great many of the polemics of other writers on the subject of air power.

In addition to his specific conclusions as to how a campaign should be conducted so as to make use of the overwhelming power of an air force, Douhet had ideas on military organization and general policy. He was an unremitting advocate of the unification of military and naval and air organization. Within a few years after 1918 he was talking of "total war." In 1921 he wrote: "The prevailing forms of social organization have given war a character of national totality—that is, the entire population and all the resources of a nation are sucked into the maw of war. And, since society is now definitely evolving along this line, it is within the power of human foresight to see now that future wars will be total in character and scope."[23]

"There are theories of land war, naval war, air war. These theories exist, evolve, develop; but a theory of *War* is almost unknown." He advocated the organization of national defense through a central ministry, and in 1927 he had the satisfaction of seeing such a ministry created in Rome.

The influence of Douhet's writings has been far-reaching. Much of what he said has been supported and documented by experience and by later writers. Much, on the other hand, has proven either overoptimistic or specifically erro-

---

[21] *The Command of the Air*, pp. 47-48.   [22] *The Command of the Air*, p. 67.
[23] *The Command of the Air*, pp. 5-6.

neous. Some of his beliefs have proved unsound during the present war; yet the trend of development has in most respects, aside from those relating to airplane design, been in the direction that Douhet foresaw. He went far astray in his specifications of the characteristics of the aircraft that would be used in future wars; but his judgment of the ways in which they should be used is more nearly valid in 1943 than it could have been in the conditions of his own time, and further development in aircraft performance is likely to bring the practice of the organizers and commanders of military effort still closer to the practices that Douhet advocated. The intense concentration of bombardment on a small area, the selection of industrial objectives, and the building of large bombers stoutly armed for their own defense (even though they do not follow Douhet to the extent of the complete abandonment of fighter protection) all seem obvious enough now; but it was Douhet who gave those ideas their most effective sponsorship during the experimental years of the 'twenties and 'thirties. The standing that Douhet's studies gained among serious military students, even in his own time, is suggested by the fact that an officer so distinguished as Colonel P. Vauthier found time to prepare an exhaustive critique of his work[24] and also by the phrasing of the introduction to that volume (dated June 7, 1934):

"The study of Douhet is an inexhaustible source for reflection. The notable doctrine that he has established may have a decisive influence on coming events. Conventional in his initial assumptions and in his methods, he shows the conclusions to which they lead. Let us take care not to treat lightly, as a Utopian dreamer, a man who may later be regarded as a Prophet."

The phrases just quoted came from no young zealot. They bear the signature of Marshal Pétain.

Bombing aircraft are not immune from reprisal either from the ground or in the air, but they are relatively less vulnerable than they were 20 years ago; and the basic assumption of the invincibility of the airplane and its inherent superiority in power and economy to all other weapons, which found little to justify it when applied to the aircraft of Douhet's own day, is much closer to the mark now. With another 20 years of development in aeronautical science it will be closer still. The men responsible for the organization of national security in Douhet's day had to depend for the time being upon the weapons that currently existed. In that sense Douhet's theories were drawn for the future, and were much less applicable with the means existing in his own time than they have subsequently become.

Even students of air power in Douhet's time took exception to the military assumptions underlying his conclusions, and to the conclusions themselves as constituting an adequate guide to air operations. General Golovine,[25] among others, challenged Douhet's overestimation of the destructive effect of bombing on civilian morale. Experience since 1939 has supported Douhet's critics on these points, rather than Douhet himself. On general principle, time works

[24] *La Doctrine de Guerre du General Douhet* (Paris, 1935).
[25] N. N. Golovine, "Air Strategy," *Royal Air Force Quarterly* (April 1936), p. 169.

with Douhet; but his computations of the amount of effort required to produce a given effect were characteristically oversanguine, as those of other advocates of the supremacy of air power very often have been.

Douhet's theory became the proclaimed lodestone of Italian policy. In no other country of Europe was it fully adopted. Germany, whose air force held Europe terrorized from 1938 until the Battle of Britain disproved the legend of its invincibility, followed Douhet in concentrating upon the destruction of enemy air forces on the ground and the wiping out of enemy bases; but departed in maintaining constant collaboration between air and surface forces, rather than relegating surface forces to an auxiliary defensive line, and also in using light bombers rather than heavy ones. Where the Luftwaffe came closest to the Douhet theory, over England in 1940, it failed—although certainly the attack could have succeeded if the Germans had had the means to conduct it on a very much larger scale than they did, and maintain it over a much longer period.

It is not within the proper scope of this study to argue the question of whether the evolution of the airplane and its armament has yet reached the stage where wars can be successfully conducted by the Douhet method. Would the Germans have done better, from their own point of view, to build a still larger aircraft industry and a still larger air force, devoting to airplanes most of the material and manpower that actually went into tanks and artillery and submarines and surface ships, transferring part of their ground troops and naval personnel to the Luftwaffe and putting most of the rest into the aircraft factories? Would it be wise policy for us to follow the same course at the present time? The point is hotly debated; but at least it is certain that the case for such a unique concentration of effort upon air power can be stronger in 1943 than it could have been ten or twenty years ago, and it is altogether probable that the passage of another decade will make it stronger still.

III

General Mitchell's activities were contemporaneous and roughly coextensive with Douhet's. They had much in common; and the conventional assignment of Douhet's name rather than Mitchell's to much of their common belief is due in part to chance, in part to Douhet's clearer and more systematic literary development of his conclusions, and partly to the livelier interest in military studies in Europe, as compared with the United States.

There was, however, between the two men an enormous difference in temperament. The difference is of some importance, for the temper in which controversies over the place of air power have been conducted have been the reflection of the tactics of those who were most actively engaged on both sides of the argument. Mitchell wrote and spoke as an intense partisan, becoming more and more impatient of opposition and increasingly disposed to denounce it as stupidly reactionary, blinded by self-interest, or dishonest. His opponents replied in kind, and the controversy grew bitter. Douhet maintained the attitude of a student and a dispassionate seeker of the truth, even in actual controversial exchanges. "It is not for me," he wrote, "nor for General Bastico, nor for anyone else, to assign the predominant role in war to a particular arm. If indeed any one arm has an overshadowing importance, it is not by any human desire, but because the discoverable facts so indicate. If the aero-chemical arm is to be decisive in future war, it will not be my doing, and I shall deserve neither praise nor blame."[26]

Mitchell shared in some degree, although without Douhet's complete confidence, the conviction of the prime efficiency of attack on the enemy's economic and industrial structure. He shared the belief in the comparative fragility of civilian morale. Like Douhet, he believed in the possibility of paralyzing civilian and industrial activities through a relatively modest volume of bombardment. Speaking of a possible enemy attack upon the major cities of the United States, he says: "It is unnecessary that these cities be destroyed, in the sense that every house be levelled with the ground. It will be sufficient to have the civilian population driven out so that they cannot carry on their usual vocations. A few gas bombs will do that."[27]

Several thousand tons of bombs fell on London in 1940 and 1941—yet the life of the city went on throughout, and only a very minor fraction of its buildings became unusable.

"In future the mere threat of bombing a town by an air force will cause it to be evacuated, and all work in factories to be stopped. To gain a lasting victory in war, the hostile nation's power to make war must be destroyed,—this means the factories, the means of communication, the food producers, even the farms, the fuel and oil supplies, and the places where people live and carry on their daily lives. Aircraft operating in the heart of an enemy's country will accomplish this object in an incredibly short space of time."[28]

"The advent of air power, which can go straight to the vital centers and either neutralize or destroy them, has put a completely new complexion on the old system of making war. It is now realized that the hostile main army in the field is a false objective, and the real objectives are the vital centers. . . . The result of warfare by air will be to bring about quick decisions. Superior air power will cause such havoc or the threat of such havoc in the opposing country that a long-drawn-out campaign will be impossible."[29]

In Mitchell's first writings after 1918 he gave much attention to the collaboration of air with surface forces; but as time passed surface forces receded more and more completely into a secondary position in his estimation, and his confidence in the technical capacities of the airplane grew. He continued, however, to attach a great importance to the use of air power for the destruction of enemy surface forces. In this respect he differed from Douhet, who was

---

[26] Vauthier, *op. cit.*, p. 121, quoting a letter from General Douhet. See also *The Command of the Air*, pp. 251-262.

[27] William Mitchell, *Skyways—A Book on Modern Aeronautics* (Philadelphia and London: J. B. Lippincott Co., 1930), p. 262.

[28] Mitchell, *Winged Defense* (New York and London: G. P. Putnam's Sons, 1925), pp. 126-127.

[29] *Skyways*, pp. 255-256.

willing virtually to ignore the surface forces while destroying the nation and the resources behind them. The difference was partly a reflection of the difference in nationality and in the geographical outlook of the two men.

In particular, the ability of aircraft to obliterate every sort of surface vessel, leaving surface vessels no military function whatever, became a veritable article of faith with Mitchell.

"If this projectile [a 2,000-lb. air bomb] hits in the vicinity of a ship within a couple of hundred feet, the underwater mining effect is so great that it will cave in the bottom of the ship, causing it to sink."[30] Many ships have been sunk by attack from the air during the present war; but with much less facility than that.

"It is probable that future wars again will be conducted by a special class, the air force, as it was by the armored knights in the Middle Ages. Again, the whole population will not have to be called out in the event of a national emergency, but only enough of it to man the machines that are the most potent in national defense."[31]

As a military pilot personally experienced in the actual command of air forces in battle, General Mitchell was much more intimately acquainted than Douhet with the tactical problems. He employed his inventive talent to the full in the improvement of existing tactical methods. One of his most strikingly farsighted proposals took the form of a plan for using parachute troops behind the enemy lines in 1918.

Mitchell avoided the Douhet error of assuming an all-purpose airplane. Fighting aircraft retained a large place in his scheme of war, wherever conditions would bring hostile air forces in direct contact with one another.

"It was proved in the European war," he wrote, "that the only effective defense against aerial attack is to whip the enemy's air forces in air battles."[32]

In all matters relating to the technical characteristics of aircraft and to the detail of their operation, Mitchell was incomparably more expert than Douhet. The intensity of his enthusiasm, however, led him on occasion greatly to overestimate the rate at which technical progress would be made in the immediate future, and even to exaggerate the immediately realizable possibilities as of the time when he wrote. He wrote, for example: "I can say now, definitely, that we can encircle the globe in a very short time on a single charge of gasoline."[33] The prediction is now seventeen years old, but it is still far short of being realized.

The most important of the differences between the two men as contributors to military thought, however, was the difference in geographical outlook.

Douhet wrote as an Italian, and he tested his theories by applying them to Italy, a nation with its principal potential enemies situated within a short distance by air, and with its land frontiers defended by mountain barriers against rapid development of a surface attack.

"Naturally, my first thought is of our own situation and the eventuality of

---

[30] *Ibid.*, p. 267.   [31] *Winged Defense*, p. 19.   [32] *Ibid.*, p. 199.
[33] *Ibid.*, p. 139.

a possible conflict between Italy and some one of her possible enemies. I admit that the theories I expound have that in the background, and therefore should not be considered applicable to all countries. In all probability, if I were specifically considering a conflict between Japan and the United States, I would not arrive at the same conclusions. To offer a general recipe for victory, applicable to all nations, would be downright presumption on my part. My intention is simply to point out the best and most efficient way for our country to prepare for a probable future war."[34]

Many American advocates of air power as the primary element in our military organization have written of the defense of the continental United States as our primary military concern; but Mitchell never accepted any such limitations, and he was first to discuss the application of air power—with a minimum of support by surface forces—in global terms. He was a tireless advocate of the arctic air routes between the continents, which have recently become the subject of so much popular interest and have led to so generous a displacement of mercator by polar-projection maps. He dwelt endlessly upon the value of a transatlantic route by way of Greenland and Iceland, and the feasibility of its military use, and upon the corresponding value and feasibility of movement between the United States and Asia by Alaska and Siberia or by the Aleutian and Kurile island chains. Very early in his military career he marked Alaska as the key to the military supremacy of the Pacific; and the introduction of the airplane, and its growing power, strengthened his conviction on that point. It was while General Mitchell was assistant chief of the Air Corps that an army squadron of three airplanes was flown around the world, crossing both the Pacific and the Atlantic by the island routes that he had portrayed as both an opportunity for the United States and a threat to America's safety. The route by way of Greenland and Iceland is in active service now. In one respect the Japanese success in establishing themselves in the Aleutians in the summer of 1942, where no bases had been laid down in time of peace, is a remarkable confirmation of General Mitchell's anticipation of the course of events to come, although the operation appears to have involved a larger use of naval vessels and to have been less concentrated upon aerial objectives than he would have expected.

Many of his predictions have come true. Others will be realized in years to come. But many of the technical developments which seemed to him to be on the horizon, or even within immediate reach at the time when he was writing, are still, after 20 years and despite continued intensive research, far from being in practical use. General Mitchell was immensely imaginative, both in technical matters and in tactical ones. He was an originator, and he was impatient of obstacles, however genuine and solid, that stood in the way of translating the image into reality. It was even more true of him than of Douhet that he characteristically foresaw the direction in which progress would be made, but he

[34] *The Command of the Air*, pp. 252-253.

was often oversanguine about its rapidity. Thus he alienated some of his supporters, and made himself vulnerable.

IV

Alexander de Seversky, a Russian military pilot during the first World War and subsequently an inventor, airplane designer, and highly skilled demonstrator of his own products in flight, consolidated his views on air power, its potentialities, and the defects in its present use, in his book *Victory Through Air Power*.[85] His views in large part follow those of Mitchell, whose disciple Seversky proclaims himself to be in the dedication of his book. He brings Mitchell down to date with fresh illustrations and in particular with an explanation of the failure of the German air force to succeed against Britain.

In Seversky's own presentation of the case for air power the outstanding feature, which characterizes him above all other students of the subject, is an insistence on the vital importance of large radius of action and the possibility of increasing radius to figures far beyond comparison with any heretofore available. He emancipates the air force of the future from the concern with extensive ground organization which Douhet conceded, and from dependence on the island stepping stones of the Arctic routes in the Atlantic and the Pacific to which Mitchell attached such importance.

He prophesies the early realization of nonstop flight around the world, in terms startlingly similar to those which Mitchell had used on the same subject 17 years earlier:

"In five years at the outside, the ultimate round-the-world range of 25,000 miles becomes inevitable."[36]

He may prove to be right; but in order that he may become so, technical progress in airplane design and improvement in powerplant economy will have to be much more rapid during the next five years than at any time in the past twenty. For an airplane to circle the world without stop at the present time, it would have to have substantially more than 75 per cent of the total load at the time in the form of fuel, leaving less than 25 per cent for structure, engines, crew, military equipment, and everything else that has to be carried.

Seversky has no doubt of the rapid future increase of range; and is equally convinced of the folly of accepting the consequences of the range limitations of existing types of aircraft, and certain of the military disaster that awaits those who neglect this paramount factor.

"Range deficiency has been the curse on Hitler's aviation."[37]

"It is sheer waste to maintain advance bases instead of hurling the full aerial potential directly against the adversary. The entire logic of aerial warfare makes it certain that ultimately war in the skies will be conducted from the

[85] Alexander P. de Seversky, *Victory Through Air Power* (New York: Simon and Schuster, 1942).
[36] *Ibid.*, p. 14.
[37] Seversky, *op. cit.*, p. 136.

home grounds, with everything in between turned into a no-man's-land."[38]

Assuming the ability to operate to enormous distances from shore bases, the naval aircraft carrier becomes unnecessary. Seversky dismisses it, much more confidently and conclusively than any of his predecessors.

The idea of operating from "the home grounds" without the burden of establishment and maintenance of advanced and intermediate bases would be welcomed by every air force officer, if it could be realized without paying a prohibitive price. From the inherent characteristics of the airplane as developed over the last 40 years, however, it appears probable that the price of such a method of operation will continue extremely high. Even if aircraft had attained the range necessary to launch bombing raids from a distance of 6000 or 8000 miles it would be likely to remain much more economical of materiel, and therefore more efficient, to operate from closer bases wherever they could be obtained, with fuel supplies secured locally or brought in by tanker at a fraction of the cost in man power and material of transporting them by air.

The advocates of air power have been the object of constant attack by conservative critics on the ground that they propose to win wars with imaginary instruments, reaching their conclusions by assuming aircraft of potentialities far exceeding those actually available. Against Douhet the charge has only a very limited pertinence; against Mitchell much more; and against Seversky much more still. Much of Seversky's book is frankly prophecy, but even when he writes in the future tense he leaves himself open to the charge of being unduly optimistic about the rapidity with which his visions of future aircraft and their performance will be realized. When he portrays the war of the future he drops into a pattern made familiar by many of his predecessors.

"From every point of the compass—across the two oceans and across the two Poles—giant bombers, each protected by its convoy of deadly fighter planes, converge upon the United States of America. There are thousands of these dreadnoughts of the skies. Each of them carries at least 50 tons of streamlined explosives and a hailstorm of light incendiary bombs. . . . With the precision of perfect planning, the invading aerial giants strike at the nerve centers of a great nation. Unerringly they pick their objectives. . . . The havoc they wreak is beyond description. New York, Detroit, Chicago and San Francisco are reduced to rubble heaps in the first twenty-four hours. Washington is wiped out before the Government has a chance to rescue its most treasured records."[39]

Despite the reminiscent quality of Seversky's forecast, it would be foolish to take comfort in the frequency with which such prophecies have been made in the past and in their failure to gain anything approaching complete realization up to the present time. They are nearer to being realized at the moment of writing this chapter than they ever have been before; and sooner or later, if the process of recurrent trial by world war continues, the development of the implements of aerial warfare and the technique of using them is altogether

[38] Seversky, *op. cit.*, p. 139.   [39] Seversky, *op. cit.*, pp. 7-8.

likely to catch up with the prophets. Military men have to remain in continuous readiness for battle with the instruments that can be produced at the time; but they must also plan to meet the problems that will be presented by the instruments that are likely to be available in the future, so that the consequences of technical development may not take them by surprise. The pioneer advocates of air power, and especially Douhet, can be read to greater advantage now than when their books were published.

EPILOGUE. Hitler: The Nazi Concept of War

BY EDWARD MEAD EARLE

IN JULY 1940 Adolf Hitler was master of western and central Europe from the North Cape to the Mediterranean, from the Channel coast to the borders of the Soviet Union. He had at his disposal a military machine of unprecedented striking power, which was stronger in numbers, materiel, experience, and morale than it had been during the September 1939 invasion of Poland. In view of the apparent invincibility of the German army and air force the rest of Europe, together with North Africa and the Middle East, seemed to be in the hollow of Hitler's hand. Not since Napoleon, if then, had any man so completely dominated the European scene or come so close to laying firm foundations for world empire. It is well not to forget these facts in less perilous days.

How did Hitler, at the head of a nation which had been crushingly defeated less than a quarter century previously, arrive at such eminence? Was he merely the tool of the German army and German heavy industry, or was he a strategist in his own right—a latter day Caesar or Bonaparte?

I

It is probably futile to argue the question whether the Führer was more indebted to his high command than they were to him. It seems reasonable to suppose that they were complementary. Hitler hardly could have built the *Wehrmacht* and the *Luftwaffe* without technical assistance of high order, and he alone could not have solved the tactical and logistical problems of the campaigns in Poland, the Low Countries, and France. The head of any state is dependent in such matters upon his general staff. On the other hand, it is highly improbable that the high command could have effected the moral, psychological, and emotional mobilization of the German nation which was so essential to its plans. Nor could it have waged the political, economic, and ideological war— the "white war"—of 1933-1939 which with termite-like thoroughness undermined the structure of international society, prevented formation of a firm anti-German coalition, and spread disunity and treason among the prospective victims of German aggression. These things were the work of the Nazi party, which was so largely Hitler's own creation and which he so completely dominated. German successes against Poland, Norway, the Low Countries, and France were largely the result of an extraordinarily effective combination of imaginative and daring *military* strategy and imaginative and daring *political* strategy. In other words, new military techniques were combined with revolutionary audacity to produce a devastating force which crushed the defenses of western Europe almost as easily as Joshua razed the walls of Jericho.[1]

It was no mean achievement, either, for Hitler to arrive at a *modus operandi*

---

[1] See Introduction for the significance of revolution to offensive warfare.

between the essentially conservative officer class and the dynamic radicalism of the Nazi extremists and "upstarts." The coordination of army and party was, in fact, achieved only by the expenditure of much blood in the purge of 1934 and by continuous political tight-rope walking thereafter. A high standard of military discipline was attained without depriving Nazi youth of the fanaticism which played so large a part in the morale of the German army. The crusading spirit of Nazism, combined with technical competence of the *Wehrmacht*, swept all before it (except the Royal Air Force) until it met the Red Army, which had been similarly conditioned by communism.[2]

Furthermore, there are careful observers of the German scene who are convinced that Hitler made definite military contributions to the armed forces of the Third Reich. The German general staff was fully determined that the nation should not succumb, as in 1918, to a war of position and attrition, and it therefore had evolved entirely new concepts of weapons, strategy, and tactics.[3] Hitler lent his enthusiastic and determined support to mechanized warfare. He may, indeed, have initiated some of its tactics, for as early as 1925 he had gone on record as believing that "motorization will be overwhelmingly decisive" in the next war.[4] His passion for high-speed motor cars, express highways (*Reichsautobahnen*), and airplanes was a factor in his determination to build, primarily for war purposes, a German motor industry without parallel in Europe and perhaps the world. Not even Goering was more convinced about air power as a weapon of war and terror. In short, blitzkrieg was thoroughly congenial to a man of Hitler's character. The measure of democratization which Hitler introduced in the army was responsible, among other things, for a high degree of competence among officer personnel.[5] Above all, however, was the controlling will, the centralized authority, and the driving power which Hitler contributed to the armed effort of the state in all its many phases. The unity, ingenuity, and daring of the high command were centered in and personified by Hitler, whose totalitarian powers guaranteed the perfect coordination of all arms, regarded as essential to success in modern war. Hitler so firmly believed in his destiny that in both the military and political spheres he took risks from which most generals would shrink.[6] No other chancellor in modern German history has been so thoroughgoing a militarist and hence so willing to sacrifice virtually

[2] There is an able discussion of the relations between the dictatorship and the army in Sigmund Neumann, *Permanent Revolution: The Total State in a World at War* (New York, 1942), pp. 174-182. It should be noted that the army leader General Schleicher and the Nazi leader Captain Roehm alike perished before Hitler's firing squads in 1934.

[3] Herbert Rosinski, *The German Army* (New York, 1940), Chaps. IV-V.

[4] *Mein Kampf*, first German edition of 1925, translated into English and published in full by Reynal and Hitchcock (New York, 1939), p. 958. All references are to this translation. Quotations reproduced here through the courtesy of Houghton Mifflin Company, owners of the basic copyright. *Ibid.*, for the importance of establishing a German motor industry.

[5] Joseph C. Harsch, *Pattern for Conquest* (New York), Chap. V. One of the best available discussions on the personnel of the *Wehrmacht*.

[6] Otto D. Tolischus, *They Wanted War* (New York, 1940), Chaps. IV, XII, especially pp. 68-69. The lack of unity of command, and the cautious, conservative, and defense-minded direction of the Allied effort during 1939-1940 is in striking contrast to the state of affairs prevailing in Germany.

what he always contended was essential to the success of his plans: the neu-machine. And among Hitler's contemporaries, only Stalin showed, before the collapse of France, the same single-mindedness of purpose, which perhaps no democratic statesman can or should emulate except in time of war and impend-ing disaster.

As Hitler until recently has been one of the most prolific and least inhibited writers and speakers of our time, he has given us a relatively clear picture of the strategy which he intended to pursue and to which for a time he adhered with remarkable consistency and pertinacity. There were deviations from the main line of action, dictated by momentary expediency, for Hitler understood the necessity of sacrificing, temporarily, a lesser to a greater objective. But no would-be conqueror has furnished his prospective victims with so clear a blue-print of his plans, and it is a tribute to our political ineptitude and our capacity for self-deception that we failed to be forewarned. It was not Hitler's master plan but the major deviations from that plan which led to the desperate situa-tion in which the Reich found itself in the summer of 1943.

In *Mein Kampf*, published in 1925, Hitler made it evident that under him Germany was destined to become the first military power of the world. In a bitter castigation of the Second Reich he denounced what he believed to be the essentially peaceful policies of the kaiser's government—industrialization, over-seas trade, and colonialism. Industrialization, said Hitler, was a form of inten-sive colonization of Germany itself and had the effect of reducing the empire to the status of a colony. Overseas trade, so-called "peaceful economic con-quest," was a monumental folly, for trade depends on peace and the pursuit of peace could have only one effect—to nullify all hope for an effective German policy; with "such an attitude toward world affairs," the future of Germany could "be looked upon as dead and buried." Marxism and Judaism—the eternal enemies of the master race—were at the root of this philosophy of peace and commerce and threatened to emasculate the German nation; hence they must be ruthlessly extirpated. A proper territorial policy "cannot find its fulfillment in the Cameroons"; the true sphere for German colonization is not Africa but the continent of Europe. The seat of world power is the Old World, and only in the Old World is to be found the *Lebensraum* which the dynamic and vital German nation requires. Territorial expansion on the European continent is not merely an end of German policy but a natural law, for "Nature reserves land and soil for that people which has the energy to take it and the industry to cultivate it." German eyes, therefore, should turn eastward, where lie the great fertile lands of the Ukraine, the mineral resources of the Urals, and the traditional field for German conquest. To the east, also, lay the enemy of all civilization, the Soviet Union, the destruction of whose bolshevist program was a special mission of a rejuvenated Germany. The resemblance to Pan-German-ism of much of this phase of Hitler's program is obvious.[7]

The conquest of territory in eastern Europe would be impossible, however,

[7] The indictment of the kaiser's policies is in *Mein Kampf*, Chap. IV. On Pan Germanism see Mildred S. Wertheimer, *The Pan-German League, 1890-1914* (New York, 1924).

if Germany were compelled, as in 1914-1918, to fight a two-front war. Therefore the first and most fundamental consideration for Germany was never to
"tolerate the establishment of two continental powers in Europe." Ultimately,
of course, Germany must be the only European military power, and the German people must "regard it not only as a right, but a duty, to prevent the establishment of [any other such] state by all means including the application of
armed force, or, in the event that such a state be already founded, to repress
it." The "mortal enemy" of Germany in this respect is France, which "relentlessly throttles us," so that "we must undertake every sacrifice which may help
bring about a nullification of the French drive for European hegemony." The
"eternal conflict" between France and Germany cannot be resolved except by
offensive action, leading to the destruction of the former. "Only when this is
fully understood in Germany, so that the German nation's will to live is no
longer allowed to waste itself in purely passive defensiveness, but is drawn
together for a decisive, active settlement with France, and is thrown into a final,
decisive battle for the vastest German final goals: Only then will it be possible
to bring to a conclusion our eternal struggle with France, in itself so fruitless;
*on condition, of course, that Germany really sees in France's destruction a
means of subsequently and finally giving our nation a chance to expand
elsewhere.*"[8] To bring about the destruction of France, it is necessary, first of
all, to isolate her—to nullify French alliances in eastern Europe and to obtain
for Germany the allies essential to protect her exposed flanks in the west.
Therefore, "in Europe, there can be for Germany in the predictable future only
two allies: England and Italy."[9]

This, then, was the essence of Hitler's strategy. France must be destroyed,
because the historic policy of France—no matter who ruled, "whether Bourbons or Jacobins, Bonapartists or bourgeois democrats, clerical republicans or
red Bolsheviks"—was to prevent Germany from waxing fat. If this central and
controlling aim of Hitler's policy—the total and permanent elimination of
France as a great power—be kept in mind, most other things become clear:
intervention in Spain, the alliance with Italy, the campaign for recovery of the
Saar, the reoccupation of the Rhineland, the rape of Czechoslovakia, the construction of the Westwall, the nonaggression pact with the Soviet Union, and
the partition of Poland. Only by the destruction of France could Germany be
assured uninterrupted conquest of eastern Europe. The vast project for a
Germanized Europe failed to materialize because of two formidable obstacles
which stood in the way: the Royal Air Force and the Red Army. The first of
these Hitler had hoped to have on his side; the latter he gravely underestimated.
And the Führer was to assure his eventual downfall by bringing into the coalition against him what he himself described as the "unheard of internal strength"
of the United States, "the gigantic American State Colossus, with its enormous wealth."[10]

[8] Hitler, *op. cit.*, pp. 963-966, 978-979. I have supplied italics. For further reference to
France as the "irreconcilable" and "mortal enemy" of Germany, which must be destroyed
see *ibid.*, pp. 902, 907-908.
[9] *Ibid.*, p. 908.   [10] *Ibid.*, pp. 180, 928.

Hitler understood clearly enough that the key to the success of all his plans was an understanding with England. In castigating the kaiser for failing to win the friendship of Britain, Hitler said, "the English people must be looked upon as the most valuable ally in the world as long as its leaders and the spirit of its masses permit us to expect that brutality and toughness which is determined to fight out, by all means, to the victorious end a struggle once started, without considering time and sacrifices. . . ."[11] Germany could not hope to defeat France or to conquer eastern Europe unless assured by agreement with Italy and England that her rear and flanks were protected. By such arrangements "Germany would be freed from its adverse strategic situation at one blow. The most powerful protection of the flank on one side, the complete guaranty of our supply of the necessities of life and raw materials on the other side, would be the blessed effect of the new order of states."[12] Germans may be skeptical of the value of alliances because of their unfortunate experiences with the "mummy" state Austria-Hungary and the moribund Turkish Empire of the First World War. But "the greatest world power of the earth [Britain] and a youthful national state [Italy] would constitute different premises for a struggle in Europe than did the putrid state corpses with which Germany had allied herself in the last war."[13] To "gain England's favor, no sacrifice should have been too great" in the years before 1914, and none was too great for the Third Reich if such sacrifice led to unchallenged control of the continent. Colonies and sea power would have to be renounced, and the industrial challenge to Britain in the world's markets would have to be withdrawn. Germany could build a fleet more easily than a continental empire, "but it can also be destroyed more quickly"; therefore naval competition with Britain was not to be recommended for any reason. The result of a British alliance "would certainly have been a momentary restriction, but a powerful and great future."[14]

Hitler understood, too, that his relations with Britain would play a large part in determining the attitude of the United States to Nazi Germany. "Because of the British Empire," he wrote, "one too easily forgets the Anglo-Saxon world as such. England cannot be compared with any other state in Europe, if only because of her linguistic and cultural communion with the American Union." He had some appreciation of the power of the United States, based upon the continental proportions of its territory and its relative invulnerability to attack.[15] But he hoped to neutralize American power by the alliance with Japan as well as by the bitter struggle within the United States between "isolationists" and "interventionists." He was also under some illusions concerning racial factors as contributing to disunity and even impotence in the United States.

Despite his apparent understanding of the vital importance to Germany of

---

[11] *Ibid.*, p. 461.
[12] It is interesting to note that Hitler considered 'Germany's central situation in Europe, with its "interior lines," a source not of strength but of weakness. And had he foreseen the devastating Anglo-American air offensive, he would have realized even more keenly that Germany's situation in the heart of Europe is no strategic paradise, as is so often presumed.
[13] *Ibid.*, pp. 964-965.   [14] *Ibid.*, pp. 183-184.   [15] *Ibid.*, pp. 180, 928-929.

British support, Hitler nevertheless pursued policies which were almost certain sooner or later to involve him in war with the British Empire. Perhaps he thought that the toughness of British character had been softened, that British leadership was not what it had been; certainly British behavior under Baldwin and Chamberlain was such that he had some warrant for underestimating England. Perhaps he thought that German air power would supplant British sea power in determining the destiny of the world. Clearly he seems not to have understood the strength of British determination to prevent Germany or any other nation from achieving undisputed control of Europe—the traditional policy of the balance of power, which became the primary concern even of Mr. Chamberlain after the occupation of Prague. He seems not to have understood that the more powerful, the more efficient, and the more ambitious a state, the greater the ultimate British resistance to its pretensions, however reluctant the British people might be to become involved in war.[16] He permitted the colonial department of the Nazi party and numerous private or quasi-official societies to conduct a belligerent propaganda campaign for the restoration of Germany's former colonies, and he built a navy with full knowledge that its military usefulness could hardly be commensurate with the damage it would do Anglo-German comity. Above all, he overlooked the moral factor in the relations between states, which sooner or later asserts itself in British and American policy. He came perilously close to destroying Britain; but since he did not, he must plead guilty to the same crime for which he indicted William II—failure to win British friendship or, at least, assure British neutrality. The alliance with Italy was poor compensation for alienation of Britain.

In isolating France, Hitler pursued a well thought out policy of whittling away, rather than demolishing at one blow, the sources of her strength. The pact of 1934 with Pilsudski cost Hitler nothing—it was simply *de jure* acknowledgment of the *de facto* situation that at the moment Poland was stronger than Germany—but it cost France a great deal, for it was the first serious breach in the bloc of French allies in eastern Europe. The reoccupation of the Rhineland, the *Anschluss* with Austria, and the support of Henlein in Czechoslovakia were still further breaches in the Versailles system of French security. Hitler was always careful before Munich, however, not to demand so much at any one time as to run serious risk of war. The destruction of smaller states was to proceed one at a time, and the sapping of French power was to be piecemeal, always avoiding a *casus belli*. "An intelligent victor will, whenever possible," wrote Hitler, "present his demands to the vanquished in installments. He can then be sure that a nation which has become characterless—and such is every one which voluntarily submits—will no longer find any sufficient reason in each of these detailed oppressions to take to arms once more. The more extortions thus cheerfully accepted, however, the more unjustified does it seem to

---

[16] On this point Sir Eyre Crowe's famous memorandum of January 1, 1907, on Anglo-German relations is as pertinent as if written today. See G. P. Gooch and H. Temperley, *British Diplomatic Documents on the Origins of the War*, III (1928), 397-420, especially p. 403 regarding the balance of power.

people finally to set about defending themselves against some new, apparently isolated, although really constantly recurring, oppression, especially if, taking everything together, so much more and greater misfortune has been borne silently and tolerantly without doing so."[17]

Hitler succeeded in isolating and destroying France. Just as his *Wehrmacht* sought a Cannae which would annihilate French armies, so Hitler sought a political Cannae which would permanently remove France and her allies from the ranks of the powers. In this respect he was breaking with the European tradition and with the German policies of his predecessors Frederick the Great and Bismarck. "France's continued existence as a great power is just as needful to us as that of any other of the great powers," wrote Bismarck in 1887. "If we should be attacked by France and emerge victorious, we would nevertheless not even consider it a possibility to annihilate a nation of forty million Europeans so highly endowed and sensitive as the French."[18] But the Nazis think otherwise and have been determined almost from the beginning to reduce France to the status of a German colony.

II

To the Nazis the armed forces of the Reich were only the cutting edge of the war machine. In their totalitarian strategy, military operations and war were not a first step against the enemy but a regrettable and unavoidable last resort, after all other methods of conquest had been exhausted. Hitler's greatest period of success, therefore, was his long series of bloodless victories up to and including Munich. Thereafter the battle rested largely with the soldiers, and there is not much evidence available to show that Hitler's military strategy will compare favorably with his triumphs in the realm of psychological and political warfare.

An indispensable first step in the war of nerves was the welding of the German people into a unity which was terrifying to the outside world. This was accomplished partly by the ruthless suppression of all internal dissent from the Nazi program—particularly by the destruction or the "coordination" of the Jews, the churches, the universities, the trade unions, the socialists, the communists, and all others who were suspected of internationalism or pacifism. It was done partly by skillful propaganda of press and radio, reinforced by party discipline and appeals to national pride. Militarism, Pan-Germanism, anti-Semitism, racial superiority, worship of the state, and other features of Hitler's program were deeply rooted in German history and were exploited by the Nazis to their own ends. "Any resurrection of the German people," wrote Hitler, "can take place only by way of regaining external power. But the prerequisites for this are not arms, as our bourgeois 'statesmen' always babble, but the forces of will power. . . . The best arms are dead and useless material as long as the spirit is missing which is ready, willing, and determined to use them." Therefore, "the question of regaining Germany's power is not, perhaps,

---

[17] *Mein Kampf*, pp. 968-969.
[18] *Die grosse Politik der europäischen Kabinette*, VII, 177-178.

How can we manufacture arms, but, How can we produce that spirit which enables the people to bear arms."[19]

It is not necessary to describe in detail the manner in which Hitler created the will to fight by bringing about the psychological and emotional mobilization of the German nation. He took over from Hindenburg and other army leaders the stab-in-the-back legend, designed to show that the German army had not been defeated in 1918 but had been betrayed on the home front, and gave the legend a popularity and degree of credence it could not otherwise have attained.[20] He led the German people to believe that in November 1918, they had voluntarily laid down their arms on promises from Woodrow Wilson of a just and generous peace and that Wilson had defaulted on his promises in the "greatest betrayal in all history." By these and other devices he bred in the hearts of all classes a consuming sense of injustice over the *Diktat* of Versailles, and among large numbers of Germans he aroused a spirit of vengeance. He revived the pre-1914 theory that a peaceful Germany was being encircled by jealous and hostile states. Among German youth he developed a cult of Spartanism, a fanatical German nationalism, and an unquestioned loyalty to the Führer which boded no good for the peace of the world.[21] By maintaining the largest private army in modern history, the S.A., and a corps of janissaries, the S.S., Hitler even before he came to power as chancellor spread the spirit of militarism throughout the land. The Brown Shirts' war cry of "Deutschland Erwache" became a national slogan. The function of the state, said Hitler, is to unite all Germans, and "to lead them, gradually and safely, to a dominating position" in the world.[22]

The story of German economic mobilization for war, *Wehrwirtschaft* and *Kriegswirtschaft*, has been told too often to need repetition here. It must be noted, however, that from the general staff the Nazis took over the idea that the "total" war of 1914-1918 was not total enough; that adequate measures—such as the development of synthetic raw materials and the building up of stock piles of critical minerals—must be taken against blockade; that the home front must be solidified economically, as well as psychologically, in support of the war effort which was to come. The Four Year Plans, under the general administration of Goering with the assistance of Major Generals Fritz Loeb, Georg Thomas, and Hermann von Hanneken, among others, brought about the complete militarization of German economy. As a result, the German army entered the war in 1939 better equipped, and with larger reserves of materiel,

[19] *Mein Kampf*, pp. 459-460, and on the same theme *ibid.*, p. 624.
[20] Hans Fried, *The Guilt of the German Army* (New York, 1942) gives an important, but much neglected, account of the stab-in-the-back legend.
[21] Concerning Spartanism see Henri Lichtenberger, *The Third Reich* (New York, 1937), Book V; S. H. Roberts, *The House That Hitler Built* (New York, 1938), Part IV, Sec. II. In the moral mobilization of Germany Hitler received support from surprising sources, as is shown in an important article by Oscar J. Hamman, "German Historians and the Advent of the National Socialist State," *Journal of Modern History*, XIII (1941), 161-188.
[22] *Mein Kampf*, p. 601. On the quasi-military organizations of Nazi Germany see a series of articles by Alfred Vagts in the *Infantry Journal*, May-September 1943.

than any other modern army. The Nazi concept of total war was inherent in their theories of the totalitarian state.[23]

Force alone, however, was never considered by Hitler to be an effective weapon. Force and the threat of force must be supplemented by words, slogans, ideas, which are among the most powerful of all weapons, as was demonstrated by the French Revolution, by Woodrow Wilson, and by the bolsheviks. Therefore the National Socialist movement was offered the Germans and the world as a basis for a New Order, which would replace the old "chaos" and "inefficiency." Because it had about it a character of inevitability—because, in the words of an American, it was the "wave of the future"—it was on the offensive, the rest of the world on the defensive. Only the ideological offensive—a fanatical belief in one's own view of life—can give victory, said Hitler. Thus the Nazi revolution was to serve not merely as an instrumentality for unifying the German people but a means, as well, of disunifying those countries which stood in the way of German expansion. To paraphrase Lenin, the revolutionary struggle in Germany was to be transformed by Hitler into a European and world-wide civil war.[24]

Hitler was a master at throwing apples of discord among other nations. He knew of the critical differences of opinion in France, Great Britain, and America and exploited them to the full. He always succeeded in discussing the issues of European politics not in the terms in which he really viewed them—as questions of power—but in the terms which would cause the maximum division in public opinion abroad. "Mental confusion, contradiction of feeling, indecisiveness, panic: these are our weapons," he told Rauschning.[25] Thus the alliance with Japan was first presented to a credulous world as the Anti-Comintern Pact and hence as an attack on bolshevism. The fear of bolshevism, Hitler correctly suspected, was so strong among the conservatives of Britain, France, and the United States that they altogether overlooked the portent to their own security in the Pacific of the camouflaged alliance. To the conservatives, also, Hitler was one who had "solved the labor problem," not one who conscripted the labor force for the manufacture of armaments. Liberals in Britain and America were befuddled by attacks on Versailles, by appeals to the sacred right of Germans in Czechoslovakia and elsewhere to be "reunited" with the motherland, and by the claims of Germany to "fair play." Anti-Semitism was bait for so many different classes in so many different places as to be a catch-all for the unwary everywhere. Those who advocated resistance to Hitler were represented to peaceful peoples as "war mongers."

In this manner, so much confusion was spread abroad that the statesmen of Europe seemed unable to see their interest and act on it. Of this, of course,

---

[23] On the economic preparation for total war see A. T. Lauterbach, *Economics in Uniform* (Princeton, 1943) ; Tolischus, *op. cit.*, Chaps. III-V.

[24] *Mein Kampf*, pp. 220-224, 784, for particularly striking passages on this theme. All of Hitler's speeches are a gold mine on the same subject. See N. H. Baynes (ed.), *The Speeches of Adolf Hitler, 1922-1939* (2 vols., Oxford, 1942) and R. de Roussy de Sales, (ed.) *My New Order* (New York, 1941), both with valuable commentaries.

[25] Hermann Rauschning, *The Voice of Destruction* (New York, 1940), p. 9.

Spain was the classic example. Ciano might realize in regard to Spain that "there no longer are frontiers, only strategic positions," but non-fascists could be persuaded by fascist propaganda that the struggle for control of Spain involved only bolshevism and catholicism. In short, the Nazis undermined morale and the will to resist everywhere, except at home, by fostering pacifism, defeatism, and "corroding uneasiness, doubt, and fear." By these methods Hitler softened up his victims, lulled them into a false sense of security, and ultimately rendered them incapable of successful armed resistance. Thus Czechoslovakia, at the beginning of 1938, was a relatively prosperous state, secure behind a bulwark of powerful fortifications, possessed of a magnificent army, buttressed by powerful allies in both east and west; nine months later she found herself unable to stave off disintegration, in which her friends lent her enemies a helping hand. If he had accomplished nothing but the conquest of Czechoslovakia without firing a shot, Hitler would have to be recorded on the pages of history as a master of political warfare. Although Goebbels and Goering played their parts, and although the *Wehrmacht* always stood in the foreground, it was Hitler who set the pace, called the tune, and reaped the rewards.

Nazi strategy, indeed, drew no clearly defined line between war and peace and considered war not peace the normal state of society. Since war to the Nazis no longer consisted solely, or even primarily, of military operations, the policy of the state in time of so-called peace was only a "broadened strategy" involving economic, psychological, and other nonmilitary weapons. Political warfare was constantly carried on, writes a former member of Hitler's entourage, "not only to render the tactical situation favorable to a succession of bloodless victories, but also to determine the particular issues which the general political situation may make ripe for settlement in accordance with the aims of National Socialism. These political activities find their explanation in the novel character of the important moves to come—pressure combined with sudden threats, now at one point and now at another, in an unending activity that tires out opponents, enabling particular questions to be isolated, divisions to be created in the opposing camp, and problems to be simplified until they become capable of solution without complications [*i.e.*, without war]." The military preparations of Nazi Germany were only one aspect of its revolutionary activities, which were designed to make armed aggression superfluous or, if necessary, certain of success. "The aim is not simply the expansion of frontiers and the acquisition of new territory, but at the same time the extension of the totalitarian revolutionary movement into other countries. All this is virtually the transfer [to the international sphere] of the modern technique of the coup d'état," with the Nazi military establishment having the same function as armed revolutionaries in a domestic insurrection. By these means Hitler found it possible to bring about far-reaching political changes without bloodshed, with his enormous armaments intended to be a threat of war, rather than an instrument of actual combat.[26]

[26] Hermann Rauschning, *The Voice of Destruction* (New York, 1939), pp. 139-140.

Sir Eric Phipps, the British ambassador in Berlin, warned his government in November 1933 that the conditions in Germany under Hitler "are not those of a normal civilized country, and the German Government is not a normal civilized Government and cannot be dealt with as if it were." But Sir Eric did not find a sympathetic ear in Whitehall, where the belief prevailed that Nazism was "a healthy national revival" rather than armed Jacobinism.[27]

Hitler's strategy, both in peace and war, was a strategy of terror. In order to come to power in Germany he won the "battle for the streets"; in order to stay in power he tortured and imprisoned and murdered his opponents; in order to have his way in Europe he projected the same methods beyond the boundaries of the Reich. The *Luftwaffe*, especially, was intended to be a means of terrorization and perhaps more than any other single weapon at Hitler's command was responsible for the Munich capitulation of Britain and France. The widest possible publicity was given abroad to the strength or the reputed strength of the German Air Force, and distinguished visitors were shown ostentatiously not only Germany's planes but her capacity to produce planes. The Douhet theory of air warfare, which Mr. Warner discusses in Chapter 20, was the accepted doctrine of the German high command. "To those who say that air power and the threat of air power is a two-edge sword which cuts both ways, the reply is plain," wrote Major George Fielding Eliot in 1939. "Not while Europe is divided ideologically between dictators and democracies, for dictators may threaten, dictators may blackmail, but free peoples cannot. Blackmail, as a weapon in the international struggle, is a weapon which may be wielded effectively only by a dictator whose purposes are obscure, whose intentions are shrouded in mystery, whose name is synonymous with ruthlessness and terrorism. . . . It is blackmail which rules Europe today, and nothing else: blackmail made possible only by the existence of air power."[28] But this very emphasis upon the terroristic possibilities of air power was no service to the *Luftwaffe* when it had to meet its greatest test over England in 1940-1941.[29]

Looking back over the years 1933-1939, one may come hastily to the conclusion that Hitler's political strategy was one long succession of triumphs, that he made no mistakes. But more mature consideration of the facts will lead to the conclusion that the whole was different from the sum of its parts, that the totality of no mistakes was one gigantic mistake. For the time was certain to come when the world would no longer be deceived and blackmailed, whatever the cost of resistance. At long last, the British people came regretfully to the decision that peace was no longer to be bought except with "blood, sweat, and tears." The British decision to guarantee the independence of Poland, followed by the scrupulous fulfillment of that pledge, marked the failure of Hitler to win

[27] G. P. Gooch, *Studies in Diplomacy and Statecraft* (London, 1942), Chap. VI, especially pp. 211 ff.

[28] *Bombs Bursting in Air* (New York, 1939), pp. 80-81. See also Edmond Taylor, *The Strategy of Terror* (Boston, 1940). For a very remarkable passage on the blackmail tactics of the Imperial German government before 1914 and the futility of appeasing the blackmailer see Sir Eyre Crowe, *loc. cit.*, p. 416.

[29] A. P. de Seversky, *Victory Through Air Power* (New York, 1942), Chap. III.

what he always contended was essential to the success of his plans: the neutrality or active friendship of the British Empire.

III

Munich and its aftermath, the occupation of the rump of Czechoslovakia in 1939, were the last of Hitler's bloodless victories. Although the events of 1939-1941 still gave him ample opportunities to demonstrate the effectiveness of political and psychological warfare, the day of exclusively nonmilitary action was past.

It is too early to write any final appraisal of Hitler as a military strategist, although the indications are that history will not speak any too well of him. We know too little of the determining factors and the controlling voices in the campaigns in Poland, Norway, the Low Countries, and France to apportion the praise or blame for German strategy and tactics. Assuming, however, that in a totalitarian country the dictator actually dictates, Hitler deserves credit for astute conduct of the war until July 1940. He took his enemies one at a time, conquering Poland before France and overrunning France before attacking England. The campaign against Norway was one of the most brilliant military operations in modern history, designed not only to protect Germany's northern flank but to strike a serious blow at British sea power. Sir Halford Mackinder had always contended that control by a military state of long stretches of the European coast line would largely nullify British control of the seas—a contention which was enthusiastically taken over by Haushofer and the geopoliticians. The conquest by Germany of the Atlantic coast of Europe from Narvik to the Spanish border was not a mortal wound to British sea power, but it made infinitely more difficult the battle against the submarine and thereby placed heavy strains upon the British people and their power to resist. The defeat of France was blitzkrieg at its maximum of efficiency and terror, and it marked the essential first step in any German conquest of Europe, as Hitler had said in 1925.

Thereafter, things went badly with Hitler and the Nazi war plan. Too cautious or too methodical to attack Britain immediately after Dunkirk, when she was badly disorganized and almost defenseless, Hitler lost his greatest single opportunity to win the war. The subsequent defeat of the Luftwaffe by the Royal Air Force was the turning point of the war. Simultaneously, Hitler made bad errors in political judgment. Dealing through traitors like Quisling and Laval, he imposed upon the occupied territories an intolerable burden of "collaboration" and—after a brief period of "correctness"—a reign of terror.[30] The New Order, which might conceivably have been given a trial by the disillusioned peoples of Europe, consequently became a synonym for exploitation and oppression.

[30] On the character of German economic operations in occupied territories see Thomas Reveille (pseudonym for Rifat Tirana), *The Spoil of Europe: The Nazi Technique in Political and Economic Conquest* (New York, 1941), and T. D. Kernan, *France on Berlin Time* (New York, 1942). Mr. Tirana's book is indispensable to an understanding of the nature of the German military occupation.

The attack on the Soviet Union meant that in 1941-1943, as in 1914-1918, Germany had to fight a two-front war and, as events proved, a war of attrition as well. Hitler, then, finds himself as these words are written in much the same situation as the Emperor William II, whom he so thoroughly berated in the pages of *Mein Kampf*. Hitler is learning what so many other conquerors learned before him, that only the final battles decide wars.

Who's Who of Contributors

CRANE BRINTON. Author of *The Lives of Talleyrand, A Decade of Revolution, 1789-1799,* and other historical works. Professor of history, Harvard University. Now overseas in government service. A.B. Harvard, D.Phil. Oxford.

GORDON A. CRAIG. Assistant professor of history, Princeton University. A.B. and Ph.D. Princeton; B.Litt. Oxford.

HARVEY A. DEWEERD. Captain, Army of the United States. Associate editor, *The Infantry Journal.* Author of *Great Soldiers of Two World Wars.* A.B. Hope College, Ph.D. Michigan.

EDWARD MEAD EARLE. Special consultant, Army Air Forces. Lecturer, Army War College. Professor in the Institute for Advanced Study at Princeton, and chairman Princeton military studies group. B.S. and Ph.D. Columbia.

IRVING M. GIBSON. Pseudonym for an American scholar who saw active service in the infantry in 1917-1918. For several years has been engaged in the study of military history, especially of period immediately preceding and following the First World War.

FELIX GILBERT. Formerly assistant to the editors of the German diplomatic documents on the origins of the war of 1914. A student of political ideas and international affairs. Now in government service. Ph.D. University of Berlin.

JEAN GOTTMANN. A geographer formerly associated with Institute of Geography at the Sorbonne. Now teaching in the Army Specialized Training Program at Princeton. Lecturer in geography, the Johns Hopkins University. *Licencié ès Lettres,* University of Paris.

HENRY GUERLAC. Chairman of the department of the history of science, University of Wisconsin. Now in government service. A.B. Cornell, Ph.D. Harvard. Author of unpublished volume *Science and War in the Old Regime.*

HAJO HOLBORN. Author of *Bismarck's European Policy in the Seventies; Germany and Turkey, 1878-1890,* and other historical works. Professor of history, Yale University. Now in government service. Ph.D. University of Berlin.

ALEXANDER KIRALFY. Author of *Victory in the Pacific.* A lifelong student of military and naval affairs and a regular contributor to professional military journals. Military analyst, *Asia* magazine.

ETIENNE MANTOUX. In active service with the Fighting French forces. Lieutenant French air force, 1939-1940. Economist and lawyer. *Docteur en Droit*, University of Lyon.

SIGMUND NEUMANN. Author of *Permanent Revolution* and of numerous articles on totalitarianism. Associate professor of government and social sciences, Wesleyan University. Research associate, Institute of International Studies, Yale University. Ph.D. University of Leipzig.

ROBERT R. PALMER. Author of *Twelve Who Ruled: The Committee of Public Safety during the Terror*. Assistant professor of history, Princeton University. Now on duty with the Historical Section, Army Ground Forces. Ph.B. Chicago, Ph.D. Cornell.

STEFAN T. POSSONY. Author of *Tomorrow's War: Its Planning, Management and Cost*. Co-author, *The Axis Grand Strategy*. Associated with the French ministries of Information and Air, 1939-1940. Now in government service in the United States. Ph.D. University of Vienna.

THEODORE ROPP. A student of naval strategy and technology, having studied in French Ministry of Marine and French Naval War College. Member of department of history, Duke University, now teaching in the naval training program. A.B. Oberlin, Ph.D. Harvard.

HANS ROTHFELS. Editor of the letters and political writings of Clausewitz and of a notable collection of the works of Bismarck. Author of *Carl von Clausewitz, Politik und Krieg*. Visiting professor of history, Brown University. Ph.D. Heidelberg.

HANS SPEIER. Co-author and co-editor, *War in Our Time*. Co-author of a forthcoming volume *The German Radio and the German People*. Professor of sociology, The New School for Social Research. Ph.D. Heidelberg.

MARGARET TUTTLE SPROUT. Co-author (with her husband Harold Sprout) of *The Rise of American Naval Power* and *Toward a New Order of Sea Power*. A.B. Oberlin, M.A. Wisconsin.

EDWARD WARNER. Vice chairman, Civil Aeronautics Board. Assistant Secretary of the Navy for aeronautics, 1926-1929. Author of *Airplane Design: Performance*. Former editor *Aviation*. Former professor of aeronautical engineering, Massachusetts Institute of Technology. A.B. Harvard, M.S. Massachusetts Institute of Technology.

DERWENT WHITTLESEY. Author of *The Earth and the State*, and *German Strategy of World Conquest*. Consultant, War Department. Professor of geography, Harvard University. Ph.B. and Ph.D. Chicago.

Editor's Note and Acknowledgments

THIS volume originated in the discussions of the seminar in military affairs of the Institute for Advanced Study and Princeton University about eighteen months ago. Its completion is largely due to the willingness of very busy men to lay aside other urgent tasks in order to write the several chapters. Therefore, the largest measure of credit must go to the twenty contributors, who have been willing and helpful co-workers. Mr. Gordon Craig and Mr. Felix Gilbert have not been assistants; they have been collaborators in every sense of the word—intelligent, cooperative, critical, and painstaking. Captain Harvey DeWeerd has had a friendly interest in the volume from its inception and has been of invaluable assistance at various stages of its preparation. The Carnegie Corporation of New York has generously supported in Princeton broader studies which have contributed to this book.

Several members of the staff of the Institute for Advanced Study have contributed to the book in various ways: Dr. Robert Kann, who read or participated in the research for several chapters and prepared the index; Dr. Edmond Silberner and Mrs. V. T. Bill, who assisted with the research; Miss Barbara Harper, Miss Hattie M. Wise, Miss Margot Cutter, and Mrs. Marion G. Hartz, who helped with the preparation of the manuscript. Without the intelligent and devoted work of Mrs. Hartz, the completion of the volume would have been delayed long beyond the present date of publication.

Mr. Datus Smith, director of Princeton University Press, and the members of his competent staff have been invaluable collaborators. Enthusiastic, resourceful, and patient, they have been helpful in almost innumerable ways. Without their friendly criticism this would have been a less satisfactory book.

The editors, authors, and publishers acknowledge with gratitude the willingness of other authors and publishers to permit us to draw upon their work, as is evidenced throughout the footnotes. The Office of Population Research of Princeton University, the American Geographical Society, and others mentioned in the bibliographies, have helped with maps and charts. Admiral William V. Pratt, Admiral Harry E. Yarnell, Mr. Max Werner, and Professors Frank A. Fetter and Harold Sprout have read portions of the manuscript and have offered helpful criticism, although they are in no way responsible for the opinions or conclusions of the authors.

Several contributors are in the service of the United States. In most instances they had completed their manuscripts before taking official positions. In every case, however, they are expressing their own personal opinions and judgments and are in no way speaking for any government agency with which they are associated.

A portion of Chapter 6 appeared in *The American Scholar*, of Chapter 8 in *Military Affairs*, and of Chapter 19 in *Foreign Affairs* by arrangement with Princeton University Press. No other portion of the manuscript has heretofore been published.

E. M. E.

Bibliographical Notes

CHAPTER I. MACHIAVELLI: THE RENAISSANCE OF THE ART OF WAR

The latest critical edition of Machiavelli is *Tutte le opere storiche e letterarie di Niccolò Machiavelli*, G. Mazzoni and M. Casella, ed. (Florence, 1929). This edition has been used throughout. Machiavelli's main works—the *Principe* and the *Discorsi*—have been frequently translated into English. They are available in an edition with an introduction by M. Lerner (Modern Library, New York, 1940). An English translation of Machiavelli's *Istorie Fiorentine* by W. K. Marriott, has been published in Everyman's Library (London, New York, 1909). A selection of Machiavelli's letters and lesser works will be found in N. Machiavelli, *The Prince and Other Works*, new translation, introduction and notes by A. H. Gilbert (Chicago, 1941). Machiavelli's *Arte della Guerra* was first translated into English by Peter Whitehorn in 1560; several other translations followed, particularly in the eighteenth century, but no modern translation of the work is available.

Among the general works which form a basis for the study of the military history of the period the following are outstanding: Hans Delbrück, *Geschichte der Kriegskunst im Rahmen der politischen Geschichte* (Berlin, 1900-1936); Sir Charles Oman, *A History of the Art of War in the Middle Ages*, 2 vols. (London, 1924); Sir Charles Oman, *A History of the Art of War in the 16th Century* (New York, 1937). M. Jähns, *Geschichte der Kriegswissenschaften*, Vol. I (Berlin, 1889), is invaluable for a study of the military literature of the past. W. Sombart, *Krieg und Kapitalismus* (Volume II in W. Sombart, *Studien zur Entwicklungsgeschichte des modernen Kapitalismus*, Munich and Leipzig, 1913—), represents a pioneer work with regard to the relation between the rise of capitalism and the changes in military organization, but its material has to be used with circumspection.

An interesting account of the decline of the feudal ideology in military matters is J. Huizinga, *The Waning of the Middle Ages*, English translation, (London, 1927). A more specialized work on the military developments in Italy during the fifteenth and sixteenth centuries is W. Bock, "Die Condottieri. Studien über die sogenannten unblutigen Schlachten," *Ebering's Historische Studien*, 110 (Berlin, 1913). The author makes a very convincing case for the Condottieri. Another brilliant survey is F. L. Taylor, *The Art of War in Italy, 1484-1529* (Cambridge, 1921). The most detailed scholarly study of the military history of the Italian fifteenth century is P. Pieri, *La crisi militare Italiana-nel Rinascimento* (Naples, 1934). This work is written with a nationalistic bias and overstates the importance of the Italian contributions to modern military development.

The two basic biographies of Machiavelli upon which all later authors are dependent are P. Villari, *Life and Times of Niccolò Machiavelli*, English translation (London, 1898), and O. Tommasini, *La vita e gli scritti di Niccolò Machiavelli*, 2 vols. (Rome, 1887 and 1911). Special sections on Machiavelli's military theory are to be found in the more specifically military works listed above. The authors differ considerably in their evaluation of Machiavelli as military theorist and an entirely satisfactory treatment of this problem is lacking. W. Hobohm, *Machiavelli's Renaissance der Kriegskunst*, 2 vols. (Berlin,

NOTE: Each author in this symposium prepared a bibliography for his own chapter, and the books cited here are those selected by the author himself. In a few cases, however, the actual form of the bibliography was recast to achieve an approximate similarity in the method of presentation.—EDITORS

1913), is exclusively devoted to a discussion of Machiavelli's "military science." Hobohm's book has been severely criticized; it is vitiated by a rather unusual approach to his subject. The author is not so much interested in establishing Machiavelli's importance in the development of military thought as he is in the extent to which Machiavelli can be regarded as a reliable source for the military history of his time.

CHAPTER 2. VAUBAN: THE IMPACT OF SCIENCE ON WAR

The plate in this chapter illustrating siegecraft is from *Traité de l'attaque et de la défense des places*, by M. le Maréchal de Vauban, 2 vols. (Pierre de Hondt, The Hague, 1742). The one of Belfort is taken from Nicolas de Fer, *Introduction à la fortification dediée à Monseigneur le Duc de Bourgogne* (Paris, 1693). Both plates appear by courtesy of the Harvard Library.

Useful for the early period are: F. R. Taylor, *The Art of War in Italy, 1484-1529* (Cambridge, 1921), and Sir Charles Oman, *The History of the Art of War in the Sixteenth Century* (London, 1937).

The best general survey of the development of French military institutions down to the time of Louis XIV is the old work of Edgard Boutaric, *Institutions militaires de la France avant les armées permanentes* (1863). For the reforms of Le Tellier and the organization of the French army in the seventeenth century the indispensable study is Louis André, *Michel le Tellier et l'organization de l'armée monarchique* (1906); and for the period of Louis XIV's personal reign an important work is Camille Rousset, *Histoire de Louvois et de son administration politique et militaire*, 4 vols. (1862-1864). Among works which describe the general characteristics of the French royal army may be mentioned Albert Duruy, *L'armée royale en 1789* (1888) and Albert Babeau's two volumes, *La vie militaire sous l'ancien régime* (1889-1890).

For the histories of the separate arms one is still obliged to consult General Susane's *Histoire de l'ancienne infanterie française* (1849), his *Histoire de la cavalerie française* (1874), and his parallel study of the French artillery arm. The standard work on the French engineers is of about the same vintage. It is Lieutenant Colonel Augoyat's *Aperçu historique sur les fortifications, les ingénieurs, et sur le corps du génie* (3 vols., 1860-1864). There is valuable information in E. Legrand-Girade, "Étude historique sur le corps du génie," *Revue du génie militaire* (1897-1898), and in C. Lecomte's *Les ingénieurs militaires en France sous le règne de Louis XIV* (1904).

Nothing is more conspicuously lacking in the field of military studies than a well-illustrated history of the arts of fortification and siegecraft. The best studies all date from the period of the Second Empire: J. Tripier, *La fortification déduite de son histoire* (1866); Cosseron de Villenoisy, *Essai historique sur la fortification* (1869); Prévost du Vernois, *De la fortification depuis Vauban* (1861). But the only work generally available in American libraries is A. de Zastrow, *Histoire de la fortification permanente* (3rd ed., 1856, trans. from the German by Ed. de la Barre-Duparcq).

A vivid way to grasp in its entirety the scope of the fortification system of France under Louis XIV is to consult a contemporary *atlas des places fortes* such as the beautifully illustrated work of Nicolas de Fer, *Introduction à la fortification*. The best statement of the prevailing view as to the role of fortresses in national defense is given by the little-known work which was used as the official text during the eighteenth century in the Ecole de Mézières: *Traité de la Sureté et Conservation des Etats, par le moyen des Forteresses, par M. Maigret, Ingénieur en Chef* (1725). Harvard and the New York Public

Library both have the De Fer, and the Maigret can be consulted in the Library of Congress.

The most recent work on Vauban in English is Sir Reginald Blomfield's *Sébastien le Prestre de Vauban, 1633-1707* (London, 1938) ; though mention should be made of E. M. Lloyd's *Vauban, Montalembert, Carnot, Engineer Studies* (London, 1887). Blomfield's book relies heavily upon the best French work available : Colonel du génie P. Lazard's *Vauban* (1934), which is a brilliant study covering all aspects of Vauban's career. Both works are admirably illustrated, Blomfield's with his own drawings of Vauban forts in their present state of preservation, and Lazard with sketches by Vauban and with photographs of models from the *Musée des plans reliefs* of the French army's geographical service. George Michel's *Histoire de Vauban* (1897) remains a useful connected narrative, while Daniel Halévy's *Vauban* (1923) is a readable but uncritical work available in English translation. Fontenelle's *Eloge du Maréchal de Vauban* and Voltaire's *Siècle de Louis XIV* are classics that may be consulted with profit. Despite its title, A. Allent's *Histoire du corps impériale de génie* (1st and only volume 1805), deals largely with Vauban, and is the earliest attempt at a serious study.

There is important material in H. Chotard: "Louis XIV, Louvois, Vauban et les fortifications du Nord de la France, d'après les lettres inédites de Louvois adressées a M. de Chazerat, Gentilhomme d'Auvergne," *Annales du Comité Flamand de France*, XVIII (1889-1890). An important critical study on Vauban's strategic contributions is the work of Gaston Zeller, *L'organization défensive des frontières du Nord et de l'Est au XVIIe siècle* (1928).

There is regrettably no standard edition of Vauban's collected works. The best editions of the two main treatises, the *Traité des sièges et de l'attaque des places* are those published by Latour-Foissac in 1795 and by Augoyat and Valazé in 1829. Both editions are rare and many libraries have only the imperfect editions of the eighteenth century, of which there are several. Of the mass of Vauban manuscripts a large part remains unpublished and even those that were published in the eighteenth or nineteenth centuries are not found in most libraries. These are Vauban's autobiographical fragment, the *Abrégé des services du Maréchal de Vauban*, written in 1703, published by Augoyat in 1839; his earliest published work, the *Directeur général des fortifications*, published in The Hague in 1685, reprinted in Paris, 1725; the *Mémoire pour servir d'instruction dans la conduite des sièges*, written in 1669, printed Leyden, 1740; his *De l'importance dont Paris est à la France, et le soin que l'on doit prendre de sa conservation* (Paris, 1825) ; *Mémoires militaires de Vauban . . . Précédés d'un avant-propos par M. Favé* (Paris, 1847-1854) ; *Mémoires inédites de Vauban sur Landau, Luxembourg et divers sujets*, ed. by Augoyat (Paris, 1841) ; *Mémoire du Maréchal de Vauban sur les fortifications de Cherbourg, 1686* (Paris, 1851). The first four volumes of the *Oisivetés* were published by Augoyat in 1842-1845.

For personalia and much else the best assemblage of materials is the work of Colonel de Rochas, *Vauban, sa famille, et ses écrits*, 2 vols. (1910). This work is extremely rare. Much of Vauban's correspondence, including letters reprinted in Rochas, has been published in the *Revue de génie militaire* in the numbers from 1897 to 1901. Vauban's *Lettres intimes inédites adressées au Marquis de Puyzieulx* have been printed with introduction and notes by Hyrvoix de Landosle (1924).

CHAPTER 3. FREDERICK THE GREAT, GUIBERT, BÜLOW: FROM DYNASTIC TO
NATIONAL WAR

The cuts used in this chapter are by the courtesy of *Military Engineer.*

The fullest repository of materials on military writers before 1800 is M.
Jähns, *Geschichte der Kriegswissenschaften vornehmlich in Deutschland,* 3
vols. (Munich, 1889-1891). This work, despite its title, contains abundant information on other than German writers. It is a huge compendium of biographical data, extensive analyses and long quotations from often inaccessible
sources. The standard over-all narrative, emphasizing strategy and policy
rather than technology, is H. Delbrück, *Geschichte der Kriegskunst im Rahmen
der politischen Geschichte,* 7 vols. (Berlin, 1900-1936).

For the authors treated in this chapter there is no substitute for their own
writings. Frederick's two Testaments give the essence of his military thought
in brief form. The Testaments were made public by the Hohenzollern family
at the close of the nineteenth century. They were described and criticized by
O. Hintze, "Das politische Testament Friedrichs des Grossen von 1752," in
Historische und Politische Aufsätze, Vol. III (Deutsche Bucherei 98/99),
p. 3-28. The Testaments are unfortunately available only in German, and may
be found in the *Werke Friedrichs des Grossen,* 10 vols. (Berlin, 1912-1914).
Much of the rest of the *Werke* consists in translation of works written by
Frederick in French, and first assembled in the *Oeuvres de Frédéric le Grand,*
30 vols. (Berlin, 1846-1856). There is a translation of the *Principes généraux
de la guerre* in T. R. Phillips, *Roots of Strategy* (Harrisburg, 1940), pp. 301-
400.

Guibert's chief military writings were reprinted in *Oeuvres militaires du
comte de Guibert,* 5 vols. (Paris, 1803). The *Essai* of 1772 was translated into
English, *A general essay on tactics, with an introductory discourse upon the
present state of politics* (London, 1781).

The writings of Bülow were never collected. Their titles may be found in
Jähns, with a statement of their contents. His most famous but by no means
most significant book was translated, *The Spirit of the Modern System of War*
(London, 1806). It is to be noted that these British translations, both of
Guibert and of Bülow, were published in both cases at a time when the doctrine
of the translated book had been recanted by the author.

CHAPTER 4. JOMINI

Jomini's main historical works are *Traité des grandes opérations militaires*
(8 vols., Paris, 1804-1816); *Histoire critique et militaire des guerres de la
Révolution* (5 vols. and atlas, Paris, 1806; 15 vols. and 4 atlases, Paris, 1819-
1824); *Vie politique et militaire de Napoléon* (4 vols., Paris, 1827); and
Précis politique et militaire de la campagne de 1815 (Paris, 1839). Of these
works the *Traité* has the most analytical approach to the subject; and it and
the *Précis de l'art de la guerre* (2 vols., Paris, 1838) contain the bulk of
Jomini's theoretical writings. The *Précis de l'art de la guerre,* Jomini's most
important work, grew out of his *Introduction à l'étude des grandes combinai-
sons de la stratégie et de la tactique* (Paris, 1829). The *Précis* and the *Traité*
have appeared in several editions which differ in detail.

Of the biographies of Jomini, the following should be noted: F. Lecomte,
Le général Jomini, sa vie et ses écrits (Lausanne, 1860; and Paris, 1894);
C. A. Ste. Beuve, *Le général Jomini* (Paris, 1869); and Xavier de Courville,
Jomini ou le devin de Napoléon (Paris, 1935). See also the important review
of Lecomte's *Jomini* by Georges Gilbert in *La nouvelle Revue* (December 1,
1888), 674-685.

Appraisals of Jomini's theories will be found in Edouard Guillon, *Nos écrivains militaires* (2 vols., Paris, 1898-1899), II, 207-219; Rudolf von Cämmerer, *The Development of Strategical Science during the Nineteenth Century* (London, 1905), chapter III; Henri Bonnal, *De la méthode dans les hautes études militaires en Allemagne et en France* (Paris, 1902), 15-17; and Spenser Wilkinson, *The French Army before Napoleon* (Oxford, 1915), 13-15. An excellent brief treatment of Jomini is that of Dallas D. Irvine, in "The French Discovery of Clausewitz and Napoleon," *Journal of the American Military Institute*, IV (1940), 145-146.

For the general background of the period of Napoleonic wars Crane Brinton's *A Decade of Revolution, 1789-1799* (New York, 1934) and Geoffrey Bruun's *Europe and the French Imperium, 1799-1814* (New York, 1938) may be consulted. These two volumes in the *Rise of Modern Europe* series, edited by W. L. Langer, summarize modern research and contain brief critical bibliographies. The latest scholarly work on Napoleon at this period is that of F. M. Kircheisen, *Napoleon I, sein Leben und seine Zeit*, 9 vols. (Munich and Leipzig, 1911-1934). Other valuable studies are: A. Chuquet, *Les guerres de la révolution*, 11 vols. (Paris, 1886-1896); R. W. Phipps, *The Armies of the First French Republic*, 3 vols. (Oxford, 1926-1929); A. Fournier, *Napoleon I, Emperor of the French* (New York, 1903), which is still the best short treatment; T. A. Dodge, *Napoleon; A History of the Art of War*, 4 vols. (Boston and New York, 1904-1907).

CHAPTER 5. CLAUSEWITZ

Clausewitz's military writings have been collected under the title *Hinterlassene Werke des Generals Carl v. Clausewitz über Krieg und Kriegführung* (Berlin, 1832-1837). The first three volumes of this collection contain the work *Vom Kriege*, the fifteenth separate edition of which appeared in 1937. A new translation of the work *On War* by O. J. Matthijs Jolles of the Institute of Military Studies, with a foreword by Richard McKeon, has been published by Modern Library (New York, 1943). Before the appearance of this translation a standard translation was Carl von Clausewitz, *On War*, translated by Colonel J. J. Graham, new and revised edition with introduction and notes by Colonel F. N. Maude, 3 vols. (London, 1918). (On the weaknesses of this translation see chapter 5, footnote 9). Another selection from Clausewitz's writings is *War According to Clausewitz*, edited with commentary by Major General T. D. Pilcher (London, New York, 1918). Volume VII of the *Hinterlassene Werke* has been translated under the title *The Campaign of 1812 in Russia* (London, 1843). In addition to this there is a translation of the memorandum written by Clausewitz in 1812 for the instruction of the Prussian Crown Prince: Carl von Clausewitz, *Principles of War*, translated and edited with an introduction by Hans W. Gatzke (Harrisburg, 1942). This last work was first published as an appendix to the book *On War*. (See above chapter 5, note 7.)

There are two collections of Clausewitz's letters and other writings: Karl Linnebach, *Karl und Marie von Clausewitz: Ein Lebensbild in Briefen und Tagebuchblättern* (Berlin, 1916) and Hans Rothfels, *Carl von Clausewitz, Politische Schriften und Briefe* (Munich, 1922). There are several notable general studies of Clausewitz. Karl Schwartz, *Leben des Generals Carl von Clausewitz . . .*, 2 vols. (Berlin, 1878), is a biography in the style of the standard "Life and Letters" biographies. Hans Rothfels, *Carl von Clausewitz, Politik und Krieg* (Berlin, 1920) uses a biographical basis and presents an interpretation of the interplay of military and political thought in Clausewitz.

Walter Malsten Schering, *Die Kriegsphilosophie von Clausewitz* (Hamburg, 1935) represents an attempt to analyze Clausewitz's specific philosophical method.

There are numerous special studies on Clausewitz which stress his influence upon military thought. R. v. Cämmerer, *Entwicklung der strategischen Wissenschaft* (Berlin, 1904), pp. 59-102, gives an evaluation of Clausewitz's contribution to the development of strategical conceptions, especially in Germany. Clausewitz's analysis of personal factors in generalship is treated in v. Freytag-Loringhoven, *Die Macht der Persönlichkeit in Kriege. Studien nach Clausewitz* (Berlin, 1905). Captain B. H. Liddell Hart, *The Ghost of Napoleon* (London, 1933); Alfred Vagts, *The History of Militarism* (New York, 1937); and Hoffman Nickerson, *The Armed Horde* (New York, 1940) are stimulating critical studies which emphasize particular aspects of Clausewitz's thought.

CHAPTER 6. ADAM SMITH, ALEXANDER HAMILTON, FRIEDRICH LIST: THE
ECONOMIC FOUNDATIONS OF MILITARY POWER

The map in this chapter is taken from Friedrich Lenz, *Friedrich List* (Munich and Berlin, 1936).

On mercantilism see the comprehensive work of Eli F. Heckscher, *Merkantilismen* (Stockholm, 1931), translated into English by M. Shapiro, *Mercantilism* (2 vols., London, 1935); J. W. Horrocks, *A Short History of Mercantilism* (London, 1925); P. W. Buck, *The Politics of Mercantilism* (New York, 1942); Gustav Schmoller, *The Mercantile System and Its Historical Significance* (translation by W. J. Ashley, London and New York, 1896); Jacob Viner, "English Theories of Foreign Trade before Adam Smith," in *Journal of Political Economy*, XXXVIII (1930), 249-301, 404-457; C. W. Cole, *Colbert and a Century of French Mercantilism* (New York, 1939).

Adam Smith's works relevant to this study are *An Inquiry into the Nature and Causes of the Wealth of Nations*, originally published in 1776 but in many subsequent editions of which the best were edited by Edwin Cannan, and are now available in the Modern Library; also *Lectures on Justice, Police, Revenue and Arms*, edited by Edwin Cannan as reported by a student in 1763 (Oxford, 1896); and *The Theory of Moral Sentiments* (London, 1759). John Rae's *Life of Adam Smith* (London, 1895) is the most satisfactory biography. The best general commentary is J. M. Clark and others, *Adam Smith, 1776-1926* (Chicago, 1928), a collection of lectures to commemorate the sesquicentennial of the publication of "The Wealth of Nations." For a German view, see A. Oncken, *Adam Smith in der Kulturgeschichte* (Vienna, 1874).

The Federal Edition of the *Works of Alexander Hamilton*, edited by Henry Cabot Lodge (12 vols., New York, 1904) is the best source of information concerning Hamilton. *The Federalist* (sesquicentennial edition, Washington, 1937, with an introduction by E. M. Earle) is available in the Modern Library. Another convenient volume is Samuel McKee, ed., *Papers on Public Credit, Commerce, and Finance by Alexander Hamilton* (New York, 1934). The best interpretations are W. S. Culbertson, *Alexander Hamilton* (New Haven, 1911), friendly to Hamilton; W. G. Sumner, *Alexander Hamilton* (New York, 1890), hostile to Hamilton. Ugo Rabbeno, *Protezionismo Americano: Saggi Storizi di Politico Comerciale* (Milan, 1893), translated into English under the title *American Commercial Policy* (London, 1895), Part III, Chap. I, is the best general criticism.

A labor of admiration and scholarship was performed by the Friedrich List

Gesellschaft, in cooperation with the German Academy, in publishing the complete works of Friedrich List: *Schriften, Reden, Briefe* (9 vols. and an index volume, Berlin, 1931-1935). This great work includes everything which List is known to have written, including some manuscripts not previously available. The best English edition of *The National System of Political Economy* is the translation by S. S. Lloyd, with an introduction by J. S. Nicholson (London, 1904). For biographical data concerning List see the introductory essay in each of the volumes of his collected works; also L. Häusser, *Friedrich Lists gesammelte Schriften* (3 vols., Stuttgart, 1850-1851), Vol. I; Karl Jentsch, *Friedrich List* (Berlin, 1910); Friedrich Lenz, *Friedrich List, der Mann und das Werk* (Munich and Berlin, 1936). For List's experiences in America: William Notz, "Friedrich List in Amerika," in *Weltwirtschaftliches Archiv* (April-July, 1925), pp. 199-293, and the same author's "Friedrich List in America," in the *American Economic Review*, XVI (1926), 249-265. A convenient summary in English of some of List's work is in Margaret E. Hirst, *Life of Friedrich List and Selections from Some of His Writings* (New York and London, 1909). For an estimate of List's part in American protectionism see Rabbeno, *op. cit.*, pp. 325-354.

On neo-mercantilism: C. J. H. Hayes, *A Generation of Materialism, 1871-1900* (New York, 1941), Chap. VI; J. Marchand, *La renaissance du mercantilisme à l'epoque contemporaine* (Paris, 1937); E. M. Earle, "The New Mercantilism," in the *Political Science Quarterly*, XL (1925), 594-600.

CHAPTER 7. ENGELS AND MARX: MILITARY CONCEPTS OF THE SOCIAL REVOLUTIONARIES

The authoritative collection *Karl Marx-Friedrich Engels. Historisch Kritische Gesamtausgabe* (Moscow, 1927-1935) includes the writings of Marx and Engels only to 1848 and is, therefore, of very limited value for the studies on military strategy, all of which were written after that date. The edition does, however, contain a complete collection of the most pertinent Marx-Engels correspondence (4 vols.). For an excellent English selection with commentary and notes see *Karl Marx and Friedrich Engels: Correspondence 1846-1895* (New York, 1934).

Important incomplete collections, including writings of the later period not otherwise available are: *Aus dem Literarischen Nachlass von Karl Marx, Friedrich Engels und Ferdinand Lasalle*, edited by Franz Mehring, 4 vols. (2nd ed., Stuttgart, 1913); *Gesammelte Schriften von Karl Marx und Friedrich Engels 1852 bis 1862*, edited by D. Ryazanov, 2 vols. (2nd ed., Stuttgart, 1920), which reprints a number of important articles from the *New York Tribune*, 1852-1855; *Notes on the War*, articles reprinted from the *Pall Mall Gazette*, 1870-1871, edited by Friedrich Adler (Vienna, 1923).

Most of the writings of Marx and Engels are available in new editions or recent reprints as well as in English translation. (See lists of International Publishers, New York.) A good collection of extracts from the writings of Marx and Engels will be found in Emil Burns, ed., *A Handbook of Marxism* (New York, International Publishers, 1935); V. Adoratsky, ed., *Karl Marx: Selected Works*, 2 vols. (Moscow, 1935).

These popular editions contain such valuable historical essays of Marx and Engels as: *Germany: Revolution and Counter-Revolution* (1852); *The Class Struggles in France 1848-1850* (1850), with Engels' significant introduction of 1895; *The Eighteenth Brumaire* (1852); *The Civil War in France* (1871), with Engels' introduction of 1891; and the prefatory notes of 1870 and 1874 to Engels' *Peasant War in Germany* (1850). Unfortunately they do not reprint the most important studies by Engels dealing specifically with military strategies, namely: *Po und Rhein* (Berlin, 1859); *Savoyen, Nizza und der Rhein* (Berlin, 1860); *Die Preussische Militärfrage und die Deutsche Arbeiterpartei* (Hamburg, 1865); and *Kann Europa abrüsten?* (Berlin, 1893). These writings are available only in the German original.

For a select bibliography of Engels' work see the compilation by Ernst Drahn in *Handwörterbuch des Staatswissenschaften*, II, 727-730 (4th ed., Jena, 1926). For biographical evaluation and critical analysis consult the standard work of Gustav Mayer, *Friedrich Engels, eine Biographie*, 2 vols. (Haag, 1934), and the shorter English biography by the same author *Friedrich Engels, a Biography* (London, 1936). Compare also D. Ryazanov, *Karl Marx and Friedrich Engels* (New York, 1927) and Franz Mehring, *Karl Marx, Geschichte seines Lebens* (Leipzig, 1918). The latter work has been translated by Edward Fitzgerald, *Karl Marx: the Story of his Life* (New York, 1935).

For Marxism and foreign politics in general see Hans Rothfels, "Marxismus und Auswärtige Politik," *Meinecke Festschrift: Deutscher Staat und Deutsche Parteien* (Munich, 1922); Oskar Blum, "Die Weltpolitischen Lehrjahre von Marx und Engels," *Archiv für Sozialwissenschaft und Sozialpolitik*, XLIV (1917-1918), 530-566; Hermann Oncken, *Historisch-politische Aufsätze*, Vol. II; and Hertneck, *Die deutsche Sozialdemokratie und die orientalische Frage im Zeitalter Bismarcks* (Diss., Berlin, 1927).

On Engels as a military leader consult August Happich, *Friedrich Engels als Soldat der Revolution* (*Hessische Beiträge zur Staat und Wirtschaftskunde*, 1931); Ernst Drahn, *Friedrich Engels als Kriegswissenschaftler* (*Kultur und Fortschrift*, Nos. 524-525); Max Schippel, "Friedrich Engels als Militärpolitischer Führer," *Sozialistische Monatshefte*, XXI (1915), 1222-1227; Max Schippel, "Die Miliz und Friedrich Engels," *Sozialistische Monatshefte*, XX (1914), 20-27; and Hugo Schulz, "Der General," *Der Kampf*, XVIII, 352-357 (Vienna, 1925). With the exception of the above mentioned biography by Gustav Mayer, which has many useful comments in this respect, the military strategy of Marx and Engels has been completely ignored in English literature.

CHAPTER 8. MOLTKE AND SCHLIEFFEN: THE PRUSSIAN-GERMAN SCHOOL

Map 1 in this chapter was adapted by G. F. Bush from F. W. Putzger, *Historischer Schul-Atlas*, Grosse Aufgabe, 52 Auflage (Bielefeld and Leipzig, 1935). Map 2 is from H. von Kuhl, *Der deutsche Generalstab in Vorbereitung und Durchführung des Weltkrieges* (Berlin, 1920). Maps 3 and 4 are from Bernhard Poll, *Deutsches Schicksal, 1914-1918* (Berlin, 1937).

The literature on the general history of the Prussian army during the nineteenth century is too large to be enumerated here. Any historical study of Prussian military legislation still has to start with the standard works on the military reforms after 1806: M. Lehmann, *Scharnhorst* (1886-1887); H. Delbrück, *Gneisenau* (3rd ed., 1908); F. Meinecke, *Boyen* (1896-1899). For the general military history, C. Frh. von der Goltz, *Kriegsgeschichte Deutschlands im 19. Jahrhundert* (1914), should be consulted as well as Volume V of H. Delbrück, *Geschichte der Kriegskunst* (1928). Although this section, written by E. Daniels, does not reach the level of Delbrück's earlier four volumes it constitutes a useful compilation. The best introduction to the specialized study of nineteenth century strategy is: R. von Cämmerer, *Entwicklung der strategischen Wissenschaft im 19. Jahrhundert* (1904) (English translation, London, 1905). A more recent essay on modern strategy is to be found in the article "Kriegskunst" by T. von Schaefer in the military dictionary *Handbuch der neuzeitlichen Wehrwissenschaften*, ed. by H. Franke (1936), Vol. I, pp. 180-227. Two books apparently devoted to the study of modern generalship *Führertum* and *Heerführer des Weltkrieges*, both ed. by F. von Cochenhausen before 1939, could not be located. General Wetzell reviewed the second book in three articles in the *Militär-Wochenblatt* (1939), pp. 2257-2263, 2329-2338, 2406-2409, under the title "Das Bild des modernen Feldherrn." Although concentrating upon the first World War these articles contain a competent treatment of nineteenth century strategy as well.

No special history of the Prussian general staff has been written. A group of older officers tried to fill the gap, following a suggestion by General Groener. The symposium was published in 1933 under the title, *Von Scharnhorst zu Schlieffen*, ed. by F. von Cochenhausen. The articles are uneven and not very critical, but contain some information not available elsewhere. The constitutional problem of the general staff was studied by G. Wohlers, *Die staatsrechtliche Stellung des Generalstabes in Preussen und Deutschland* (1920).

The writings of Moltke were collected after his death in two large editions: H. von Moltke, *Gesammelte Schriften und Denkwürdigkeiten*, 8 vols. (1891-1894); and *Militärische Werke*, ed. by the Prussian general staff, 13 vols. (1892-1912). These editions do not contain his memoranda on the problems of a two-front war during 1871-1890. They were edited by F. von Schmerfeld: H. Graf von Moltke, *Die deutschen Aufmarschpläne 1871-1890, Forschungen und Darstellungen aus dem Reichsarchiv*, Vol. VII (1928). (A brief analysis and description will be found in P. Rassow, *Der Plan des Feldmarschalls Grafen Moltke für den Zweifronten-Krieg, 1871-1890* [1936].) Additional material on Moltke's thought about the two-front war is to be found in Vol. VI of the German publication on the origins of the World War, *Die Grosse Politik der Europäischen Kabinette, 1871-1914*. Of some documentary value is the study prepared by the historical section of the Prussian general staff: *Moltke in der Vorbereitung und Durchführung der Operationen* (*Kriegsgeschichtliche Einzelschriften*, XXXVII, 1905).

The existing biographical studies of Moltke are only of modest value. The most suggestive is the one by W. von Blume (2nd edition, 1907). Others were written by M. Jähns (1894-1900), W. Bigge (1900), F. Freiherr von der Goltz (1903), Spencer Wilkinson (*The Early Life of Moltke*, 1913), A. von Janson (1915), and H. von Seeckt (1931).

For the study of Moltke's strategy the military histories of the wars of 1866 and 1870-1871 should be consulted. They are enumerated in Dahlmann-Waitz, *Quellenkunde der deutschen Geschichte*, 9th edition, ed. by H. Haering (Leipzig 1931), nos. 14382-14405; 14440-14491. Among the general historical works most useful for the understanding of Moltke's strategy are H. Friedjung, *Der Kampf um die Vorherrschaft Deutschlands* (1st edition, Stuttgart, 1896; 10th edition, 1916-1917), and O. von Lettow-Vorbeck, *Geschichte des Krieges von 1866 in Deutschland* (1896-1902).

The monographical studies of Moltke's strategy are even more important. First place should be given to General von Schlichting's monograph *Moltke und Benedek* (1900), one of the classics of modern strategy. Schlichting's study was written as a critique of the military chapters of Friedjung's historical work and arrived through its higher understanding of the military and strategic problems at a fairer historical judgment on both victors and vanquished of 1866. A long debate developed which is summed up in later editions of Friedjung. Out of the literature the following articles should be noted: A. Krauss, *Moltke, Benedek und Napoleon* (1901); H. Delbrück, "Moltke,"

Erinnerungen, Aufsätze und Reden (1902) ; A von Bogulawski, *Strategische Erörterungen* (1901), Frh. von Freytag-Loringhoven, *Die Heerführung Napoleons in ihrer Bedeutung für unsere Zeit,* 1909. Of high interest is, of course, Schlieffen's treatment of Moltke's strategy in his *Cannae* articles. (See below.)

The impact of railroad building on modern strategy was treated by E. A. Pratt, *The Rise of Rail-Power in War and Conquest, 1833-1914* (1915). For the history of the German railroads as means of warfare see H. von Staabs, *Aufmarsch nach zwei Fronten, auf Grund der Operationspläne von 1871-1914* (1925). His successor as chief of the railroad section of the German general staff, W. Groener, contributed an article on the railroad mobilization in 1914 to the work *Die deutschen Eisenbahnen der Gegenwart* (new edition 1923, ed. by Prussian Ministry of Public Works). Since then the subject has found a full treatment in the official German history of the World War: Reichsarchiv, *Der Weltkrieg: Das deutsche Feldeisenbahnwesen*; Vol. I: *Die Eisenbahnen zu Kriegsbeginn,* 1928.

A good many studies deal with the relationship between politics and strategy. Out of them the following may be mentioned: W. von Blume, "Politik und Strategie. Bismarck und Moltke," *Preussische Jahrbücher,* Vol. CXI (1903) ; W. Busch, *Bismarck und Moltke* (1916) ; H. von Haeften, "Bismarck und Moltke," *Preussische Jahrbücher,* Vol. CLXXVII (1919) ; P. Schmitthenner, *Politik und Kriegsführung in der neuesten Geschichte* (1937).

Shortly after Schlieffen's death his published articles and public speeches were collected under the title: Graf Alfred von Schlieffen, *Gesammelte Schriften,* 2 vols. (Berlin, 1913). An abbreviated edition of these collected writings appeared in 1925 under the title *Cannae.* The bulk of both editions is formed by the series of studies which Schlieffen devoted to the encirclement battles from Cannae to Sedan. An abbreviated English translation of the Cannae articles was published in 1931 at Fort Leavenworth, Kansas. The most important addition to the writings of Schlieffen is contained in the luxurious edition of his official writings started by the German general staff six years ago: *Dienstschriften des Chefs des Generalstabes der Armee, Generalfeldmarschall Graf von Schlieffen*; Vol. I: *Die taktisch-strategischen Aufgaben aus den Jahren 1891-1905* (Berlin, 1937) ; Vol. II: *Die Grossen Generalstabsreisen Ost aus den Jahren 1891-1905* (Berlin, 1938). The present war apparently interrupted the progress of the edition. Though the two volumes afford a full opportunity for the study of the Tannenberg idea and very intimate glimpses of the growth of Schlieffen's strategic conceptions with regard to a war against France, they do not contain the official plans for operations in the west which were apparently reserved for publication in subsequent volumes. The best historical sources for the "Schlieffen plan" today are still H. von Kuhl, *Der deutsche Generalstab in Vorbereitung und Durchführung des Weltkrieges,* 2nd edition (Berlin, 1920) ; W. Foerster, *Graf Schlieffen und der Weltkrieg* (Berlin, 1921) ; the official German history of the first World War: Reichsarchiv, *Der Weltkrieg* (Berlin, 1925) ; R. von Collenberg, "Graf Schlieffen und die deutsche Mobilmachung," *Wissen und Wehr* (1927) ; W. Foerster, *Aus der Gedankenwerkstatt des deutschen Generalstabs* (Berlin, 1931).

Biographical sketches of Schlieffen were written by the director of the historical section of the general staff, General Freiherr von Freytag-Loringhoven (1920), by H. Rochs (1921), W. Elze (1928), F. von Boetticher (1933), E. Bircher (1937). H. A. DeWeerd's article on Schlieffen in his *Great Soldiers of the Two World Wars* (1941), is a well balanced portrait and critical study of his strategical ideas.

The discussion of Schlieffen's strategic ideas runs like a red thread through all modern German books on strategy. It plays the greatest part in the German critique of the operations of the first World War. In addition to the above mentioned studies by H. von Kuhl and W. Foerster and the official German history of the World War, which was written chiefly under the direction of General H. von Haeften, the outstanding work came from the pen of General W. Groener, who was chief of the railroad section of the general staff in 1914, and succeeded Ludendorff in the fall of 1918. As minister of war under the Republic he became one of the chief fathers of the modern German army and its strategy. His *Das Testament des Grafen Schlieffen* (Berlin, 1927) is the most distinguished and profound study of Schlieffen. Groener supplemented it later by his *Der Feldherr wider Willen* (Berlin, 1931), a study of the strategy of the younger Moltke. The veneration enjoyed by Schlieffen in German military circles is almost general. A good expression of it is found in a special issue of the *Militärwissenschaftliche Rundschau* in 1938: Lieutenant General von Zoellner, *Schlieffens Vermächtnis*. The chief opponent of Schlieffen before 1914, General F. von Bernhardi, has found no followers in the younger generation. However, there has been a school of military thought which placed Moltke the Elder above Schlieffen, criticizing either the rigidity of Schlieffen's operational schemes or recommending Moltke's idea of an offensive in the east as the better solution of the two-front war. The best representative of this school is probably General E. Buchfinck. See his article "Moltke und Schlieffen," *Historische Zeitschrift*, Vol. CLVIII (1938). General Wetzell's articles, mentioned above, seem partly motivated by a desire to justify Ludendorff's approval of Moltke's deployment of the German forces (see in addition his article in *Militär-Wochenblatt*, 1934, pp. 843 ff.). Ludendorff had been chief of the mobilization section of the general staff prior to August 1914. Ludendorff himself defended the change of the Schlieffen plan by the younger Moltke in an article in *Deutsche Wehr* (1930).

The French study, J. Courbis, *Le comte Schlieffen, organisateur et stratège* (Paris, 1938), did not use the new Schlieffen sources of 1937-1938, quoted above, nor did it penetrate deeply into the Schlieffen problem. The introduction by General Daille dwelt upon the irrational philosophy of the German military school, which seemed unbearable to the logical Latin mind, and Daille assured the French that they had nothing to fear from a repetition of the Schlieffen plan.

J. V. Bredt, *Die belgische Neutralität und der Schlieffensche Feldzugplan* (1929), is the chief source for the treatment of the Belgian question in German military and political circles before 1914. Special volumes of the official German history of the World War show the influence of Schlieffen's concept of modern war upon the economic and financial preparations in Germany: Reichsarchiv, *Der Weltkrieg, Kriegsrüstung und Kriegswirtschaft*, Vol. I and Vol. I, Annexes.

CHAPTER 9. DU PICQ AND FOCH: THE FRENCH SCHOOL

The basic studies of the development of the French general staff are those by Dallas D. Irvine, "The French and Prussian Staff Systems before 1870," *Journal of the American Military History Foundation*, II (1938), 192-203; "The French Discovery of Clausewitz and Napoleon," *Journal of the American Military Institute*, IV (1941), 143-161; and "The Origin of Capital Staffs," *Journal of Modern History*, X (1938), 161-179. On French military institutions in general see Joseph Monteilhet, *Les Institutions Militaires de la France 1814-1932*, second edition (Paris, 1917).

Du Picq's writings are few in number, and practically nothing has been written on him. The French edition of his *Etudes sur le Combat* (Paris, 1880) contains a preface which is the only available biographical source. This work has been translated into English: C. Ardant Du Picq, *Battle Studies Ancient and Modern Battle*, translated from the eighth edition in the French by Colonel John N. Greely and Major Robert C. Cotton (New York, 1921). Unfortunately this work is not complete.

On Du Picq's intellectual background and the times in which he wrote, see Auguste Frederick Marmont, *Esprit des Institutions Militaires* (Paris, 1845); Count Henri Amedée le Lorgne Ideville, *Memoirs of Marshal Bugeaud* (based on Bugeaud's private correspondence and original documents 1784-1849, edited from the French by Charlotte M. Yonge); and Louis Jules Trochu, *L'Armée Française en 1867* (Paris, 1867). The military events of the period leading up to the Franco-Prussian war and of the war itself will be found in Emile Ollivier, *L'Empire Libéral* (Paris, 1898-1916), volumes XIV-XVII. See also Pierre de la Gorce, *Histoire du Second Empire*, 7 vols. (Paris, 1908-1911). For the popularity of Du Picq's work during the First World War see Jean Norton Cru, *Témoins. Essai d'analyse et de critique des souvenirs de combattants* (Paris, 1929).

Foch's writings are readily available and the following works may be consulted for the significant aspects of his theory: *Des Principes de la Guerre* (*The Principles of War*), translated by Hilaire Belloc (New York, 1920); *De la Conduite de la Guerre* (third edition, Paris, 1915); *The Memoirs of Marshal Foch*, translated by Colonel T. Bentley-Mott (New York, 1931); and *Morale in War*, under "Army" in *Encyclopaedia Britannica*. There have been in addition many interpretations of Foch's theory and many first-hand accounts of Foch the soldier. Perhaps the most notable of these are Raymond Recouly, *Marshal Foch, His Own Words on Many Subjects* (London, 1929); Charles Bugnet, *Foch Speaks* (London, 1929); C. Le Goffic, *Mes entretiens avec Foch* (Paris, 1929); Louis Madelin, *Le Maréchal Foch* (Paris, 1929); H. Bidou, "Le Maréchal Foch, écrivain militaire," *Minerve Française* (February 1920); Major General Sir G. Aston, *The Biography of the Late Marshal Foch* (New York, 1929); B. H. Liddell Hart, *Foch, the Man of Orleans* (London, 1931); and André Tardieu, *Avec Foch* (Paris, 1939).

Interesting works on the question of morale and the role of the battle in warfare are Gustave Le Bon, *Hier et Demain* (Paris, 1918); Charles Coste, *La Psychologie du Combat*, preface by Jacques Chevalier (second edition, Paris, 1929); Louis Huot and Paul Voivenel, *Le Courage* (Paris, 1917); General Percin, *Le Combat* (Paris, 1914); and G. Stanley Hall, *Morale, the Supreme Standard of Life and Conduct* (New York, 1920).

CHAPTER 10. BUGEAUD, GALLIÉNI, LYAUTEY: THE DEVELOPMENT OF FRENCH COLONIAL WARFARE

Acknowledgment is made to the *Geographical Review* for the use of the two maps in this chapter.

The most authoritative works on Bugeaud are those of General Paul Azan. See his *Bugeaud et l'Algérie* (Paris, 1930); *Conquête et pacification de l'Algérie* (Paris, 1931); *L'Armée d'Afrique de 1830 à 1852* (Paris, 1936); and *Les grands soldats de l'Algérie* (*Cahiers du Centenaire de l'Algérie*, No. 4, Paris, 1930). Other valuable studies by the same author are *L'expédition d'Alger* (Paris, 1929); *L'expédition de Fez*, introduction by Lyautey (Paris, 1924); and *L'Armée indigène nord-africaine* (Paris, 1925). Bugeaud's theories of mountain warfare are treated systematically in Général Huré, "Stratégie et tactique Marocaines," *Revue des Questions de Défense Nationale*, I, No. 3 (July 1939), pp. 397-412. For his administrative system, see Albert Ringel, *Les "bureaux arabes" de Bugeaud et les "cercles militaires" de Galliéni* (Paris, 1903).

Galliéni's writings are readily accessible. See especially *La Pacification de Madagascar, 1896-1899* (Paris, 1900) ; *Neuf ans à Madagascar* (Paris, 1908) ; "Lettres de Madagascar," *Revue des Deux Mondes*, XLIV (1928), 776-809, and XLV (1928), 63-86; and *Les Carnets de Galliéni* (Paris, 1932). Among secondary works on Galliéni, those of Emile-Felix Gautier are of special interest. Gautier, a famous geographer, worked with both Galliéni and Lyautey in Madagascar. See his "Le Général Galliéni," *Annales de Géographie* (Paris, 1916), 310-313; "Documents d'archives soundanais concernant le Général Galliéni," *La Géographie*, XLII (1924), 133-146; and *La Conquête du Sahara, Essai de Psychologie Politique* (Paris, 1909). Albert Ringel's work, cited above, throws light upon Galliéni's theories of administration. Of the many other treatments of Galliéni's achievement, the following should be noted: Captain B. H. Liddell Hart, "Galliéni," *Atlantic Monthly*, CXL (1927), 354-364; Lallier du Coudray, "Galliéni et Lyautey," *Bulletin de la Société de Géographie et d'études coloniales de Marseille*, XLVIII (1925), 5-21; René Musset, "Galliéni et Madagascar" in *Mélanges de Géographie et d'orientalisme offerts à E. F. Gautier* (Tours, 1937), pp. 388-390; and Jerome et Jean Tharaud, "Galliéni et Lyautey," *Revue Hebdomadaire*, XII (1920), 125-137.

Lyautey's literary remains are more extensive than those of either of his predecessors. In the order of their publication, they are: *Du rôle social de l'officier*; *Du rôle colonial de l'Armée* (Paris, 1900) ; *Dans le Sud de Madagascar* (Paris, 1903) ; *Paroles d'Action* (Paris, 1927) ; *Lettres du Tonkin et de Madagascar* (Paris, 1920) ; *Intimate letters from Tonquin* (London, 1932) ; *Lettres de Jeunesse* (Paris, 1931) ; *Vers le Maroc. Lettres du Sud-Oranais* (Paris, 1937) ; *Lettres du Sud de Madagascar* (Paris, 1935) ; and "Lyautey d'après lui-même," *Revue Universelle*, LXIII (1935), 129-146. Secondary works which discuss the various phases of Lyautey's career are numerous. In addition to those which treat both Lyautey and Galliéni and which are cited above, the following are most noteworthy: Louis Barthou, *La Bataille du Maroc* (Paris, 1919), *Lyautey* (Paris, 1929), and "Le sourire de Lyautey au Maroc," *Revue des Deux Mondes*, LX (1930), 580-590; Augustin Bernard, "L'oeuvre du Maréchal Lyautey," *La Geographie* (1920), pp. 337-360; Henri Bidou, "Réception du Général Lyautey à l'Académie Française," *Revue des Deux Mondes*, LVIII (1920), 639-644; Général Gouraud, "Le Maréchal Lyautey," *Revue des Deux Mondes*, XXII (1934), 925-931; and André Maurois, *Lyautey* (Paris, 1931).

On the general question of the pacification of Morocco, see Lieutenant Colonel Bernard, "Les étapes de la Pacification Marocaine (1907-34)." *Renseignements Coloniaux*, suppl. to the *Bulletin du Comité de l'Afrique Française* (1936), No. 8, pp. 113-150; Général Catroux, "L'Achèvement de la Pacification Marocaine," *Revue Politique et Parlementaire*, CLXI (1934), No. 479, pp. 24-46; Capitaine P. Vallerie, "La penetration militaire au Maroc, 1934," in *Revue de l'Infanterie* (March 1, 1935) (the whole issue is on the pacification of Morocco) ; and René Pinon, *Au Maroc: la fin des temps héroiques* (Paris, 1935). From the purely military point of view, two works by General Fabre are valuable studies: *Le Bataillon au Combat* and *La Tactique au Maroc*. Of special interest, in view of North Africa's role in the present war are General Catroux, "La Position Stratégique de l'Italie en Afrique du Nord,"

Politique Etrangère (Paris, June 1939), pp. 271-281; General Armengaud, "La Securité de notre Afrique du Nord," in *Revue des Deux Mondes*, XLIX (1939), 550-564; Ignotus, "Henri Giraud, Général d'Armée," *Revue de Paris* (1939), No. 6, pp. 777-786; General Paul Azan, "La Bataille de Tunisie," *L'Illustration*, CCIII (July 8, 1939), 361-364; and G. L. Steer, *A Date in the Desert* (London, 1939).

CHAPTER II. DELBRÜCK: THE MILITARY HISTORIAN

Delbrück's first major work was *Das Leben des Feldmarschalls Grafen Neidhardt von Gneisenau* (Berlin, 1882). This work, which has gone through four editions since its initial publication, remains the standard biography of the Prussian general. In *Die Perserkriege und die Burgunderkriege: zwei kombinierte kriegsgeschichtliche Studien* (Berlin, 1887), Delbrück first clearly outlined his method of approaching military history and his conception of the importance of reconstructing single battles. Early full scale presentations of his strategical theories will be found in *Die Strategie des Perikles erläutert durch die Strategie Friedrichs des Grossen: mit einem Anhang über Thucydides und Kleon* (Berlin, 1890); and *Friedrich, Napoleon, Moltke* (Berlin, 1892).

The first volume of the *Geschichte der Kriegskunst im Rahmen der politischen Geschichte* appeared in 1900; the second in 1902; and the third and fourth in 1907 and 1920 respectively. A second edition of the first two volumes (Berlin, 1908) and a third edition of the first volume (Berlin, 1920) contain additional notes and answers to critics but are otherwise unchanged. The fourth volume of the *Geschichte*, the last which Delbrück wrote, ends with an account of the wars of Liberation. The work was continued by Emil Daniels; the fifth and sixth volumes, covering the period between the Crimean and Franco-Prussian wars, appearing in 1928 and 1932. In 1936, a seventh volume, which discusses the American Civil War and the Boer and Russo-Japanese wars, was published under the joint authorship of Daniels and Otto Haintz.

Hans Delbrück, *Numbers in History* (London, 1913) is a reprint of two lectures delivered by the historian at the University of London in 1913. This volume, in brief compass and in Delbrück's own words, surveys the first three volumes of the *Geschichte der Kriegskunst* and outlines the main themes.

Delbrück's shorter military writings are scattered through the pages of the *Preussische Jahrbücher* and other publications. There are, however, four collections of the articles which Delbrück himself considered most important. *Historische und politische Aufsätze* (Berlin, 1886; second ed., 1907) contains an important essay, "Über die Verschiedenheit der Strategie Friedrichs und Napoleons." *Erinnerungen, Aufsätze und Reden* (Berlin, 1902; third ed., 1905) includes an article on the work of the general staff in the Danish War of 1864, in addition to a notable essay on Moltke. Delbrück's World War writings have been collected in the three volumes of *Krieg und Politik* (Berlin, 1917-1919). A final collection appeared in 1926 under the title *Vor und nach dem Weltkrieg* and includes Delbrück's most important articles for the periods 1902-1914 and 1919-1925.

For Delbrück's position during the World War, see the collections cited above, and also the pamphlet *Bismarcks Erbe* (Berlin, 1915), which is perhaps his most impassioned plea for a negotiated peace with the Allies. Delbrück's masterly critique of Ludendorff's strategy in 1918 is printed in *Das Werk des Untersuchungsausschusses der Deutschen Verfassunggebenden Nationalversammlung und des Deutschen Reichstages 1919-1926. Die Ursachen des Deutschen Zusammenbruchs im Jahre 1918* (Vierte Reihe im Werk des

Untersuchungsausschusses) (Berlin, 1925), III, 239-373. Selections from Delbrück's testimony will be found also in R. H. Lutz, ed., *The Causes of the German Collapse in 1918*, Hoover War Library Publications, No. 4 (Stanford, 1934).

Many of Delbrück's afterthoughts on the war exist only in pamphlet form. See, for example, *Ludendorff, Tirpitz, Falkenhayn* (Berlin, 1920); *Ludendorffs Selbstporträt* (Berlin, 1922), an answer to Ludendorff's *Kriegführung und Politik* (Berlin, 1922); *Kautsky und Harden* (Berlin, 1920); and *Der Stand der Kriegsschuldfrage* (Berlin, 1925). The last two works are largely concerned with the question of war guilt.

Even an incomplete listing of Delbrück's works must also include his *Regierung und Volkswille* (Berlin, 1914), a series of lectures on the imperial government and constitution; and his five volume *Weltgeschichte* (Berlin, 1924-1928). The former work has been translated into English by Roy S. MacElwee under the title *Government and the Will of the People* (New York, 1923).

No full scale biography of Delbrück has yet been written. For biographical details, consult the introductions to volumes I and IV of the *Geschichte der Kriegskunst* and the epilogue to *Krieg und Politik*; and see also Johannes Ziekursch in *Deutsches biographisches Jahrbuch* (Berlin, 1929) and Friedrich Meinecke in *Historische Zeitschrift*, CXL (1929), 703. Richard H. Bauer's article in Bernadotte Schmitt, ed., *Some Historians of Modern Europe* (Chicago, 1942), pp. 100-127, is a careful account of Delbrück's life and work although Delbrück's military writings are treated only in a general manner. F. J. Schmidt, Konrad Molinski and Siegfried Mette in *Hans Delbrück: der Historiker und Politiker* (Berlin, 1928) discuss the philosophical basis of Delbrück's writings and his importance as a historian and a politician. The historian's political and military ideas are also treated fully in *Am Webstuhl der Zeit: eine Erinnerungsgabe Hans Delbrück dem Achtzigjährigen von Freunden und Schülern dargebracht* (Berlin, 1928), a collection of essays by Emil Daniels, Paul Rohrbach, Generals Groener and Buchfinck, and others. See also Arthur Rosenberg, "Hans Delbrück, der Kritiker der Kriegsgeschichte," *Die Gesellschaft*, VI (1921), 245; Franz Mehring, "Eine Geschichte der Kriegskunst," *Die Neue Zeit*, Erganzungsheft, no. 4 (October 16, 1908); and V. Marcu, *Men and Forces of Our Time*, translated by Eden and Cedar Paul (New York, 1931), pp. 201 ff.

Delbrück's strategical theories gave rise to a flood of controversial literature. The most important articles appearing before 1920 are listed in *Geschichte der Kriegskunst*, IV, 439-444. For a more recent and thorough appraisal of Delbrück's strategical concepts see Otto Hintze, "Delbrück, Clausewitz und die Strategie Friedrichs des Grossen," *Forschungen zur Brandenburgischen und Preussischen Geschichte*, XXXIII (1920), 131-177.

CHAPTER 12. CHURCHILL, LLOYD GEORGE, AND CLEMENCEAU: THE EMERGENCE OF THE CIVILIAN

On Churchill, see his own four volume personal history, *The World Crisis* (London and New York, 1923-1928). Not only do these volumes contain some of the finest prose in recent British literature, but they are a valuable source of information on many phases of the First World War. They include the principal papers and memoranda written by Churchill as First Lord of the Admiralty and Minister for Munitions. An appraisal of Churchill's work in the Admiralty is found in R. M. Dawson, "The Cabinet Minister and Administration: Winston S. Churchill at the Admiralty, 1911-1915," *Canadian Journal of Economics*

and Science, VI (August 1940), 325-358. For Churchill's role in the development of the tank, see Colonel Ernest D. Swinton, *Eyewitness: Being Personal
Reminiscences of Certain Phases of the Great War, Including the Genesis of
the Tank* (London, 1933).

David Lloyd George's *War Memoirs* (London and New York, 1933-1937),
in six volumes, is the personal record of the wartime prime minister. Though
condemned by many writers as biased and presenting a distorted picture of
Haig and Robertson, this memoir must be regarded as an important part of the
literature on the First World War. The War Office point of view in the controversy with Lloyd George will be found in Field Marshal Sir William Robertson, *Soldiers and Statesmen* (London, 1926) and *From Private to Field
Marshal* (London and New York, 1921). See also R. M. Dawson, "The Cabinet Minister and Administration: Asquith, Lloyd George, Curzon," *Political
Science Quarterly*, LV (September 1940), 348-377.

The problem of preparing for the war of materiel, with special reference to
British experience, is ably discussed in Charles W. Baker, *Government and
Operation of Industry in Great Britain and the United States during the World
War* (New York, 1921) and in Brigadier General Sir J. E. Edmonds, *Military
Operations, France and Belgium* (London, 1930-1941). The latter work, in the
volumes for 1915 and 1916, discusses the conditions of trench warfare and the
need for materiel. For general background material on the problem of British
military organization, see R. B. Haldane, *Richard Burdon Haldane: An Autobiography* (London, 1929).

The best available biography of Rathenau is H. Kessler, *Walther Rathenau,
sein Leben und sein Werk* (Berlin, 1928). An English translation was pub-
lished in New York in 1929. Rathenau's contribution to the organization of
German war economy in the First World War is also treated in Ernest Juenger,
Die totale Mobilmachung (Berlin, 1931). For general information on the
German experience in the First World War see also Albrecht Mendelssohn-
Bartholdy, *The War and German Society* (New Haven, 1937) ; and Frank P.
Chambers, *The War Behind the War: A History of Political and Civilian
Fronts* (New York, 1939).

On Clemenceau, see his *Grandeurs et misères d'une victoire* (Paris, 1930).
This was Clemenceau's reply to Foch's posthumous attack on him made through
the pages of R. Recouly's *Mémorial de Foch* (Paris, 1930). The latter work
is based on a series of interviews with Foch in his last years, and its publication
let loose a flood of controversial literature. Clemenceau's military adviser, J. H.
Mordacq, gives a general account of Clemenceau's war service in his *Le
ministère Clemenceau: journal d'un témoin* (Paris, 1930), a work which should
be used with caution.

S. C. Davis, *The French War Machine* (London and New York, 1937) is
an examination of French military institutions in general, the early chapters
dealing with the problems of 1914-1918. The relation of the French government to the conduct of the war is discussed in Part III of the third chapter
of Pierre Renouvin's *The Forms of War Government in France* (New Haven,
1927). The most valuable single work on the relations of the French government and the high command is Lieutenant Colonel Charles Bugnet, *Rue St.
Dominique et G.Q.G.* (Paris, 1937). See also J. M. Bourget, *Gouvernment et
commandement: les leçons de la guerre mondiale* (Paris, 1930), a valuable
addition to the series *Collection de memoires, études et documents pour servir
a l'histoire de la guerre mondiale*.

The problem of unified command during the First World War is discussed
in Tasker H. Bliss, "The Unified Command," *Foreign Affairs*, I (December

1922), 1-30 and Brigadier General E. L. Spears, *Prelude to Victory* (London, 1939). The latter is a brilliant account of the difficulties of waging coalition warfare under the conditions of 1917-1918, written by the British liaison officer with the French G.Q.G.

On United States military policy see J. H. Mordacq, *Politique et stratégie dans une démocratie* (Paris, 1912), a study of civil and military relations in democratic states which devotes a long chapter to Union and Confederate practices in the Civil War. A sound treatment of American military policy through the Spanish American War is F. V. Greene, *The Revolutionary War and the Military Policy of the United States* (New York, 1911). A striking example of the capacity of a first rate civilian mind to influence military institutions is Elihu Root's *Military and Colonial Policy of the United States* (Cambridge, Mass., 1916), a collection of the most important papers dealing with Root's reforms in the War Department. H. White, *Executive Influence in Determining Military Policy in the United States* (Urbana, Illinois, 1925) is a doctoral thesis covering the influence of the executive in American military policy through the Defense Act of 1920.

One of the most valuable sources of information on the civilian control of war economy in the United States is Bernard Baruch's *American Industry in the War: A Report of the War Industries Board* (New York, 1941). This work, the final report of the chairman of the War Industries Board of 1918, should be considered in connection with Grosvenor Clarkson's *Industrial America in the World War: the Strategy Behind the Line, 1917-1918* (New York, 1923). Other treatments of the problem of military policy in the United States are Brigadier General John McAuley Palmer, *America in Arms: The Experience of the United States with Military Organization* (New Haven, 1941); Pendleton E. Herring, *The Impact of War: Our American Democracy under Arms* (New York, 1941), an examination of the civil-military relationship in the United States; and the articles of Edward Mead Earle, "Military Policy and Security," *Political Science Quarterly*, LIII (March 1938), 4-12; "National Security and Foreign Policy," *Yale Review*, XXIX (Spring 1940), 444-460; and "Political and Military Strategy for the United States," *Proceedings of the Academy of Political Science*, XIX (January 1941), 112-119. A recent article on the general question of the role of the civilian is Lindsay Rogers, "Civilian Control of Military Policy," *Foreign Affairs*, XVIII (January 1940), 280-291.

Among the general works, Lewis Mumford's *Technics and Civilization* (New York, 1934) is a brilliant discussion of the impact of technical developments on society, including the military. Jesse D. Clarkson and Thomas C. Cochran (eds.), *War as a Social Institution* (New York, 1941) is a collection of essays showing the historian's perspective on wartime problems. Chapters bearing on the relation of the civilian to modern war include: "Civilian and Military Elements in Modern War" by H. A. DeWeerd; "The Social and Political Aspects of Conscription: Europe's Experience" by Colonel Herman Beukema; "War and Modern Dictatorships" by Arthur Rosenberg; and "War and Economic Institutions" by Charles E. Rothwell. Another valuable symposium is Sir George Aston (ed.), *The Study of War for Statesmen and Citizens* (London, 1927), a work based on the assumption that civilians must have some knowledge of the basic factors in war. Major General Sir F. K. Maurice, *Government and War: A Study of the Conduct of War* (London, 1926) is a study of the relationship of government in a democracy to the conduct of war written by a distinguished British soldier and writer. Major General J. F. C. Fuller, *War and Western Civilization: A Study of War as a*

Political Instrument and the Expression of Mass Democracy (London, 1932) is the comment of a brilliant military thinker on the problems of war and the democratic form of government, some of the political and social views of which need not be taken at their face value.

CHAPTER 13. LUDENDORFF: THE GERMAN CONCEPT OF TOTAL WAR

Ludendorff's theories of total war are presented most systematically in his *Der Totale Krieg* (Munich, 1935). He has left a record of his war experiences and the conclusions drawn therefrom in *Meine Kriegserinnerungen* (Munich, 1919) and *Kriegführung und Politik* (Berlin, 1922). The former work has been published in an English edition: *My War Memories: Ludendorff's Own Story, August 1914-November 1918* (New York, 1919). Ludendorff's ideas on religion and on the race question may be found in the pamphlet *Friedrich der Grosse auf Seiten Ludendorffs. Friedrichs des Grossen Gedanken über Religion. Aus seinen Werken* (Munich, 1935) and in E. Ludendorff, *Die Judenmacht, ihr Wesen und Ende* (Munich, 1939). Interesting for an understanding of Ludendorff's connection with National Socialism in the early years of the movement is E. Ludendorff, *Auf dem Weg zur Feldherrnhalle. Lebenserinnerungen aus der Zeit des 9 November 1923; mit Dokumenten in 6 Anlagen* (Munich, 1938).

Innumerable pamphlets and books have been written attacking or defending Ludendorff. Perhaps the most scathing is Hans Delbrück, *Ludendorffs Selbstporträt* (Berlin, 1922), an answer to the general's *Kriegführung und Politik*. For a more temperate appraisal of Ludendorff's career, see Karl Tschuppik, *Ludendorff: the Tragedy of a Military Mind*, translated by W. H. Johnston (Boston, New York, 1932).

For military questions in general in the pre-first World War period, see Herbert Rosinski, *The German Army* (London, New York, 1939) and R. Schmidt-Bückeburg, *Das Militär-Kabinett der preussischen Könige und deutschen Kaiser* (Berlin, 1933), and consult bibliography to chapter 8. Among the general works, Carl Schmitt, *Staatsgefüge und Zusammenbruch des zweiten Reiches* (Hamburg, 1934) and Arthur Rosenberg, *Die Entstehung der deutschen Republik* (Berlin, 1928) are especially valuable. Competent discussions of events on the home front and of the impact of the war will be found in Albrecht Mendelssohn-Bartholdy, *The War and German Society* (New Haven, 1937) ; and Frank P. Chambers, *The War Behind the War, 1914-1918: A History of the Political and Civilian Fronts* (New York, 1939). A basic source of materials for the study of Germany's economic position in the war is Reichsarchiv, *Der Weltkrieg 1914 bis 1918. Kriegsrüstung und Kriegswirtschaft* (Berlin, 1930). See also Germany: Reichstag: *Untersuchungsausschuss: Die Ursachen des deutschen Zusammenbruchs im Jahre 1918* (Berlin, 1922-) and R. H. Lutz, ed., *The Causes of the German Collapse in 1918*, Hoover War Library Publications, No. 4 (Stanford, 1934). The works of Ernst Juenger are valuable for an understanding of the lessons drawn by the Germans themselves from their experience in the years 1914-1918. See his *Die totale Mobilmachung* (Berlin, 1931) and *Der Arbeiter* (Hamburg, 1932).

National Socialist literature on the question of total war is voluminous. See especially Guido Fischer, *Wehrwirtschaft* (Leipzig, 1936) ; R. von Schumacher and Hans Hummel, *Vom Kriege zwischen den Kriegen* (Stuttgart, 1937) ; Kurt Hesse, ed., *Schriften zur kriegswirtschaftlichen Forschung und Schulung*, and his *Die kriegswirtschaftliche Gedanke* (Hamburg, 1935) ; as well as the articles in *Kriegswirtschaftliche Jahresberichte* (Hamburg) and Hermann Franke, ed., *Handbuch der neuzeitlichen Wehrwissenschaften* (Leipzig, 1936). The question of technological progress and its relations to war is treated at length in Karl Justrow, *Der technische Krieg*, 2 vols. (Berlin, 1938-1939); while the problem of industrial mobilization is discussed in Institut für Konjunkturforschung, *Industrielle Mobilmachung* (Hamburg, 1936). *Wehrwirtschaft* is one of the primary interests of the geopoliticians. See Karl Haushofer, *Wehr-Geopolitik* (Berlin, 1932) and Ewald Banse, *Raum und Volk im Weltkriege: Gedanken über eine nationale Wehrlehre* (Oldenburg i.O., 1932), translated by Alan Harris, *Germany Prepares for War* (New York, 1934). Consult also bibliography to chapter 16.

The following works may be cited as of general interest: B. von Volkmann-Leander, *Soldaten oder Militärs?* (Munich, 1935); Franz Neumann, *Behemoth: the Structure and Practice of National Socialism* (New York, 1942); Hans Gerth, "The Nazi Party: its Leadership and Composition," *American Journal of Sociology*, XLV (1940); Eric Voegelin, "Extended Strategy," *Journal of Politics*, II (1940), 189-200; Albert T. Lauterbach, "Roots and Implications of the Idea of Military Society," *Military Affairs*, V (1941), 1-20; and Hans Speier and Alfred Kahler, eds., *War in Our Time* (New York, 1939).

Chapter 14. Lenin, Trotsky, Stalin: Soviet Concepts of War

The map in this chapter is taken from *The Red Army* by Erich Wollenberg, (London, 1940). The two charts were used through the courtesy of the Office of Population Research, Princeton, New Jersey.

On the Russian Revolution insofar as it is dealt with in this book see William H. Chamberlin, *The Russian Revolution 1917-1921*, 2 vols. (New York, 1935); John Wheeler-Bennett, *Brest-Litovsk, the Forgotten Peace* (London, 1938); John Reed, *Ten Days That Shook the World* (New York, 1935); Winston S. Churchill, *The Aftermath: The World Crisis 1918-1928* (New York, 1929); C. K. Cumming and W. W. Pettit, *Russian-American Relations* (New York, 1920); Michael T. Florinsky, *The End of the Russian Empire* (New Haven, 1931); T. A. Taracouzio, *War and Peace in Soviet Diplomacy* (Cambridge, Mass., 1940).

A collection of documents and materials on the Russian Revolution has been prepared by James Bunyan and H. H. Fisher, *The Bolshevik Revolution 1917-1918* (Stanford University, California, 1934); see also James Bunyan, *Intervention, Civil War and Communism in Russia, April-December 1918* (Baltimore, 1936); Publications of the Department of State, *Papers Relating to the Foreign Relations of the United States, 1918, Russia*, 3 vols. (Washington, 1932).

On the leaders of the Russian Revolution see V. I. Lenin, *Works* (English translation, New York, 1929); D. S. Mirsky, *Lenin* (Boston, 1931); Harold J. Laski, "Vladimir Ilich Ulyanov (Nikolay Lenin)," in *Encyclopedia of the Social Sciences* (1937); Leon Trotsky, *My Life* (New York, 1930); Leon Trotsky, *The History of the Russian Revolution* (English translation, 3 vols. in one, New York, 1937).

For eyewitness accounts of representatives of the Allies, Germany and of Imperial Russia see Major General Sir Alfred Knox, *With the Russian Army 1914-1917*, 2 vols. (New York, 1921); Major General Max Hoffmann, *War Diaries and Other Papers*, 2 vols. (English translation, London, 1929); Bruce Lockhart, *Memoirs of a British Agent* (London, 1932); General A. A. Brussilov, *A Soldier's Note Book 1914-18* (London, 1930); Major General W. S. Graves, *America's Siberian Adventure* (New York, 1931). Also on intervention see L. I. Strakhovsky, *The Origins of American Intervention in North Russia* (Princeton, 1937).

On the Red Army see, for the controversies on problems of military organization during the early period of Soviet rule, Michail Tukhachevsky, *Die Rote Armee und die Miliz* (German translation, Leipzig, 1921), advocating a regular army; M. Frunze, *Works*, in Russian (Moscow, 1927), favoring a militia system; S. Gusev, "Our Disagreements in Questions of War," in Russian (*The Bolshevik*, No. 15, 1924, pp. 34-49).

On the modern Red Army see the brilliant analysis by Max Werner, *The Military Strength of the Powers* (New York, 1939) and the comprehensive and thorough study by D. F. White, *The Growth of the Red Army* (Princeton, 1943; Erich Wollenberg, *The Red Army, a Study of the Growth of Soviet Imperialism* (London, 1940); Nicolaus Basseches, *The Unknown Army* (New York, 1943); Michel Berchin and Eliahu Ben-Horin, *The Red Army* (New York, 1942); N. Vishniakov and F. Arkhipov, *The Structure of the Armed Forces of the U.S.S.R.*, in Russian (Moscow, 1930); N. V. Piatnitskii, *The Red Army of the USSR*, in Russian (Paris, 1932); Leonard Engel, "The Red Officer Corps," *Infantry Journal*, LII (1943), 18-24; T. Adamheit, *Rote Armee, Rote Welt-Revolution, Roter Imperialismus* (Berlin, 1935).

The military doctrine and strategy of the Red Army are analyzed by Max Werner in two books: *The Great Offensive* (New York, 1942) and *Attack Can Win in '43* (Boston, 1943); Triandafilov, *The Character of Operations of Modern Armies*, in Russian (Moscow, 1929); Michail Tukhachevsky, "War as a Problem of Military Struggle," in Russian (*Great Soviet Encyclopedia*, Vol. XII); Boris Shaposhnikov, *The Brain Trust of the Army*, in Russian (Moscow, 1927); A. Svechin, *Strategy*, in Russian (Moscow, 1927); L. S. Amiragov, "On the Character of Future Wars," in Russian, in *War and Revolution*, July-August 1934; D. F. White, "Soviet Philosophy of War," *Political Science Quarterly*, LI (1936), 321-353; *The Field Service Regulations of the Red Army*, in Russian (Moscow, 1936) which replaced the Field Service Regulations issued in 1929. The latter are not available in American libraries, but an interesting comparison between the new and the old Field Service Regulations has been written by Michail Tukhachevsky, "On the New Field Service Regulations of the Red Army," in Russian (in *The Bolshevik*, May-August 1937). A wealth of material on strategic and tactical questions is to be found in the daily Soviet newspaper *Krasnaya Svesda* (The *Red Star*) and the monthly periodical *Voyennaia Mysl* (*Military Thought*).

Analyses of Soviet Russia's war economy and military potential relevant to this study are William H. Chamberlin, *Russia's Iron Age* (Boston, 1934); Ambassador Joseph E. Davies, *Mission to Moscow* (New York, 1941); Albert Rhys Williams, *The Russians* (New York, 1943); H. C. Cassidy, *Moscow Dateline 1941-1943* (Boston, 1943); K. Voroshilov, *The Defense of the USSR*, in Russian (Moscow, 1927); *The Land of Socialism Today and Tomorrow; Reports at the 18th Congress of the Communist Party of Soviet Russia, March 10-21, 1939* (Moscow, 1939); A. Baikov, "The Economic Basis of the Defense System of the U.S.S.R.," in Russian, in *Voyennaia Mysl* (November 1939), pp. 21-36; "Soviet War Economy," *The Economist*, CXLI (London, 1941), 3, 17-18.

CHAPTER 15. MAGINOT AND LIDDELL HART: THE DOCTRINE OF DEFENSE

Important sources for the history of the debate on French military organization in the post World War period are *Projet de loi . . . sur le recrutement de l'armée présenté par M. Maginot, ministre de la guerre* (Paris, 1923) and

Paul-Bernier (ed.), *Rapport fait au nom de la commission de l'armée chargée d'examiner le projet de loi relatif au recrutement de l'armée* (Paris, 1927). The records of the Riom trial, where the whole question was thoroughly aired, are unfortunately not yet available. Excellent commentaries of the proceedings at Riom can be found, however, in the *New York Times*, February 19 to April 3, 1942.

The literature on the French army before the present war is extensive and controversial. One of the most important prewar authorities was General Debeney, chief of staff of the French army. See his *Sur la sécurité militaire de la France* (Paris, 1930) and *La guerre et les hommes, reflexions d'après guerre* (Paris, 1937). A work with considerable inside information is P. Bénazet, *Défense nationale, notre sécurité* (Paris, 1938). The author was, as early as 1913, chairman of the military committee of the Chamber of Deputies and later occupied a similar position in the Senate. Outstanding works of the "rightist" school are General H. Mordacq, *Les leçons de 1914 et la prochaine guerre* (Paris, 1934) and L. Souchon, *Feue l'armée Française* (Paris, 1939), the latter being a violent attack upon those groups which advocated a reduction of the term of military service. An earlier work by an exponent of the militia system is that of Lieutenant Colonel E. Mayer, *La guerre d'hier et l'armée demain* (Paris, 1921). A more technical discussion by a member of the defensive school is General Maurin's *L'armée moderne* (Paris, 1938). P. Reynaud, *Le problème militaire Française* (Paris, 1937) is of considerable interest in view of the fact that Reynaud was the parliamentary champion of General De Gaulle's ideas. Perhaps the most widely read book on military tactics before 1940 was General L. Chauvineau, *Une invasion, est-elle encore possible?* (Paris, 1939). At Riom this book was declared to be one of the factors in the fall of France.

On Maginot and the construction of the famous line, see P. Belperron, *Maginot of the Line* (London, 1940). Both Ph. Barrès, *Charles De Gaulle* (New York, 1941) and H. A. DeWeerd, "De Gaulle as a Soldier," *Yale Review*, XXXII (1943), No. 4, pp. 760-776, are valuable studies of De Gaulle's career and his efforts to modernize the French army. "Pertinax" (A. Géraud), *Les Fossoyeurs*, 2 vols. (New York, 1943) is of special interest as a study of leading French figures in the army debate and French political life in general. The author attempts to assess the responsibility of Gamelin, Reynaud, Daladier and Pétain for the weakness of France in 1940. On Pétain, see also J. M. Bourget, "La légende de Maréchal Pétain," *Revue de Paris* (January 1, 1931), pp. 57-70. The political background to the army problem is discussed in detail in H. Pol, *Suicide of a Democracy* (New York, 1940).

Two special studies of French military organization should be noted: J. Monteilhet, *Les institutions militaires de la France* (Paris, 1934), a study of the French army since the Restoration and S. C. Davis, *The French War Machine* (London, 1927). Interesting French appraisals of Italian and German preparations for war are A. Vauthier, *La doctrine de guerre de Général Douhet* (Paris, 1935) ; and General Nollet, *Une expérience de désarmement; cinq ans de contrôle militaire en Allemagne* (Paris, 1932). The student of French military questions in this period should consult also *Revue Militaire Générale,* the most important of the military periodicals, and *Revue des Deux Mondes,* which contains many valuable contributions to the study of the question.

The basic official sources for material on British military legislation in the period are Hansard's *Parliamentary Debates* and Great Britain, Parliament, *Accounts and Papers.*

Of the leading commentators on military affairs, the most prolific was Major

General J. Fuller, who published about one hundred titles. His most representative works are *The Reformation of War* (London, 1923) ; *The Foundation of the Science of War* (London, 1926) ; *Lectures on Field Service Regulations III: Operations Between Mechanized Forces* (London, 1932) ; *The Army in My Time* (London, 1935) ; *Memoirs of an Unconventional Soldier* (London, 1936). The first two of these supplied the intellectual foundation for the so-called "Fuller or mechanized" school. His *Lectures on Field Service Regulations III* is his most important scientific contribution. General H. Rowan-Robinson, *Imperial Defense: A Problem of Four Dimensions* (London, 1938) is an outstanding work by a brilliant member of the Fuller school, the more so because of its careful discussion of air and naval problems.

Of Liddell Hart's many works, the most significant are *The Remaking of Modern Armies* (London, 1927) ; "Grave Deficiencies of the Army," *English Review*, LVI (February 1933) ; *The Defense of Britain* (London, 1939) ; *Dynamic Defense* (London, 1941) ; and *The Current of War* (London, 1941). Liddell Hart, the author of the "limited liability" theory of war, was an unofficial spokesman for the War Office before 1939.

V. W. Germains' *The Mechanization of War* (London, 1927) represents the most thoughtful criticism of the Fuller school. L. Clive, *The People's Army* (London, 1938) presents the views of the Labor Party in the army controversy. P. R. C. Groves, *Behind the Smoke Screen* (London, 1934) ; J. R. Kennedy, *Modern War and Defense Reconstruction* (London, 1936) ; and Lord G. N. Strabolgi, *New Wars and Weapons* (London, 1930) are rather immoderate attacks upon the high command and the "Colonel Blimps" of the War Office and the Admiralty.

Major E. W. Sheppard's *A Short History of the British Army* (London, 1940) and *Tanks in the Next War* (London, 1938) are both short accounts, written in an even temper. Questions of naval and air power are treated systematically in G. C. Slessor, *Air Power and Armies* (London, 1936) and B. Acworth, *Britain in Danger* (London, 1937). Special studies of value will be found in *Royal United Service Institution Quarterly*, an important semiofficial publication, and *Army Quarterly*, one of the best periodicals of its kind in the English language.

Among the general works, A. Wolfers, *Britain and France Between Two Wars* (New York, 1940) ; R. A. Chaput, *Disarmament in British Foreign Policy* (London, 1935) ; and J. F. Kennedy, *Why England Slept* (New York, 1940) should be noted, the last two being, indeed, indispensable for an understanding of British unpreparedness in the period.

With regard to both Britain and France, the student of military affairs should consult *Command and General Staff School Quarterly*, the official magazine of the U.S. Army Staff School at Leavenworth, Kansas. Its methodical reviews of foreign military periodicals make it an indispensable guide.

CHAPTER 16. HAUSHOFER: THE GEOPOLITICIANS

The base of the map entitled "A Construction of Principal Panregions," used in this chapter was furnished by courtesy of McKnight and McKnight. The map of "The Heartland and World Island vis-à-vis the Insular Continents" is from H. J. Mackinder, "The Geographical Pivot of History," *Geographical Journal*, XXIII (1904), 435. It is reproduced by the courtesy of the Royal Geographical Society.

The classical treatise on political geography is Friedrich Ratzel, *Politische Geographie* (Munich and Berlin, 1897; 3rd edition, revised, 1923). Halford J. Mackinder's *Democratic Ideals and Reality* (New York, 1919) has been reprinted in a new edition (1942) with an introduction by Edward Mead Earle and a preface by Major George Fielding Eliot. This work represents the final formulation by its author of the theory of the Heartland, one of the principal space concepts borrowed by the geopoliticians.

The first inclusion and definition of the term "geopolitics" as an integral part of a system of politics is to be found in Rudolf Kjellén, *Staten som Lifsform* (Stockholm, 1916), translated as *Der Staat als Lebensform* (Leipzig, 1917; reprinted, 1924). See also Kjellén's *Die Grossmächte vor und nach dem Weltkriege* (Berlin and Leipzig, 1930). This is the twenty-second edition in German translation of the author's *Die Grossmächte der Gegenwart*, which was first published in 1914. To it are appended studies of postwar changes by three of the earlier and least unsound geopoliticians. This work is edited by Karl Haushofer.

The bulk of Haushofer's writings has appeared in successive issues of the *Zeitschrift für Geopolitik* as "Reports on the Indo-Pacific Area" (*Berichte über den indo-pazifischen Raum*). Of his many publications the following have been more discussed than others and may be taken as a sampling of his significant works. Karl Haushofer, Erich Obst, Hermann Lautensach, and Otto Maull, *Bausteine zur Geopolitik* (Berlin-Grünewald, 1928). This work reprints articles by four early editors of the *Zeitschrift für Geopolitik*, with Haushofer contributing the lion's share. Haushofer's *Geopolitik des pazifischen Ozeans* (Berlin, 1924; 2nd edition, 1939) is an application of geopolitics to the area of Haushofer's prime interest. The geopolitician's viewpoint of boundaries, one of the most discussed topics of political geography, is presented in his *Grenzen in ihrer geographischen und politischen Bedeutung* (Berlin-Grunewald, 1927; 2nd edition, 1939). *Weltpolitik von Heute* is a presentation of geopolitics for popular consumption. One of Haushofer's many works on Japan is *Japan und die Japaner; eine Landeskunde und Volkskunde* (Leipzig and Berlin, 1923; 2nd edition, 1933): while his pervading interest in military affairs is represented by his *Wehr-Geopolitik* (Berlin, 1932).

An outspoken statement of the ways in which a knowledge of geography can be used in war will be found in Ewald Banse, *Raum und Volk im Weltkriege: Gedanken über eine nationale Wehrlehre* (Oldenburg i.O., 1932). This has been translated by Alan Harris as *Germany Prepares for War* (New York, 1941). Johannes Kühn, "Über den Sinn des gegenwärtiges Krieges," *Zeitschrift für Geopolitik*, XVII (1940), 57-62, 105-112, 156-1571, is a defense of the present war from the standpoint of geopolitics.

Of current works on geopolitics see the two scholarly and brief essays on the theory and meaning of geopolitics by Johannes Mattern, *Geopolitik* (Baltimore, 1942); Hans W. Weigert, *Generals and Geographers* (New York, 1942), a philosophical treatment which centers mainly on the writings of Haushofer; Andreas Dorpalen, *The World of General Haushofer* (New York, 1942), an intellectual biography of Haushofer with long quotations from his own and other works; and Robert Strausz-Hupé, *Geopolitics* (New York, 1942), which is a discussion of the entire movement.

Geographic aspects of geopolitics and its precursors, illustrated with quotations, will be found in Derwent Whittlesey, *German Strategy of World Conquest* (New York, 1942). The student should consult also Alfred Vagts, "Geography in War and Geopolitics," in *Military Affairs*, VII (1943), 79-88, and Jean Gottmann, "The Background of Geopolitics," in *Military Affairs*, VI (1942), 197-205.

CHAPTER 17. MAHAN: EVANGELIST OF SEA POWER

The map in this chapter is from *Toward a New Order of Sea Power*, by Harold and Margaret Sprout (Princeton University Press, 1940).

There are three works by Admiral Mahan which constitute a unified series dealing with his concept of sea power. They are: *The Influence of Sea Power upon History, 1660-1783* (Boston, 1890); *The Influence of Sea Power upon the French Revolution and Empire, 1793-1812* (Boston, 1892); and *Sea Power in its Relation to the War of 1812* (Boston, 1905).

While writing his large work on sea power, Mahan prepared a short life of Admiral Farragut (1892). Farragut's son, then living in New York, made available a great quantity of materials in addition to the admiral's journals and correspondence. This book, however, is in no way comparable to *The Life of Nelson*, published in 1897. In this second biography Mahan undertook to write a definitive life of the man who was the "embodiment of British sea power," presenting both the "milieu" and the man in their proper relations, a task which he found difficult, but accomplished with some success.

Two books, *From Sail to Steam: Recollections of Naval Life* (1907) and *The Harvest Within* (1909), are autobiographical. The other large group of Mahan's writings consists of collections of essays on current subjects. The first of these volumes, including eight essays written between 1890 and 1897, appeared in 1897 under the title, *The Interest of America in Sea Power, Present and Future*. A second, entitled *Lessons of the War with Spain and Other Articles*, made its appearance two years later. In 1900 there came a third, *The Problem of Asia and Its Effect upon International Policies*.

The War in South Africa, which also appeared in 1900, was followed by two more books of essays. *Types of Naval Officers Drawn from the History of the British Navy* (1901) is a series of sketches of typical naval officers. *Retrospect and Prospect, Studies in International Relations, Naval and Political* (1902) contained two essays which, like the two volumes just cited, were of special interest to British readers.

Some Neglected Aspects of War (1907) embodied Mahan's views on some of the questions that came before the Second Hague Conference. A collection of articles on naval administration which appeared in 1908, and *Naval Strategy, Compared and Contrasted with the Principles and Practice of Military Operations on Land*, published in 1911, are of special interest to members of the naval profession. *The Interest of America in International Conditions* (1910) has more appeal for civilian readers.

The Major Operations of the Navies in the War of American Independence, published separately in 1913, first appeared as a chapter in Clowes' *History of the Royal Navy* in 1897. Mahan's last published works were articles and letters which appeared in the latter part of 1914, after the issue between land power and sea power was once more joined in the First World War.

The accepted biography of Mahan, based upon his papers, is Captain W. D. Puleston's *The Life and Work of Captain Alfred Thayer Mahan* (New York, 1940).

For Mahan's influence on Great Britain see A. J. Marder, *The Anatomy of British Sea Power* (New York, 1940) and R. H. Heindel, *The American Impact on Great Britain* (Philadelphia, 1940). For Mahan's influence on Germany: A. Vagts, *Deutschland und die Vereinigten Staaten in der Weltpolitik*, 2 vols. (New York, 1935); E. Kehr, *Schlachtflottenbau und Parteipolitik 1894-1901* (Berlin, 1930); E. von Halle, *Die Seemacht in der deutschen Geschichte* (Leipzig, 1907); H. Hallman, *Krügerdepesche und Flottenfrage* (Stuttgart, 1927); Hallman, *Der Weg zum deutschen Schlachtflottenbau*

(Stuttgart, 1933) ; Admiral Tirpitz, *My Memoirs* (English translation, New York, 1919) ; H. Rosinski, "German Theories of Sea Warfare" in *Brassey's Naval Annual* (1940), p. 89; H. Rosinski, "Mahan and the Present War" in *Brassey's Naval Annual* (1941), pp. 9-11.

On American naval policy see H. and M. Sprout, *The Rise of American Naval Power* (rev. ed., Princeton, 1942) ; H. and M. Sprout, *Toward a New Order of Sea Power* (rev. ed., Princeton, 1943) ; and G. T. Davis, *A Navy Second to None: the Development of Modern American Naval Policy* (New York, 1940). Allan Westcott, *Mahan on Naval Warfare* (Boston, 1941) is a brief and not altogether satisfactory collection of some of Mahan's writings.

On geopolitics see Chapter 16 on Haushofer; Robert Strausz-Hupé, *Geopolitics: The Struggle for Space and Power* (New York, 1942) ; Hans Weigert, *Generals and Geographers* (New York, 1942). For Britain's geopolitical position: Sir Halford Mackinder, *Britain and the British Seas* (2nd ed., New York, 1930).

CHAPTER 18. CONTINENTAL DOCTRINES OF SEA POWER

On the general background of the *Jeune Ecole* see Jean Grivel, *De la marine militaire considerée dans ses rapports avec le commerce et avec la défense du pays* (Paris, 1837) and Richard Grivel, *De la guerre maritime, avant et depuis les nouvelles inventions* (Paris, 1869). The Grivels, father and son, were among the most interesting forerunners of this French school of naval strategy. The most important work of Admiral Théophile Aube, the leader of the *Jeune Ecole*, is his *A terre et à bord, notes d'un marin* (Paris, 1884). Gabriel Charmes, *La reforme de la marine* (Paris, 1886) is the work of one of the chief publicists of the school. The leading French followers of Mahan were Gabriel Darrieus, *La guerre sur mer* (Paris, 1907) and René Daveluy, *L'esprit de la guerre navale* (3 vols., Paris, 1909-1910).

Other valuable studies of French naval strategy in the period before the first World War are A. Gougeard, *La marine de guerre, son passé et son avenir* (Paris, 1884) ; and Commandant Z. (Paul Fontin) and H. Montechant (J. H. Vignot), *Les guerres navales de demain* (Paris, 1891). The most notable modern French work on naval strategy is Raoul Castex, *Théories stratégiques* (second edition, 5 vols., Paris, 1937).

Among the German followers of the *Jeune Ecole*, see Lothar Persius, *Warum die Flotte versagte* (Leipzig, 1925). The most important German works on the strategy of the first World War are Wolfgang Wegener, *Die Seestrategie des Weltkrieges* (Berlin, 1929) and Paul Sethe, *Die ausgebliebene Seeschlacht: die englische Flottenführung, 1911-1915* (Berlin, 1932). The latter is a keen analysis of the reasons for the inactivity of the two great battle fleets during the war.

The most interesting Italian works on the subject are Domenico Bonamico, "La strategia navale nel secolo XIX" and "Il dominio del mare del punto di vista italiano," *Rivista Marittima* (1899 and 1900) ; Giovanni Sechi, *Elementi di arte militare marittima* (2 vols., Leghorn, 1903-1906) ; Romeo Bernotti, *La guerra marittima, studio critico sull' impiego dei mezzi nella guerra mondiale* (Florence, 1923) ; and Oscar di Giamberardino, *L'arte della guerra in mare* (2 vols., Rome, 1937). See also G. Fioravanzo, *La guerra sul mare et la guerra integrale* (2 vols., Turin, 1930-1931).

CHAPTER 19. JAPANESE NAVAL STRATEGY

Reliable English materials on Japanese naval strategy are few and far between. The student should consult Captain Viscount Ogasawara, "Historical Essay on the Japanese Navy (to 1893)" in *Revue Maritime*, CXCVIII (Paris, September 1913), 257-266, 381-392, CC (January-March 1914), 93-124, as one of the few Japanese accounts in a western language. Special aspects of Japanese strategy are treated in "The Battle of the Yellow Sea," *United States Naval Institute Proceedings* (September-October 1914), pp. 1283-1299; and "The Battle of the Sea of Japan," *United States Naval Institute Proceedings* (July-August 1914), pp. 961-1007.

Fred T. Jane, *The Imperial Japanese Navy* (London, 1904), is still a standard account, and valuable material may be found in Vice Admiral G. A. Ballard, C. B., *The Influence of the Sea on the Political History of Japan* (New York, 1921). For Togo's career, there are two main sources of information: Vice Admiral Viscount Nagayo Ogasawara, *Life of Admiral Togo*, translated by J. and T. Inouys (Tokyo, 1934) and Edwin A. Falk, *Togo and the Rise of Japanese Sea Power* (New York, 1936). See Kenneth Colegrove, *Militarism in Japan* (Boston, 1936) for relations of politics and the armed forces.

Chapters XIII and XIV of Captain A. T. Mahan, *Naval Strategy* (Boston, 1918) are useful; and the whole question of the balance of naval power in the Far East is treated in Chapter V of Archibald S. Hurd, *Naval Efficiency* (London, 1902).

Among the general works, see Captain Gabriel Darrieus, *War on the Sea*, translated by Professor Philip R. Alger, U.S.N. (Annapolis, 1908); H. W. Wilson, *Battleships in Action*, I (London, n.d.); and Brigadier-General G. G. Aston, C.B., *Letters on Amphibious Wars* (London, 1911), pp. 162-360.

CHAPTER 20. DOUHET, MITCHELL, SEVERSKY: THEORIES OF AIR WARFARE

Giulio Douhet's first presentation of the theory which bears his name was his *Il Dominio dell' Aria: saggio sul' arte della guerra aerea* (Rome, Stab. Poligr. Per l'Amministrazione della guerra, 1921). This work outlines his ideas in their first and limited form. For a more complete exposition see his *Il Dominio del' Aria* (second ed., Rome, Instituto Nazionale Fascista di Cultura, 1927). An earlier work was his *La Possibilita dell' Aeronavigazione* (Rome, Tip. Unione ed., 1910). "La Guerra del' 19—," Douhet's fictional account of a future war, will be found in *Rivista Aeronautica* (March 1930), pp. 409-502. Douhet's principal military writings have been translated by Dino Ferrari, *The Command of the Air* (New York, 1942).

A comprehensive French appraisal of the Douhet theory is Arsene M. P. Vauthier, *La Doctrine de Guerre du Général Douhet* (Paris, 1935). See also the same author's *La Danger Aerien et l'Avenir du Pays* (Paris, 1930).

Mitchell's chief writings are *Our Air Force: The Keystone of National Defense* (New York, 1921); *Winged Defense: the Development and Possibilities of Modern Air Power—Economic and Military* (New York, 1925); and *Skyways: a Book on Modern Aeronautics* (London and Philadelphia, 1930). The basic source for the controversy which led to Mitchell's court martial in 1925 is: *U.S.: 68th Congress: House: Select Committee on Inquiry into the Operation of the United States Air Service* (Florian Lampert, chairman). *Hearings* (Washington, Government Printing Office, 1925). See references under the name of General Mitchell in *Index*, VI, 367-384. Consult also *U.S.: President's Aircraft Board* (Dwight W. Morrow, chairman). *Hear-*

ings, September 29-30, October 1-2, 1925 (Washington, Government Printing Office, 1925), II, 475-908. A recent biography of Mitchell is Isaac Don Levine, *Mitchell: Pioneer of Air Power* (New York, 1943).

The modern successors of Douhet and Mitchell are Alexander de Seversky, *Victory Through Air Power* (New York, 1942) and William Bernard Ziff, *The Coming Battle of Germany* (with an introduction by William E. Gillmore, New York, 1942). Among the many other titles, the following should be noted: William Carrington Sherman, *Air Warfare* (New York, 1926); Alford Joseph Williams, *Airpower* (New York, 1940); and Allan Michie, *Air Offensive against Germany* (New York, 1943).

Index

Proper names of persons of relatively minor significance to the topic of this volume, as well as merely casual references, are not included in the Index. Subject entries are not always to be found under the literal terms used by individual authors. Such terms are sometimes replaced by broader terms which deal with general concepts. Footnotes, except in so far as purely bibliographical data are concerned are included in the Index.

Abd-el-Kader, 236, 237
Abd-el-Krim, 252
absolute war, 28, 102-106, 111, 221-223, 306
administration, colonial-military, 240-242, 244, 247, 257
advance guard, task of, 224, 225
air-defense, 490, 494
air force, 374, 494; British, 377, 379; German, 312, 494, 497, 514, 515; Russian, 357, 359
air power, 485-503
air power vs. sea power, 499
air warfare, 374, 485-503
Alexander the Great, 273
anti-aircraft defense, 490, 494
Ariosto, 4
armies, size of, 57, 64, 77, 78, 100, 151, 208, 215, 264-267, 274, 346, 357, 369, 380; size of, in battle line, see particularly 264-267
artillery, 7-9, 13-15, 17, 20, 27, 28, 30-32, 38, 57, 58, 62, 64, 73, 186, 226, 227, 373
artillery schools, 38
attack, 110, 190, 229, 232, 383, 384; counter-attack, 374, 375, 383
Aube, 447-449, 453, 455
autarky, see particularly 398, 399, 511

Bacon, 33, 124, 125
Bagdad Railway, 151
balance of power, 73, 99, 145, 392, 427, 450
ballistics, 31, 32, 38
barracks, 38, 39
"base of operations," see particularly 70
bastions, 30, 31, 41, 42, 65
"battaglione," 18
battle, nature of, 22, 24, 60, 66, 93, 103-105, 178, 210, 211, 222, 225, 226, 268, 293
battle line, 17, 18, 39, 56, 58, 66, 178, 190, 195, 267, 269; see also tactics
battle (air) plane, 490, 492
battleship, 328, 379, 434-437, 443, 447, 455
Bavaria, general staff, 394, 397
bayonet, 38, 212
Belidor, 42
Belisarius, 273, 274
"bellum justum," 22
Benedek, 184
Bentham, 127, 128
Bernhardi, 274
Berthier, 67, 82
Bethmann-Hollweg, 314
Biringuccio, 31
Bismarck, see particularly 203, 279, 314, 450, 451

"blitzkrieg," 9, 53, 67, 98, 360, 361, 369, 370, 371, 384, 505, 515
blockade, 110, 315, 425, 427, 432, 433, 435, 436, 446, 447, 453, 455; counterblockade, 110, 428
Blücher, 96
Blume, 312
bomber, 491-493, 495, 496, 502
Bonaparte, see Napoleon I
Boyen, 172
Brest-Litovsk, peace negotiations of, 325-328, 347
British military organization, 375-377, 379; see also Cardwell system and Haldane system
Brussilov, 337
Budenny, 329, 340
Bugeaud, 207, 215, 234-259; see particularly 236-238
Bülow, Dietrich von, 49-73; see particularly 69-73; 80, 84, 87, 90, 101

Caesar, 19, 32, 266, 273
canals and waterways, France, 36, 46-48
Cannae, battle of, 189, 190, 205, 269
Cardwell system, 375, 376; see also British military organization
Carey, 138, 141, 142
Carnot, 37, 40, 68
casernes, see barracks
Castex, 441, 442, 455, 456
Castracani, 14
Catroux, 235, 254-257
Cavaignac, 119
cavalry, 13, 17, 22, 26-28, 57, 60
Channing, 79
Charles VIII of France, 8, 9, 26
Charles the Bold of Burgundy, 6
Chauvineau, 374, 375
Churchill, Winston, vii, viii, 287-304, see particularly 292-296; 331, 334, 335, 364, 386, 387
citizen army, see particularly 49, 63, 64, 67, 172, 173; see also conscription, general; democratic army; "nation in arms"; national army
Civil war, American, see particularly 166, 167, 288, 289
Clausewitz, 25, 79, 80, 88, 89, 92, 93-113, 158, 172, 173, 175, 178, 203, 218, 219, 221, 222, 260, 261, 272, 273-275, 282, 306, 307, 317, 318, 323, 334, 336, 450
Clemenceau, 232, 233, 258, 280, 287-305; see particularly 302-305; 368, 371
Clerville, 35, 39

"Coalition, wars of," 107
coast defense, 415, 416, 425-427, 434, 438, 447
Colbert, 32, 39, 117, 420
"colonel général de l'infanterie," 27
colonial warfare (general principles), 234, 235, 241-243, 245, 246
colonial warfare, French, 234-259
Colonna, Egidio, 21
"command of the air," 490, 491
commerce raiding (guerre de course), 39, 415, 416, 431-433, 443, 444, 446, 448, 449-451, 456
Commissaires (French Army), 28
commissary general of artillery, 218
commissary general of engineering, 35
communications, line of, see particularly 97, 190; sea lines of, see particularly 424, 430, 431; aerial, see particularly 500
Commynes, 19
"concentration of force" (naval), 431, 433, 450
Condé, 35, 43
condottieri, see particularly 12, 13
Consalvo, 9
conscription, general, see particularly 10, 16, 20, 73, 138, 168, 169, 170-173, 208, 319, 355, 380
continental system, 78, 119
convoy system, 449, 454
Corbett, 453, 454
Cornazzano, 22
counterattack, see attack
counterblockade, see blockade
Craig, Malin, 382
"culminating point," concept of, 111

Darrieus, 441, 455, 469
defense, see strategy of defensive
defense in depth, 41, 359, 360
defensive, see strategy of defensive
De Gaulle, see Gaulle, de
Delbrück, 52, 69, 109, 189, 260-283
democratic army, see particularly, 168-171; see also citizen army; conscription general; "nation in arms"; and national army
Denikin, 329, 331, 333
destroyer, 447, 453
discipline, military, 18, 55, 210-213, 332, 339, 357
division, 62, 64, 66
Douhet, 320, 453, 455, 485-503, see particularly 487-500
Du Picq, see Picq, du

economic warfare and war economics, see particularly 15, 117-154, 290-291, 312, 315, 320, 350-353, 511; see also Wehrwirtschaft
Eliot, G. F., 514
"enceinte" (main enclosure of a fortress), 30, 41, 42
Engels, Friedrich, 104, 154-171, 288, 323, 336, 337, 345
engineering, military, 27, 30, 31, 38, 42, 186, 313, 319; see also Mézières, Ecole de
England, invasion of, 166

Ermattungsstrategie, see strategy of exhaustion
Errard, 31, 44
Eugene, Prince of Savoy, 204
Eurafrica, see Panregions

Fabry, 366, 368
Fairgrieve, 390
Falkenhayn, 291, 313, 360
feudal army, 5-7
fighter plane, 493, 495
firearms, see particularly 7, 26, 38, 167, 183-184, 226, 227, 271; development of, 296, 297
firepower, see particularly 227, 373, 374, 378, 379, 381; naval, 478
Fisher, Lord John, 295
Five Year Plans (Russian), see particularly 350-353, 358
"fleet in being," 434
Florence, campaign against Pisa, 10, 11; military organization, 11
Foch, 93, 206-233, see particularly 218-233; 258, 300, 303, 304
Foix, 9
Fontenelle, 29, 34, 37
fortifications, 21, 30, 39-48, 59, 61, 64, 65, 70, 163, 164, 165, 371-373; in depth, 41; see also defense in depth
"fortified zone," 48, 373
fortresses, 30, 33, 39, 40-41, 43, 44, 45-48, 59, 65, 97
forts, 43, 65
Four Year Plan, 511
Francis I, of France, 20, 44
Frederick, "the great elector" of Brandenburg, 53, 54
Frederick the Great (Frederick II, King of Prussia), 19, 49-74, see particularly 53-62; 80-81, 83, 88, 91, 92, 112, 163, 172, 189, 190, 193, 204, 273-275, 276, 278, 307
Frederick III, Emperor of Germany, 176, 183, 261, 262
Frederick Charles, Prince of Prussia, 183, 184
Frederick William, King of Prussia, 54
Frederick William, Crown Prince of Prussia, see Frederick III, Emperor of Germany
French, Academie Royale des Sciences, 32, 35, 37
French, Sir John, 290, 293
French army organization, 27 f., 62 f., 166, 208, 215, 217, 218, 365-370
French invasion of Italy (1494), see Charles VIII, of France
French Revolution (1789—), military influence of, 26, 49, 50, 52, 66, 68, 77, 97, 161, 512
Fritsch, 188, 201
frontiers, 405, 406, 409
Frontinus, 14, 32
Frunze, 329, 342, 343-346
Fuller, 376, 377, 379, 381, 383

Galileo, 30-32
Galliéni, 200, 234-259, see particularly 238-245; 302
Gamelin, 200, 366

Gaulle, de, 215, 257, 350, 372-373, 375, 379, 383
Gautier, 241, 248, 249
general staff system, see particularly 289, 298, 317
geography, political, 388, 389, 391
geopolitics, 69, 318, 319, 388-411, see particularly 396-411, 444
Germains, V. W., 382, 383
German army organization, see Prussian and German army organization
German general staff, see Prussian and German general staff and also Bavaria, general staff
German-Japanese alliance, 512
German-Japanese naval war 1914, 478, 479
German military cabinet, 308-310
German Navy League, 308-310, 392, 442
German, Reichstag peace resolution of July 1917, 278, 279
German, Reichswehr, see particularly 369
German, secretaries of war, 309, 310
Giorgio Martini, Francesco di, 7
Giraud, 235, 254
Gneisenau, 67, 68, 73, 98, 172-176, 203, 263, 279
Goebbels, 317, 320, 513
Goering, 117, 291, 321, 505, 511, 512, 513
Gorlice-Tarnow, battle of, 201
Grand master of artillery, French, 28
Grandmaison, 195, 217, 229
"grand strategy," see strategy
"Greater Germany," see particularly 141, 144, 145, 153, 165, 395, 506, 507
Gribeauval, 62, 91
Groener, 201, 264, 267
Grotius, 33
"guerra corte e grosso," see blitzkrieg
guerre de course, see commerce raiding
guerrilla warfare, 342, 344, 361-363
Guibert, 49-73, see particularly 62-69; 91, 207, 210
Guicciardini, 8, 9, 15
gunpowder, invention of, 6
Gustavus Adolphus, 273, 274

Haeften, 320
Haig, 297-299, 300, 303
Halle, Ernst von, 442
Hamilton, 117-154, see particularly 128-138
Hannibal, 189, 190, 205, 269, 417
Hart, see Liddell Hart
Haushofer, 388-411, 444, 445, see particularly 394-411; 515
"heartland" concept, 404, 405, 445, see also geopolitics
Henry II, of France, 28, 31
Henry IV, of France, 30, 31
Hentsch, 195, 199
Herodotus, 266-268
Hess, Rudolf, 408
Hideyoshi, 465, 467, 468, 478
Hindenburg, 279, 281, 313, 314, 326, 393
Hitler, xi, 205, 307-309, 316, 319, 354, 358, 360, 361, 382, 393, 408, 409, 454, see partic-

ularly 504-516; Mein Kampf, 354, 393, 408, 409, 454, 505-512, 516
Hoffmann, 189, 325, 326
Hore-Belisha, 380, 384
Huré, 235, 253, 254
Hutcheson, 126

infantry, 8, 9, 13, 16-18, 20, 23, 26, 27, 57, 59, 67, 70, 183, 184, 226, 227, 377, 382
initiative (strategical) see strategy, initiative in
intendants d'armée French, 28
international law, 33
isolationism, military, 106
Isthmian canal, see Panama canal
Ito, 462, 469, 470, 472, 478, 482, 483

Jähns, 52
Jane, F. T., 453, 455
Japan, geopolitical studies, 394, 395, 409
Japan, naval literature, 460-463
Jaurès, 171, 345, 366, 368
Jefferson, vii, 19, 135, 136-139
Jeune Ecole (naval), 446, 447, 449, 450, 452-456
Joffre, 200, 216, 235, 302
Jomini, 77-93, 101, 173, 224, 430, 431, 450
Juenger, Ernst, 316

Kanimura, 473, 476, 478
Kautsky, 335
Kitchener, 290, 293
Kjellén, 390, 391
"knights," military organization, 4, 5, 7
Kolchak, 329, 331, 333, 343
Krafft von Delmensingen, 202
Kutuzov, 356, 361

land bases, see particularly 455
"Landsknechte," 9
Landwehr, see territorials (national guards)
Lazard, 40, 41, 44, 48
"Lebensraum," see "living space"
Leeb, Wilhelm v., 201, 360, 361, 373
Leipzig, battle of, 78, 180, 189
Lenin, 160, 170, 322-364, see particularly 322-334
Leonardo da Vinci, 30
Le Tellier, 27
Leuthen, battle of, 190
levée en masse, 68, 77, 97, 333, 368
Liddell Hart, 93, 94, 219, 350, 365-387, see particularly 376-386
limited warfare, see particularly 99, 273
lines of operations, 20, 83, 84, 86, 87, 88, 89, 92; exterior and interior, 83-87, 92, 109, 179, 185, 189; single, 86; double, 86, 87; naval, 430; see also zones of operations
List, Friedrich, 117-154, see particularly 139-154
"living space" (Lebensraum), 389-400, 408, 506
Lloyd George, 280, 287-305, see particularly 292-307
logistics, 68, 89, 164
London, Royal society of, 32

Louis XIV of France, 26-29, 33, 35, 36, 38, 42-45, 48, 49
Louvois, 26, 27, 35, 36, 38, 43, 45
Ludendorff, 93, 189, 202, 205, 231, 261, 279, 281, see particularly 306-321; 326, 397
Luftwaffe, see Airforce, German
Lyautey, 234-259, see particularly 239-255, 257-259; 302 ·

Machiavelli, 3-25, 26, 30, 33, 49
Mack, 71, 110, 190
Mackinder, 148, 390, 391, 404, 405, 444, 445, 452, 515
Madison, 138, 139, 287
Maginot, 365-387; Maginot line, 365-387, see particularly 371-375
Mahan, ix, 146, 390, 415-445, 446, 447, 449-455, 461
Maistre, 227, 228
Marathon, battle of, 266-269
Marengo, battle of, 70, 88, 91
Mareth line, 256
Marlborough, 204
Marx, 154-171, 323, 345
mass armies, 26, 77, 93-94, 168, 190, 208, 215, 275, 288, 346, 357
materiel, war of, see particularly 290, 293, 350, 351-353; see also economic warfare and war economics and Wehrwirtschaft
Maurin, 372
Mazarin, 35, 44
mechanized and motorized forces, 202, 253-254, 255, 378, 379, 380, 384, 505
mechanized warfare, see particularly 253, 254, 255, 257, 288, 289, 350, 354, 355, 505
medieval army, 5, 7
medieval warfare, 14, 270
Mehring, 264
mercantilism, 117-122, 127-131, 140, 142, 153, 154, 418
mercenaries, see particularly 5, 11, 172
Mézières, Ecole de (school of military engineering), 41, 46
Michel, 200
military potential, see war potential
military science, see particularly 72, 73, 80, 84, 100, 101, 342, 343
militia, 11, 27, 126, 169, 345, 346, 366, 367-369
Miltiades, 268, 270
Mitchell, 378, 485-503, see particularly 487-488, 497-501
mobilization, see particularly 177, 181, 369, 370
Moltke, Helmuth Karl v., 1800-1891, 166, 167, 172-205, see particularly 174-185; 212, 220, 260, 279, 312, 396
Moltke, Helmuth Johannes v., 1848-1916, 195-199, 204, 291
Mommsen, 417, 418
Montalembert, 41
Montesquieu, 63, 91, 136
morale, 112, 113, 210-215, 217, 228, 313, 489, 490, 491, 496, 498
Morat and Nancy, battles of, 6, 270

Mordacq, 303, 304
movement, war of, see particularly 52, 58, 60, 63, 64, 66, 85, 177-205, 236, 237, 290, 344, 359

Napoleon I, ix, 60, 62, 66, 71, 72, 77, 78, 79, 81, 83, 88, 91, 96-99, 101, 110, 163, 172-174, 180, 187, 189, 190, 193, 204, 207, 208, 212, 215, 219-223, 273, 277, 361, 368
Napoleon III, 82, 164, 166, 185, 208, 368
"narrow seas," control of, 421, 423, 426, 429, 435, 483
Nassau, Maurice of, 20
national army, see particularly 64, 67; see also citizen army; conscription, general; democratic army; "nation in arms"
"nation in arms," 168-170; see also citizen army; democratic army; conscription, general
National Socialist Party, 318-321; see particularly 504, 505, 511, 512
naval bases and stations, 421, 426-428, 438, 447, 450
navigation acts (British), 122, 123
Nelson, 470
Neuf-Brisach, fortifications of, 41
Ney, 81, 82
Niedermeyer, 396
Niederwerfungsstrategie, see strategy of annihilation
Niel, 208, 215
Nivelle, 300, 302, 371

occupation, military, colonial, 241-244, 246, 253, 254
Oman, 282
operations, see lines of operations and zones of operations
"Ordinanza" of 1506, 10
Osoaviakhim (Russian organization for the mobilization of civilians), 354, 355

Pagan, 31
Painlevé, 300, 371
Panama canal, 146, 426, 438, 439 (Isthmian canal)
Pan-German League, 310, 392
Panregions, 400-403, 404; Pan-America, 402, 403, 404; Pan-Asia, 402, 403, 404; Eurafrica (German panarea), 403; see also geopolitics
parachute troops, 499
Paris "commune" (1871), 170
Paris, fortifications of, 47
Pascal, viii, 29
Patricius, 21, 22
"perfect war," see "absolute war"
Pericles, 273-275, 279
Pershing, 304
Pétain, 302, 303, 361, 370, 372, 374, 496
Picq, Ardant du, 206-233, see particularly 206-218
Pisa, see Florence, campaign against Pisa
Pisan, C. de, 7, 21
"pivot area," 404, see also geopolitics

planes vs. battleships, 435, 436
Poincaré, 304, 367
Polybius, 16, 19, 210
position, war of, see particularly 52, 59, 60, 61, 276, 290
professional army, see particularly 5, 6, 27, 49-50, 67, 96, 97, 126-127, 135, 136, 208, 345, 346, 369
propaganda, 313, 315-317, 320, 324, 325, 328, 337, 510, 511-513; see also psychological warfare
Prussian and German army organization, 54-56, 71, 72, 172, 173, 174, 176, 177, 203, 308, 311, 319; Prussian and German general staff, 174, 179, 180, 185, 186, 187, 195, 310, 312, 313, 393, 396, 407, see also Bavaria, general staff; Prussian and German officers corps, see particularly 311
psychological warfare, 14, 107, 112, 214, 315-317, 320, 324, 325, 328, 510, 511, 513, 514
pyrotechnics, military, 31

R.A.F., see air force, British
railways, military importance of, 148-152, 166, 177, 181, 183, 187, 199, 200, 201, 288, 289
range of airplanes, 501
Rathenau, 290-291, 316
Ratzel, 388, 389, 391, 395, 397, 452
Rauschning, 512
Red Army, see particularly 322, 327, 329, 332, 333, 335, 338, 339-347, 349, 351, 353, 355-357, 359-362; Red Army, political commissars, 341, 357, 358; Red air force, 357, 379
Reyher, 176-177
Richelieu, 26-30, 44
ricochet bombardment, 40
Riff war, 252
Ritter, Karl, 396
Robertson, Sir William, 295, 297-299, 300
Rohan, Henri de, 32
Rojdestvensky, 474, 475, 476, 477, 482
Roman, military institutions, 16, 20, 23, 24, 211, 238, 269-270; see also strategy, ancient
Roon, 176, 396
Roosevelt, Theodore, 134, 431, 436, 439, 440
Rowan-Robinson, 381, 382
rules of warfare, 33
Russian army organization, see Red Army
Russian civil war and wars of foreign intervention in Russia, 328-335
Russo-Japanese naval war, 1904-1905, 470-478

Sadowa, battle of, 167, 181, 184, 186, 212
Saxe, marshal de, 19, 84
Scharnhorst, 69, 95-96, 172-176, 180, 188, 193, 203
Schlieffen, 93, 172-205, see particularly 186-205; 260, 267, 312; "Schlieffen Plan," see particularly 191-203, 221, 451
science and war, see particularly 29, 30, 99, 100

sea power, 415-456, see also air power vs. sea power
sea power, naval organization and strategy, Great Britain, 145-148, 378, 379, 390, 419, 420, 422, 423, 424, 429, 431-433, 435, 439, 440, 444, 445, 446-449, 450, 453; France, 39, 419-421, 431-433, 446-449, 455, 484; Germany, 204, 423, 434, 439, 442, 443, 444, 445, 446, 447, 451, 453, 454; Italy, 434, 448, 451, 455, 484; Japan, 423, 448, 457-484, see also transport fleet (Japanese); Russia, 447, 449, 451, 452, 470, 471, 472-475; United States, 133, 134, 415, 423, 425-428, 434, 436, 437-439, 440, 449, 453
Sebastopol, siege of, 163, 164
security, military and national, 124-125, 127, 130-132, 134-137, 142, 224, 232, 233, 370; security, collective, 358
Sedan, battle of, 185
Seeckt, 201, 215, 360, 369, 370, 396
Servan, 63
Seversky, 485-503, see particularly 501-503
Shaposhnikov, 329, 340, 349, 358, 359
Sheppard, 380, 382
Shotwell, 291
siege warfare, 8, 27, 30, 33, 34, 39, 40, 46, 163
signal corps, 186
Sino-Japanese naval war (1894-1895), 468-470
Slessor, J. C., 380
Smith, Adam, 79, 117-154, see particularly 117-128
Soderini, 10
Sokolovsky, 363
Sorel, Georges, 156
Spears, 300
Spee, 478, 479
speed of planes, 494, 495
Spengler, 288, 318, 319, 397
Stalin, 322-364, see particularly 348-364
Stalingrad, siege of, 329
standing army, see particularly 124, 125, 126, 127, 375, see also professional army
Steuben, vii
strategy, definition of, viii
strategy, see particularly vii-xi, 24, 63, 70, 72, 85-90, 92, 94, 100, 103-105, 109, 110, 165, 173-180, 183, 184, 187, 189, 190, 223-227, 274, 275; see also battle, nature of; movement, war of; position, war of; lines of operations; zones of operations
  aerial, see particularly, 494
  "extended strategy," 513
  ancient, 210, 211, 236, 238
  of annihilation, 109, 110, 187, 272-275, 280, 281, 350
  of attrition, 109, 110, 187, 344, 359, 361
  of the Austrian-Prussian War 1866, 181, 183-185
  of colonial warfare, 236, 237, 241, 248, 250-256
  defensive, x, 48, 60, 61, 67, 89, 110-112, 160, 188, 201, 231, 276, 318, 334, 359-363, 372-374, 381, 382, 384-386; "active defense," strategy of, 361
  of exhaustion, 273, 274, 276-278

of the Franco-German War 1870-1871,
see particularly 185, 220
"Grand Strategy," viii, 393
initiative in, 88, 201, 359
naval, *see* particularly 415, 416, 418, 420,
426, 427, 430-439, 442-448, 450, 452,
453-456; naval defensive, 420, 434,
435, 447-449; Japanese naval, 463-
484; Japanese naval in World War
II, 481-484; naval offensive, 433-455
offensive, *see* particularly x, 61, 88, 89,
110, 111, 158, 188-190, 192-194, 217,
218, 225, 276, 280-282, 344, 345, 349,
350, 361, 363, 382-384
à outrance, 216, 217, 231, 232, 373
political, *see* particularly 261, 277-280,
282, 301, 346, 358, 393, 394, 407, 450,
504, 506-511, 513, 514
revolutionary, 158-162, 167-170, 341, 342,
344, 346
of World War I, *see* particularly 189-
205, 221, 228, 229, 236, 275-277, 279-
281, 282, 290, 294-299, 303, 515
of World War II, *see* particularly 202,
359-363, 384-385
Strausz-Hupé, 444-445
submarines, 451, 452, 453, 455, 456
Suez, 146
supply system, *see* particularly 65, 97, 100,
294
Suvorov, 338, 356
Svechin, 344, 345

tactics, definition of, viii
tactics, 63, 67, 70, 72, 85, 89, 90, 93, 100, 103,
110, 178, 180, 185, 202, 267, 268, 269, 281,
see also battle line; aerial, 499; in colonial
warfare, 236-238, 252, 253; naval, *see* particularly, 418; Japanese, naval, 464, 468,
469, 475-477; revolutionary, 170, 331, 341,
342; shock, 211, 212
Talleyrand, 33
tanks, 202, 295, 296, 359, 374, 376, 377, 380,
381
Tannenberg, battle of, 189, 201
target, selection, aerial, 490, 496, 498
Tartaglia, 31, 32
technical troops, *see* particularly, 313
technology, military, *see* particularly ix, 32,
33, 38, 186, 312, 313, 317, 319
territorials (national guards), 126
Thomas, Georg, 291
Timoshenko, 329, 357
Ting, 468-470, 472, 473, 483
Tirpitz, 203, 308, 443, 444, 449, 451
Tocqueville, 287
Togo, 461, 462, 471, 472, 473-475, 476-478,
483
Tolstoy, 338
torpedo boat, 447, 453
total war, 13, 14, 49, 156, 158, 296, 350, 355,

see particularly 306-321; 393, 407, 488,
497, 511, 512
Tracy, R. F., 436
transport fleet (Japanese), 463, 464, 466;
scheme, 470
trenches, 39, 40
trench warfare, 226, 290, 292, 293, 294
Trochu, 207, 215
Trotsky, 322-364, *see* particularly 325-329,
331-333, 335-345, 361
Tsushima, naval battle of, 474-477
Tukhachevsky, 329, 340, 344, 345, 346, 349,
350, 358, 359
Turenne, 29, 43
two front war, *see* particularly 166, 188, 189,
507
"two power standard" (naval), 440

Ulm, battle (siege) of, 71, 72, 190
Urbino, duke of, 7

Vauban, 26-48, 49, 59, 65, 165
Vegetius, 16, 21, 32
Verdun, battle of, 371-373
Verdy du Vernois, 219, 220
Vinci, Leonardo da, *see* Leonardo da Vinci
Voroshilov, 329, 340, 342, 344, 348, 349, 353,
357

Wallenstein, 273
war as social phenomenon, *see* particularly
318, 319
war potential, 118, 129, 144, 351, 353, 364,
498
Washington, George, 128, 130, 132
Washington, Treaty of Naval Limitations,
480
Wegener, 454
Wehrwirtschaft, see particularly 117, 153,
320, 511; *see also* economic warfare and
war economics
Weygand, 329, 332, 361
William I, Emperor of Germany, 176, 181,
183
William II, Emperor of Germany, 181, 203,
204, 307, 308, 309, 392, 442, 443, 451, 509,
516
Willisen, 164, 165
Wilson, Woodrow, 304, 324, 325, 440, 511,
512
"world island," 404, 405, *see also* geopolitics
Wright, Peter, 297, 298, 299
Wright, Quincy, 288

Yudenitch, 329

Zeller, 44
Ziff, 486
Zollverein, German, 139, 141, 147, 152
zones of operations, 87, 88, 89, *see also* lines
of operations